Universitext

Eberhard Freitag · Rolf Busam

Complex Analysis

Second Edition

 Springer

Professor Dr. Eberhard Freitag
Universität Heidelberg
Inst. Mathematik
Im Neuenheimer Feld 288
69120 Heidelberg
Germany
freitag@mathi.uni-heidelberg.de

Dr. Rolf Busam
Universität Heidelberg
Inst. Mathematik
Im Neuenheimer Feld 288
69120 Heidelberg
Germany
busam@mathi.uni-heidelberg.de

ISBN 978-3-540-93982-5   e-ISBN 978-3-540-93983-2

Library of Congress Control Number: 2008943985

Mathematics Subject Classification (2000): 30-01, 11-01, 11F11, 11F66, 11M45, 11N05, 30B50, 33E05

© 2009, 2005 Springer-Verlag Berlin Heidelberg

Cover design: WMXDesign GmbH, Heidelberg

Printed on acid-free paper

9 8 7 6 5 4 3 2 1

springer.com

In Memoriam
Hans Maaß
(1911–1992)

Preface to the Second English Edition

Meanwhile the fourth edition of the German version appeared. In this second English edition we adapt the content closely to this German edition. We also followed several suggestions and tried to improve the language and typography. We thank Shari Scott for her support.

Heidelberg, August 2008

Eberhard Freitag
Rolf Busam

Preface to the English Edition

This book is a translation of the forthcoming fourth edition of our German book "Funktionentheorie I" (Springer 2005). The translation and the LaTeX files have been produced by Dan Fulea. He also made a lot of suggestions for improvement which influenced the English version of the book. It is a pleasure for us to express to him our thanks. We also want to thank our colleagues Diarmuid Crowley, Winfried Kohnen and Jörg Sixt for useful suggestions concerning the translation.

Over the years, a great number of students, friends, and colleagues have contributed many suggestions and have helped to detect errors and to clear the text.

The many new applications and exercises were completed in the last decade to also allow a partial parallel approach using computer algebra systems and graphic tools, which may have a fruitful, powerful impact especially in complex analysis.

Last but not least, we are indebted to Clemens Heine (Springer, Heidelberg), who revived our translation project initially started by Springer, New York, and brought it to its final stage.

Heidelberg, Easter 2005

Eberhard Freitag
Rolf Busam

Contents

Introduction

The complex numbers have their historical origin in the 16th century when they were created during attempts to solve *algebraic equations*. G. CARDANO (1545) has already introduced formal expressions as for instance $5 \pm \sqrt{-15}$, in order to express solutions of quadratic and cubic equations. Around 1560 R. BOMBELLI computed systematically using such expressions and found 4 as a solution of the equation $x^3 = 15x + 4$ in the disguised form

$$4 = \sqrt[3]{2 + \sqrt{-121}} + \sqrt[3]{2 - \sqrt{-121}} \,.$$

Also in the work of G.W. LEIBNIZ (1675) one can find equations of this kind, e.g.

$$\sqrt{1 + \sqrt{-3}} + \sqrt{1 - \sqrt{-3}} = \sqrt{6} \,.$$

In the year 1777 L. EULER introduced the notation $i = \sqrt{-1}$ for the *imaginary unit*.

The terminology "complex number" is due to C.F. GAUSS (1831). The rigorous introduction of complex numbers as pairs of real numbers goes back to W.R. HAMILTON (1837).

Sometimes it is already advantageous to introduce and make use of complex numbers in real analysis. One should for example think of the integration of rational functions, which is based on the partial fraction decomposition, and therefore on the Fundamental Theorem of Algebra:

Over the field of complex numbers
any polynomial decomposes as a product of linear factors.

Another example for the fruitful use of complex numbers is related to FOURIER series. Following EULER (1748) one can combine the real angular functions sine and cosine, and obtain the "exponential function"

$$e^{\mathrm{i}x} := \cos x + \mathrm{i} \sin x \,.$$

Then the addition theorems for sine and cosine reduce to the simple formula

E. Freitag and R. Busam, *Complex Analysis*,
DOI: 10.1007/978-3-540-93983-2_Intro, © Springer-Verlag Berlin Heidelberg 2009

$$e^{i(x+y)} = e^{ix}e^{iy} \ .$$

In particular,

$$\left(e^{ix}\right)^n = e^{inx} \text{ holds for all integers } n \ .$$

The FOURIER series of a sufficiently smooth function f, defined on the real line with period 1, can be written in terms of such expressions as

$$f(x) = \sum_{n=-\infty}^{\infty} a_n e^{2\pi inx} \ .$$

Here it is irrelevant whether f is real or complex valued.

In these examples the complex numbers serve as useful, but ultimately dispensable tools. New aspects come into play when we consider complex valued functions depending on a *complex variable*, that is when we start to study functions $f : D \to \mathbb{C}$ with two-dimensional domains D systematically. The dimension two is ensured when we restrict to *open domains of definition* $D \subset \mathbb{C}$. Analogously to the situation in real analysis one introduces the notion of complex differentiability by requiring the existence of the limit

$$f'(a) := \lim_{z \to a} \frac{f(z) - f(a)}{z - a}$$

for all $a \in D$. It turns out that this notion has a drastically different behavior than real differentiability. We will show for instance that a (first order) complex differentiable function automatically is complex differentiable arbitrarily often. We will see more, namely that complex differentiable functions can always be developed locally as power series. For this reason, complex differentiable functions (defined on open domains) are also called *analytic functions*.

"Complex analysis" is the theory of such analytic functions.

Many classical functions from real analysis can be analytically extended to complex analysis. It turns out that these extensions are unique, as for instance in the case

$$e^{x+iy} := e^x e^{iy} \ .$$

It follows from the relation

$$\boxed{e^{2\pi i} = 1}$$

that the complex exponential function is periodic with the *purely imaginary* period $2\pi i$. This observation is fundamental for the complex analysis. As a consequence one can observe further phenomena:

1. The complex logarithm cannot be introduced as the unique inverse function of the exponential function in a natural way. It is *a priori* determined only up to a multiple of $2\pi i$.

2. The function $1/z$ $(z \neq 0)$ does not have any primitive in the punctured complex plane. A related fact is the following: the line integral of $1/z$ with respect to a circle line centered at the origin and oriented anticlockwise yields the non-zero value

$$\oint_{|z|=r} \frac{1}{z} \, dz = 2\pi i \qquad (r > 0) \, .$$

Central results of complex analysis, like e.g. the *Residue Theorem*, are nothing but a highly generalized version of these statements.

Real functions often show their true nature only if one considers also their analytic extensions. For instance, in real analysis it is not directly transparent why the power series representation

$$\frac{1}{1+x^2} = 1 - x^2 + x^4 - x^6 \pm \cdots$$

is valid only for $|x| < 1$. In the complex theory this phenomenon becomes understandable, simply because the considered function has singularities at $\pm i$. Then its power series representation is valid in the biggest open disk excluding the singularities, namely the unit disk.

In real analysis it is also hard to understand why the TAYLOR series of the $\mathcal{C}^\infty$ function

$$f(x) = \begin{cases} e^{-1/x^2} \, , & x \neq 0 \, , \\ 0 \, , & x = 0 \, , \end{cases}$$

around 0 converges for all $x \in \mathbb{R}$, but does not represent the function at any point other than zero. In the complex theory this phenomenon becomes understandable, since the function e^{-1/z^2} has an *essential singularity* at zero.

Less trivial examples are even more impressive. Here one should mention the RIEMANN ζ-function

$$\zeta(s) = \sum_{n=1}^{\infty} n^{-s} \, ,$$

which will be extensively studied in the last chapter of the book as a function of the *complex variable* traditionally denoted by s using the methods of complex analysis, which will be presented throughout the preceding chapters. From the analytical properties of the ζ-function we deduce the *Prime Number Theorem*.

RIEMANN's celebrated work on the ζ function [Ri2] is a brilliant example for the thesis he presented already eight years in advance in his dissertation [Ri1]

"Die Einführung der complexen Grössen in die Mathematik hat ihren Ursprung und nächsten Zweck in der Theorie einfacher durch Grössenoperationen ausgedrückter Abhängigkeitsgesetze zwischen veränderlichen Grössen. Wendet man nämlich diese Abhängigkeitsgesetze in einem erweiterten Umfange an, indem man den veränderlichen Grössen, auf welche sie sich beziehen, complexe Werthe giebt, so tritt eine sonst versteckt bleibende Harmonie und Regelmäßigkeit hervor."

In translation:

"The introduction of complex variables in mathematics has its origin and its proximate purpose in the theory of simple dependency rules for variables expressed by variable operations. If one applies these dependency rules in an extended manner by associating complex values to the variables referred to by these rules, then there emerges an otherwise hidden harmony and regularity."

Complex numbers are not only useful auxiliary tools, but in some respect are indispensable in many applications like e.g physics and other sciences: The commutation relations in quantum mechanics for impulse and coordinate operators

$$PQ - QP = \frac{h}{2\pi i} I,$$

and respectively the SCHRÖDINGER equation for the HAMILTON operator H,

$$H \, \Psi(x,t) = i \frac{h}{2\pi} \, \partial_t \Psi(x,t)$$

contain the imaginary unit i.

Since there exist several textbooks on complex analysis, a new attempt in this direction needed a special justification. The main idea of this book, and of a second forthcoming volume was to give an extensive description of classical complex analysis, where "classical" means that sheaf theoretical and cohomological methods are omitted. Obviously, it is not possible to include all material that can be considered as classical complex analysis. For instance, if somebody is especially interested in value distribution theory, or in applications of conformal maps, then one will be quickly disappointed and might put this book aside. The line pursued in this text can be described by keywords as follows:

The first four chapters contain an introduction to complex analysis, roughly corresponding to a course "complex analysis I" (four hours each week). Here, the fundamental results of complex analysis are treated.

After the foundations of the theory of analytic functions have been laid, we proceed to the theory of *elliptic functions*, then to *elliptic modular functions* —and after some excursions to analytic number theory— in a second volume we move on to *Riemann surfaces, the local theory of analytic functions of several variables*, to *abelian functions*, and finally we discuss *modular functions for several variables*.

Lot of emphasis is put on completeness of presentation, in the sense that all required notions and concepts are carefully developed. Except for basics in real analysis and linear algebra, as they are nowadays taught in standard introductory courses, we do not want to assume anything else in this first book. In a second volume only some simple topological concepts will be compiled without proof and subsequently used.

We made efforts to introduce as few notions as possible in order to quickly advance to the core of the studied problem. A series of important results will have several proofs. If a special case of a general proposition will be used in an important context, we strived to give a simpler proof for this special case as well. This is in accordance with our philosophy, that a thorough understanding can only be achieved if one turns things around and over and highlights them from different points of view.

We hope that this comprehensive presentation will convey a feeling for the way, in which the topics are related with each other, and for their origin.

Attempts like this are not new. Our text was primarily modelled on the lectures of H. MAASS, to whom we both owe our education in complex analysis. In the same breath, we would also like to mention the lectures of C.L. SIEGEL. Both sources are attempts to trace a great historical epoch, which among others is connected with the names of A.L. CAUCHY, N.H. ABEL, C.G.J. JACOBI, B. RIEMANN and K. WEIERSTRASS, and to present the results, which they developed.

Our objectives are very similar to both mentioned examples. Methodically however, our approach differs in many aspects. This will emerge especially in the second volume, where we will again dwell on the differences.

The present volume presents a comparatively simple introduction to the complex analysis in one variable. The content corresponds to a two semester course with accompanying seminars.

The first three chapters contain the standard material including the Residue Theorem, which should be covered in any introduction. In the fourth chapter —we rank it among the introductory lectures— we treat problems that are less obligatory. We present the Gamma function in detail to illustrate the methods by a beautiful example. We further focus on the Theorems of WEIERSTRASS and MITTAG-LEFFLER on the construction of analytic functions with prescribed zeros and poles. Finally, as a highlight, we prove the *Riemann Mapping Theorem* which states that any proper sub-domain of the complex plane $\mathbb{C}$ "without holes" is conform equivalent to the unit disk.

Only now, in an appendix to chapter IV we will treat the question of *simply connectedness* and we will give different equivalent characterizations for simply connected domains. Roughly speaking, they are domains without holes. In this context different versions, namely the homotopical and homological versions, of the CAUCHY Integral Theorem will be deduced.

These equivalences are useful for a good insight into the theory, and they are important for further developments. But nevertheless they are of minor importance for the development of the standard repertoire of complex analysis. Among simply connected domains we will only need *star-shaped domains* (and some domains that can be constructed from star-shaped domains). Consequently one needs the CAUCHY Integral Theorem only for star-shaped domains, which can be reduced to triangular paths by an idea of A. DINGHAS without any topological complications.

Therefore we will deliberately deal with star-shaped domains for a longer time to avoid the notion of simply connectedness. The price to be paid for this approach is that we have to introduce the concept of an *elementary domain*. By definition this is a domain where the CAUCHY Integral Theorem holds without exception. For this it is enough to know that star-shaped domains are elementary domains, and we postpone their final topological identification to the appendix of the fourth chapter. For the sake of a lucid methodology we have postponed this to a later point. In principle we could do without it completely in the first volume.

The subject of the fifth chapter is the theory of *elliptic functions*, i.e. meromorphic functions with two linearly independent periods. Historically these functions appeared as inverse functions of certain elliptic integrals, as for example the integral

$$y = \int_*^x \frac{1}{\sqrt{1 - t^4}} \, dt .$$

It is easier to do the converse, namely to obtain the elliptic integrals as a byproduct of the impressively beautiful and simple theory of elliptic functions. One of the great achievements of complex analysis is the elegant and transparent construction of the theory of elliptic integrals. As quite usual nowadays, we will choose the WEIERSTRASS approach to the $\wp$-function.

In connection with ABEL's Theorem we will also give a short account of the historically older approach using the JACOBI theta function. We finish the fifth chapter by proving that any complex number is the absolute invariant of a period lattice. This fact implies that one indeed obtains any elliptic integral of the first kind as the inverse function of an elliptic function. At this point the elliptic modular function $j(\tau)$ appears.

As simple as this theory appears nowadays, it remains highly obscure how an elliptic integral gives rise to a period lattice, and thus to an elliptic function. In the second volume, the more complicated theory of RIEMANN surfaces will allow deeper insight into these questions.

In the sixth chapter we will systematically introduce —as a continuation of the end of fifth chapter— the theory of modular functions and modular forms.

Of central interest will be *structural results*, the determination of all modular forms for the full modular group, and for certain subgroups.

Other important examples of modular forms like EISENSTEIN and theta series have arithmetical significance.

One of the most beautiful applications of complex analysis can be found in analytical number theory. For instance, the FOURIER coefficients of modular forms have arithmetic meaning: The FOURIER coefficients of the theta series are representation numbers associated to quadratic forms, those of the EISENSTEIN are sums of divisor powers. Identities between modular forms obtained from complex analysis give rise to number theoretical applications. Following JACOBI, as an example we determine the *number of representations* of a natural number as a sum of four and respectively eight squares of integers. The necessary identities from complex analysis will be deduced independently from the structure theorems for modular forms.

A special section was dedicated to HECKE's theory on the connection between FOURIER series satisfying a transformation rule with respect to the transformation and DIRICHLET series satisfying a functional equation. This theory is a bridge between modular functions and DIRICHLET series. However, the theory of HECKE operators will not be discussed, merely in the exercises we will go into it. Afterwards we will concentrate in detail on the most famous among the DIRICHLET series, the RIEMANN ζ-function. As a classical application we will give a complete proof of the *Prime Number Theorem* with a weak estimate for the error term.

In all chapters there are numerous exercises, easy ones at the beginning, but with increasing chapter number there will also be harder exercises complementing the main text. Occasionally the exercises will require notions from topology or algebra not introduced in the text.

The present material originates in the standard lectures for mathematicians and physicists at the Ruprecht-Karls University of Heidelberg.

Heidelberg, Easter   *Eberhard Freitag*
April 2005   *Rolf Busam*

I

Differential Calculus in the Complex Plane $\mathbb{C}$

In this chapter we shall first give an introduction to *complex numbers* and their *topology*. In doing so we shall assume that this is not the first time the reader has encountered the system $\mathbb{C}$ of complex numbers. The same assumption is made for topological notions in $\mathbb{C}$ (*convergence, continuity* etc.). For this reason we shall not dwell on these matters. In Sect. I.4 we introduce the notion of *complex derivative*. One can begin reading directly with this section if one is already sufficiently familiar with the complex numbers and their topology. In Sect. I.5 the *relationship* between *real differentiability* and *complex differentiability* will be treated (*the Cauchy-Riemann differential equations*).

The story of the complex numbers from their early beginnings in the 16th century until their eventual full acceptance in the course of the 19th century —probably by the scientific authority of C.F. GAUSS— as well as the rather lengthy period of uncertainty and unclarity about them, is an impressive example in the history of mathematics. The historically interested reader should read [Re2]. For more historical remarks about complex numbers see also [CE].

I.1 Complex Numbers

It is well known that not every polynomial with real coefficients has a real root, e.g. the polynomial

$$P(x) = x^2 + 1 \ .$$

There is no real number x with $x^2 + 1 = 0$. If, nonetheless, one wishes to arrange that this and similar equations have solutions, this can only be achieved if one constructs an extension of $\mathbb{R}$, in which such solutions exist. One extends the field $\mathbb{R}$ of real numbers to the field $\mathbb{C}$ of the complex numbers. In fact, in this field, *every polynomial equation*, not just the equation $x^2 + 1 = 0$, has solutions. This is the statement of the *"Fundamental Theorem of Algebra"*.

E. Freitag and R. Busam, *Complex Analysis*,
DOI: 10.1007/978-3-540-93983-2_I, © Springer-Verlag Berlin Heidelberg 2009

Theorem I.1.1 *There exists a field $\mathbb{C}$ with the following properties:*

(1) *The field $\mathbb{R}$ of real numbers is a subfield of $\mathbb{C}$, i.e. $\mathbb{R}$ is a subset of $\mathbb{C}$, and addition and multiplication in $\mathbb{R}$ are the restrictions to $\mathbb{R}$ of the addition and multiplication in $\mathbb{C}$.*

(2) *The equation*

$$X^2 + 1 = 0$$

has exactly two solutions in $\mathbb{C}$.

(3) *Let i be one of the two solutions; then $-\mathrm{i}$ is the other. The map*

$$\mathbb{R} \times \mathbb{R} \longrightarrow \mathbb{C} \ ,$$
$$(x, y) \mapsto x + \mathrm{i}y \ ,$$

is a bijection.

We call $\mathbb{C}$ the field of **complex numbers**. *(Strictly speaking one can not speak of "the" field of real numbers and of "the" field of complex numbers, because by the axioms they only are determined up to isomorphism.)*

Proof. The proof of existence is suggested by (3). One defines on the set $\mathbb{C} := \mathbb{R} \times \mathbb{R}$ the following *composition laws*,

$$(x, y) + (u, v) := (x + u, y + v),$$
$$(x, y) \cdot (u, v) := (xu - yv, xv + yu)$$

and then first shows that the *field axioms* hold. These are:

(1) *The associative laws*

$$(z + z') + z'' = z + (z' + z'') \ ,$$
$$(zz')z'' = z(z'z'') \ .$$

(2) *The commutative laws*

$$z + z' = z' + z \ ,$$
$$zz' = z'z \ .$$

(3) *The distributive laws*

$$z(z' + z'') = zz' + zz'' \ ,$$
$$(z' + z'')z = z'z + z''z \ .$$

(4) *The existence of neutral elements*

    (a) There exists a (unique) element $\underline{0} \in \mathbb{C}$ with the property

$$z + \underline{0} = z \text{ for all } z \in \mathbb{C} .$$

    (b) There exists a (unique) element $\underline{1} \in \mathbb{C}$ with the property

$$z \cdot \underline{1} = z \text{ for all } z \in \mathbb{C} \text{ and } \underline{1} \neq \underline{0} .$$

(5) *The existence of inverse elements*

    (a) For each $z \in \mathbb{C}$ there exists a (unique) element $-z \in \mathbb{C}$ with the property

$$z + (-z) = \underline{0} .$$

    (b) For each $z \in \mathbb{C}$, $z \neq \underline{0}$, there exists a (unique) element $z^{-1} \in \mathbb{C}$ with the property

$$z \cdot z^{-1} = \underline{1} .$$

Verification of the field axioms

The axioms (1) – (3) can be verified by direct calculation.

(4) (a) $\underline{0} := (0,0)$.
     (b) $\underline{1} := (1,0)$.
(5) (a) $-(x,y) := (-x,-y)$.
     (b) Assume $z = (x,y) \neq (0,0)$. Then $x^2 + y^2 \neq 0$. A direct calculation shows that

$$z^{-1} := \left(\frac{x}{x^2 + y^2} , -\frac{y}{x^2 + y^2} \right)$$

    is the inverse of z.

Obviously

$$(a,0)(x,y) = (ax, ay) ,$$

and therefore, in particular,

$$(a,0)(b,0) = (ab,0) .$$

In addition, we have

$$(a,0) + (b,0) = (a+b,0) .$$

Therefore

$$\mathbb{C}_{\mathbb{R}} := \{ (a,0) ; a \in \mathbb{R} \}$$

is a subfield of $\mathbb{C}$, in which the arithmetic is just the same as in $\mathbb{R}$ itself.

More precisely: The map

$$\iota : \mathbb{R} \longrightarrow \mathbb{C}_{\mathbb{R}} \ , \qquad a \longmapsto (a,0) \ ,$$

is an isomorphism of fields.

Thus we have constructed a field $\mathbb{C}$, which does not actually contain $\mathbb{R}$, but a field $\mathbb{C}_{\mathbb{R}}$ which is isomorphic to $\mathbb{R}$. One could then easily construct by set-theoretical manipulations a field $\widetilde{\mathbb{C}}$ isomorphic to $\mathbb{C}$ which actually does contain the given field $\mathbb{R}$ as a subfield. We will skip this construction and simply identify the real number a with the complex number $(a,0)$.

To simplify matters further we shall use the

Notation $\mathrm{i} := (0,1)$ and call i the *imaginary unit* (L. EULER, 1777).

Obviously then

(a) $\mathrm{i}^2 = \mathrm{i} \cdot \mathrm{i} = (0,1) \cdot (0,1) = (0 \cdot 0 - 1 \cdot 1, 0 \cdot 1 + 1 \cdot 0) = (-1,0)$,
(b) $(x,y) = (x,0) + (0,y) = (x,0) \cdot (1,0) + (y,0) \cdot (0,1)$

or, written more simply,

(a) $\mathrm{i}^2 = -1$,   (b) $(x,y) = x + y\,\mathrm{i} = x + \mathrm{i}y$.

Thus each complex number can be written *uniquely* in the form $z = x + \mathrm{i}y$ with real numbers x and y. Therefore we have proved Theorem I.1.1.   □

It can be shown that a field $\mathbb{C}$ is "essentially" uniquely determined by properties (1) – (3) in Theorem I.1.1 (cf. Exercise 13 in I.1).

In the unique representation $z = x + \mathrm{i}y$ we say

x is the *real part* of z and

y is the *imaginary part* of z.

Notation. $x = \mathrm{Re}\,(z)$ and $y = \mathrm{Im}\,(z)$.

If $\mathrm{Re}\,(z) = 0$, then z is said to be *purely imaginary*.

Remark. Note the following essential difference from the field $\mathbb{R}$ of real numbers: $\mathbb{R}$ is an *ordered field,* i.e. there is in $\mathbb{R}$ a special subset P of the so-called "positive elements", such that the following holds:

(1) For each real number a exactly one of the following cases occurs:

(a) $a \in P$ (b) $a = 0$ or   (c) $-a \in P$.

(2) For arbitrary $a, b \in P$,

$$a + b \in P \quad \text{and} \quad ab \in P \ .$$

However, it is easy to show that $\mathbb{C}$ cannot be ordered, i.e. there is no subset $P \subset \mathbb{C}$, for which (1) and (2) hold for any $a, b \in P$ (because of $\mathrm{i}^2 = -1$).

Passing to the *conjugate complex* is often useful in working with complex numbers:

Let $z = x + \mathrm{i}y$, $x, y \in \mathbb{R}$. We put $\overline{z} = x - \mathrm{i}y$ and call $\overline{z}$ the *complex conjugate* of z. It is easy to check the following arithmetical rules for the conjugation map

$$^{-} : \mathbb{C} \longrightarrow \mathbb{C} \ , \quad z \longmapsto \overline{z} \ .$$

Remark I.1.2 *For $z, w \in \mathbb{C}$ holds:*

(1) $\qquad\qquad\qquad \bar{\bar{z}} = z$,

(2) $\qquad\qquad \overline{z \pm w} = \bar{z} \pm \bar{w}$, $\qquad\qquad \overline{zw} = \bar{z} \cdot \bar{w}$,

(3) $\qquad\qquad Re\, z = (z + \bar{z})/2$, $\qquad\qquad Im\, z = (z - \bar{z})\,/\,2i$,

(4) $\qquad\qquad z \in \mathbb{R} \Longleftrightarrow z = \bar{z}$, $\qquad z \in i\mathbb{R} \Longleftrightarrow z = -\bar{z}$.

The map $^- : \mathbb{C} \to \mathbb{C}$, $z \mapsto \bar{z}$, is therefore an involutive field automorphism with $\mathbb{R}$ as its invariant field.

Obviously

$$z\bar{z} = x^2 + y^2$$

is a nonnegative real number.

Definition I.1.3 *The **absolute value** or **modulus** of a complex number z is defined by*

$$|z| := \sqrt{z\bar{z}} = \sqrt{x^2 + y^2} .$$

Clearly $|z|$ is the EUCLIDean distance of z from the origin. We have

$$|z| \geq 0$$

and

$$|z| = 0 \quad \Longleftrightarrow \quad z = 0 .$$

Remark I.1.4 *For $z, w \in \mathbb{C}$ we have:*

(1) $\qquad\qquad |z \cdot w| = |z| \cdot |w|$,

(2) $\qquad\qquad |Re\, z| \leq |z|, \quad |Im\, z| \leq |z|$,

(3) $\qquad\qquad |z \pm w| \leq |z| + |w|$ $\qquad$ *(triangle inequality)* ,

(4) $\qquad\qquad \big|\, |z| - |w| \,\big| \leq |z \pm w|$ $\qquad$ *(triangle inequality)* .

By using the formula $z\bar{z} = |z|^2$ one also gets a simple expression for the inverse of a complex number $z \neq 0$:

$$\boxed{z^{-1} = \frac{\bar{z}}{|z|^2} .}$$

Example.

$$(1 + i)^{-1} = \frac{1 - i}{2} .$$

Geometric visualization in the Gaussian number plane

(1) The addition of complex numbers is just the vector addition of pairs of real numbers:

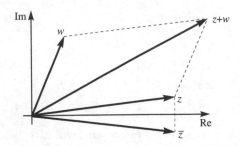

(2) $\bar{z} = x - iy$ is obtained from $z = x + iy$ by reflection along the real axis.

(3) A geometrical meaning for the *multiplication* of complex numbers can be found by means of *polar coordinates*. It is known from real analysis that any point $(x, y) \neq (0, 0)$ can be written in the form

$$(x, y) = r(\cos \varphi, \sin \varphi), \quad r > 0 .$$

In this expression r is uniquely determined,

$$r = \sqrt{x^2 + y^2} ,$$

however, the angle φ (measured in radians) is only fixed up to the addition of an integer multiple of 2π.[1] If we use the notation

$$\mathbb{R}_+^{\bullet} := \{ x \in \mathbb{R}; \quad x > 0 \}$$

for the set of positive real numbers, and

$$\mathbb{C}^{\bullet} := \mathbb{C} \setminus \{0\}$$

for the complex plane with the origin removed, then we have:

Theorem I.1.5 *The map*

$$\mathbb{R}_+^{\bullet} \times \mathbb{R} \longrightarrow \mathbb{C}^{\bullet} ,$$
$$(r, \varphi) \mapsto r(\cos \varphi + i \sin \varphi) ,$$

is surjective.

Supplement. From

$$r(\cos \varphi + i \sin \varphi) = r'(\cos \varphi' + i \sin \varphi'),$$

$$r, r' > 0,$$

it follows that

$$r = r' \quad \text{and} \quad \varphi - \varphi' = 2\pi k , \quad k \in \mathbb{Z} .$$

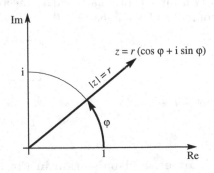

---

[1] One also says: modulo 2π.

Remark. In the *polar coordinate representation* of $z \in \mathbb{C}^{\bullet}$,

$$(*) \qquad\qquad z = r(\cos\varphi + \mathrm{i}\sin\varphi) ,$$

the r is therefore uniquely determined by z ($r = \sqrt{z\bar{z}}$), but φ is only determined up to an integer multiple of 2π. Each $\varphi \in \mathbb{R}$, for which (*) holds, is called *an argument* of z. Therefore if φ_0 is a fixed argument of z, then any other argument φ of z has the form

$$\varphi = \varphi_0 + 2\pi k , \ k \in \mathbb{Z} .$$

The *uniqueness* of the polar coordinate representation can be achieved, if, for example, one demands that φ lies in the interval $]-\pi, \pi]$. In other words, the map

$$\mathbb{R}_+^{\bullet} \times \,]-\pi, \pi] \longrightarrow \mathbb{C}^{\bullet} , \qquad (r, \varphi) \mapsto r(\cos\varphi + \mathrm{i}\sin\varphi) ,$$

is bijective. We call $\varphi \in \,]-\pi, \pi]$ the *principal value* of the argument and sometimes denote it by $\mathrm{Arg}(z)$.
Examples. $\mathrm{Arg}(1) = \mathrm{Arg}(2005) = 0$, $\mathrm{Arg}(\mathrm{i}) = \pi/2$, $\mathrm{Arg}(-\mathrm{i}) = -\pi/2$, $\mathrm{Arg}(-1) = \pi$.

Theorem I.1.6 *We have*

$$(\cos\varphi + \mathrm{i}\sin\varphi)(\cos\varphi' + \mathrm{i}\sin\varphi') \;=\; \cos(\varphi + \varphi') + \mathrm{i}\sin(\varphi + \varphi')$$

or

$$\boxed{\begin{aligned} &\cos(\varphi + \varphi') = \cos\varphi \cdot \cos\varphi' - \sin\varphi \cdot \sin\varphi' \\ &\sin(\varphi + \varphi') = \sin\varphi \cdot \cos\varphi' + \cos\varphi \cdot \sin\varphi' \\ &\textit{(addition theorem for circular functions)} \end{aligned}}$$

Theorems I.1.5 and I.1.6 give a geometrical meaning to the multiplication of complex numbers. Namely, when

$$z = r(\cos\varphi + \mathrm{i}\sin\varphi) , \qquad z' = r'(\cos\varphi' + \mathrm{i}\sin\varphi') ,$$

then the product is

$$zz' = rr'\big(\cos(\varphi + \varphi') + \mathrm{i}\sin(\varphi + \varphi')\big) .$$

Therefore rr' is the absolute value of zz' and $\varphi + \varphi'$ is an argument for zz', which one can express neatly, but not quite precisely, as:

$$\boxed{\begin{aligned} &\text{Complex numbers are multiplied} \\ &\text{by multiplying their absolute values} \\ &\text{and adding their arguments.} \end{aligned}}$$

If $z = r(\cos \varphi + \mathrm{i} \sin \varphi) \neq 0$, then

$$\frac{1}{z} = \frac{\bar{z}}{z\bar{z}} = \frac{1}{r}(\cos \varphi - \mathrm{i} \sin \varphi) ,$$

from which one may similarly read off a simple geometrical construction for $1/z$.

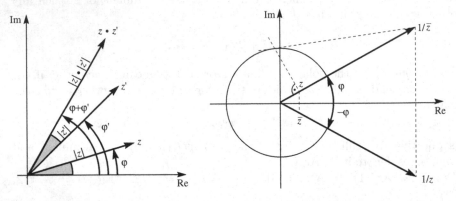

Let $n \in \mathbb{Z}$ be an integer. As usual we define a^n for complex numbers a by

$$a^n = \overbrace{a \cdots\cdots a}^{n \text{ times}} , \qquad \text{if } n > 0 ,$$
$$a^0 = 1 ,$$
$$a^n = (a^{-1})^{-n} , \qquad \text{if } n < 0 \text{ and } a \neq 0.$$

We have the rules:

$$a^n \cdot a^m = a^{n+m} ,$$
$$(a^n)^m = a^{nm} ,$$
$$a^n \cdot b^n = (a \cdot b)^n .$$

Of course, the *binomial formula also holds*:

$$(a + b)^n = \sum_{\nu=0}^{n} \binom{n}{\nu} a^\nu \, b^{n-\nu}$$

for complex numbers $a, b \in \mathbb{C}$ and $n \in \mathbb{N}_0$. The involved binomial coefficients are defined as

$$\binom{n}{0} := 1 \quad \text{and} \quad \binom{n}{\nu} := \frac{n(n-1)\cdots(n-\nu+1)}{\nu!} , \ 1 \leq \nu \leq n.$$

A complex number a is called an nth **root of unity** $(n \in \mathbb{N})$, if $a^n = 1$.

Theorem I.1.7 *For each $n \in \mathbb{N}$ there are exactly n different n^{th} roots of unity, namely*

$$\zeta_\nu := \cos \frac{2\pi\nu}{n} + \mathrm{i} \sin \frac{2\pi\nu}{n} \,, \quad 0 \leq \nu < n \,.$$

Proof. Using I.1.6 it is easy to show by induction on n, that

$$(\cos\varphi + \mathrm{i}\sin\varphi)^n = \cos n\varphi + \mathrm{i}\sin n\varphi$$

(L. EULER, 1748, 1749, A. DE MOIVRE, 1707, 1730)

for arbitrary natural n. Since roots of unity are of absolute value 1, they can be written in the form

$$\cos\varphi + \mathrm{i}\sin\varphi \,.$$

This number is an nth root of unity iff $n\varphi$ is an integer multiple of 2π, i.e. $\varphi = 2\pi\nu/n$. Then it follows from Theorem I.1.5, that one can restrict to $0 \leq \nu \leq n-1$. Thus the n numbers

$$\zeta_\nu := \zeta_{\nu,n} := \cos \frac{2\pi\nu}{n} + \mathrm{i}\sin \frac{2\pi\nu}{n} \,, \quad \nu = 0 \,, \ldots, n-1 \,,$$

give the n different n^{th} roots of unity.   □

Remark. For $\zeta_1 = \zeta_{1,n} = \cos \dfrac{2\pi}{n} + \mathrm{i}\sin \dfrac{2\pi}{n}$ we have

$$\zeta_\nu = \zeta_1^\nu \,, \quad \nu = 0, 1, \ldots, n-1 \,.$$

Examples of nth roots of unity:

$n = 1$ $\{1\}$.

$n = 2$ $\{1, -1\} = \{ (-1)^\nu ; \quad \nu = 0, 1\}$.

$n = 3$ $\left\{ 1, \; -\dfrac{1}{2} + \dfrac{\mathrm{i}}{2}\sqrt{3} \,, \; -\dfrac{1}{2} - \dfrac{\mathrm{i}}{2}\sqrt{3} \right\}$
$= \left\{ \left(-\tfrac{1}{2} + \tfrac{\mathrm{i}}{2}\sqrt{3}\right)^\nu ; \quad 0 \leq \nu \leq 2 \right\} = \left\{ \zeta_{1,3}^\nu ; \quad 0 \leq \nu \leq 2 \right\}$.

$n = 4$ $\{ 1, \mathrm{i}, -1, -\mathrm{i} \} = \{ \mathrm{i}^\nu ; \quad 0 \leq \nu \leq 3 \} = \left\{ \zeta_{1,4}^\nu ; \quad 0 \leq \nu \leq 3 \right\}$.

$n = 5$ $\left\{ \zeta_{1,5}^\nu ; \quad 0 \leq \nu \leq 4 \right\} \,, \quad \zeta_{1,5} = \dfrac{\sqrt{5}-1}{4} + \dfrac{\mathrm{i}}{4}\sqrt{2(5+\sqrt{5})}$.

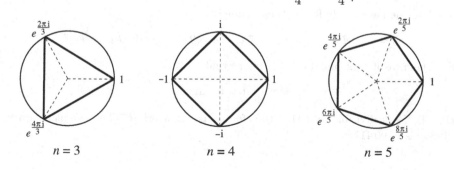

$n = 3$ $n = 4$ $n = 5$

All n^{th} roots of unity lie on the boundary of the unit disk, the *unit circle*
$S^1 := \{ z \in \mathbb{C}; \quad |z| = 1 \}$. They are the vertices of an regular n-gon inscribed
in S^1 (one vertex is always $(1,0) = 1$). Because of this, one also calls the
equation

$$z^n = 1$$

the *cyclotomic equation.* We have, as we shall see,

$$z^n - 1 = (z - \zeta_0) \cdot (z - \zeta_1) \cdot \ldots \cdot (z - \zeta_{n-1})$$

with

$$\zeta_\nu = \cos \frac{2\pi}{n}\nu + \mathrm{i} \sin \frac{2\pi}{n}\nu , \quad 0 \leq \nu \leq n - 1 .$$

The ζ_ν are the zeros of the polynomial

$$P(z) := z^n - 1 .$$

The polynomial P thus has n different zeros. This is a special case of the
Fundamental Theorem of Algebra. It asserts:

> *Each nonconstant complex*
> *polynomial has*
> *as many zeros as its degree.*

In this statement we must, of course, count the zeros with their multiplicities.
We shall encounter several proofs of this important theorem.

Remark. The regular n-gon is constructible with ruler and compass, if the nth roots
of unity can be obtained by repeated extraction of square roots and ordinary arithmetical operations from rational numbers. According to a theorem due to C.F.
GAUSS this is only the case when n has the form

$$n = 2^l F_{k_1} \ldots F_{k_r} ,$$

where $l, k_j \in \mathbb{N}_0$ and the F_{k_j}, $j = 1, \ldots, r$ are different so-called *Fermat primes.*
The latter are primes of the form

$$F_k = 2^{2^k} + 1 , \quad k \in \mathbb{N}_0 .$$

To date one knows only five of these, namely

$$F_0 = 3 , \quad F_1 = 5 , \quad F_2 = 17 , \quad F_3 = 257 , \quad \text{and} \quad F_4 = 65537 .$$

For the next of these numbers it happens that

$$F_5 = 2^{32} + 1 \equiv 0 \mod 641 ,$$

that is, F_5 is divisible by 641 —therefore it is not a prime. (The complementary
divisor is $6\,700\,417$.)

Exercises for I.1

1. Find the *real* and *imaginary parts* of each of the following complex numbers:

$$\frac{i-1}{i+1} \; ; \quad \frac{3+4i}{1-2i} \; ; \quad i^n, \, n \in \mathbb{Z} \; ; \quad \left(\frac{1+i}{\sqrt{2}}\right)^n, \, n \in \mathbb{Z} \; ;$$

$$\left(\frac{1+i\sqrt{3}}{2}\right)^n, \, n \in \mathbb{Z} \; ; \quad \sum_{\nu=0}^{7}\left(\frac{1-i}{\sqrt{2}}\right)^\nu \; ; \quad \frac{(1+i)^4}{(1-i)^3} + \frac{(1-i)^4}{(1+i)^3} \; .$$

2. Calculate the *absolute value* and an *argument* for each of the following complex numbers:

$$-3+i \; ; \quad -13 \; ; \quad (1+i)^{17} - (1-i)^{17} \; ; \quad i^{4711} \; ; \quad \frac{3+4i}{1-2i} \; ;$$

$$\frac{1+ia}{1-ia} \, , \; a \in \mathbb{R} \; ; \quad \frac{1-i\sqrt{3}}{1+i\sqrt{3}} \; ; \quad (1-i)^n \, , \; n \in \mathbb{Z} \; .$$

3. Prove the *"Triangle Inequality"*

$$|z+w| \le |z| + |w| \, , \quad z, w \in \mathbb{C} \, ,$$

and discuss when it becomes an equality; also prove the *"Triangle Inequality"*

$$\big||z| - |w|\big| \le |z-w| \, , \quad z, w \in \mathbb{C} \, .$$

4. For $z = x + iy$, $w = u + iv$, with $x, y, u, v \in \mathbb{R}$, the standard scalar product in the $\mathbb{R}$-vector space $\mathbb{C} = \mathbb{R} \times \mathbb{R}$ with respect to the basis $(1, i)$ is defined by

$$\langle z, w \rangle := \mathrm{Re}\,(z\overline{w}) = xu + yv \, .$$

Verify by direct calculation that, for $z, w \in \mathbb{C}$

$$\langle z, w \rangle^2 + \langle iz, w \rangle^2 = |z|^2 \, |w|^2$$

and infer from this the CAUCHY-SCHWARZ inequality in $\mathbb{R}^2$:

$$|\langle z, w \rangle|^2 = |xu + yv|^2 \le |z|^2 \, |w|^2 = (x^2 + y^2)(u^2 + v^2) \, .$$

In addition, show the following identities for $z, w \in \mathbb{C}$ by direct calculation:

$$|z+w|^2 = |z|^2 + 2\langle z, w \rangle + |w|^2 \qquad \text{(cosine law)} \, ,$$
$$|z-w|^2 = |z|^2 - 2\langle z, w \rangle + |w|^2 \, ,$$
$$|z+w|^2 + |z-w|^2 = 2(|z|^2 + |w|^2) \qquad \text{(parallelogram law)} \, .$$

Further, show that for each pair $(z, w) \in \mathbb{C}^{\bullet} \times \mathbb{C}^{\bullet}$ there is a unique real number $\omega := \omega(z, w) \in \,]-\pi, \pi]$ with

$$\cos \omega = \cos \omega(z, w) = \frac{\langle z, w \rangle}{|z|\,|w|}$$

and

$$\sin \omega = \sin \omega(z, w) = \frac{\langle iz, w \rangle}{|z|\, |w|} \ .$$

$\omega = \omega(z, w)$ is called the *oriented angle* between z and w and will often be denoted by $\angle(z, w)$.

Show: $\angle(1, i) = \pi/2$, $\angle(i, 1) = -\pi/2 = -\angle(1, i)$.

5.   Suppose $n \in \mathbb{N}$ and $z_\nu, w_\nu \in \mathbb{C}$ for $1 \le \nu \le n$. Prove

$$\left| \sum_{\nu=1}^{n} z_\nu w_\nu \right|^2 = \sum_{\nu=1}^{n} |z_\nu|^2 \cdot \sum_{\nu=1}^{n} |w_\nu|^2 - \sum_{1 \le \nu < \mu \le n} |z_\nu \overline{w}_\mu - z_\mu \overline{w}_\nu|^2$$

(the LAGRANGE Identity) and conclude from this the CAUCHY-SCHWARZ Inequality in $\mathbb{C}^n$:

$$\left| \sum_{\nu=1}^{n} z_\nu w_\nu \right|^2 \le \sum_{\nu=1}^{n} |z_\nu|^2 \cdot \sum_{\nu=1}^{n} |w_\nu|^2 \ .$$

6.   Sketch the following subsets of $\mathbb{C}$ in the complex plane:

(a)   Assume $a, b \in \mathbb{C}$, $b \neq 0$;

$$G_0 := \left\{ z \in \mathbb{C} \ ; \quad \mathrm{Im}\left(\frac{z-a}{b}\right) = 0 \right\} \ ,$$

$$G_+ := \left\{ z \in \mathbb{C} \ ; \quad \mathrm{Im}\left(\frac{z-a}{b}\right) > 0 \right\} \qquad \text{and}$$

$$G_- := \left\{ z \in \mathbb{C} \ ; \quad \mathrm{Im}\left(\frac{z-a}{b}\right) < 0 \right\} \ .$$

(b)   Consider $a, c \in \mathbb{R}$ and $b \in \mathbb{C}$ with $b\overline{b} - ac > 0$,

$$K := \left\{ z \in \mathbb{C} \ ; \quad az\overline{z} + \overline{b}z + b\overline{z} + c = 0 \right\} \ .$$

(c) $L := \left\{ z \in \mathbb{C} \ ; \quad \left| z - \frac{\sqrt{2}}{2} \right| \cdot \left| z + \frac{\sqrt{2}}{2} \right| = \frac{1}{2} \right\} \ .$

7.   **Square roots and the solvability of quadratic equations in $\mathbb{C}$**

Let $c - a + ib \neq 0$ be a given complex number. By splitting it into its real and imaginary part show that there are exactly two complex numbers z_1 and z_2 such that

$$z_1^2 = z_2^2 = c \ . \text{ One has } z_2 = -z_1 \ .$$

(z_1 and z_2 are called the *square roots* of c.) For example, determine the square roots of

$$5 + 7i \ , \quad \text{and} \quad \sqrt{2} + i\sqrt{2} \ .$$

Use also polar coordinates for this exercise. Furthermore, show that a quadratic equation

$$z^2 + \alpha z + \beta = 0 \ , \quad \alpha, \beta \in \mathbb{C} \text{ arbitrary} \ ,$$

always has at most two solutions $z_1, z_2 \in \mathbb{C}$.

8.   **Existence of nth roots**

Assume $a \in \mathbb{C}$ and $n \in \mathbb{N}$. A complex number z is called (an) n^{th} root of a if $z^n = a$.

Show: If $a = r(\cos \varphi + i \sin \varphi) \neq 0$, then a has exactly n (different) n^{th} roots, namely the complex numbers

$$z_\nu = \sqrt[n]{r}\left(\cos\frac{\varphi+2\pi\nu}{n} + i\sin\frac{\varphi+2\pi\nu}{n}\right)\,,\quad 0\le\nu\le n-1\,.$$

In the special case $a=1$ (thus $r=1$, $\varphi=0$), we get Theorem I.1.7.

9. Determine all $z\in\mathbb{C}$ such that $z^3-i=0$.

10. Let P be a polynomial with complex coefficients:

$$P(z):=a_n z^n + a_{n-1}z^{n-1}+\cdots+a_0 \text{ with } n\in\mathbb{N}_0\,,\ a_\nu\in\mathbb{C}\,,\ \text{for } 0\le\nu\le n\,.$$

A real or complex number ζ is called a *root* or a *zero* of P, if $P(\zeta)=0$.
Show: If all the coefficients a_ν are real, then we have

$$P(\zeta)=0 \implies P(\overline{\zeta})=0\,.$$

In other words, if the polynomial P has only *real coefficients* then the roots of P which are not real occur as pairs of complex conjugate numbers.

11. (a) Let $\mathbb{H}:=\{z\in\mathbb{C}\,;\ \operatorname{Im}z>0\}$ be the *upper half-plane.*
Show: $z\in\mathbb{H}\iff -1/z\in\mathbb{H}$.

(b) Assume $z,a\in\mathbb{C}$.
Show: $|1-z\bar a|^2-|z-a|^2=(1-|z|^2)(1-|a|^2)$.
Deduce: If $|a|<1$, then

$$|z|<1\iff\left|\frac{z-a}{\bar a z-1}\right|<1\quad\text{and}\quad |z|=1\iff\left|\frac{z-a}{\bar a z-1}\right|=1\,.$$

12. Verify for $z=x+iy\in\mathbb{C}$ the inequalities

$$\frac{|x|+|y|}{\sqrt2}\le|z|=\sqrt{x^2+y^2}\le|x|+|y|$$

and

$$\max\{|x|\,,\,|y|\}\le|z|\le\sqrt2\max\{|x|\,,\,|y|\}\,.$$

13. Let $\widetilde{\mathbb{C}}$ be another field of complex numbers. Determine all mappings $\varphi:\mathbb{C}\to\widetilde{\mathbb{C}}$ with the following properties:
(a) $\varphi(z+w)=\varphi(z)+\varphi(w)$ for all $z,w\in\mathbb{C}$,
(b) $\varphi(zw)=\varphi(z)\varphi(w)$ for all $z,w\in\mathbb{C}$,
(c) $\varphi(x)=x$ for all $x\in\mathbb{R}$.
Remark. It turns out that such mappings exist, and they are *automatically bijective*; thus they give isomorphisms $\mathbb{C}\to\widetilde{\mathbb{C}}$ that leave $\mathbb{R}$ elementwise fixed. *The field of complex numbers is therefore essentially uniquely determined.* In the special case $\mathbb{C}=\widetilde{\mathbb{C}}$ we get automorphisms of $\mathbb{C}$ with the fixed field $\mathbb{R}$.

Remark. What automorphisms (i.e. isomorphisms onto itself) admits the field of real numbers $\mathbb{R}$? *Hint.* Such an automorphism of $\mathbb{R}$ must preserve the ordering of $\mathbb{R}$!

14. Each $z \in S^1 \setminus \{-1\}$,
     $S^1 := \{z \in \mathbb{C} ; \quad |z| = 1\}$,
     can be uniquely represented in the form

$$z = \frac{1 + i\lambda}{1 - i\lambda} = \frac{1 - \lambda^2}{1 + \lambda^2} + \frac{2\lambda}{1 + \lambda^2} i$$

with $\lambda \in \mathbb{R}$.

15. (a) Consider the map

$$f : \mathbb{C}^{\bullet} \longrightarrow \mathbb{C} \text{ with } f(z) = 1/\bar{z} .$$

Give a geometrical construction (with ruler and compass) for the image $f(z)$ and justify calling this map *"reflection at the unit circle"*. Find the image under f of each of

$(\alpha) \quad D_1 := \{z \in \mathbb{C} ; \quad 0 < |z| < 1\}$,

$(\beta) \quad D_2 := \{z \in \mathbb{C} ; \quad |z| > 1\}$,

$(\gamma) \quad D_3 := \{z \in \mathbb{C} ; \quad |z| = 1\}$.

(b) Now consider the map

$$g : \mathbb{C}^{\bullet} \longrightarrow \mathbb{C} \text{ with } g(z) = 1/z \ (= \overline{f(z)})$$

and give a geometrical construction for the image $g(z)$ of z. Why is this map called *"inversion at the unit circle"*? What are the fixed points of g, i.e. for which $z \in \mathbb{C}^{\bullet}$ is it true that $g(z) = z$?

16. Assume $n \in \mathbb{N}$ and let $W(n) = \{z \in \mathbb{C} ; \ z^n = 1\}$ be the set of nth roots of unity.

Show:

(a) $W(n)$ is a subgroup of $\mathbb{C}^{\bullet}$ (and so is a group itself).

(b) $W(n)$ is a cyclic group of order n, i.e. there is a $\zeta \in W(n)$ such that

$$W(n) = \{\zeta^{\nu}; \quad 0 \le \nu < n\} .$$

Such a root of unity ζ is called a *primitive root of unity*.

Deduce that $W(n) \simeq \mathbb{Z}/n\mathbb{Z}$.

For which $d \in \mathbb{N}$ with $1 \le d \le n$ is the power ζ^d again a primitive n^{th} root of unity? Therefore how many primitive n^{th} roots of unity are there?

Other introductions of the complex numbers

In section I.1 the complex numbers were introduced as pairs of real numbers (following C. WESSEL, 1796, J.R. ARGAND, 1806, C.F. GAUSS, 1811, 1831, and W.R. HAMILTON, 1835). From considering the geometry of $\mathbb{R}^2$ (rotations and dilatations!) the following approach to the complex numbers is plausible:

17. Let

$$\mathcal{C} := \left\{ \begin{pmatrix} a & -b \\ b & a \end{pmatrix}; \quad a, b \in \mathbb{R} \right\} \subset M(2 \times 2; \mathbb{R})$$

with ordinary addition and multiplication of (real) 2×2 matrices.

Show: $\mathcal{C}$ is a field, which is isomorphic to $\mathbb{C}$, the field of complex numbers.

18. As remarked during the introduction of the complex numbers, the polynomial $P = X^2 + 1 \in \mathbb{R}[X]$ has no roots in $\mathbb{R}$, in particular it does not decompose into polynomials of smaller degrees, so P is *irreducible* in $\mathbb{R}[X]$. In algebra (see, for instance, [La2]) it is shown how one constructs for each irreducible polynomial P in the polynomial ring $K[X]$ over a field K a minimal extension field E in which the given polynomial does have a root. In our special case ($K = \mathbb{R}$, $P = X^2 + 1$), this means that one takes the factor ring of $\mathbb{R}[X]$ with respect to the ideal $(X^2 + 1)$. This is isomorphic to $\mathbb{C}$.

19. **Hamilton's Quaternions** (W. R. HAMILTON, 1843)

We consider the following map

$$H : \mathbb{C} \times \mathbb{C} \longrightarrow M(2 \times 2; \mathbb{C}) ,$$

$$(z, w) \mapsto H(z, w) := \begin{pmatrix} z & -w \\ \overline{w} & \overline{z} \end{pmatrix}$$

and denote its image by

$$\mathcal{H} := \{ H(z, w); \quad (z, w) \in \mathbb{C} \times \mathbb{C} \} \subset M(2 \times 2; \mathbb{C}) .$$

Show that $\mathcal{H}$ is a *skew field*, i.e. in $\mathcal{H}$ all the field axioms hold with the exception of the commutativity law for multiplication.

Remark. The notation $\mathcal{H}$ is intended to remind us of Sir WILLIAM ROWAN HAMILTON (1805-1865). One calls $\mathcal{H}$ the HAMILTON quaternions.

20. **Cayley Numbers** (Λ. CAYLEY, 1845)

Let

$$\mathcal{C} := \mathcal{H} \times \mathcal{H} .$$

Consider the following composition law:

$$\mathcal{C} \times \mathcal{C} \longrightarrow \mathcal{C} ,$$

$$((H_1, H_2), (K_1, K_2)) \mapsto (H_1 K_1 - \bar{K}_2' H_2, H_2 \bar{K}_1' + K_2 H_1) .$$

Here $\bar{H}'$ denotes the conjugate transposed matrix of $H \in \mathcal{H} \subset M(2 \times 2; \mathbb{C})$.

Show that this defines on an $\mathbb{R}$-bilinear map, which has no zero divisors, i.e. the "product" of two elements in $\mathcal{C}$ is zero, iff one of the two factors vanishes. This "CAYLEY multiplication" is, in general, neither commutative nor associative.

A deep theorem (M. A. KERVAIRE (1958), J. MILNOR (1958), J. BOTT (1958)) says that on an n-dimensional ($n < \infty$) real vector space V a bilinear form free of zero divisors can only exist when $n = 1, 2, 4$ or 8. Examples of such structures are the "real numbers", the "complex numbers", the "HAMILTON quaternions" and the "CAYLEY numbers". Compare with the article of F. HIRZEBRUCH, [Hi].

I.2 Convergent Sequences and Series

We assume that the reader is familiar with the topology of $\mathbb{R}^p$ from real analysis of several variables. In the case $\mathbb{C} = \mathbb{R}^2$, the fundamental definitions and properties will be briefly recalled.[2]

Definition I.2.1 *A sequence $(z_n)_{n\geq 0}$ of complex numbers is called a **null sequence** if for each $\varepsilon > 0$ there is a natural number N such that*

$$|z_n| < \varepsilon \quad \text{for all} \quad n \geq N .$$

Definition I.2.2 *A sequence*

$$z_0, \ z_1, \ z_2, \ \ldots$$

of complex numbers converges to the complex number z if the sequence of differences $z_0 - z, z_1 - z, \ldots$ is a null sequence.

It is well-known that the limit z is uniquely determined, and we write

$$z = \lim_{n\to\infty} z_n \quad \text{or} \quad z_n \to z \text{ as } n \to \infty .$$

From the equivalence of the Euclidean metric and the maximum metric for $\mathbb{R}^2$, or simply from

$$|\text{Re } z| , \ |\text{Im } z| \ \leq \ |z| \ \leq \ |\text{Re } z| + |\text{Im } z| ,$$

there follows:

Remark I.2.3 *Let (z_n) be a sequence of complex numbers, and z be another complex number. The following statements are equivalent:*

(1) $z_n \to z$ *for $n \to \infty$.*

(2) $\text{Re } z_n \to \text{Re } z$ *and* $\text{Im } z_n \to \text{Im } z$ *for $n \to \infty$.*

Remark I.2.4 *From $z_n \to z$ and $w_n \to w$ as $n \to \infty$ follows:*

(1) $z_n \pm w_n \to z \pm w$,

(2) $z_n \cdot w_n \to z \cdot w$,

(3) $|z_n| \to |z|$,

(4) $\overline{z_n} \to \overline{z}$,

(5) $z_n^{-1} \to z^{-1}$ *in case of $z \neq 0$, $z_n \neq 0$ for all n.*

---

[2] For topological purposes we shall always identify $\mathbb{C}$ with $\mathbb{R}^2$:

$$\mathbb{C} \ni z \longmapsto (\text{Re } z, \text{Im } z) \in \mathbb{R}^2 .$$

One can prove this either by splitting the involved complex numbers into real and imaginary parts, or just by translating the usual proofs from real analysis.

Example.

$$\lim_{n \to \infty} z^n = 0 \quad \text{for} \quad |z| < 1 .$$

The assertion follows from the corresponding theorem for real z by using

$$|z^n| = |z|^n .$$

Infinite Series of Complex Numbers

Let z_0, z_1, z_2, ... be a sequence of complex numbers. One can associate to it a new sequence, the *sequence of its partial sums* S_0, S_1, S_2, ... with

$$S_n := z_0 + z_1 + \cdots + z_n .$$

The sequence (S_n) is also called the *series associated to the sequence* (z_n). For this one uses the notation

$$\sum_{n=0}^{\infty} z_n = z_0 + z_1 + z_2 + \cdots .$$

If the sequence (S_n) converges, then one calls

$$S := \lim_{n \to \infty} S_n$$

the *value* or the *sum* of the series. In this case, one also writes

$$S = \sum_{n=0}^{\infty} z_n = z_0 + z_1 + z_2 + \cdots .$$

We follow here the usual but not entirely precise traditional notation: The symbol $\sum_{n=0}^{\infty} z_n$ will be used with *two* meanings:

(1) *On the one hand* as a synonym for the sequence (S_n) of partial sums of the sequence (z_n).
(2) *On the other hand* (if (S_n) converges) for their sum, i.e. the limit $S = \lim_{n \to \infty} S_n$. Thus S is a number in this case.

Which of the two meanings is intended is usually clear from the context. For this, see also Exercise 9 in section I.2.

Example. The *geometric series* converges for all $z \in \mathbb{C}$ with $|z| < 1$:

$$\frac{1}{1-z} = 1 + z + z^2 + \cdots \quad \text{for } |z| < 1 .$$

The proof of this follows from the formula (proved, for instance, by induction on n)

$$\frac{1 - z^{n+1}}{1 - z} = 1 + z + \cdots + z^n \qquad \text{for } z \neq 1 .$$

A series

$$z_0 + z_1 + z_2 + \cdots$$

is called *absolutely convergent,* if the series of absolute values

$$|z_0| + |z_1| + |z_2| + \cdots$$

converges.

Theorem I.2.5 *An absolutely convergent series converges.*

Proof. We assume that the corresponding theorem for the real case is known. The assertion then follows from Remark I.2.3. $\qquad \square$

Using Theorem I.2.5 one may extend many elementary functions into the complex plane.

Remark I.2.6 *The series*

$$\sum_{n=0}^{\infty} \frac{z^n}{n!} , \qquad \sum_{n=0}^{\infty} \frac{(-1)^n}{(2n+1)!} z^{2n+1} \qquad and \qquad \sum_{n=0}^{\infty} \frac{(-1)^n}{(2n)!} z^{2n}$$

are absolutely convergent for all $z \in \mathbb{C}$.
We define

$$
\begin{aligned}
\exp(z) &:= \sum_{n=0}^{\infty} \frac{z^n}{n!} && \text{(complex exponential function),} \\
\sin(z) &:= \sum_{n=0}^{\infty} \frac{(-1)^n}{(2n+1)!} z^{2n+1} && \text{(complex sine),} \\
\cos(z) &:= \sum_{n=0}^{\infty} \frac{(-1)^n}{(2n)!} z^{2n} && \text{(complex cosine).}
\end{aligned}
$$

Lemma I.2.7 (Cauchy Multiplication Theorem) *Let*

$$\sum_{n=0}^{\infty} a_n \qquad and \qquad \sum_{n=0}^{\infty} b_n$$

be absolutely convergent series. Then we have

$$\sum_{n=0}^{\infty} \left(\sum_{\nu=0}^{n} a_\nu b_{n-\nu} \right) = \left(\sum_{n=0}^{\infty} a_n \right) \cdot \left(\sum_{n=0}^{\infty} b_n \right) ,$$

where the series on the left-hand side is also absolutely convergent.

The proof is literally as in the real case. From the Multiplication Theorem I.2.7 follows

$$\exp(z)\exp(w) = \sum_{n=0}^{\infty}\sum_{\nu=0}^{n}\frac{z^\nu w^{n-\nu}}{\nu!(n-\nu)!} = \sum_{n=0}^{\infty}\frac{(z+w)^n}{n!} = \exp(z+w) \ .$$

Theorem I.2.8 *For arbitrary complex numbers z and w*

$$\boxed{\begin{array}{c} \exp(z+w) = \exp(z)\cdot\exp(w) \\ \textit{Addition theorem or functional equation} \end{array}}$$

Corollary I.2.8$_1$ In particular, we have $\exp(z) \neq 0$ for all $z \in \mathbb{C}$ because of $\exp(z)\exp(-z) = 1$, and

$$\exp(z)^n = \exp(nz) \qquad \text{for } n \in \mathbb{Z} \ .$$

The function $\exp(z)$ coincides for real z with the real exponential function. For complex z *we define*

$$e^z := \exp(z) \ .$$

In this way the functional equation in I.2.8 becomes a *power law:*

$$e^{z+w} = e^z e^w \ .$$

However, note the remark at the end of the paragraph about the problems of taking complex powers. We will use both the notations e^z and $\exp(z)$.

Remark I.2.9 *We have*

$$\boxed{\begin{array}{c} \exp(\mathrm{i}z) = \cos z + \mathrm{i}\sin z \ , \\[2mm] \cos(z) = \dfrac{\exp(\mathrm{i}z) + \exp(-\mathrm{i}z)}{2} \ , \\[2mm] \sin(z) = \dfrac{\exp(\mathrm{i}z) - \exp(-\mathrm{i}z)}{2\mathrm{i}} \ . \end{array}}$$

Corollary I.2.9$_1$ Let $z = x + \mathrm{i}y$. Then we have

$$e^z = e^x(\cos y + \mathrm{i}\sin y) \ , \qquad \text{and therefore}$$
$$\mathrm{Re}\ e^z = e^x \cos y \ ,$$
$$\mathrm{Im}\ e^z = e^x \sin y \ ,$$
$$|e^z| = e^x \ .$$

Corollary I.2.9$_2$ For arbitrary complex numbers $z, w \in \mathbb{C}$ we have the

---

Addition theorems

$$\cos(z + w) = \cos z \cos w - \sin z \sin w \ ,$$

$$\sin(z + w) = \sin z \cos w + \cos z \sin w \ .$$

The complex exponential function is *not injective*. We have

$$e^{2\pi i k} = 1 \text{ for any } k \in \mathbb{Z} \ .$$

From the supplement to Theorem I.1.5 there follows a more precise result:

Remark I.2.10 *We have for all* $z, w \in \mathbb{C}$

$$\exp(z) = \exp(w) \Longleftrightarrow z - w \in 2\pi i \mathbb{Z} \ ,$$

and, in particular,

$$\text{Ker } \exp := \left\{ \ z \in \mathbb{C} \ ; \quad \exp(z) = 1 \ \right\} = 2\pi i \ \mathbb{Z} \ .$$

For $w \in \mathbb{C}$, because of the functional equation for the exponential

$$\exp(z + w) = \exp(z)\exp(w) \ ,$$

we have the equation

$$\exp(z + w) = \exp(z) \qquad \text{for all } z \in \mathbb{C}$$

if and only if

$$\exp(w) = 1 \Longleftrightarrow w \in \text{Ker } \exp = 2\pi i \ \mathbb{Z} \ .$$

The equation

$$\text{Ker } \exp = 2\pi i \ \mathbb{Z}$$

can be interpreted, because of this, as a *periodicity property* of exp:

The complex exponential function is periodic, and has as periods the numbers and only the numbers

$$2\pi i \ k \ , \ k \in \mathbb{Z} \ .$$

Corollary I.2.10$_1$ For all $z \in \mathbb{C}$ and all $k \in \mathbb{Z}$ we have

$$\sin z = 0 \quad \Longleftrightarrow \quad z = k\pi \ ,$$

$$\cos z = 0 \quad \Longleftrightarrow \quad z = \left(k + \frac{1}{2}\right)\pi \ .$$

This is because, for example, $\sin z = (\exp(iz) - \exp(-iz))/2i = 0$ means nothing else but $\exp(2iz) = 1$, i.e. $z = k\pi$, $k \in \mathbb{Z}$. The complex sine and cosine functions therefore have only the roots, which are known for the real functions.

Its periodicity causes difficulties in inverting the complex exponential function to define a complex logarithm. To attack this problem we restrict the domain of definition of exp in a suitable way.

Principal Branch of the Logarithm

We shall denote by S the strip
$S = \{\, w \in \mathbb{C}\, ;\quad -\pi < \operatorname{Im} w \leq \pi \,\}$.
The restriction of exp to S is injective by I.2.10. Each value of exp is taken already in S. The image of exp is, by I.1.5, $\mathbb{C}^\bullet$, the plane punctured at 0. Because of this the complex exponential function gives a *bijective map*

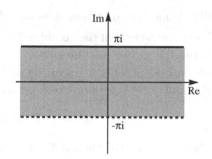

$$S \xrightarrow{\ \exp\ } \mathbb{C}^\bullet \ ,$$
$$w \longrightarrow e^w \ .$$

Therefore, to each point $z \in \mathbb{C}^\bullet$ there corresponds a uniquely determined number $w \in S$ with the property $e^w = z$. We call this number w the *principal value* of the logarithm of z and denote[3] it by

$$w = \operatorname{Log} z \ .$$

Therefore we have proved:

Theorem I.2.11 *There exists a function —the so-called **principal branch of the logarithm**—*

$$\operatorname{Log} : \mathbb{C}^\bullet \longrightarrow \mathbb{C} \ ,$$

which is uniquely determined by the following two properties:

(a) $\qquad\qquad \exp(\operatorname{Log} z) = z$,

(b) $\qquad\qquad -\pi < \operatorname{Im} \operatorname{Log} z \leq \pi \qquad$ *for all $z \neq 0$.*

Supplement. *From the equation*

$$\exp(w) = z$$

follows

$$w = \operatorname{Log} z + 2\pi i k \ , \quad k \in \mathbb{Z} \ .$$

Only if w is contained in S, one can conclude

$$w = \operatorname{Log} z \ .$$

---

[3] The notations $w = \log z$ and $w = \ln z$ are also common in the literature.

In particular, $\operatorname{Log} z$ coincides for positive real z with the usual real (natural) logarithm:

$$\operatorname{Log} z = \log z \ .$$

Remark I.2.12 *To each complex number $z \neq 0$ there corresponds a real number φ, which is uniquely determined by the following two properties:*

$$(a) \qquad\qquad -\pi < \varphi \leq \pi \ ,$$

$$(b) \qquad\qquad \frac{z}{|z|} = \cos\varphi + \mathrm{i}\sin\varphi \quad (= e^{\mathrm{i}\varphi}) \ .$$

This is an immediate consequence of I.1.5 and a special case of I.2.11.

The construction of the complex logarithm therefore contains a generalization of the representation of a complex number in polar coordinates.

We call the number φ occurring in I.2.12 the *principal value of the argument* of z and write (cf. the remark before I.1.6)

$$\varphi = \operatorname{Arg} z \ .$$

Theorem I.2.13 *For $z \in \mathbb{C}^{\bullet}$ one has*

$$\operatorname{Log} z = \log|z| + \mathrm{i}\operatorname{Arg} z \ .$$

Here $\log|z|$ is the usual real natural logarithm of the positive number $|z|$.

Proof. By Theorem I.2.11 it is sufficient to show:

$$\exp\bigl(\log|z| + \mathrm{i}\operatorname{Arg} z\bigr) = z \ ;$$

this follows immediately from I.2.9$_1$ and I.2.12.   □

We close this paragraph with a *warning about calculations with complex powers.*
If $a \in \mathbb{C}^{\bullet}$, $b \in \mathbb{C}$, then one can define $a^b := \exp(b \operatorname{Log} a)$. This definition is, however, artificially, since if b does not lie in $\mathbb{Z}$, then

$$\exp(\,b \operatorname{Log} a\,) \neq \exp\bigl(\,b(\operatorname{Log} a + 2\pi\mathrm{i}\,k)\,\bigr) \ , \quad k \in \mathbb{Z} \ .$$

Each number in

$$\bigl\{\exp(\,b(\log|a| + \mathrm{i}\operatorname{Arg} a)\,)\ \exp(2\pi\mathrm{i}\,bk) \ ; \quad k \in \mathbb{Z}\bigr\}$$

can be considered to be a^b in its own right.[4] Using the *principal value of the logarithm* one has, for instance,

$$\mathrm{i}^{\mathrm{i}} = \exp(\,\mathrm{i}\operatorname{Log}\mathrm{i}\,) = \exp(\,\mathrm{i}(\log|\mathrm{i}| + \mathrm{i}\operatorname{Arg}\mathrm{i})\,)$$

$$= \exp\left(\,\mathrm{i}\Bigl(0 + \mathrm{i}\frac{\pi}{2}\Bigr)\,\right) = \exp\Bigl(-\frac{\pi}{2}\Bigr) \approx 0.20787957635076190854\ldots \ .$$

---

[4] Each of these numbers is *one* of the bth powers of a.

All possible values of i^i lie, because of the equality

$$\exp\big(\, i(\log|i| + i\operatorname{Arg} i)\,\big)\, \exp\big(\, 2\pi i\, ik\,\big) = \exp\big(\, -(4k+1)\pi/2\,\big)\,,$$

in the following set of positive real numbers:

$$\left\{\, \exp\!\left(-(4k+1)\frac{\pi}{2}\right)\, ;\quad k \in \mathbb{Z}\,\right\}.$$

The nth roots of unity, i.e. the solutions of $z^n = 1$, are now exactly the $1/n$-th powers of 1. More general, the nth roots of a number $a \in \mathbb{C}^*$ are just the $1/n$-th powers of a.

Remark. The number $e^z := \exp(z)$ is *one* of the zth powers of e.

Care should be taken when one formally uses the exponentiation laws, which hold for real numbers. For example, in general it is not true that

$$(a_1 a_2)^b = a_1^b a_2^b\,.$$

Example. Set $a_1 = a_2 = -1$, and $b = 1/2$. Then (using the principal value) we have
$(a_1 a_2)^{1/2} = \big((-1)(-1)\big)^{1/2} = 1^{1/2} = 1 \neq -1 = i \cdot i = (-1)^{1/2}(-1)^{1/2} = a_1^{1/2} a_2^{1/2}\,.$
Which of the rules known for real numbers carry over to complex numbers has to be checked checked case by case. There is no difficulty if one defines $a^b = \exp(\, b \log a\,)$ for real and *positive* a, because one can take the usual real logarithm. Then the rules

$$(a_1 a_2)^b = a_1^b a_2^b \qquad (a_1 > 0,\; a_2 > 0)$$

are also valid for complex b.

Exercises for I.2

1. Let $z_0 = x_0 + iy_0 \neq 0$ be a given complex number. Define the sequence $(z_n)_{n\geq 0}$ recursively by

$$z_{n+1} = \frac{1}{2}\left(z_n + \frac{1}{z_n}\right)\,,\quad n \geq 0\,.$$

   Show:
   If $x_0 > 0$, then $\lim_{n\to\infty} z_n = 1$.
   If $x_0 < 0$, then $\lim_{n\to\infty} z_n = -1$.
   If $x_0 = 0$, $y_0 \neq 0$, then $(z_n)_{n\geq 0}$ is undefined or divergent.
   Hint. Consider $w_{n+1} = \dfrac{z_{n+1} - 1}{z_{n+1} + 1}$.

2. Let $a \in \mathbb{C}^*$ be given. For which $z_0 \in \mathbb{C}$ converges the sequence (z_n), which is recursively defined by

$$z_{n+1} = \frac{1}{2}\left(z_n + \frac{a}{z_n}\right) \qquad \text{for } n \geq 0\,?$$

3. A sequence $(z_n)_{n\geq 0}$ of complex numbers is called a *Cauchy sequence,* if for each $\varepsilon > 0$ there is an index $n_0 \in \mathbb{N}_0$, such that for all $n, m \in \mathbb{N}_0$ with $n, m \geq n_0$

$$|z_n - z_m| < \varepsilon\,.$$

   Show: A sequence $(z_n)_{n\geq 0}$, $z_n \in \mathbb{C}$ is convergent if and only if it is a Cauchy sequence.

4.   Prove the following inequalities.
     (a) For all $z \in \mathbb{C}$ we have

$$|\exp(z) - 1| \;\leq\; \exp(|z|) - 1 \;\leq\; |z| \exp(|z|) \;.$$

(b) For all $z \in \mathbb{C}$ with $|z| \leq 1$ we have

$$|\exp(z) - 1| \;\leq\; 2|z| \;.$$

5.   Determine, in each case, all the $z \in \mathbb{C}$ with

$$\exp(z) = -2 , \qquad \exp(z) = i , \qquad \exp(z) = -i ,$$
$$\sin z = 100 , \qquad \sin z = 7i , \qquad \sin z = 1 - i ,$$
$$\cos z = 3i , \qquad \cos z = 3 + 4i , \qquad \cos z = 13 .$$

6.   The (complex) hyperbolic functions cosh and sinh are defined similarly to the real ones. For $z \in \mathbb{C}$ let

$$\cosh z := \frac{\exp(z) + \exp(-z)}{2} \qquad \text{and} \qquad \sinh z := \frac{\exp(z) - \exp(-z)}{2} \;.$$

Show:
(a) $\sinh z = -i \sin(iz)$, $\cosh z = \cos(iz)$ for all $z \in \mathbb{C}$.
(b) Addition theorems

$$\boxed{\begin{aligned} \sinh(z + w) &= \sinh z \cosh w + \cosh z \sinh w \;, \\ \cosh(z + w) &= \cosh z \cosh w + \sinh z \sinh w \;. \end{aligned}}$$

(c) $\cosh^2 z - \sinh^2 z = 1$ for all $z \in \mathbb{C}$.
(d) sinh and cosh have the *period* $2\pi i$, i.e.

$$\begin{aligned} \sinh(z + 2\pi i) &= \sinh z \\ \cosh(z + 2\pi i) &= \cosh z \end{aligned} \qquad \text{for all } z \in \mathbb{C} .$$

(e) For all $z \in \mathbb{C}$ the series $\displaystyle\sum \frac{z^{2n}}{(2n)!}$ and $\displaystyle\sum \frac{z^{2n+1}}{(2n+1)!}$ are absolutely convergent, and one has

$$\cosh z = \sum_{n=0}^{\infty} \frac{z^{2n}}{(2n)!} \qquad \text{and} \qquad \sinh z = \sum_{n=0}^{\infty} \frac{z^{2n+1}}{(2n+1)!} \;.$$

7.   For all $z = x + iy \in \mathbb{C}$ one has:

(a) $\qquad \overline{\exp(z)} = \exp(\bar{z}) , \quad \overline{\sin(z)} = \sin(\bar{z}) , \quad \overline{\cos(z)} = \cos(\bar{z}) .$

(b)
$$\boxed{\begin{aligned} \cos z &= \cos(x + iy) = \cos x \cosh y - i \sin x \sinh y \;, \\ \sin z &= \sin(x + iy) = \sin x \cosh y + i \cos x \sinh y \;. \end{aligned}}$$

In the special case $x = 0$, $y \in \mathbb{R}$ we have

$$\cos(iy) = \frac{1}{2}(e^y + e^{-y}) = \cosh y \quad \text{and} \quad \sin(iy) = \frac{i}{2}(e^y - e^{-y}) = i \sinh y .$$

Determine all the $z \in \mathbb{C}$ with $|\sin z| \leq 1$, and find an $n \in \mathbb{N}$ such that

$$|\sin(in)| > 10\,000 .$$

8. **Definition of the tangent and cotangent**

For $z \in \mathbb{C} \setminus \{ (k + 1/2)\pi; \ k \in \mathbb{Z} \}$ let

$$\tan z := \frac{\sin z}{\cos z} ,$$

and for $z \in \mathbb{C} \setminus \{ k\pi; \ k \in \mathbb{Z} \}$ let

$$\cot z := \frac{\cos z}{\sin z} .$$

Show:

$$\tan z = \frac{1}{i} \frac{\exp(2iz) - 1}{\exp(2iz) + 1} , \quad \cot z = i \frac{\exp(2iz) + 1}{\exp(2iz) - 1} ,$$
$$\tan(z + \pi/2) = -\cot z, \quad \tan(-z) = -\tan z, \quad \tan z = \tan(z + \pi),$$
$$\tan z = \cot z - 2\cot(2z), \quad \cot(z + \pi) = \cot z.$$

9. Let $\mathrm{Maps}(\mathbb{N}_0, \mathbb{C})$ be the set of all maps of $\mathbb{N}_0$ into $\mathbb{C}$ (= the set of all complex sequences).

Show: The map

$$\sum : \mathrm{Maps}(\mathbb{N}_0, \mathbb{C}) \longrightarrow \mathrm{Maps}(\mathbb{N}_0, \mathbb{C}) ,$$
$$(a_n)_{n \geq 0} \longmapsto (S_n)_{n \geq 0} \text{ with } S_n := a_0 + a_1 + \cdots + a_n ,$$

is bijective (telescope trick). The theories of sequences and of infinite series are therefore in principle the same.

10. Let $(a_n)_{n \geq 0}$ and $(b_n)_{n \geq 0}$ be two sequences of complex numbers such that $a_n = b_n - b_{n+1}$, $n \geq 0$.

Show: The series $\sum_{n=0}^{\infty} a_n$ is convergent if and only if the sequence (b_n) is convergent, and then

$$\sum_{n=0}^{\infty} a_n = b_0 - \lim_{n \to \infty} b_{n+1} .$$

Example: $\displaystyle\sum_{n=0}^{\infty} \frac{1}{(n+1)(n+2)} = 1.$

11. **Binomial series**

For $\alpha \in \mathbb{C}$ and $\nu \in \mathbb{N}$ let

$$\binom{\alpha}{0} := 1 \quad \text{and} \quad \binom{\alpha}{\nu} := \prod_{j=1}^{\nu} \frac{\alpha - j + 1}{j} .$$

Show: $\sum_{\nu=0}^{\infty} \binom{\alpha}{\nu} z^{\nu}$ is absolutely convergent for all $z \in \mathbb{C}$ with $|z| < 1$.

Let $b_{\alpha}(z) := \sum_{\nu=0}^{\infty} \binom{\alpha}{\nu} z^{\nu}$.

Show: For all $z \in \mathbb{C}$ with $|z| < 1$ and arbitrary $\alpha, \beta \in \mathbb{C}$ we have

$$b_{\alpha+\beta}(z) = b_{\alpha}(z)\, b_{\beta}(z) \ .$$

Remark. We shall see later that for $z \in \mathbb{C}$ with $|z| < 1$ there holds

$$b_{\alpha}(z) = (1+z)^{\alpha} := \exp\big(\alpha \log(1+z)\big) \ .$$

For $\alpha = n \in \mathbb{N}_0$, one obtains the binomial formula

$$(1+z)^n = \sum_{\nu=0}^{n} \binom{n}{\nu} z^{\nu} \ .$$

12.   For $k \in \mathbb{N}_0$, and $z \in \mathbb{C}$ with $|z| < 1$, show

$$\frac{1}{(1-z)^{k+1}} = \sum_{n=0}^{\infty} \binom{n+k}{k} z^n = \sum_{n=0}^{\infty} \binom{n+k}{n} z^n \ .$$

13.   Let $(a_n)_{n \geq 0}$ and $(b_n)_{n \geq 0}$ be two sequences of complex numbers and

$$A_n := a_0 + a_1 + \cdots + a_n \ , \qquad n \in \mathbb{N}_0 \ .$$

Show: For each $m \geq 0$ and each $n \geq m$ we have

$$\sum_{\nu=m}^{n} a_{\nu} b_{\nu} = \sum_{\nu=m}^{n} A_{\nu}(b_{\nu} - b_{\nu+1}) - A_{m-1} b_m + A_n b_{n+1}$$

(ABEL's partial summation , N. H. ABEL, 1826)

where in the case $m = 0$ we set $A_{-1} = 0$ (empty sum).

14.   *Show:* Under the conditions of (13) a series of the form $\sum a_n b_n$ is always convergent if

  (a) the series $\sum a_n(b_n - b_{n+1})$ and

  (b) the sequence $(a_n b_{n+1})$ are convergent (N. H. ABEL, 1826).

15.   If $\sum a_n$ is convergent and $(b_n)_{n \geq 0}$ is a sequence of real numbers, which is monotone and bounded, then the series $\sum a_n b_n$ is convergent (P.G.L. DIRICHLET, 1863).

16.   Assume that the series $\sum a_n$ is absolutely convergent, and let $a := \sum_{n=0}^{\infty} a_n$. Suppose that the series $\sum b_n$ is convergent, and assume $b := \sum_{n=0}^{\infty} b_n$.

Show: if $c_n := \sum_{\nu=0}^{n} a_{\nu} b_{n-\nu}$, then the series $\sum c_n$ is convergent, and for $c := \sum_{n=0}^{\infty} c_n$ one has

$$c = ab \qquad \text{(MERTENS' theorem, F. MERTENS, 1875).}$$

17. Let $(a_n)_{n\geq 0}$ be a sequence of complex numbers, and let $(S_n) = (\sum_{\nu=0}^{n} a_\nu)$ be the associated sequence of partial sums. Set

$$\sigma_n := \frac{S_0 + S_1 + \cdots + S_n}{n+1} \ , \quad n \geq 0 \ .$$

Show: if (S_n) is convergent and $S := \lim_{n \to \infty} S_n$, then (σ_n) is also convergent, and

$$\lim_{n \to \infty} \sigma_n = S \ .$$

Show by a counterexample that, in general, one cannot deduce from the convergence of (σ_n) the convergence of (S_n).

18. Show that for $\varphi \in \mathbb{R} - 2\pi\mathbb{Z}$ and for all $n \in \mathbb{N}$ one has

$$\frac{1}{2} + \sum_{\nu=1}^{n} \cos \nu\varphi = \frac{\sin\big((n+1/2)\varphi\big)}{2\sin(\varphi/2)} \qquad \text{and}$$

$$\sum_{\nu=1}^{n} \sin \nu\varphi = \frac{\sin(n\varphi/2)\sin\big((n+1)\varphi/2\big)}{\sin(\varphi/2)} \ .$$

19. Show that for all $n \in \mathbb{N}$ one has

$$\boxed{\prod_{\nu=1}^{n-1} \sin \frac{\nu\pi}{n} = \frac{n}{2^{n-1}} \ .}$$

Hint. $z^n - 1 = \prod_{\nu=1}^{n}(z - \zeta^\nu), \quad \zeta := \cos\dfrac{2\pi}{n} + \mathrm{i}\sin\dfrac{2\pi}{n} \ .$

20. (a) For each of the following complex numbers calculate the principal value of the logarithm:

$$\mathrm{i} \ ; \quad -\mathrm{i} \ ; \quad -1 \ ; \quad x \in \mathbb{R} , \ x > 0 \ ; \quad 1 + \mathrm{i} \ .$$

(b) Calculate the principal values of the following numbers and compare them.

$$\big(\mathrm{i}(\mathrm{i} - 1)\big)^{\mathrm{i}} \text{ and } \mathrm{i}^{\mathrm{i}} \cdot (\mathrm{i} - 1)^{\mathrm{i}}$$

(c) Calculate

$$\{a^b\} := \big\{ \ \exp(\ b\log|a| + \mathrm{i}b\operatorname{Arg} a \ ) \ \exp(\ 2\pi\mathrm{i}\,bk \ ) \ ; \ k \in \mathbb{Z}\big\}$$

for

$$(a,b) \in \Big\{ (-1,\mathrm{i}), \ (1,\sqrt{2}), \ (-2,\sqrt{2}) \Big\}$$

and the corresponding principal values.

21. **Connection of *Arg* with *arccos***

Recall the definition of the *real* arccos: arccos is the inverse function of cos restricted to $[0, \pi]$, thus

$$\arccos t = \varphi \qquad \Longleftrightarrow \qquad 0 \leq \varphi \leq \pi \text{ and } \cos\varphi = t \ .$$

Show: For $z = x + \mathrm{i}y \neq 0$ we have

$$\operatorname{Arg} z = \begin{cases} \pi & \text{if } y = 0 \text{ and } x < 0 \ , \\ \operatorname{sign}(y) \arccos \dfrac{x}{\sqrt{x^2 + y^2}} & \text{otherwise} \ . \end{cases}$$

22. For $z, w \in \mathbb{C}^{\bullet}$ show

$$\mathrm{Log}(zw) = \mathrm{Log}(z) + \mathrm{Log}(w) + 2\pi i k(z,w) \qquad \text{where}$$

$$k(z,w) = \begin{cases} 0, & \text{if } -\pi < \mathrm{Arg}\, z + \mathrm{Arg}\, w \leq \pi, \\ +1, & \text{if } -2\pi < \mathrm{Arg}\, z + \mathrm{Arg}\, w \leq -\pi, \\ -1, & \text{if } \pi < \mathrm{Arg}\, z + \mathrm{Arg}\, w \leq 2\pi. \end{cases}$$

23. There is a problem posed by TH. CLAUSEN in *Journal für reine and angewandte Mathematik* (CRELLE's Journal) , Band 2 (1827), pages 286-287:

"If e is the base for the hyperbolic (= natural) logarithms, π denotes half of the perimeter of the unit circle and n is a positive or negative number, then it is well-known that

$$e^{2n\pi\sqrt{-1}} = 1,$$
$$e^{1+2n\pi\sqrt{-1}} = e,$$

and thus also that

$$e^{(1+2n\pi\sqrt{-1})^2} = e = e^{1+4n\pi\sqrt{-1}-4n^2\pi^2}.$$

However, since $e^{1+4n\pi\sqrt{-1}} = e$, it would follow from this that $e^{-4n^2\pi^2} = 1$, which is absurd. Find the mistake in the derivation of this result."

I.3 Continuity

Definition I.3.1 *A function*

$$f : D \longrightarrow \mathbb{R}^q, \quad D \subset \mathbb{R}^p,$$

*is called **continuous at a point** $a \in D$, if for each $\varepsilon > 0$ there exists $\delta > 0$ with the property* [5]

$$|f(z) - f(a)| < \varepsilon \quad \text{if} \quad |z - a| < \delta, \quad z \in D.$$
$$(\varepsilon\text{-}\delta\text{-Definition of Continuity})$$

An equivalent formulation is as follows: One has

$$f(a_n) \to f(a) \quad \text{for} \quad n \to \infty$$

for each sequence (a_n) converging to a (Sequence Criterion).

The function f is called continuous, if it is continuous at each point of D.

We are primarily interested in the case $p = q = 2$, i.e.

$$f : D \longrightarrow \mathbb{C}, \quad D \subset \mathbb{C}.$$

Then, from I.2.4 there follows:

---

[5] We denote by $|\cdot|$ the EUCLIDean norm (in $\mathbb{R}^p$ and $\mathbb{R}^q$).

Remark I.3.2 *The sum, difference and product of two continuous functions are continuous.*

Remark I.3.3 *The function*

$$\mathbb{C}^{\bullet} \longrightarrow \mathbb{C} , \quad z \longmapsto \frac{1}{z} ,$$

is continuous.

Let

$$f : D \longrightarrow \mathbb{C} \quad \text{and} \quad g : D' \longrightarrow \mathbb{C}$$

be two functions. If the image of f is contained in the domain of definition of g, $f(D) \subset D'$, then one can define the composition

$$g \circ f : D \longrightarrow \mathbb{C} ,$$
$$z \longmapsto g\big(f(z)\big) .$$

Remark I.3.4 *The composition of two continuous functions is continuous.*

If $f : D \to \mathbb{C}$ is a continuous function without zeros, then by I.3.3 and I.3.4 the following function is also continuous:

$$\frac{1}{f} : D \longrightarrow \mathbb{C} .$$

Remark I.3.5 *A function $f : D \to \mathbb{C}$, $D \subset \mathbb{C}$, is continuous if and only if the real and imaginary parts of f are continuous.*

$$(Re \ f)(z) := Re \ f(z) ,$$
$$(Im \ f)(z) := Im \ f(z) .$$

In particular the absolute value of a continuous function is continuous:

$$|f| = \sqrt{(Re \ f)^2 + (Im \ f)^2} .$$

Examples.

(1) Each polynomial

$$P(z) = a_0 + a_1 z + \cdots + a_n z^n , \quad n \in \mathbb{N}_0 , \ a_{\nu} \in \mathbb{C} , \ 0 \le \nu \le n ,$$

is continuous on $\mathbb{C}$.

(2) The functions

$$\exp , \ \sin \ \text{and} \ \cos : \mathbb{C} \longrightarrow \mathbb{C}$$

are continuous (since the real and imaginary parts are).

Let
$$f : D \longrightarrow \mathbb{C} , \qquad D \subset \mathbb{C} ,$$
be an injective function. Then the *inverse function*
$$f^{-1} : f(D) \longrightarrow D \subset \mathbb{C}$$
is well-defined. It is characterized by the properties
$$f\big(f^{-1}(w)\big) = w \qquad \text{for all } w \in f(D) ,$$
$$f^{-1}\big(f(z)\big) = z \qquad \text{for all } z \in D .$$

Remark I.3.6 *The inverse of a continuous function is not necessarily continuous.*

Example. We consider the principal value of the argument, restricted to the circle
$$S^1 := \{ z \in \mathbb{C} ; \quad |z| = 1 \} .$$
By definition, this function is the inverse function of the continuous function
$$] - \pi, \pi] \longrightarrow S^1, \quad x \longmapsto \cos x + \mathrm{i} \sin x ,$$
but itself it is not continuous. For this notice:

Remark I.3.7 *The function*
$$S^1 \longrightarrow \mathbb{C} , \quad z \longmapsto \operatorname{Arg} z ,$$
is discontinuous at $z = -1$.

Corollary I.3.7$_1$ The principal value of the logarithm is discontinuous along the negative real axis.

Proof of the Remark. Let
$$a_n = e^{(\pi - 1/n)\mathrm{i}} \text{ and } b_n = e^{(-\pi + 1/n)\mathrm{i}} , \quad n \in \mathbb{N} .$$
On the one hand
$$\operatorname{Arg} a_n = \pi - \frac{1}{n} \text{ and } \operatorname{Arg} b_n = -\pi + \frac{1}{n} ,$$
$$\Longrightarrow \qquad \lim_{n \to \infty} \operatorname{Arg} a_n = \pi \text{ and } \lim_{n \to \infty} \operatorname{Arg} b_n = -\pi ,$$
but, on the other hand, $\lim_{n \to \infty} a_n = \lim_{n \to \infty} b_n = -1 = e^{\pi \mathrm{i}} = e^{-\pi \mathrm{i}}$. Therefore Arg is discontinuous at $z = -1$. $\square$

One can also see that the restriction of Arg to S^1 is discontinuous in the following way: The set S^1 is compact (see I.3.10). If Arg were continuous then $] - \pi, \pi] = \operatorname{Arg}(S^1)$ would also be compact. However, this is not the case.

We quickly recall the usual topological notions in $\mathbb{R}^p$ (where we are especially interested in the case $p = 2$).

Definition I.3.8 *A subset $D \subset \mathbb{R}^p$ is called* **open**, *if for each $a \in D$ there exists an $\varepsilon > 0$ such that the ε-neighborhood (for $p = 2$ a disk)*

$$U_\varepsilon(a) := \left\{ \, z \in \mathbb{R}^p \, ; \quad |z - a| < \varepsilon \, \right\}$$

is contained in D.

Definition I.3.9 *A set $A \subset \mathbb{R}^p$ is called* **closed**, *if one of the two following equivalent conditions is satisfied.*

(a) *The complement*

$$\mathbb{R}^p - A = \left\{ \, z \in \mathbb{R}^p \, ; \quad z \notin A \, \right\}$$

   is open.
(b) *The limit of any convergent sequence of points in A is also in A.*

Definition I.3.10 *A set $A \subset \mathbb{R}^p$ is called* **compact**, *if for each covering*

$$A \subset \bigcup_{\lambda \in \Lambda} U_\lambda \qquad (\Lambda \text{ an arbitrary index set})$$

by a family $(U_\lambda)_{\lambda \in \Lambda}$ of open sets $U_\lambda \subset \mathbb{R}^p$ there is a finite sub-covering, i.e. there is a finite subset $\Lambda_0 \subset \Lambda$ with the property

$$A \subset \bigcup_{\lambda \in \Lambda_0} U_\lambda \, .$$

The following theorems are known from real analysis:

Theorem I.3.11 (Heine-Borel) *A set $A \subset \mathbb{R}^p$ is compact if and only if it is bounded and closed.*

Theorem I.3.12 *The image of a compact set $A \subset \mathbb{R}^p$ under a continuous map $f : \mathbb{R}^p \to \mathbb{R}^q$ is also compact. In particular, a continuous real-valued function i.e. $(q = 1)$ on A is bounded and takes its minimum and its maximum.*

Theorem I.3.13 *The inverse of a continuous injective function $f : A \to \mathbb{C}$ on a compact domain $A \subset \mathbb{C}$ is also continuous.*

Exercises for I.3

1.   Prove the equivalence of the ε-δ-continuity and sequence continuity in I.3.1.
2.   Using Exercise 21 in I.2 show that $\mathrm{Arg} : \mathbb{C}_- \to \mathbb{R}$ is continuous. Here $\mathbb{C}_-$ is the complex plane slit along the negative real axis :

$$\mathbb{C}_- := \mathbb{C} \setminus \left\{ \, t \in \mathbb{R} \, ; \, t \leq 0 \, \right\} .$$

   Deduce that the principal value of the logarithm is also continuous on $\mathbb{C}_-$.

3. Set $D \subset \mathbb{R}^p$.
    A point $a \in D$ is called an *interior point* (of D)
    if together with a there exists a ε-ball $U_\varepsilon(a) :=$
    $\{x \in \mathbb{R}^p ; \quad |x - a| < \varepsilon\}$ which is contained in D.
    Show: D is open $\iff$ each point of D is an interior
    point. A subset $U \subset \mathbb{R}^p$ is called a *neighborhood of*
    $a \in \mathbb{R}^p$ if U contains an ε-ball $U_\varepsilon(a)$.
    Show: D is open $\iff$ D is a neighborhood of each
    point $a \in D$.

    Let $\overset{\circ}{D} := \{ x \in D ; \quad D \text{ neighborhood of } x \}$

    Show: D is open $\iff$ $D = \overset{\circ}{D}$.
    $\overset{\circ}{D}$ is always open, and for each open subset $U \subset \mathbb{R}^p$
    with $U \subset D$ we have $U \subset \overset{\circ}{D}$.

4. Let $M \subset \mathbb{R}^p$. A point $a \in \mathbb{R}^p$ is called an *accumulation point* of M if for each
    ε-ball $U_\varepsilon(a)$ there holds

    $$U_\varepsilon(a) \cap (M \setminus \{a\}) \neq \emptyset .$$

    In each ε-ball for a there is therefore a point of M different from a.

    Notation. $M' := \{x \in \mathbb{R}^p ; \quad x \text{ is an accumulation point of } M\}$.

    Show: For a subset $A \subset \mathbb{R}^p$ the following are equivalent:

    (a) A is closed, i.e. $\mathbb{R}^p - A$ is open.
    (b) For each convergent sequence (a_n), $a_n \in A$ we have $\lim_{n \to \infty} a_n \in A$.
    (c) $A \supset A'$.

    Show that in addition:
    $$\bar{A} := A \cup A'$$

    is always closed, and for each closed set $B \subset \mathbb{R}^p$ with $B \supset A$ we have $B \supset \bar{A}$.
    $\bar{A}$ is called the *closure* (or *closed hull*) of A.

5. Let $(x_n)_{n \geq 0}$ be a sequence in $\mathbb{R}^p$. A point $a \in \mathbb{R}^p$ is called an *accumulation
    value* of the sequence (x_n) if for each ε-ball $U_\varepsilon(a)$ there are infinitely many
    indices n such that $x_n \in U_\varepsilon(a)$.

    Show (BOLZANO-WEIERSTRASS Theorem): Any bounded sequence (x_n), $x_n \in$
    $\mathbb{R}^p$ has an accumulation value.

    A subset $K \subset \mathbb{R}^p$ is called *sequence compact* if each sequence $(x_n)_{n \geq 0}$ with
    $x_n \in K$ has (at least) one accumulation value in K.

    Show: For a subset $K \subset \mathbb{R}^p$ the following are equivalent:

    (a) K is compact,
    (b) K is sequence compact.

    Remark. This equivalence holds for any metric space.

6. For all $z \in \mathbb{C}$
    $$\lim_{n \to \infty} (1 + z/n)^n = \exp(z) .$$

    More generally: For each sequence (z_n), $z_n \in \mathbb{C}$ with $\lim_{n \to \infty} z_n = z$ we have
    $$\lim_{n \to \infty} (1 + z_n/n)^n = \exp(z) .$$

7. Prove HEINE's theorem (E. HEINE, 1872):

    If $K \subset \mathbb{C}$ is compact and $f : K \to \mathbb{C}$ is continuous then f is *uniformly continuous* on K, i. e. for each $\varepsilon > 0$ there exists a $\delta > 0$ so that for all $z, z' \in K$ with $|z - z'| < \delta$,
    $$|f(z) - f(z')| < \varepsilon .$$

8. For any subsets $A, B \subset \mathbb{C}$,
    $$d(A, B) := \inf \{ |z - w| \; ; \quad z \in A , \; w \in B \}$$

    is called the *distance between A and B*. If $B = \{w\}$, then one simply writes $d(A, w)$ instead of $d(A, \{w\})$.

    Show:

(a) If $A \subset \mathbb{C}$ is a closed subset and $b \in \mathbb{C}$ is arbitrary, then there is an $a \in A$ with
    $$d(A, b) = |a - b| .$$

(b) If $A \subset \mathbb{C}$ is a closed subset and $B \subset \mathbb{C}$ is compact, then there are elements $a \in A$ and $b \in B$ such that
    $$d(A, B) = |a - b| .$$

9. There does not exist a function $f : \mathbb{C}^{\bullet} \to \mathbb{C}^{\bullet}$ with both of the following properties

    (a) $f(zw) = f(z)f(w)$ for all $z, w \in \mathbb{C}^{\bullet}$, and

    (b) $\left(f(z) \right)^2 = z$ for all $z \in \mathbb{C}^{\bullet}$.

10. *Show:*

    (a) There is no continuous function $f : \mathbb{C}^{\bullet} \to \mathbb{C}^{\bullet}$ such that
    $$\left(f(z) \right)^2 = z \text{ for all } z \in \mathbb{C}^{\bullet} .$$

    (b) There is no continuous function $q : \mathbb{C} \to \mathbb{C}$ such that
    $$\left(q(z) \right)^2 = z \text{ for all } z \in \mathbb{C} .$$

11. There is no continuous function $\varphi : \mathbb{C}^{\bullet} \to \mathbb{R}$ such that
    $$z = |z| \exp\left(i\varphi(z) \right) \text{ for all } z \in \mathbb{C}^{\bullet} .$$

12. There is no continuous function $l : \mathbb{C}^{\bullet} \to \mathbb{C}$ such that
    $$\exp\left(l(z) \right) = z \text{ for all } z \in \mathbb{C}^{\bullet} .$$

13. Let $n \geq 2$ be a natural number. There is no function $f : \mathbb{C}^{\bullet} \to \mathbb{C}^{\bullet}$ with the two properties

    (a) $f(zw) = f(z)f(w)$ for all $z, w \in \mathbb{C}^{\bullet}$,   and

    (b) $\left(f(z) \right)^n = z$ for all $z \in \mathbb{C}^{\bullet}$ $(n \in \mathbb{N} , \; n \geq 2)$.

14. Let $n \geq 2$ be a natural number. There is no continuous function $q_n : \mathbb{C} \to \mathbb{C}$ such that
    $$\left(q_n(z) \right)^n = z \text{ for all } z \in \mathbb{C} .$$

I.4 Complex Derivatives

Let $D \subset \mathbb{C}$ be a set of complex numbers. A point $a \in \mathbb{C}$ is called an *accumulation point* of D, if for each $\varepsilon > 0$ there exists a point

$$z \in D \quad \text{with} \quad 0 < |z - a| < \varepsilon \, .$$

Let $f : D \to \mathbb{C}$ be a function and $l \in \mathbb{C}$ a complex number. The statement

$$f(z) \to l \quad \text{for} \quad z \to a$$

by definition means:

(a) a is an accumulation point of D.
(b) The function

$$\tilde{f} : D \cup \{a\} \longrightarrow \mathbb{C} \, ,$$

$$z \longmapsto \tilde{f}(z) = \begin{cases} f(z) & \text{for } z \neq a \, , \ z \in D \, , \\ l & \text{for } z = a \, , \end{cases}$$

is continuous at a, therefore:
For any $\varepsilon > 0$ there exists a $\delta > 0$ such that

$$|f(z) - l| < \varepsilon \text{ for all } z \in D \, , \ z \neq a \text{ with } |z - a| < \delta \, .$$

It is easy to see that the limit l is unique.
We say: l is the limit of f at a (or approaching a). The notation

$$l = \lim_{\substack{z \to a \\ z \neq a}} f(z) \quad \text{or} \quad l = \lim_{z \to a} f(z)$$

is therefore justified. Notice that different notions of a limit are used in the literature, which differ in whether the point a is included or not.

Definition I.4.1 *A function*

$$f : D \longrightarrow \mathbb{C} \, , \quad D \subset \mathbb{C} \, ,$$

*is called **complex differentiable**, or does have a **complex derivative**, at the point $a \in D$ iff the following limit exists:*

$$\lim_{z \to a} \frac{f(z) - f(a)}{z - a} \, .$$

We denote this limit, if it exists, by $f'(a)$. (The function $z \mapsto \frac{f(z)-f(a)}{z-a}$ is defined in $D \setminus \{a\}$. By assumption a is an accumulation point of $D - \{a\}$ (and hence of D).

If f is differentiable at *each* point of D, then one can consider the complex derivative again as a function on D

$$f' : D \longrightarrow \mathbb{C} \, ,$$
$$z \longmapsto f'(z) \, .$$

Special Case. Let D be an interval in the real line, say

$$D = [a, b] \, , \quad a < b \, .$$

Decompose f into real and imaginary parts

$$f(x) = u(x) + iv(x) \, .$$

Here u and v are ordinary real functions of one real variable.

Obviously, f is complex differentiable if and only if the functions u and v are differentiable, and we have

$$f'(x) = u'(x) + iv'(x) \, .$$

Complex differentiability is therefore a generalization of real differentiability. We shall see, however, that the situation for *open* domains of definition $D \subset \mathbb{C}$ is completely different.

Sometimes another formulation of differentiability is useful:

Remark I.4.2 *Assume that $a \in D$ is an accumulation point of $D \subset \mathbb{C}$ and $f : D \to \mathbb{C}$ a function and $l \in \mathbb{C}$. Then the following statements are equivalent:*

(a) *f is complex differentiable at a, and there has the derivative l.*

(b) *There exists a function $\varphi : D \to \mathbb{C}$ which is continuous at a such that*

$$f(z) = f(a) + \varphi(z) \, (z - a) \quad and \quad \varphi(a) = l \, .$$

(c) *There exists a function $\rho : D \to \mathbb{C}$ which is continuous at a such that*

$$f(z) = f(a) + l \, (z - a) + \rho(z) \, (z - a) \quad and \quad \rho(a) = 0 \, .$$

(d) *If one defines $r : D \to \mathbb{C}$ by the equation*

$$f(z) = f(a) + l \, (z - a) + r(z) \, ,$$

then

$$\lim_{z \to a} \frac{r(z)}{z - a} = 0 \quad or \ equivalently \quad \lim_{z \to a} \frac{r(z)}{|z - a|} = 0 \, .$$

In each case $l = f'(a)$.

The equivalence of the assertions is obvious from their definitions.

Corollary I.4.2$_1$ A function which is differentiable at a is continuous at a.

As in the real case one shows the following *properties of permanence:*

Theorem I.4.3 *Let the functions $f, g : D \to \mathbb{C}$, $D \subset \mathbb{C}$, be complex differentiable at $a \in D$. Then the functions*

$$f + g ; \quad \lambda f , \ \lambda \in \mathbb{C} ; \quad f \cdot g ; \ \text{and} \ \frac{1}{f} , \ \text{if } f(a) \neq 0$$

are complex differentiable at a, and we have

$$(f + g)'(a) = f'(a) + g'(a) ,$$
$$(fg)'(a) = f'(a)g(a) + f(a)g'(a) ,$$
$$(\lambda f)'(a) = \lambda f'(a) ,$$
$$\left(\frac{1}{f}\right)'(a) = -\frac{f'}{f^2}(a) .$$

Application. The function

$$f(z) = z^n , \quad n \in \mathbb{Z} ,$$
(with domain of definition $\mathbb{C}$ if $n \geq 0$, and $\mathbb{C}^\bullet$ otherwise)

is complex differentiable, and

$$f'(z) = n \, z^{n-1} .$$

The reformulation of differentiability in Remark I.4.2 is of use in proving the Chain Rule.

Theorem I.4.4 (Chain Rule) *Assume that the functions*

$$f : D \longrightarrow \mathbb{C} \quad and \quad g : D' \longrightarrow \mathbb{C}$$

can be composed, i.e. $f(D) \subset D'$. In addition, assume that

$$f \ at \ a \in D \ and \ g \ at \ f(a) \in D'$$

are complex differentiable. Then their composition

$$g \circ f : D \longrightarrow \mathbb{C} ,$$
$$z \longmapsto g(f(z)) ,$$

is differentiable at $z = a$, and we have

$$(g \circ f)'(a) = g'(f(a)) \cdot f'(a) .$$

Proof. By assumption

$$f(z) - f(a) = \varphi(z)\,(z - a), \quad \varphi \text{ continuous at } a \text{ and } \varphi(a) = f'(a) \,,$$
$$g(w) - g(b) = \psi(w)\,(w - b), \quad \psi \text{ continuous at } b = f(a) \text{ and } \psi(b) = g'(b) \,.$$

Because of this (for $z \neq a$)

$$\frac{g\big(f(z)\big) - g\big(f(a)\big)}{z - a} \;=\; \psi\big(f(z)\big) \cdot \frac{f(z) - f(a)}{z - a} \,.$$

Passing to the limit we get

$$(g \circ f)'(a) \;=\; \psi\big(f(a)\big) f'(a) \;=\; g'\big(f(a)\big) f'(a) \,.$$

$\square$

Examples.

(1) By repeated application of the rules in I.4.3 one obtains that each polynomial

$$P(z) = \sum_{\nu=0}^{n} a_\nu \, z^\nu \qquad \text{with } n \in \mathbb{N}_0 \text{ and } a_\nu \in \mathbb{C} \text{ for } 0 \leq \nu \leq n \,,$$

is complex differentiable for all $z \in \mathbb{C}$ and that

$$P'(z) = \sum_{\nu=1}^{n} \nu a_\nu z^{\nu-1}.$$

(2) If $P, Q : \mathbb{C} \to \mathbb{C}$ are polynomials and $N(Q) = \{\, z \in \mathbb{C} ; \quad Q(z) = 0 \,\}$ is the set of roots of Q, then the rational function

$$f : \mathbb{C} \setminus N(Q) \longrightarrow \mathbb{C} \,,$$

$$z \longmapsto f(z) := \frac{P(z)}{Q(z)} \,,$$

is complex differentiable. These results immediately follow from Example (1) and the rules in I.4.3.

(3) We use the observation from the following section that the complex exponential function is complex differentiable and has itself as derivative (cf. Example 4):

$$\exp' = \exp \,,$$

and that the principal branch of the logarithm, Log, in the slit plane

$$\mathbb{C}_- := \mathbb{C} - \{ t \in \mathbb{R}; \ t \leq 0 \}$$

is complex differentiable with (cf. I.4, Exercise 6)

$$\mathrm{Log}'(z) = \frac{1}{z} \,.$$

Using the Chain Rule I.4.4 we then obtain that for $s \in \mathbb{C}$ the function

$$f : \mathbb{C}_- \longrightarrow \mathbb{C} \,,$$
$$z \longmapsto z^s := \exp(s \,\mathrm{Log}\, z) \,,$$

is complex differentiable and we have

$$f'(z) = \exp(s \,\mathrm{Log}\, z) \cdot s \frac{1}{z} = s z^{s-1} \,.$$

(4) Let $a \in \mathbb{C}$ and (c_ν) be a sequence of complex numbers. A series of the type

$$\sum_{\nu=0}^{\infty} c_\nu \,(z - a)^\nu$$

is called a *power series centered at* a with coefficients c_ν.

We assume that the power series

$$\sum_{\nu=0}^{\infty} c_\nu (z - a)^\nu$$

converges in the disk

$$U_R(a) = \{ z \in \mathbb{C} \,;\quad |z - a| < R \} \qquad (R > 0),$$

and, for $z \in U_R(a)$, we define

$$f(z) := \sum_{\nu=0}^{\infty} c_\nu (z - a)^\nu \,.$$

We shall later show (section III.2), that $f(z)$ is complex differentiable for all $z \in U_R(a)$ and that we have

$$f'(z) = \sum_{\nu=1}^{\infty} \nu c_\nu \,(z - a)^{\nu-1} \quad \text{(termwise differentiation of a power series)} \,.$$

From these results, for instance, the formulas $\exp' = \exp$, and also $\sin' = \cos$ and $\cos' = -\sin$ follow.

In the following sections we shall meet some other methods for checking complex differentiability.

Exercises for I.4

1. Prove the differentiation rules in Theorem I.4.3 using property (b) in I.4.2.

2. Investigate the continuity and complex differentiability of the following functions f. Find the derivatives at points where they exist.

   (a) $f(z) = z\mathrm{Re}\,(z)$, $f(z) = \bar{z}$,
       $f(z) = z\bar{z}$, $f(z) = z/\,|z|$, $z \neq 0$.

   (b) The exponential function exp is differentiable, and we have $\exp' = \exp$.

3. If the function $f : \mathbb{C} \to \mathbb{C}$ is complex differentiable at all points $z \in \mathbb{C}$ and takes only real or pure imaginary values, then f is constant.

4. Let $f : D \to \mathbb{C}$ be complex differentiable at $a \in D$ and $D^* := \{\, z \; ; \; \bar{z} \in D \,\}$. Then the function $g : D^* \to \mathbb{C}$ defined by

$$g(z) = \overline{f(\bar{z})}$$

   is complex differentiable at $\bar{a}$, and we have

$$g'(\bar{a}) = \overline{f'(a)} \ .$$

5. Prove the following *variant of the theorem of invertible functions*: Let D and $D' \subset \mathbb{C}$ be open and $f : D \to \mathbb{C}$ and $g : D' \to \mathbb{C}$ continuous functions with $f(D) \subset D'$ and $g(f(z)) = z$ for all $z \in D$.

   Show: If g is complex differentiable at $b = f(a)$ and $g'(b) \neq 0$, then f is complex differentiable at a, and we have

$$f'(a) = \frac{1}{g'(b)} \ .$$

6. By Exercise 2 in I.3 the principal value of the logarithm in the slit plane $\mathbb{C}_-$ is continuous. Show, by using Exercise 5, that it is actually complex differentiable in $\mathbb{C}_-$ and that there $\mathrm{Log}'(z) = 1/z$.

I.5 The Cauchy-Riemann Differential Equations

The starting point of our considerations is the formal similarity of Remark I.4.2 with the notion of total differentiability in real analysis:

A map

$$f : D \longrightarrow \mathbb{R}^q , \quad D \subset \mathbb{R}^p \ open ,$$

is called **totally differentiable** *at a point* $a \in D$, *if there exists an* $\mathbb{R}$-*linear map*

$$A : \mathbb{R}^p \longrightarrow \mathbb{R}^q ,$$

such that the remainder r *introduced by the equation*

$$f(x) - f(a) = A(x - a) + r(x)$$

satisfies

$$\lim_{x \to a} \frac{r(x)}{|x - a|} = 0 .$$

Here, $|x - a|$ *denotes the Euclidean distance between* x *and* a.

The linear map A is uniquely determined and is called the *Jacobian* of f at a (also the total differential of f at a, or the tangent map to f at a).

Notation. $A = J(f; a)$.

Looking back to I.4.2 shows that any function, which is complex differentiable at a point, is also differentiable at this point in the sense of the real analysis. More precisely:

Remark I.5.1 *For a function*

$$f : D \longrightarrow \mathbb{C} , \quad D \subset \mathbb{C} \ open , \quad a \in D ,$$

the following two statements are equivalent:

(a) *f is complex differentiable at a.*
(b) *f is totally differentiable at a (in the sense of real analysis by considering $\mathbb{C} = \mathbb{R}^2$), and the Jacobian*

$$J(f; a) : \mathbb{C} \longrightarrow \mathbb{C}$$

is of the form

$$J(f; a)z = l \, z$$

with l a suitable complex number. Of course the number l is the derivative $f'(a)$.

We are immediately led to the following question:

For which $\mathbb{R}$-linear maps $A : \mathbb{R}^2 \to \mathbb{R}^2$ exists a complex number $l \in \mathbb{C} = \mathbb{R}^2$ such that

$$Az = l \, z \ ?$$

In other words: When is an $\mathbb{R}$-linear map $A : \mathbb{R}^2 \to \mathbb{R}^2$ also $\mathbb{C}$-linear?

Remark I.5.2 *For an $\mathbb{R}$-linear map*

$$A : \mathbb{C} \longrightarrow \mathbb{C}$$

the following four statements are equivalent:

(1) *There exists a complex number l with $Az = lz$.*
(2) *A is $\mathbb{C}$-linear.*
(3) *$A(i) = iA(1)$.*

(4) *The matrix with respect to the canonical basis* 1 (= (1, 0)) *and* i (= (0, 1))
     has the special form

$$\begin{pmatrix} \alpha & -\beta \\ \beta & \alpha \end{pmatrix} \qquad (\alpha, \beta \in \mathbb{R}) \ .$$

Proof. The statements (1), (2) and (3) are trivially equivalent. It remains to
prove the equivalence of (1) and (4).

First we recall how to introduce the matrix corresponding to a linear map

$$A : \mathbb{R}^2 \longrightarrow \mathbb{R}^2 \ .$$

Since A is $\mathbb{R}$-linear, we have

$$A(x, y) = (ax + by, cx + dy)$$

with certain real numbers a, b, c, d. The corresponding matrix is

$$\begin{pmatrix} a & b \\ c & d \end{pmatrix} \ .$$

If one sets $A(x, y) =: (u, v)$, then this equation simply can be written as matrix
multiplication

$$\begin{pmatrix} u \\ v \end{pmatrix} = \begin{pmatrix} a & b \\ c & d \end{pmatrix} \begin{pmatrix} x \\ y \end{pmatrix} \ .$$

In doing this, we identify $\mathbb{C}$ with $\mathbb{R}^2$ via the isomorphism

$$\mathbb{C} \xrightarrow{\sim} \mathbb{R}^2$$
$$x + \mathrm{i}y \longmapsto \begin{pmatrix} x \\ y \end{pmatrix} \ .$$

Consider now the special case

$$Az = l z \ , \qquad l = \alpha + \mathrm{i}\beta \ ,$$

and thus

$$A(x, y) = (\ \alpha x - \beta y \ , \ \beta x + \alpha y\ ) \ , \quad \text{with } z = (x, y) \ .$$

This shows (1) $\Rightarrow$ (4). The converse also follows from this formula.   □

Each nonzero complex number $l \neq 0$ can be written in the form $l = r e^{\mathrm{i}\varphi}, r > 0$,
$\varphi \in \mathbb{R}$, (Theorem I.1.5). Multiplication by r effects a *dilation* by the factor
r, and multiplication by $e^{\mathrm{i}\varphi}$ gives a *rotation* by the angle φ. A map which
is given by multiplication with a complex number is called also a similarity
transformation.

The maps of the complex plane $\mathbb{C}$ *into itself, which can be written as multiplication with a nonzero complex number, are exactly the **similarity transformations (rotation-dilations)**.*

Similarity transformations are obviously **angle-preserving** and are also **orientation-preserving**. In fact, a converse holds too; cf. Remark I.5.14.

From real analysis we know how to compute the JACOBI matrix —i.e. the matrix of the JACOBIian— of a totally differentiable function. To do this we split f into its real and imaginary part: $f(z) = u(x, y) + iv(x, y)$, $z = x + iy$.

Let the map

$$f : D \longrightarrow \mathbb{R}^2 , \quad D \subset \mathbb{R}^2 \ open \ ,$$

be totally differentiable at $a \in D$. Then the partial derivatives of u and v exist at a, and we have

$$J(f; a) \longleftrightarrow \begin{pmatrix} \dfrac{\partial u}{\partial x}(a) & \dfrac{\partial u}{\partial y}(a) \\ \dfrac{\partial v}{\partial x}(a) & \dfrac{\partial v}{\partial y}(a) \end{pmatrix}$$

(= the Jacobian matrix of f at a).

One can summarize remarks I.5.1 and I.5.2 as follows:

Theorem I.5.3 (A.L. Cauchy, 1825; B. Riemann, 1851) *For a function*

$$f : D \longrightarrow \mathbb{C} , \quad D \subset \mathbb{C} \ open , \quad a \in D ,$$

the following two statements are equivalent:

(a) *f is complex differentiable at a.*
(b) *f is totally differentiable at a in the sense of real analysis ($\mathbb{C} = \mathbb{R}^2$), and for $u := \mathrm{Re} f$ and $v := \mathrm{Im} f$ the following differential equations hold:*

$$\boxed{\begin{array}{c} \textit{Cauchy-Riemann differential equations} \\ \dfrac{\partial u}{\partial x}(a) = \dfrac{\partial v}{\partial y}(a) , \qquad \dfrac{\partial u}{\partial y}(a) = -\dfrac{\partial v}{\partial x}(a) . \end{array}}$$

In case that (a) or (b) holds one has:

$$f'(a) = \frac{\partial u}{\partial x}(a) + i\frac{\partial v}{\partial x}(a) = \frac{\partial v}{\partial y}(a) - i\frac{\partial u}{\partial y}(a) .$$

Remark on notation. Instead of

$$\frac{\partial u}{\partial x}(a) , \qquad resp. \qquad \frac{\partial u}{\partial y}(a)$$

one often writes

$$u_x(a) \ or \ \partial_1 u(a) , \qquad resp. \qquad u_y(a) \ or \ \partial_2 u(a) ,$$

and correspondingly for v. For the functional determinant of a complex differentiable function $f = u + iv$ we obtain

$$\det J(f;a) \;=\; u_x(a)^2 + v_x(a)^2 \;=\; u_y(a)^2 + v_y(a)^2 \;=\; |f'(a)|^2 \;,$$

which is therefore non-negative, and, in fact, positive when $f'(a)$ is different from 0.

It should be mentioned that the CAUCHY-RIEMANN differential equations can also be derived simply as follows:
If the function $f : D \to \mathbb{C}$, $D \subset \mathbb{C}$ open, is complex differentiable at $a \in D$, then, in particular,

$$f'(a) = \lim_{h \to 0} \frac{f(a+h) - f(a)}{h} = \lim_{h \to 0} \frac{f(a+ih) - f(a)}{ih} \;,$$

where h varies only over real numbers. If one decomposes f into real and imaginary parts,

$$f = u + iv \;,$$

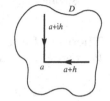

then it follows that

$$f'(a) = \partial_1 u(a) + i\partial_1 v(a) = \frac{1}{i}\big[\partial_2 u(a) + i\partial_2 v(a)\big] \;.$$

From this the CAUCHY-RIEMANN equations follow immediately. However, this proof does not provide the converse assertion without extra trouble, i.e. that the differentiability of f follows from the CAUCHY-RIEMANN equations (with total differentiability assumed).

It is well known that the existence of partial derivatives does not imply that f has a total derivative. But the following *sufficient criterion* for total differentiability is known from real analysis:

If the partial derivatives of the map

$$f : D \longrightarrow \mathbb{R}^q \;, \quad D \subset \mathbb{R}^p \ open \;,$$

exist at each point and are continuous, then f is totally differentiable.

Examples.

(1) We already know that the function f with

$$f(z) = z^2 \quad \text{or more generally} \quad f(z) = z^n \;, \; n \in \mathbb{N} \;,$$

is complex differentiable. Therefore the CAUCHY-RIEMANN equations must hold. From

$$f(z) = (x + iy)^2 = x^2 - y^2 + 2ixy$$

i.e.

$$u(x,y) = x^2 - y^2 \;, \qquad v(x,y) = 2xy \;,$$

it follows

$$\partial_1 u(x,y) = 2x , \qquad \partial_2 u(x,y) = -2y ,$$
$$\partial_1 v(x,y) = 2y , \qquad \partial_2 v(x,y) = 2x .$$

Therefore the CAUCHY-RIEMANN equations are satisfied.
(2) The function

$$\sigma(z) = \overline{z}$$

is not complex differentiable anywhere, because

$$u(x,y) = x , \qquad v(x,y) = -y ,$$

giving

$$1 = \partial_1 u \neq \partial_2 v = -1 .$$

Theorem I.5.4 *The functions* exp, sin *and* cos *are complex differentiable in the entire complex plane* $\mathbb{C}$, *and*

$$\exp' = \exp , \quad \sin' = \cos , \quad \cos' = -\sin .$$

Proof. We have for instance

$$\exp(z) = e^x(\cos y + \mathrm{i}\sin y) ,$$

i.e.

$$u(x,y) = e^x \cos y , \quad v(x,y) = e^x \sin y .$$

The CAUCHY-RIEMANN equations can be easily checked, as can then the expressions for the derivatives; the latter are continuous.   □

Remark I.5.5 (Characterization of locally constant functions)   *Let* $D \subset \mathbb{C}$ *be open, and* $f : D \to \mathbb{C}$ *a function. Then there are equivalent:*

(a) *f is locally constant in D.*
(b) *f is complex differentiable for all $z \in D$ and*

$$f'(z) = 0 \text{ for all } z \in D .$$

Supplement.   *In particular, any complex differentiable function in D with only real (or purely imaginary) values is locally constant in D.*

We are using the term *locally constant* for a function f which is constant in some neighborhood of any point. (A set $U \subset \mathbb{C}$ is called a *neighborhood* of a, if U contains a full disk around a.)

Proof. It is only necessary to show (b) $\Rightarrow$ (a):
If $f = u + \mathrm{i}v$, then $f' = u_x + \mathrm{i}v_x$; $u_x = v_y$ and $u_y = -v_x$. Therefore

$$u_x(a) = u_y(a) = 0 \text{ and } v_x(a) = v_y(a) = 0$$

for all $a \in D$. It is well known from real analysis that then u and v are locally constant in D. Therefore $f = u + \mathrm{i}v$ is locally constant in D as well.

Let f be a complex differentiable function that has only real values. It follows from the CAUCHY-RIEMANN equations that the derivative of f vanishes, and thus the function f is locally constant. $\qquad\qquad\qquad\qquad\square$

For example, the functions $f(z) = |\sin z|$ and $g(z) = \operatorname{Re} z$ are not complex differentiable in $\mathbb{C}$.

From this we see that being "complex differentiable" is a very strong restriction.

Terminology. *A function*

$$f : D \longrightarrow \mathbb{C} \ , \ D \subset \mathbb{C} \ \text{open} \ ,$$

which is complex differentiable at every point of D is also called (complex) analytic or holomorphic or regular in D.

f is called analytic at a point $a \in D$, iff there exists an open neighborhood $U \subset D$ of a such that f is analytic in U.

Example. The function $f(z) = z\bar{z}$ is complex differentiable at $a = 0$, but is not analytic at 0.

In the following, we shall use the terminology "analytic" in preference to the alternatives "complex differentiable" or "holomorphic" in D.

Definition I.5.6 *A subset $D \subset \mathbb{C}$ is called **connected** iff each locally constant function $f : D \to \mathbb{C}$ is constant.*

With this we can formulate the supplement to I.5.5 in the following way:

The real part of a function which is analytic in a connected open set $D \subset \mathbb{C}$ is uniquely determined by its imaginary part, up to an additive constant.

Namely, if f and g are two analytic functions with the same imaginary part, then $f - g$ has only real values.

We obtained the CAUCHY-RIEMANN equations as a result of the trivial remark I.5.1. As another application we prove the complex *Implicit Function Theorem* using the corresponding real theorem.

Theorem I.5.7 (Implicit Function) *Let there be given an analytic function*

$$f : D \longrightarrow \mathbb{C} \ , \ D \subset \mathbb{C} \ \text{open} \ ,$$

with continuous derivative.

Part 1 *Assume that at a point $a \in D$ we have $f'(a) \neq 0$. Then there exists an open set*

$$D_0 \ , \quad D_0 \subset D \ , \quad a \in D_0 \ ,$$

such that the restriction $f|D_0$ is injective and such that $f'(z) \neq 0$ for $z \in D_0$.

Part 2 *Assume that f is injective and $f'(z) \neq 0$ for all $z \in D$. Then its image $f(D)$ is an open subset of $\mathbb{C}$. The inverse function*

$$f^{-1} : f(D) \longrightarrow \mathbb{C}$$

is analytic, and its derivative is

$$f^{-1\prime}\big(f(z)\big) = \frac{1}{f'(z)} \ .$$

Later we will see that the derivative of an analytic function is always continuous (and, in fact, analytic); cf. II.3.4.

Proof of I.5.7. We use the analogous theorem from real analysis.

Part 1. We must know that the JACOBI map

$$J(f; a) : \mathbb{R}^2 \longrightarrow \mathbb{R}^2$$

is an isomorphism, and thus bijective. This follows from I.5.1:

$$J(f; a)z = f'(a)z \ , \quad f'(a) \neq 0 \ .$$

Part 2. The real implicit function theorem says in addition: The range of
a continuously partial differentiable (and thus totally differentiable) map is
open, if the JACOBI map is an isomorphism for all $a \in D$. If f is, in addition,
injective, then the inverse map is also totally differentiable, and the JACOBI
map of f^{-1} at $f(a)$ is exactly the inverse map to $J(f; a)$, i.e.

$$J(f; a)^{-1} = J\big(f^{-1}; f(a)\big) \ .$$

Taking into account that the inverse map of

$$\mathbb{C} \longrightarrow \mathbb{C} , \quad z \mapsto l\,z \quad (l \in \mathbb{C}^{\bullet}) \ ,$$

is given by $z \mapsto l^{-1}\,z$, we are done. $\qquad\square$

Remark. It is unsatisfying that the rather involved theorem of inverse functions
of real analysis had to be used in the proof. A simple function theoretic proof
therefore would be welcome. We shall return to find such a proof later (see
also III.7.6).

Example. The exponential function exp is complex differentiable and its
derivative does not vanish anywhere. The restriction of exp to the domain

$$-\pi < \operatorname{Im} z \leq \pi$$

is injective. However this region is not open. Thus we restrict exp to a somewhat smaller open region

$$D := \{\, z \in \mathbb{C} ; \quad -\pi < \operatorname{Im} z < \pi \,\} \ .$$

Obviously we have

$$\exp(D) = \mathbb{C}_- = \mathbb{C} \setminus \{\, x \in \mathbb{R} ; \ x \leq 0 \,\}$$

(the complex plane slit along the negative real axis).

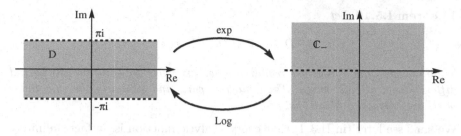

From the implicit function theorem there now follows:

Theorem I.5.8 *The principal branch of the logarithm is analytic in the plane slit along the negative real axis* $\mathbb{C}_-$, *and there we have*

$$\text{Log}'(z) = \frac{1}{z} \ .$$

We have already shown that the principal branch of the logarithm is not even continuous at the points of the negative real axis. More precisely:

Remark I.5.9 *If $a < 0$ is a negative real number then*

$$\lim_{\substack{z \to a \\ Im\ z > 0}} \text{Log}\, z = \log |a| + \pi i \qquad (= \text{Log}\, a) \ ,$$

$$\lim_{\substack{z \to a \\ Im\ z < 0}} \text{Log}\, z = \log |a| - \pi i \ .$$

The principal branch of the logarithm therefore makes a "jump of $2\pi i$" when crossing the negative real axis.

In connection with the CAUCHY-RIEMANN equations the following question arises. Suppose that we are given a "sufficiently smooth" —let us say twice continuously differentiable— function

$$u : D \longrightarrow \mathbb{R} \ , \quad D \subset \mathbb{R}^2 \text{ open} \ .$$

Can one find an analytic function $f : D \to \mathbb{C}$ whose real part is u?
If there is such a function f, then from the CAUCHY-RIEMANN differential equations we have

$$\partial_1^2 u = \quad \partial_1(\partial_2 v) \ ,$$
$$\partial_2^2 u = -\partial_2(\partial_1 v) \ .$$

By a theorem H.A. SCHWARZ, partial derivatives commute. We obtain the *Laplace differential equation*

$$\Delta u := \left(\frac{\partial^2}{\partial x^2} + \frac{\partial^2}{\partial y^2} \right) u = 0 \ .$$

Functions which satisfy this differential equation are called *potential functions* or *harmonic functions* and $\Delta = \partial_1^2 + \partial_2^2$ is called the *Laplace operator*.

Theorem I.5.10 *Let*

$$f : D \longrightarrow \mathbb{C} , \ D \subset \mathbb{C} \ open ,$$

be an analytic function whose real and imaginary parts are continuously partial differentiable at least twice. Then the real part, and similarly the imaginary part, are harmonic functions.

We shall see later (in II.3.4) that every analytic function is, in fact, infinitely often complex differentiable. So the real and imaginary parts are, in particular, infinitely often partial differentiable.

Examples of harmonic functions can therefore be obtained by considering the real and imaginary parts of analytic functions:

$$
\begin{aligned}
u(x,y) &= & x^3 - 3xy^2 & = \operatorname{Re}\left(z^3\right) , \\
v(x,y) &= & 3x^2 y - y^3 & = \operatorname{Im}\left(z^3\right) , \\
u(x,y) &= & \cos x \cosh y & = \operatorname{Re}\left(\cos z\right) , \\
v(x,y) &= - & \sin x \sinh y & = \operatorname{Im}\left(\cos z\right) .
\end{aligned}
$$

Is it true that any harmonic function is the real part of some analytic function? For some particular domains of definition the answer is yes.

Theorem I.5.11 *Let $D \subset \mathbb{C}$ be an open rectangle, whose sides are parallel to the coordinate axes. Let $u : D \to \mathbb{R}$ be a harmonic function. Then there is an analytic function $f : D \to \mathbb{C}$ with real part u.*

The function f is uniquely determined up to a purely imaginary constant. The harmonic function $v : D \to \mathbb{R}$ with $f = u + iv$ is called a conjugate harmonic function to u. It is uniquely determined up to an additive real constant. (The theorem is true more generally for "simply connected" regions $D \subset \mathbb{C}$; see also the remark at the end of II.2, and Appendix C to Chapter IV.)

Proof of I.5.11. Let

$$D =]a, b[\times]c, d[\quad \text{with} \quad a < b \text{ and } c < d .$$

We choose two points

$$x_0 \in]a, b[\quad \text{and} \quad y_0 \in]c, d[.$$

From the equation $\partial_1 u = \partial_2 v$ follows that, for each $x \in]a, b[$,

$$v(x,y) = \int_{y_0}^{y} \partial_1 u(x,t) \, dt + h(x) .$$

From LEIBNIZ's rule the integral is differentiable as a function of x and we have

$$\partial_1 v(x,y) = \int_{y_0}^y \partial_1^2 u(x,t)\,dt + h'(x) = -\int_{y_0}^y \partial_2^2 u(x,t)\,dt + h'(x)$$
$$= \partial_2 u(x,y_0) - \partial_2 u(x,y) + h'(x)\ ,$$

and therefore

$$h'(x) = -\partial_2 u(x,y_0)\ .$$

This suggests to try the following:

$$v(x,y) := \int_{y_0}^y \partial_1 u(x,t)\,dt - \int_{x_0}^x \partial_2 u(t,y_0)\,dt\ .$$

Now we have to verify the CAUCHY-RIEMANN equations using the fundamental theorem of calculus and LEIBNIZ's rule. (The LEIBNIZ rule is formulated and proved in II.3.) In the following, theorem I.5.11 is no longer used.   □

Remarkably, the function

$$u(x,y) := \log\sqrt{x^2 + y^2}$$

is a harmonic function in the entire region $\mathbb{R}^2 \setminus \{(0,0)\} = \mathbb{C}^\bullet$. However, there is no analytic function $f : \mathbb{C}^\bullet \to \mathbb{C}$ with

$$\operatorname{Re} f(z) = \log\sqrt{x^2 + y^2} = \log|z|\ ,$$

since f would have to agree in the plane slit along the negative real axis with $\operatorname{Log} z$, up to an additive constant. But then f could not be continuous at the points of the negative real axis. Theorem I.5.11 therefore is not true for arbitrary regions $D \subset \mathbb{C}$. But in the plane slit along the negative real axis the principal branch Log of the logarithm is an analytic function with $\operatorname{Re} \operatorname{Log} = u$.

Remark.

(1) The construction of the conjugate harmonic function v of u involves integration, similarly the following consideration.

   If a harmonic function u is given in D and one defines

$$g : D \longrightarrow \mathbb{C} \quad \text{by} \quad g = \partial_1 u - i\partial_2 u\ ,$$

   then g is analytic (one checks this using the CAUCHY-RIEMANN equations). If D is an open rectangle (or more generally a so-called elementary domain), then there is an analytic function $f : D \to \mathbb{C}$ with $f' = g$, as we shall show in the next chapter. If $f = U + iV$, then

$$f' = \partial_1 U + i\partial_1 V = \partial_1 U - i\partial_2 U = g = \partial_1 u - i\partial_2 u$$

   and therefore $U = u + \text{const}$. The real part U of the analytic function f agrees, up to an additive constant, with the given harmonic function

u, and for v one may choose V. The question of whether for a given harmonic function $u : D \to \mathbb{R}$ there is an analytic function $f : D \to \mathbb{C}$ with Re $f = u$ is equivalent to *find a primitive*. We shall consider the question of the existence of primitives in the next chapter.

(2) In addition, the LAPLACE equation

$$\partial_1^2 u + \partial_2^2 u = 0$$

is precisely the *"exactness condition"* of the partial differential equation system

$$\boxed{\begin{aligned} \partial_1 u &= \partial_2 v \ , \\ \partial_2 u &= -\partial_1 v \ , \end{aligned}}$$

where u is the known and v is the unknown function. In other words, this is the *"integrability condition"* for the vector field

$$D \longrightarrow \mathbb{R}^2 \ ,$$
$$(x,y) \longmapsto \left(-\partial_2 u(x,y) \ , \ \partial_1 u(x,y) \right) \ .$$

Example. We find $a \in \mathbb{R}$ so that the function defined by

$$u_a : \mathbb{R}^2 \longrightarrow \mathbb{R} \ ,$$
$$(x,y) \longmapsto x^3 + axy^2 \ ,$$

is harmonic, and also determine all the conjugate harmonic functions to u_a, i.e. the analytic functions $f : \mathbb{C} \to \mathbb{C}$ with Re $f = u_a$. From

$$0 = \Delta u_a(x,y) = 6x + 2ax \quad \text{for all} \quad x,y \in \mathbb{R}^2$$

it follows that $a = -3$, and $u := u_{-3}$ is harmonic. We find f, resp. v, using the two methods above and another method.

First method. Construction using the method of the first proof of Theorem I.5.11: We choose $(x_0, y_0) = (0,0)$ and obtain

$$u(x,y) = x^3 - 3xy^2 \quad \Longrightarrow \quad \begin{cases} \partial_1 u(x,y) = 3x^2 - 3y^2 \ , \\ \partial_2 u(x,y) = -6xy \ . \end{cases}$$

From this we have $\partial_2 u(x,0) = 0$ and so

$$v(x,y) = \int_0^y (3x^2 - 3t^2)\, dt = 3x^2 y - y^3 \ .$$

Therefore v is a conjugate harmonic function to u, and

$$f(z) = x^3 - 3xy^2 + \mathrm{i}(3x^2 y - y^3) = z^3$$

is an analytic function with $\operatorname{Re} f = u$. All other analytic functions with this property can be obtained, by Theorem I.5.11, after addition of a purely imaginary constant to f.

Second method. Define g by

$$g(z) = 3x^2 - 3y^2 + i6xy = 3(x + iy)^2 = 3z^2 .$$

Obviously g is analytic and an analytic function $f : \mathbb{C} \to \mathbb{C}$ with $f' = g$ is given by $f(z) = z^3$:

$$\operatorname{Im}(z^3) = \operatorname{Im}\left((x + iy)^3\right) = 3x^2 y - y^3 =: v(x, y) .$$

Third method. Construct

$$f(z) := 2u\left(\frac{z}{2}, \frac{z}{2i}\right) - u(0,0)$$

and obtain $f(z) = z^3$ (cf. Exercise 19 in I.5).

Elementary facts concerning conformal maps

Definition I.5.12 *A bijective $\mathbb{R}$-linear map $T : \mathbb{R}^n \to \mathbb{R}^n$ is called*

(a) **orientation-preserving**, *if* $\det T > 0$,
(b) **angle-preserving** *if, for all* $x, y \in \mathbb{R}^n$,

$$|Tx|\,|Ty|\,\langle x, y\rangle = |x|\,|y|\,\langle Tx, Ty\rangle .$$

Here $\langle\,,\,\rangle$ denotes the standard scalar product.

Remark. In the case $n = 2$ conditions (a) and (b) mean only that the *oriented angle* between z and w is preserved (cf. Exercise 4 in I.1).

Note: The $\mathbb{R}$-linear map $\mathbb{C} \to \mathbb{C}$, $z \mapsto \bar{z}$, is angle-preserving, but is not also orientation-preserving!

Definition I.5.13 *A continuously differentiable map*

$$f : D \longrightarrow D' , \quad D, D' \subset \mathbb{R}^n \text{ open} ,$$

is called (locally) **conformal** , *if the Jacobi map $J(f; a)$ is angle and orientation preserving at each point $a \in D$.*

If, in addition, f is bijective, then f is called globally conformal.

For $n = 2$ we have (cf. Exercise 18 in I.5):

Remark I.5.14 *An $\mathbb{R}$-linear map of the complex plane ($\mathbb{C} = \mathbb{R}^2$) to itself is a similarity transformation if and only if it is both angle and orientation preserving.*

Therefore we have

Theorem I.5.15 *A map*

$$f : D \longrightarrow D' , \quad D, D' \subset \mathbb{C} \ open ,$$

is locally conformal if and only if it is analytic and its derivative is analytic and does not vanish anywhere.

The geometrical significance of a conformal map is the following:

The oriented angle between two regular curves in D at an intersection point $a \in D$ is equal to the oriented angle between the image curves at their intersection $f(a)$.

(The notion " regular" will be made more precise in Exercise 11 in II.1.)

Example. The exponential function exp provides a globally conformal map of the strip $-\pi < \operatorname{Im} z < \pi$ onto the slit plane $\mathbb{C}_-$.

At points where the derivative of an analytic function vanishes, angles need not be preserved by this function. This can be seen from the example of the function $f(z) = z^n$, $n \geq 2$. The angle at the origin is obviously multiplied by n.

Geometrical visualization of complex functions

In differential calculus one often tries to visualize a function $f : D \to \mathbb{R}$, $D \subset \mathbb{R}$, by its graph:
$G(f) := \{ (x, y) \in D \times \mathbb{R} ; \quad y = f(x) \}.$

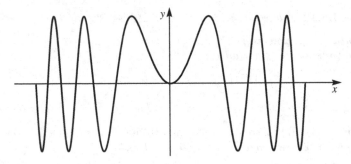

For a subset $D \subset \mathbb{R}^2$ and a function $f : D \to \mathbb{R}$ one can similarly visualize f using its graph

$$G(f) = \{ (x, y, z) \in D \times \mathbb{R} ; \quad z = f(x, y) \} \subset \mathbb{R}^3$$

as a "surface" in $\mathbb{R}^3$ (here $f(x, y) = x^3 - 3xy^2$):

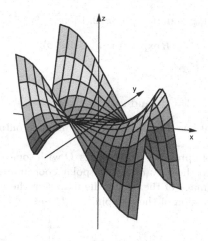

For a map $f : D \to \mathbb{C}$, $(D \subset \mathbb{C})$, the graph is a subset of $\mathbb{R}^4$, which cannot be visualized. But there is also here an adequate method to visualize such maps. For this the following point of view is very useful: One thinks of two copies of the complex plane, a z- or x-y-plane and a w- or u-v-plane:

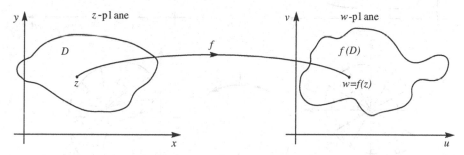

To visualize a map $f : D \to \mathbb{C}$ with Re $f = u$ and Im $f = v$ one can use various methods.

First method. The points $z \in D$ are mapped by f into which w-points? One gets a first impression if one can see both D and $f(D)$ explicitly.

For example, let $D := \{\, z \in \mathbb{C} \ ; \quad \text{Re } z > 0 \text{ and Im } z > 0 \,\}$, the so-called "first quadrant", and let $f : D \to \mathbb{C}$ be defined by $z \mapsto z^2$.

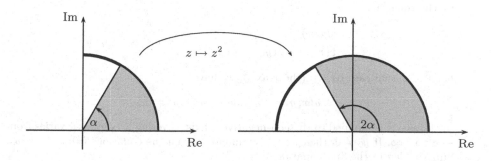

If one sets $z := r\exp(i\varphi)$, $r > 0$, $0 < \varphi < \pi/2$, then

$$z^2 = R\exp(i\psi) = r^2\exp(i2\varphi) \ ,$$

and therefore

$$R = r^2 \quad \text{and} \quad \psi \equiv 2\varphi \ (\text{mod } 2\pi) \ .$$

Clearly the first quadrant has been "opened" and mapped onto the so-called "upper half-plane" $\mathbb{H} := \{ z \in \mathbb{C} \ ; \ \operatorname{Im} z > 0 \}$.
One gets a more precise impression if one marks D with some sort of a covering net, e.g. with lines parallel to the axes or with a polar coordinate net (as we just did) and then considers the image of the net by the map f in the w-plane. The finer the net, the better the impression of the mapping $f : D \to \mathbb{C}$ will be.

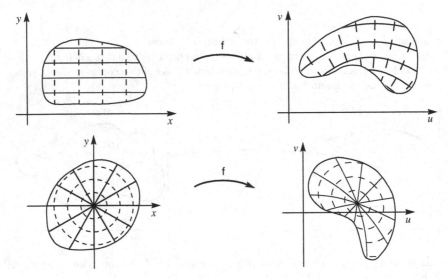

We continue with the example $f(z) = z^2$, but this time let us take as its domain of definition the whole complex plane $\mathbb{C}$. From $z = x + iy$, $w = u + iv$ and $z^2 = w$ it follows that

$$u(x,y) = x^2 - y^2 \ , \quad v(x,y) = 2xy \ .$$

The image of a line parallel to the x-axis $-\infty < x < \infty$, $y = y_0$, is therefore given by the equations

$$\left.\begin{array}{l} u(x,y) = x^2 - y_0^2 \\ v(x,y) = 2xy_0 \end{array}\right\} \quad -\infty < x < \infty \ . \tag{$*$}$$

For the special case $y_0 = 0$ (the x-axis), we have

$$u(x,y) = x^2 \quad \text{and} \quad v(x,y) = 0 \ ,$$

so the x-axis is mapped to the non-negative u-axis, covering it twice as x varies from $-\infty$ to $+\infty$. If $y_0 \neq 0$, then we can eliminate x from the equations $(*)$: $x = v/2y_0$. Substituting in the first equation we get

$$u = \frac{v^2}{4y_0^2} - y_0^2 \ .$$

This is the equation of a parabola, opened to the right, with the u-axis as a symmetry axis and the origin as focus. The points of intersection with the axis are

$$u = -y_0^2 \qquad \text{(intersection with the } u\text{-axis) and}$$

$$v = \pm 2y_0^2 \qquad \text{(intersection with the } v\text{-axis) .}$$

Thus lines parallel to the x-axis are mapped onto confocal parabolas. Since $f(z) = f(-z)$, both lines $-\infty < x < \infty$, $y = y_0$, and $-\infty < x < \infty$, $y = -y_0$, obviously have the same image. The images of the lines $x = x_0$ parallel to the y-axis $-\infty < y < \infty$, can be found in the same way, and for them one obtains a family of confocal parabolas open to the left if $x_0 \neq 0$. For $x_0 = 0$ (the imaginary axis) one has as image the negative real axis (covered twice).

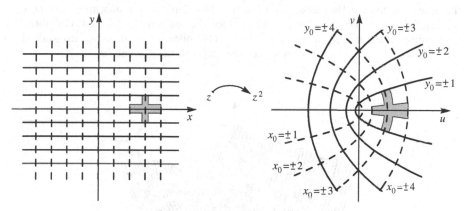

Note that, excepting the intersection point at $f(0) = 0$, the intersections in the image net are rectangular. This is because the map f is locally conformal away from the origin. At the origin the angles are doubled.

The method above is closely connected with the second method.

Second method. The *"contour line method"*:

For fixed $c \in \mathbb{R}$ we look at the level lines

$$N_u^c = \{ (x,y) \in D \ ; \quad u(x,y) = c \} \qquad \text{resp.} \qquad N_v^c = \{ (x,y) \in D \ ; \quad v(x,y) = c \} \ .$$

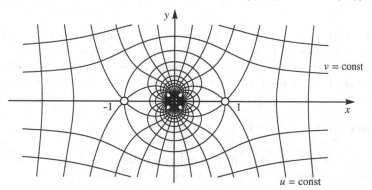

Example: $u = \operatorname{Re} w$, $v = \operatorname{Im} w$ for $w = 1/2(z + 1/z)$

Here, one can either give the "contour maps" of u and v separately or draw both families of curves on top of each other. In this way one obtains a net on D, from which one can read off $f(z) = u(x,y) + iv(x,y)$. If f has an inverse map g:

$$g : f(D) \longrightarrow D ,$$

then the lines of the image of the x-y-net are exactly the contour lines of the real and imaginary parts of g, the inverse of f,

$$g : f(D) \longrightarrow D , \quad (u,v) \longmapsto (x,y) .$$

Third method. The "analytic mountainscape" or the "analytic landscape":
If one considers

$$\{ (z, w) \in D \times \mathbb{R} ; \quad w = |f(z)| \} \subset \mathbb{R}^3 ,$$

Then one can look at this subset of $\mathbb{R}^3$ as the "function's mountainscape" over D. By adding further marker lines, e.g. lines on which the real part is constant, one obtains a so-called "relief image" of the function f. The following figure is for instance the mountainscape of the complex sine, i.e. the graph of $(x,y) \to \sin(x + iy)$.

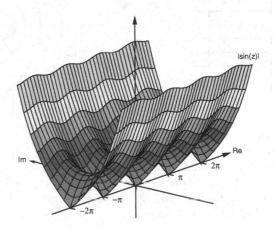

We shall see (III.3.5) that the absolute value surface has no maxima, and can have minima only at zeros of f. Thus in the "analytical landscape" there are no peaks, and the valley bottoms reach down to the complex plane (poor mountain climbers!). Just imagine that it rained in this analytical landscape. Where would the water collect?

Exercises for I.5

1. Re-examine the complex differentiability of the examples from Exercise 2 of I.4, now using the CAUCHY-RIEMANN differential equations.

2. Let $f : \mathbb{C} \to \mathbb{C}$ be defined by $f(z) = x^3 y^2 + i x^2 y^3$.
    Show: f is complex differentiable exactly on the coordinate axes, and there is
    no open subset $D \subset \mathbb{C}$ such that $f|D$ is analytic.

3. Write the following functions in the form $f = u + iv$ and give explicit formulas
    for u and v.

    $$\text{(a)} \quad f(z) = \sin z, \qquad\qquad \text{(b)} \quad f(z) = \cos z,$$
    $$\text{(c)} \quad f(z) = \sinh z, \qquad\qquad \text{(d)} \quad f(z) = \cosh z,$$
    $$\text{(e)} \quad f(z) = \exp\left(z^2\right), \qquad \text{(f)} \quad f(z) = z^3 + z.$$

    Show that in every case the CAUCHY-RIEMANN equations are satisfied (for all
    $z \in \mathbb{C}$), and conclude that these functions are analytic in $\mathbb{C}$.

4. The function $f : \mathbb{C} \to \mathbb{C}$,

    $$f(z) = \begin{cases} \exp(-1/z^4) & \text{for } z \neq 0, \\ 0 & \text{for } z = 0, \end{cases}$$

    satisfies the CAUCHY-RIEMANN equations for all $z \in \mathbb{C}$ and is complex differ-
    entiable for all $z \in \mathbb{C}^\bullet$, but not at the origin.

5. What is the maximal open set $D \subset \mathbb{C}$, such that $f : D \to \mathbb{C}$, $f(z) := \mathrm{Log}(z^5+1)$,
    is well defined and analytic.

6. If $f : D \to \mathbb{C}$ is analytic, $D \subset \mathbb{C}$ is open, and one of the following conditions
    holds:
    (a) Re $f = $ constant,
    (b) Im $f = $ constant,
    (c) $|f| = $ constant,
    then f is locally constant.

7. For each of the harmonic functions given below construct an analytic function
    $f : D \to \mathbb{C}$ with the given real part u:
    (a) $D = \mathbb{C}$ and $u : D \to \mathbb{R}$ with $u(x, y) = x^3 - 3xy^2 + 1$.
    (b) $D = \mathbb{C}^\bullet$ and $u : D \to \mathbb{R}$ with $u(x, y) = \dfrac{x}{x^2 + y^2}$.
    (c) $D = \mathbb{C}$ and $u : D \to \mathbb{R}$ with $u(x, y) = e^x \left(x \cos y - y \sin y\right)$.
    (d) $D = \mathbb{C}_-$ and $u : D \to \mathbb{R}$ with $u(x, y) = \sqrt{\dfrac{x + \sqrt{x^2 + y^2}}{2}}$.

8. **The Laplace operator in polar coordinates**
    Let $\mathbb{R}_+^\bullet \times \mathbb{R} \to \mathbb{R}^2 \setminus \{(0, 0)\}$ be the map defined by $(x, y) = (r \cos \varphi, r \sin \varphi)$. In
    addition, let $D \subset \mathbb{R}^2 \setminus \{(0, 0)\}$ be an open subset and $u : D \to \mathbb{R}$ a function
    which is twice continuously differentiable. Let $\Omega := \{ (r, \varphi) \; ; \; (x, y) \in D\}$ and

    $$U : \Omega \longrightarrow \mathbb{R}, \quad U(r, \varphi) = u(x, y).$$

    Show:
    $$(\varDelta u)(x, y) = \left(U_{rr} + \frac{1}{r} U_r + \frac{1}{r^2} U_{\varphi\varphi} \right) (r, \varphi).$$

9. Determine all harmonic functions

    $$u : \mathbb{C}^\bullet = \mathbb{R}^2 \setminus \{(0, 0)\} \longrightarrow \mathbb{R},$$

    that depend only on $r := \sqrt{x^2 + y^2}$.

10. Let $D \subset \mathbb{C}$ be open and $D' \subset \mathbb{C}$ another open subset. Let $\varphi : D \to D'$ be analytic and even twice continuously differentiable, and $\eta : D' \to \mathbb{R}$ twice continuously partial differentiable.

    Show:
    $$\Delta(\eta \circ \varphi) = ((\Delta\eta) \circ \varphi) \, |\varphi'|^2 \ .$$

    Deduce: If φ is conformal then η is harmonic if and only if $\eta \circ \varphi$ is harmonic.

11. **Characterization of the exponential function by a differential equation**

    Let $D = \mathbb{R}$ or $D = \mathbb{C}$. Let $C \in \mathbb{C}$ be a constant and $f : D \to \mathbb{C}$ differentiable with
    $$f'(z) = C \, f(z) \quad \text{for all} \quad z \in D \ .$$
    If $A = f(0)$, then
    $$f(z) = A \, \exp(Cz) \quad \text{for all} \quad z \in D \ .$$

12. Find all continuous maps
    $$\chi : \mathbb{R} \longrightarrow S^1 = \{ z \in \mathbb{C} ; \quad |z| = 1 \}$$
    which satisfy
    $$\chi(x + t) = \chi(x) \, \chi(t) \quad \text{for all} \quad x, t \in \mathbb{R} \ .$$

    Hint. Such a χ is in fact differentiable. So make use of Exercise 11.

    Result. Each such χ (i.e. each so-called *continuous character* of $(\mathbb{R}, +)$) is of the form
    $$\chi(x) = \chi_y(x) = e^{ixy} \quad (y \in \mathbb{R}) \ .$$

13. Sketch the following level lines for the map $f : \mathbb{C} \to \mathbb{C}, \ z \mapsto z^3$
    $$\{ z \in \mathbb{C} ; \ \mathrm{Re}\, f(z) = c \} , \quad \{ z \in \mathbb{C} ; \ \mathrm{Im}\, f(z) = c \} , \quad \{ z \in \mathbb{C} ; \ |f(z)| = c \}$$
    for $c \in \mathbb{Z}$ with $|c| \le 5$.
    Go on to find the images under f of these level lines and the images of the lines parallel to the real axis (resp. the imaginary axis).

14. Let $D = \{ z \in \mathbb{C} ; \ -\pi < \mathrm{Im}\, z < \pi , \ 0 < \mathrm{Re}\, z < b \}$ and $f = \exp|D$.

    Show: f defines a conformal map from D onto the set D', where $D' = f(D)$ should also be determined.

15. The *Joukowski function* —named after the Russian mathematician N.J. JOUKOWSKI (1847-1921)–
    $$f : \mathbb{C}^\bullet \longrightarrow \mathbb{C} , \quad z \mapsto \frac{1}{2} \left(z + \frac{1}{z} \right) ,$$
    is analytic, is not injective since $f(z) = f(1/z)$, but it is (locally) conformal because of $f'(z) = \frac{1}{2}(1 - 1/z^2)$ in $\mathbb{C}^\bullet \setminus \{1, -1\}$.

    Show (by introducing polar coordinates):
    (a) The image of a circle is $C_r := \{ z \in \mathbb{C} ; \ |z| = r \}$, $r > 0$, under f is
       (i) in the case $r \ne 1$, an ellipse with the foci ± 1 and semi-axes $\frac{1}{2}\left(r + \frac{1}{r}\right)$, resp. $\frac{1}{2}\left|r - \frac{1}{r}\right|$,
       (ii) and else $f(C_1) = [-1, 1]$.

(b) The image of a half-line $r \mapsto re^{i\varphi}, r > 0,$ ($\varphi \notin \{0, \pm\pi/2, \pi\}$, φ fixed) is
a branch of a hyperbola with the foci ± 1.

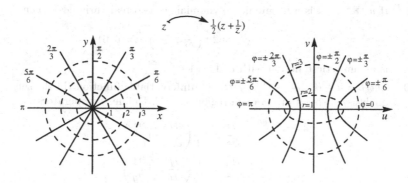

Further show: If

$$D_1 := \{ z \in \mathbb{C} ; \quad |z| > 1 \}$$

and

$$D_2 := \{ z \in \mathbb{C} ; \quad 0 < |z| < 1 \} ,$$

then the restriction of f to D_1, and respectively D_2, gives a conformal map
from these open sets onto the plane slit along the real axis from -1 to 1,
$\mathbb{C} \setminus \{ t \in \mathbb{R} ; \quad -1 \le t \le 1 \}$.

Note: For $z = x + iy \in D_1$ we have $|z|^2 = x^2 + y^2 > 1$.

The JOUKOWSKI function plays an important role in aerodynamics (for example
in describing flow around lifting surfaces —cf. the JOUKOWSKI-KUTTA profile,
W.M. KUTTA, 1902, N.J. JOUKOWSKI, 1906).

16. Let
$$D = \left\{ z \in \mathbb{C} ; \quad -\frac{\pi}{2} < \operatorname{Re} z < \frac{\pi}{2} \right\}.$$

Show:

(a) For $f(z) = \sin z$ we have $f(D) = \mathbb{C} \setminus \{ t \in \mathbb{R} ; \quad |t| \ge 1 \}$.

(b) For $f(z) = \tan z$ we have $f(D) = \mathbb{C} \setminus \{ ti ; \quad t \in \mathbb{R} , t \ge 1 \text{ or } t \le -1 \}$.
The map $\tan : D \to f(D)$ is conformal, and its inverse is

$$g(z) = \frac{1}{2i} \operatorname{Log} \frac{1 + iz}{1 - iz} .$$

17. Let us consider the upper half-plane $\mathbb{H} = \{ z \in \mathbb{C}; \quad \operatorname{Im} z > 0 \}$ and the unit
disk $\mathbb{E} = \{ q \in \mathbb{C}; \quad |q| < 1 \}$.
Show: The function

$$f(z) := \frac{z - i}{z + i}$$

provides a globally conformal map of $\mathbb{H}$ onto $\mathbb{E}$. What is its inverse map?
The map f is also called the *Cayley map* (A. CAYLEY, 1846).

18. For an $\mathbb{R}$-linear map $T : \mathbb{C} \to \mathbb{C}$ the following properties are equivalent:
    (a) T is a similarity transformation (rotation-dilation),
    (b) T is orientation- and angle-preserving .

19. If $u : \mathbb{R}^2 \to \mathbb{R}$ is a harmonic polynomial (in two real variables), then

$$f(z) = 2u\left(\frac{z}{2}, \frac{z}{2i} \right) - u(0,0)$$

is an analytic function with real part u.

20. Let $f = u + iv : D \to \mathbb{C}$ be a totally differentiable function (in the sense of real analysis), defined on an open region $D \subset \mathbb{C}$. Define the operators

$$\frac{\partial f}{\partial z} := \frac{1}{2}\left(\frac{\partial f}{\partial x} - i\frac{\partial f}{\partial y} \right),$$

$$\frac{\partial f}{\partial \bar{z}} := \frac{1}{2}\left(\frac{\partial f}{\partial x} + i\frac{\partial f}{\partial y} \right).$$

Show: f is analytic if and only if $\dfrac{\partial f}{\partial \bar{z}} = 0$, and when this is the case one has $f' = \dfrac{\partial f}{\partial z}$.

Remark. These differential operators $\partial := \dfrac{\partial}{\partial z}$ and $\bar{\partial} := \dfrac{\partial}{\partial \bar{z}}$ were originally introduced by H. POINCARÉ (1899). A systematic calculus was developed by W. WIRTINGER (1927) —the so-called *Wirtinger calculus*. However, it is not very important in the one-variable function theory; its full significance lies in many-variable function theory, for which it was originally developed by WIRTINGER.

21. Where does the function $f : \mathbb{C}^{\bullet} \to \mathbb{C}$, $f(z) = z\bar{z} + z/\bar{z}$, satisfy the CAUCHY-RIEMANN differential equations?

II

Integral Calculus in the Complex Plane $\mathbb{C}$

In Section I.5 we already encountered the problem of finding a primitive function for a given analytic function $f : D \to \mathbb{C}$, $D \subset \mathbb{C}$ open, i.e., an analytic function $F : D \to \mathbb{C}$ such that $F' = f$.

In general, one may ask: Which functions $f : D \to \mathbb{C}$, $D \subset \mathbb{C}$ open, have a primitive? Recall that in the real case any *continuous* function $f : [a, b] \to \mathbb{R}$, $a < b$, has a primitive, namely, for example the integral

$$F(x) := \int_a^x f(t)\, dt \ .$$

Whether one uses the notion of a RIEMANN integral or the integral for regulated functions is irrelevant in this connection.

In the complex case the situation however is different. We shall see that a function that has a primitive must itself already be analytic, and that is, as we already know, a much stronger condition than just continuity. To explore the similarities with and differences from real analysis we will attempt to construct a primitive using an integration process

$$F(z) = \int_{z_0}^z f(\zeta)\, d\zeta \ , \quad z_0 \text{ fixed} \ .$$

For this we first have to introduce a suitable complex integral, the *complex line integral*. In contrast to the real case this not only depends on the starting and end points, but also on the choice of the curve connecting them. One obtains a primitive only when one can prove its independence of this choice.

The *Cauchy Integral Theorem* (A.L. CAUCHY, 1814, 1825) is the main result in this direction. However, as it can be extracted from a letter of C.F. GAUSS to F.W. BESSEL sent on December 18, 1811, GAUSS already knew the statement of CAUCHY's Integral Theorem (C.F. GAUSS, Werke 8, 90-92).

An extension of the CAUCHY Integral Theorem is provided by the *Cauchy Integral Formulas* (A.L. CAUCHY, 1831), which are themselves a special case of the *Residue Theorem*, which is a powerful tool for function theory. However, we shall only get to the Residue Theorem in the next chapter.

E. Freitag and R. Busam, *Complex Analysis*,
DOI: 10.1007/978-3-540-93983-2_II, © Springer-Verlag Berlin Heidelberg 2009

II.1 Complex Line Integrals

A complex-valued function

$$f : [a, b] \longrightarrow \mathbb{C} \quad (a, b \in \mathbb{R}, \ a < b)$$

on a real interval is called *integrable*, if Re f, Im $f : [a, b] \to \mathbb{R}$ are integrable functions in the sense of real analysis. (For instance, in the RIEMANN sense or in the sense of a regulated function. Which notion of integral is to be used is not important, it is only essential that all *continuous functions* are integrable.) Then one defines the integral

$$\int_a^b f(x) \, dx := \int_a^b \operatorname{Re} f(x) \, dx + \mathrm{i} \int_a^b \operatorname{Im} f(x) \, dx$$

and furthermore

$$\int_b^a f(x) \, dx := - \int_a^b f(x) \, dx \, , \qquad \int_a^a f(x) \, dx := 0 \, .$$

The usual rules of calculation with RIEMANN integrals, or with integrals of regulated functions, then can be extended to complex-valued functions:

(1) The integral is $\mathbb{C}$-linear: For continuous functions $f, g : [a, b] \to \mathbb{C}$ the following holds:

$$\int_a^b (f(x) + g(x)) \, dx = \int_a^b f(x) \, dx + \int_a^b g(x) \, dx \, ,$$

$$\int_a^b \lambda f(x) \, dx = \lambda \int_a^b f(x) \, dx \quad (\lambda \in \mathbb{C}) \, .$$

(2) If f is continuous and F is a primitive of f, i. e. $F' = f$, then

$$\int_a^b f(x) \, dx - F(b) - F(a).$$

(3)

$$\left| \int_a^b f(x) \, dx \right| \leq \int_a^b |f(x)| \, dx \leq (b - a)C \, , \qquad \text{if } |f(x)| \leq C$$

for all $x \in [a, b]$. This inequality holds for step functions from the triangle inequality, the general case follows by approximation.

(4) *Substitution rule:* Let $M_1, M_2 \subset \mathbb{R}$ be intervals, $a, b \in M_1$ and

$\varphi : M_1 \longrightarrow M_2$ continuously differentiable and $f : M_2 \longrightarrow \mathbb{C}$ continuous.

Then

$$\int_{\varphi(a)}^{\varphi(b)} f(y) \, dy = \int_a^b f(\varphi(x)) \varphi'(x) \, dx \, .$$

Proof. If F is a primitive of f, then $F \circ \varphi$ is a primitive of $(f \circ \varphi) \varphi'$. $\square$

(5) *Partial integration*

$$\int_a^b u(x)v'(x)\,dx \;=\; uv\Big|_a^b \;-\; \int_a^b u'(x)v(x)\,dx \;.$$

Here u and $v : [a,b] \to \mathbb{C}$ are continuously differentiable functions. The proof follows from the product formula $(uv)' = uv' + u'v$. $\qquad\qquad\square$

Definition II.1.1 *A **curve** is a continuous map*

$$\alpha : [a,b] \longrightarrow \mathbb{C}\,, \quad a < b\,,$$

*from a compact real interval into the complex plane. We call $\alpha(a)$ the **starting point**, and $\alpha(b)$ the **end point** of α.*

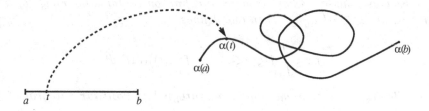

Examples.

(1) The straight line connecting $z, w \in \mathbb{C}$ is parametrized by

$$\alpha : [0,1] \longrightarrow \mathbb{C}\,, \quad \alpha(t) = z + t(w - z) \quad (\alpha(0) = z\,, \ \alpha(1) = w)\,.$$

(2) The k-fold unit circle, $k \in \mathbb{Z}$, is

$$\varepsilon_k : [0,1] \longrightarrow \mathbb{C}\,, \quad \varepsilon_k(t) = \exp(2\pi i k t)\,.$$

Definition II.1.2 *A curve is called **smooth**, if it is continuously differentiable.*

Definition II.1.3 *A curve is called **piecewise smooth**, if there is a partition*

$$a = a_0 < a_1 < \cdots < a_n = b$$

such that the restrictions

$$\alpha_\nu := \alpha \mid [a_\nu, a_{\nu+1}]\,, \quad 0 \le \nu < n\,,$$

are smooth.

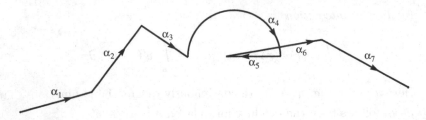

Definition II.1.4 *Let*

$$\alpha : [a,b] \longrightarrow \mathbb{C}$$

be a smooth curve and

$$f : D \longrightarrow \mathbb{C}, \quad D \subset \mathbb{C},$$

a continuous function, whose domain of definition contains the image of the curve α, i.e. $D \supset \alpha([a,b])$. Then one defines

$$\int_\alpha f := \int_\alpha f(\zeta)\, d\zeta := \int_a^b f(\alpha(t))\alpha'(t)\, dt ,$$

*and calls this complex number the **line integral** or **contour integral** of f along α.*

If α is only piecewise smooth, there exists a partition

$$a = a_0 < \cdots < a_n = b ,$$

such that the restrictions

$$\alpha_\nu : [a_\nu, a_{\nu+1}] \longrightarrow \mathbb{C}, \quad 0 \le \nu < n ,$$

are smooth. In this case we define

$$\int_\alpha f(\zeta)\, d\zeta := \sum_{\nu=0}^{n-1} \int_{\alpha_\nu} f(\zeta)\, d\zeta .$$

It is obvious that this definition does not depend on the choice of the partition. By the *arc length* of a smooth curve we mean

$$l(\alpha) := \int_a^b |\alpha'(t)|\, dt .$$

The length of a piecewise smooth curve is

$$l(\alpha) := \sum_{\nu=0}^{n-1} l(\alpha_\nu) .$$

Examples.

(1) The length of the straight line connecting z and w is

$$l(\alpha) = |z - w| \ .$$

(2) The arc length of a k-fold unit circle is

$$l(\varepsilon_k) = 2\pi \, |k| \ .$$

Now we shall list the fundamental properties of complex line integrals. The proofs all follow immediately from properties (1) – (5) of the integral $\int_a^b f(x) \, dx$.

Remark II.1.5 *The complex line integral has the following properties:*

1. *$\int_\alpha f$ is $\mathbb{C}$-linear in f.*
2. *The "standard estimate" states*

$$\left| \int_\alpha f(\zeta) \, d\zeta \right| \leq C \cdot l(\alpha), \ \text{if } |f(\zeta)| \leq C \text{ for all } \zeta \in \text{Image } \alpha \ .$$

3. *The line integral generalizes the ordinary Riemann integral (or the integral of regulated functions). If*

$$\alpha : [a, b] \longrightarrow \mathbb{C} \ , \quad \alpha(t) = t \ ,$$

then $\alpha'(t) = 1$, and for any continuous $f : [a, b] \to \mathbb{C}$ one has:

$$\int_\alpha f(\zeta) \, d\zeta = \int_a^b f(t) \, dt \ .$$

4. *Parameter invariance of the line integral:*

Let $\alpha : [c, d] \to \mathbb{C}$ be a piecewise smooth curve and

$$f : D \longrightarrow \mathbb{C}, \quad \text{Image } \alpha \subset D \subset \mathbb{C} \ ,$$

a continuous function, and

$$\varphi : [a, b] \longrightarrow [c, d] \quad (\ a < b \ , \ c < d \ )$$

a continuously differentiable function with $\varphi(a) = c$, $\varphi(b) = d$. Then we have

$$\int_\alpha f(\zeta) \, d\zeta = \int_{\alpha \circ \varphi} f(\zeta) \, d\zeta \ .$$

5. *Let*

$$f : D \longrightarrow \mathbb{C} \ , \ D \subset \mathbb{C} \ \text{open} \ ,$$

be a continuous function, which has a primitive F (i.e. $F' = f$). Then for any piecewise smooth curve α in D

$$\int_\alpha f(\zeta) \, d\zeta = F\bigl(\alpha(b)\bigr) - F\bigl(\alpha(a)\bigr) \ .$$

The last point in the remark implies:

Theorem II.1.6 *If a continuous function $f : D \to \mathbb{C}$, $D \subset \mathbb{C}$ open, has a primitive then*

$$\int_\alpha f(\zeta)\, d\zeta = 0$$

*for **any** closed piecewise smooth curve α in D.*

(A curve $\alpha : [a, b] \to \mathbb{C}$ is called *closed*, if $\alpha(a) = \alpha(b)$.)

 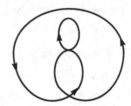

Remark II.1.7 *Let $r > 0$ and*

$$\alpha(t) = r \exp(it)\ ,\quad 0 \le t \le 2\pi\ ,$$

(a circle with the "counterclockwise" orientation). Then for $n \in \mathbb{Z}$

$$\int_\alpha \zeta^n\, d\zeta = \begin{cases} 0 & \text{for } n \neq -1\ , \\ 2\pi i & \text{for } n = -1\ . \end{cases}$$

Corollary II.1.7$_1$ In the domain $D = \mathbb{C}^\bullet$ the (continuous) function

$$f : D \longrightarrow \mathbb{C}\ ,\quad z \longmapsto \frac{1}{z}\ ,$$

does not have a primitive.

Otherwise, because of II.1.6, the integral along any closed curve in $\mathbb{C}^\bullet$ would have to vanish. However,

$$\int_\alpha \frac{1}{\zeta}\, d\zeta\ =\ 2\pi i$$

for the circle line (counterclockwise oriented)

$$\alpha : [0, 2\pi] \longrightarrow \mathbb{C}^\bullet\ ,$$
$$t \longmapsto r \exp(it)\quad (r > 0)\ .$$

Proof of II.1.7. In case of $n \neq -1$ the function $f(z) = z^n$ has the primitive $F(z) = \dfrac{z^{n+1}}{n+1}$. Therefore its integral along any closed curve vanishes. For $n = -1$, however, we have

$$\int_\alpha \zeta^{-1} d\zeta = \int_0^{2\pi} (re^{it})^{-1} \, rie^{it} \, dt = i \int_0^{2\pi} dt = 2\pi i .$$

□

A different proof of the above formula uses the principal branch of the logarithm, which makes a "jump of $2\pi i$" while crossing the negative real axis (see I.5.8).

Exercises for II.1

1. The figure on the right shows a closed curve α, Give an explicit parametrization for α and calculate

$$\frac{1}{2\pi i} \int_\alpha \frac{1}{z} \, dz .$$

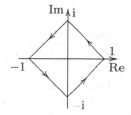

2. Let $\alpha : [0, \pi] \to \mathbb{C}$ be defined by

$$\alpha(t) := \exp(it)$$

and $\beta : [0, 2] \to \mathbb{C}$ by

$$\beta(t) = \begin{cases} 1 + t(-i - 1) & \text{for } t \in [0, 1] , \\ 1 - t + i(t - 2) & \text{for } t \in [1, 2] . \end{cases}$$

Sketch α and β, and calculate

$$\int_\alpha \frac{1}{z} \, dz \qquad \text{and} \qquad \int_\beta \frac{1}{z} \, dz .$$

3. Prove the transformation invariance of the line integral, II.1.5, (4).

4. Sketch the following curve α ("figure eight")

$$\alpha(t) := \begin{cases} 1 - \exp(\ it) & \text{for } t \in [0, 2\pi] , \\ -1 + \exp(-it) & \text{for } t \in [2\pi, 4\pi] . \end{cases}$$

5. Compute

$$\int_\alpha z \exp(z^2) \, dz ,$$

where

    (a) α is the line between the point 0 and the point $1 + i$,

    (b) α is the piece of the parabola with equation $y = x^2$, which lies between the points 0 and $1 + i$.

6. Compute

$$\int_\alpha \sin z \, dz ,$$

where α is the piece of the parabola with equation $y = x^2$, which lies between the points 0 and $-1 + i$.

7.   Let $[a,b]$ and $[c,d]$ $(a < b$ and $c < d)$ be compact intervals in $\mathbb{R}$.

Show: There is an affine map

$$\varphi : [a,b] \longrightarrow [c,d] ,$$
$$t \longmapsto \alpha t + \beta ,$$

with $\varphi(a) = c$ and $\varphi(b) = d$.

8.   Let $R > 0$ be a positive number. We consider the curve

$$\beta(t) = R \exp(it) , \qquad 0 \le t \le \frac{\pi}{4} .$$

Show:

$$\left| \int_\beta \exp(iz^2)\, dz \right| \le \frac{\pi(1 - \exp(-R^2))}{4R} < \frac{\pi}{4R} .$$

9.   Let $\alpha : [a,b] \to \mathbb{C}$ be continuously differentiable and assume that the function $f :$ Image $\alpha \to \mathbb{C}$ is continuous.

Show: For any $\varepsilon > 0$ there exists a $\delta > 0$ with the following property: If $\{a_0, \ldots, a_N\}$ and $\{c_1, \ldots, c_N\}$ are finite subsets of $[a,b]$ with

$$a = a_0 \le c_1 \le a_1 \le c_2 \le a_2 \le \cdots \le a_{N-1} \le c_N \le a_N = b$$

and

$$a_\nu - a_{\nu-1} < \delta \text{ for } \nu = 1, \ldots, N ,$$

then

$$\left| \int_\alpha f(z)\, dz - \sum_{\nu=1}^N f(\alpha(c_\nu)) \cdot (\alpha(a_\nu) - \alpha(a_{\nu-1})) \right| < \varepsilon .$$

(Approximation of the line integral by a RIEMANN sum.)

10.   By splitting f into its real and imaginary parts, represent the complex line integral $\int_\alpha f(z)\, dz$ in terms of *real* integrals.

Result: If $f = u + iv$, $\alpha(t) = x(t) + iy(t)$, $t \in [a,b]$, then

$$\int_\alpha f(z)\, dz = \int_\alpha (u\, dx - v\, dy) + i \int_\alpha (v\, dx + u\, dy)$$

$$= \int_a^b \left[u(x(t),y(t))\, x'(t) - v(x(t),y(t))\, y'(t) \right] dt$$

$$+ i \int_a^b \left[v(x(t),y(t))\, x'(t) + u(x(t),y(t))\, y'(t) \right] dt .$$

11.   A smooth curve is called *regular* if its derivative does not vanish anywhere. Assume that there are given an analytic function $f : D \to \mathbb{C}$, $D \subset \mathbb{C}$ open, and a point $a \in D$ with $f'(a) \ne 0$, and also two regular curves $\alpha, \beta : [-1,1] \to D$ with $\alpha(0) = \beta(0) = a$. One may then consider the oriented angle $\angle(\alpha'(0), \beta'(0))$ (see I.1, Exercise 4). This is the angle between the two intersecting curves. Show that the two image curves $f \circ \alpha$ and $f \circ \beta$ intersect with the same angle at their intersection point $f(a) = f(\alpha(0)) = f(\beta(0))$.

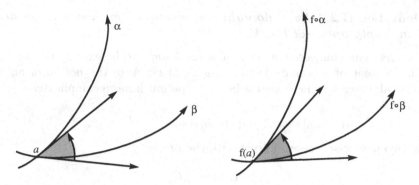

Thus an analytic function is "angle- and orientation-preserving" at any point at which its derivative does not vanish (see also Exercise 18 in I.5).

II.2 The Cauchy Integral Theorem

By an interval $[a, b]$ we will always mean a real interval. And we shall always understand, without mentioning it, that expressions like

$$a \leq b, \quad a < b, \quad [a, b]$$

imply that a and b are real.

Definition II.2.1 *A set $D \subset \mathbb{C}$ is called **arcwise connected**, if for any two points $z, w \in D$ there is a piecewise smooth curve joining z and w and lying entirely inside D, such that*

$$\alpha : [a, b] \longrightarrow D, \quad \alpha(a) = z, \quad \alpha(b) = w.$$

Remark II.2.2 *Every arcwise connected set $D \subset \mathbb{C}$ is connected, i.e. every locally constant function on D is constant.*

Proof. Let $f : D \to \mathbb{C}$ be locally constant. If f is not constant (this is an indirect proof), then there exist points $z, w \in D$ with $f(z) \neq f(w)$. Join z and w by a piecewise smooth curve within D

$$\alpha : [a, b] \longrightarrow D.$$

Since α is continuous

$$g(t) = f\big(\alpha(t)\big)$$

is locally constant. Therefore $g'(t) = 0$, and so $g = \text{const}$. But we have

$$g(a) = f(z) \neq f(w) = g(b). \qquad \square$$

It should be mentioned that for *open* sets D the converse of II.2.2 also holds, although we will not make use of this.

Definition II.2.3 *By a* ***domain*** *we understand an* ***arcwise connected non-empty open set*** $D \subset \mathbb{C}$.

Remark. The connected subsets of $\mathbb{R}$ are known to be exactly the *intervals*. The concept of a domain is thus a generalization of the notion of an open interval. However, the domains in $\mathbb{C}$ can be much more complicated.

Let
$$\alpha : [a, b] \longrightarrow \mathbb{C} \text{ and } \beta : [b, c] \longrightarrow \mathbb{C} , \qquad a \leq b \leq c ,$$

be two piecewise smooth curves with the property
$$\alpha(b) = \beta(b) .$$

Then the formula
$$\alpha \oplus \beta : [a, c] \longrightarrow \mathbb{C} ,$$

$$(\alpha \oplus \beta)(t) = \begin{cases} \alpha(t) & \text{for } a \leq t \leq b , \\ \beta(t) & \text{for } b \leq t \leq c , \end{cases}$$

also defines a piecewise smooth curve. The curve $\alpha \oplus \beta$ is called the *composition* of α and β.

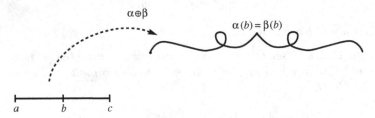

If f is a continuous function, whose domain of definition contains the images of α and β, then
$$\int_{\alpha \oplus \beta} f(\zeta) \, d\zeta = \int_{\alpha} f(\zeta) \, d\zeta + \int_{\beta} f(\zeta) \, d\zeta .$$

For any curve
$$\alpha : [a, b] \longrightarrow \mathbb{C}$$

the *reciprocal* curve is defined by
$$\alpha^- : [a, b] \longrightarrow \mathbb{C} ,$$
$$t \mapsto \alpha(b + a - t) .$$
Obviously we have the *reversal rule*

$$\int_{\alpha^-} f(\zeta) \, d\zeta = - \int_{\alpha} f(\zeta) \, d\zeta$$

for all continuous functions f with Image α in the domain of definition of f.

Convention. We shall assume, unless the contrary is explicitly mentioned, that curves are *piecewise smooth*.

Theorem II.2.4 *For a continuous function*

$$f : D \longrightarrow \mathbb{C} , \quad D \subset \mathbb{C} \text{ a domain} ,$$

the following three statements are equivalent:

(a) *f has a primitive.*
(b) *The integral of f along any closed curve in D vanishes.*
(c) *The integral f along any curve in D depends only on the beginning and end points of the curve.*

Proof.
(a) $\Rightarrow$ (b): Theorem II.1.6.
(b) $\Rightarrow$ (c): Let

$$\alpha : [a,b] \longrightarrow D \text{ and } \beta : [c,d] \longrightarrow D$$

be two curves with the same starting and end points. We have to show

$$\int_\alpha f = \int_\beta f .$$

There is no loss of generality in assuming $b = c$, since by II.1.5, (4) one may replace β by the curve

$$t \longmapsto \beta(t + c - b) , \quad b \le t \le b + (d - c) .$$

Now, we can consider the closed curve $\alpha \oplus \beta^-$, and obtain

$$0 = \int_{\alpha \oplus \beta^-} f = \int_\alpha f - \int_\beta f .$$

(c) $\Rightarrow$ (a): We fix a point $z_* \in D$ and consider

$$F(z) = \int_{z_*}^z f(\zeta) \, d\zeta$$

as the integral of f along some curve connecting z_* with z within D. The assumption ensures that the integral does not depend on the choice of the curve.

Claim. $F' = f$. For the proof, we consider an arbitrary, but for the moment fixed point $z_0 \in D$ and show $F'(z_0) = f(z_0)$. Since D is open, there is a full disk $U_\varrho(z_0)$ around z_0 in D. For $z \in U_\varrho(z_0)$, by definition, we have

$$F(z) = \int_{z_*}^z f(\zeta) \, d\zeta = \int_{z_*}^{z_0} f(\zeta) \, d\zeta + \int_{z_0}^z f(\zeta) \, d\zeta = F(z_0) + \int_{z_0}^z f(\zeta) \, d\zeta ,$$

where we can take the integral from z_0 to z along the line segment connecting them:

$$\sigma(z_0, z)(t) := z_0 + t(z - z_0) , \quad 0 \le t \le 1 .$$

Since $\int_{\sigma(z_0,z)} d\zeta = z - z_0$ we have

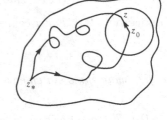

$$F(z) = F(z_0) + f(z_0)(z - z_0) + r(z) \text{ with}$$

$$r(z) = \int_{\sigma(z_0,z)} \big(f(\zeta) - f(z_0)\big) \, d\zeta .$$

By the continuity of f at z_0 there is for any $\varepsilon > 0$ a δ, $0 < \delta < \varrho$, such that for all $z \in D$ with $|z - z_0| < \delta$,

$$|f(z) - f(z_0)| < \varepsilon .$$

Therefore the usual estimate for integrals implies

$$|r(z)| \le |z - z_0| \cdot \varepsilon .$$

But this means that F is complex differentiable at z_0 and $F'(z_0) = f(z_0)$. Since $z_0 \in D$ was arbitrary, F must be a primitive for f.   □

The existence of a primitive is thus reduced to the question of the vanishing of line integrals along closed curves. In the next section we shall prove a vanishing theorem for differentiable functions and special closed curves, namely triangular paths.

Let $z_1, z_2, z_3 \in \mathbb{C}$ be three points in the complex plane. The *triangle spanned* by z_1, z_2, z_3 is the point set

$$\Delta := \{\, z \in \mathbb{C} ; \quad z = t_1 z_1 + t_2 z_2 + t_3 z_3, \ 0 \le t_1, t_2, t_3, \ t_1 + t_2 + t_3 = 1 \,\} .$$

Clearly this set is convex, i.e. with any pair of points in Δ the line segment connecting them also lies in Δ, and Δ is, in fact, the smallest convex set containing z_1, z_2 and z_3 (their convex hull).

By the *triangular path* $\langle z_1, z_2, z_3 \rangle$ we mean the closed curve

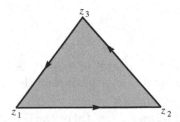

$$\langle z_1, z_2, z_3 \rangle = \alpha := \alpha_1 \oplus \alpha_2 \oplus \alpha_3 , \qquad \text{with}$$
$$\alpha_1(t) = z_1 + (t - 0)\,(z_2 - z_1) , \quad 0 \le t \le 1 ,$$
$$\alpha_2(t) = z_2 + (t - 1)\,(z_3 - z_2) , \quad 1 \le t \le 2 ,$$
$$\alpha_3(t) = z_3 + (t - 2)\,(z_1 - z_3) , \quad 2 \le t \le 3 .$$

We obviously have

$$\text{Image } \alpha \subset \Delta \qquad \text{(or precisely Image } \alpha = \text{Boundary } \Delta).$$

The following theorem is the key for solving the problem of the existence of a primitive. It is sometimes called the *Fundamental Lemma* of complex analysis.

Theorem II.2.5 (Cauchy Integral Theorem for triangular paths, E. Goursat, 1883/84, 1899; A. Pringsheim, 1901) *Let*

$$f : D \longrightarrow \mathbb{C} , \quad D \subset \mathbb{C} \text{ open },$$

be an analytic function (i.e. complex differentiable at any point $z \in D$). Let z_1, z_2, z_3 be three points in D, such that the triangle they span is also contained in D; then

$$\int_{\langle z_1, z_2, z_3 \rangle} f(\zeta) \, d\zeta = 0 .$$

Proof. We shall inductively construct a sequence of triangular paths

$$\alpha^{(n)} = \langle z_1^{(n)}, z_2^{(n)}, z_3^{(n)} \rangle , \quad n = 0, 1, 2, 3, \ldots \quad ,$$

in the following steps:

(a) $\alpha^{(0)} := \alpha = \langle z_1, z_2, z_3 \rangle$.

(b) $\alpha^{(n+1)}$ is one of the following four triangular paths

$$\alpha_1^{(n)} : \left\langle \frac{z_1^{(n)} + z_2^{(n)}}{2} , z_2^{(n)} , \frac{z_2^{(n)} + z_3^{(n)}}{2} \right\rangle ,$$

$$\alpha_2^{(n)} : \left\langle \frac{z_2^{(n)} + z_3^{(n)}}{2} , z_3^{(n)} , \frac{z_1^{(n)} + z_3^{(n)}}{2} \right\rangle , \quad z_1$$

$$\alpha_3^{(n)} : \left\langle \frac{z_1^{(n)} + z_3^{(n)}}{2} , z_1^{(n)} , \frac{z_1^{(n)} + z_2^{(n)}}{2} \right\rangle ,$$

$$\alpha_4^{(n)} : \left\langle \frac{z_1^{(n)} + z_2^{(n)}}{2} , \frac{z_2^{(n)} + z_3^{(n)}}{2} , \frac{z_1^{(n)} + z_3^{(n)}}{2} \right\rangle .$$

Thus we choose

$$\alpha^{(n+1)} = \alpha_1^{(n)} \text{ or } \alpha_2^{(n)} \text{ or } \alpha_3^{(n)} \text{ or } \alpha_4^{(n)} .$$

So we are partitioning the triangle using lines parallel to the sides and passing through their midpoints. Obviously the triangles corresponding to the triangular paths $\alpha_\nu^{(n)}$ and $\alpha^{(n)}$ are entirely contained in $\Delta = \Delta^{(0)}$, and we have

$$\int_{\alpha^{(n)}} = \int_{\alpha_1^{(n)}} + \int_{\alpha_2^{(n)}} + \int_{\alpha_3^{(n)}} + \int_{\alpha_4^{(n)}} .$$

(c) We can and do choose $\alpha^{(n+1)}$ such that

$$\left| \int_{\alpha^{(n)}} f \right| \le 4 \left| \int_{\alpha^{(n+1)}} f \right| .$$

From this follows

$$\left| \int_{\alpha} f(\zeta) \, d\zeta \right| \le 4^n \left| \int_{\alpha^{(n)}} f(\zeta) \, d\zeta \right| .$$

The closed triangles $\Delta^{(n)}$ are nested

$$\Delta = \Delta^{(0)} \supset \Delta^{(1)} \supset \Delta^{(2)} \supset \cdots$$

($\Delta^{(n)}$ is the triangle corresponding to the triangular path $\alpha^{(n)}$). By CAN-TOR's theorem for nested "intervals" there is a point z_0, which is contained in all these triangles. We then use the fact that f is complex differentiable there:

$$f(z) - f(z_0) = f'(z_0)(z - z_0) + r(z) \qquad \text{with} \qquad \lim_{z \to z_0} \frac{r(z)}{|z - z_0|} = 0 .$$

Since the affine part $z \mapsto f(z_0) + f'(z_0)(z - z_0)$ has a primitive, we have

$$\int_{\alpha^{(n)}} f(\zeta) \, d\zeta = \int_{\alpha^{(n)}} r(\zeta) \, d\zeta$$

and therefore

$$\left| \int_{\alpha} f(\zeta) \, d\zeta \right| \le 4^n \left| \int_{\alpha^{(n)}} r(\zeta) \, d\zeta \right| .$$

We shall now prove that the right-hand side converges to 0 for $n \to \infty$. Let $\varepsilon > 0$. There exists $\delta > 0$ with

$$|r(z)| \le \varepsilon |z - z_0| \text{ for all } z \in D \text{ with } |z - z_0| < \delta .$$

If n is large enough, $n \ge N$, then

$$\Delta^{(n)} \subset U_\delta(z_0) .$$

In addition,

$$|z - z_0| \le l(\alpha^{(n)}) = \frac{1}{2^n} l(\alpha) \text{ for } z \in \Delta^{(n)} .$$

We get

$$\left| \int_{\alpha} f(\zeta) \, d\zeta \right| \le 4^n \cdot l(\alpha^{(n)}) \cdot \varepsilon l(\alpha^{(n)}) = l(\alpha)^2 \cdot \varepsilon$$

for any positive ε and thus

$$\int_{\alpha} f(\zeta) \, d\zeta = 0 . \qquad\qquad \square$$

For non-analytic functions this theorem is false. For example the integral of $f(z) = |z|^2$ along a triangle path is usually different from 0, as one verifies by direct computation.

Definition II.2.6 *A **star-shaped domain** is an open set $D \subset \mathbb{C}$ with the following property: There is a point $z_* \in D$ such that for each point $z \in D$ the whole line segment joining z_* and z is contained in D:*

$$\{ z_* + t(z - z_*) \ ; \ t \in [0,1] \} \subset D .$$

The point z_* is not uniquely determined, and is called a (possible) *star center*.

Remark. Since one can join any two points through the star center, a star domain is arcwise connected, and therefore a domain.

Examples.

(1) Each convex domain is star-shaped, in particular, *any open disk is star-shaped.* Each point of the convex domain can be chosen as the star center.

(2) The plane slit along the negative real axis is star-shaped. (As star centers we can take points $x \in \mathbb{R}$, $x > 0$, and only such points.)

(3) An open disk $U_r(a)$, from which we remove finitely many line segments which join a boundary point b with a point on the straight line between a and b.

(4) $D = \mathbb{C}^{\bullet} = \mathbb{C} \setminus \{0\}$ is not star-shaped since any $z_* \in \mathbb{C}^{\bullet}$ cannot be a star center for the point $z := -z_*$ "cannot be seen from" z_*.

(5) The *annulus* $\mathcal{R} = \{ z \in \mathbb{C} ; \quad r < |z| < R \}$, $0 < r < R$, is not star-shaped.

(6) The ring sector

$$\{ z = z_0 + \zeta \varrho e^{i\varphi} \ ; \quad r < \varrho < R \, , 0 < \varphi < \beta \ \} \subset \mathcal{R} \, , \quad \zeta, z_0 \in \mathbb{C} \, , \ |\zeta| = 1 \, ,$$

is star-shaped, if $\beta < \pi$ and $\cos \frac{\beta}{2} > \frac{r}{R}$.

(7) In the following figure the three left domains are star-shaped, the right one is not.

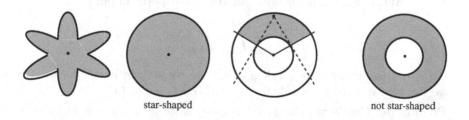

star-shaped   not star-shaped

Theorem II.2.7 (Cauchy Integral Theorem for star domains)

Version 1. Let

$$f : D \longrightarrow \mathbb{C}$$

be an analytic function on a star domain $D \subset \mathbb{C}$. Then the integral f along any closed curve in D vanishes.

Version 2. Each analytic function f defined on a star domain D has a primitive in D.

Corollary *In arbitrary domains $D \subset \mathbb{C}$ an analytic function has, at least locally, a primitive, i.e. for each point $a \in D$ there is an open neighborhood $U \subset D$ of a, such that $f \mid U$ has a primitive.*

Taking into account II.2.4, both versions of the theorem are clearly equivalent. We will proof the second version. So, let $z_* \in D$ be a star center and F be defined by

$$F(z) = \int_{z_*}^{z} f(\zeta) \, d\zeta \;,$$

where the integral is taken *along the line segment connecting z_* with z.* If $z_0 \in D$ is an arbitrary point, then the line segment connecting z_0 and z does not have to lie in D. But there does exist a disk around z_0 which is entirely contained in D. It is easy to see then that:

If z is a point in this disk, then the entire triangle spanned by z_*, z_0 and z is contained in D.

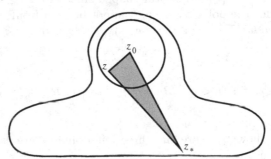

Then CAUCHY's integral theorem for triangular paths implies

$$\int_{z_*}^{z_0} + \int_{z_0}^{z} + \int_{z}^{z_*} = 0 \;.$$

(In each case integration is taken along the connecting line segments.) Now we can repeat word-for-word the proof of II.2.4, (c) $\Rightarrow$ (a).

Proof of the Corollary. The proof is clear, since for each $a \in D$ there is an open disk $U_\varepsilon(a)$ with $U_\varepsilon(a) \subset D$, and disks are convex, and thus certainly star-shaped. $\qquad \square$

Thus we have achieved a solution to our existence problem for star domains.

As an application of II.2.7 we get a new construction of the principal branch
of the logarithm as a primitive of $1/z$ in the star domain $\mathbb{C}_-$, namely

$$L(z) := \int_1^z \frac{1}{\zeta}\, d\zeta \ .$$

One integrates along some curve connecting 1 with z
in $\mathbb{C}_-$. Since the functions L and Log have the same
derivatives, and coincide at a point ($z = 1$), we have
$L(z) = \text{Log}(z)$ for $z \in \mathbb{C}_-$. If one chooses as the curve
the line segment from 1 to $|z|$ and then the arc from
$|z|$ to $z = |z|e^{\mathrm{i}\varphi}$, we obtain the form we already know

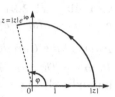

$$L(z) = \int_1^{|z|} \frac{1}{t}\, dt + \mathrm{i} \int_0^{\varphi} dt = \log|z| + \mathrm{i}\,\text{Arg}\, z \ .$$

The following variant of II.2.7 is a useful tool:

Corollary II.2.7$_1$ Let $f : D \to \mathbb{C}$ be a continuous function in a star domain
D with center z_*. If f is complex differentiable at every point $z \neq z_*$, then f
has a primitive in D.

Proof. As one can see from the proof of II.2.7,
it is enough to show that

$$\int_{z_*}^{z_0} + \int_{z_0}^{z} + \int_{z}^{z_*} = 0 \ ,$$

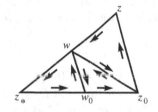

where we may assume that the triangle Δ
spanned by z_*, z_0 and z is entirely contained
within D.
Moreover, we can assume $z_* \neq z$ and $z_* \neq z_0$. Let w, resp. w_0, be an arbitrary
point different from z_* on the line segment between z_* and z, resp. z_* and
z_0. From the CAUCHY integral theorem for triangular paths (II.2.5 above) the
integrals along the paths $\langle w_0, z_0, w \rangle$ and $\langle z_0, z, w \rangle$ vanish. On the other hand,
we have

$$\int_{\langle z_*, z_0, z \rangle} = \int_{\langle z_*, w_0, w \rangle} + \int_{\langle w_0, z_0, w \rangle} + \int_{\langle z_0, z, w \rangle} = \int_{\langle z_*, w_0, w \rangle} \ .$$

The assertion now follows by passing to the limit

$$w \to z_* \ , \quad w_0 \to z_* \ . \qquad\qquad \square$$

Definition II.2.8 *A domain $D \subset \mathbb{C}$ is called an **elementary domain**, if
any analytic function defined on D has a primitive in D.*

Any star domain is thus an elementary domain. For example, $\mathbb{C}_-$, the plane
cut along the negative real axis, is an elementary domain. In this connection
it is of interest to note:

Theorem II.2.9 *Let $f : D \to \mathbb{C}$ be an analytic function on an elementary domain, let f' also be analytic[1], and $f(z) \neq 0$ for all $z \in D$. Then there exists an analytic function $h : D \to \mathbb{C}$ with the property*

$$f(z) = \exp\big(h(z)\big) \ .$$

The function h is called an *analytic branch of the logarithm of f.*

Corollary II.2.9₁ *Under the assumptions in II.2.9, there exists for any $n \in \mathbb{N}$ an analytic function $H : D \to \mathbb{C}$ with $H^n = f$.*

Proof of the Corollary. Set $H(z) = \exp\left(\frac{1}{n}h(z)\right)$. □

Proof of Theorem II.2.9. Let F be a primitive of f'/f. Then one can check immediately that, with

$$G(z) = \left(\frac{\exp\big(F(z)\big)}{f(z)}\right) \ ,$$

one has $G'(z) = 0$ for all $z \in D$. Therefore

$$\exp\big(F(z)\big) = C \, f(z) \text{ for all } z \in D$$

with some nonzero constant C. Since $\exp : \mathbb{C} \to \mathbb{C}^\bullet$ is surjective one can write this in the form $C = \exp(c)$. The function

$$h(z) = F(z) - c$$

has the desired property. □

Since the function $f(z) = 1/z$ does not have a primitive in the punctured plane $\mathbb{C}^\bullet$, we see that $\mathbb{C}^\bullet$ is not an elementary domain; however it is not true that any elementary domain must be star-shaped, as the following construction shows:

Remark II.2.10 *Let D and D' be two elementary domains. If $D \cap D'$ is non-empty and connected, then $D \cup D'$ is also an elementary domain.*

Corollary. *Slitted annuli are elementary domains.*

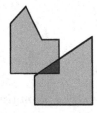

←— elementary domain
non-elementary domain —→

Proof of II.2.10. Let $f : D \cup D' \to \mathbb{C}$ be analytic. By assumption there exist primitives

---

[1] Actually, this assumption is unnecessary by II.3.4.

$$F_1 : D \longrightarrow \mathbb{C}, \quad F_2 : D' \longrightarrow \mathbb{C}.$$

The difference $F_1 - F_2$ must be locally constant in $D \cap D'$, and therefore constant since $D \cap D'$ is connected. By addition of a constant if necessary, one may assume

$$F_1 \mid D \cap D' = F_2 \mid D \cap D'.$$

The functions F_1 and F_2 now glue to a single function

$$F : D \cup D' \longrightarrow \mathbb{C}. \qquad \qquad \square$$

The following is also immediately clear:

Remark II.2.11 *Let*

$$D_1 \subset D_2 \subset D_3 \subset \cdots$$

be an increasing sequence of elementary domains. Then their union

$$D = \bigcup_{n=1}^{\infty} D_n$$

is also an elementary domain.

It can be shown (in a non-trivial way) that with the two constructions above give all elementary domains starting from disks.

We shall later obtain a simple topological characterization of elementary domains (see Appendix C of Chapter IV):

*Elementary domains are precisely the so-called **simply connected domains**. (Intuitively these are the domains "without holes").*

For practical purposes this characterization of elementary domains is not so important. For this reason we postpone the proof of this theorem. More elementary domains can be obtained by means of *conformal mappings* (cf. I.5.13).

Remark II.2.12 *Let $D \subset \mathbb{C}$ be an elementary domain and*

$$\varphi : D \longrightarrow D^*$$

a globally conformal mapping of D onto the domain D^*. We assume that its derivative is analytic. Then D^* is also an elementary domain.

Proof. We have to show: Any analytic function $f^* : D^* \to \mathbb{C}$ has a primitive F^*. This can naturally be reduced to checking the corresponding statement for D.

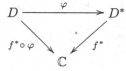

For if $f^* : D^* \to \mathbb{C}$ is analytic, then so is $f^* \circ \varphi : D \to \mathbb{C}$ analytic. But then

$$(f^* \circ \varphi)\varphi' : D \longrightarrow \mathbb{C}$$

is analytic, and so has a primitive F. (Here we have to assume that φ' is also analytic. This condition is, as we shall see in following sections, automatically satisfied.) In fact, $F^* := F \circ \varphi^{-1}$ is analytic (φ^{-1} is analytic too!) and $F^{*\prime} = f^*$.   □

Exercises for II.2

1. Which of the following subsets of $\mathbb{C}$ are domains?

   (a) $\{z \in \mathbb{C};\ \ \left|z^2 - 3\right| < 1\}$,
   (b) $\{z \in \mathbb{C};\ \ \left|z^2 - 1\right| < 3\}$,
   (c) $\{z \in \mathbb{C};\ \ \left||z|^2 - 2\right| < 1\}$,
   (d) $\{z \in \mathbb{C};\ \ \left|z^2 - 1\right| < 1\}$,
   (e) $\{z \in \mathbb{C};\ \ z + |z| \neq 0\}$,
   (f) $\{z \in \mathbb{C};\ 0 < x < 1,\ 0 < y < 1\} - \bigcup\limits_{n=2}^{\infty} \{x + iy;\ x = 1/n,\ 0 < y \leq 1/2\}$.

2. Let $z_0, \ldots, z_N \in \mathbb{C}$ ($N \in \mathbb{N}$). Define the line segments connecting z_ν with $z_{\nu+1}$ ($\nu = 0, 1, \ldots, N-1$) by

   $$\alpha_\nu : [\nu, \nu+1] \longrightarrow \mathbb{C} \text{ with } \alpha_\nu(t) = z_\nu + (t - \nu)(z_{\nu+1} - z_\nu) .$$

   Then $\alpha := \alpha_1 \oplus \alpha_2 \oplus \cdots \oplus \alpha_{N-1}$ defines a curve $\alpha : [0, N] \to \mathbb{C}$. One calls α the *polygonal path*, which joins z_0 with z_N (along $z_1, z_2, \ldots, z_{N-1}$).

   Show: An open set $D \subset \mathbb{C}$ is connected (and thus a domain) if and only if any two points of D can be connected by a polygonal path α inside D (i.e. Image $\alpha \subset D$).

3. Let $a \in \mathbb{C}$, $\varepsilon > 0$. The punctured disk

   $$\overset{\bullet}{U}_\varepsilon(a) := \{z \subset \mathbb{C};\ \ 0 < |z - a| < \varepsilon\} ,$$

   is a domain.

   Deduce: If $D \subset \mathbb{C}$ is a domain and $z_1, \ldots, z_m$ are finitely many points, then the set $D' := D \setminus \{z_1, \ldots, z_m\}$ is also a domain.

4. Let $\emptyset \neq D \subset \mathbb{C}$ be open. The continuous function

   $$f : D \longrightarrow \mathbb{C},\quad z \longmapsto \bar{z},$$

   has no primitive in D.

5. For $\alpha : [0, 1] \to \mathbb{C}$ with $\alpha(t) = \exp(2\pi i t)$ compute

   $$\int_\alpha 1/|z|\ dz , \qquad \int_\alpha 1/(|z|^2)\ dz , \qquad \bullet \quad \text{and show} \qquad \left|\int_\alpha 1/(4 + 3z)\ dz\right| \leq 2\pi .$$

6. Let
$$D := \{ z \in \mathbb{C}; \quad 1 < |z| < 3 \}$$
and $\alpha : [0,1] \to D$ be defined by $\alpha(t) = 2\exp(2\pi i t)$. Calculate
$$\int_\alpha \frac{1}{z}\, dz .$$

7. For $a, b \in \mathbb{R}_+^\bullet$, let $\alpha, \beta : [0,1] \to \mathbb{C}$ be defined by
$$\alpha(t) := a\cos 2\pi t + i a \sin 2\pi t ,$$
$$\beta(t) := a\cos 2\pi t + i b \sin 2\pi t .$$

(a) *Show:*
$$\int_\alpha \frac{1}{z}\, dz = \int_\beta \frac{1}{z}\, dz .$$

(b) *Show using* (a)
$$\int_0^{2\pi} \frac{1}{a^2 \cos^2 t + b^2 \sin^2 t}\, dt = \frac{2\pi}{ab} .$$

8. Let $D_1, D_2 \subset \mathbb{C}$ be star domains with the common star center z_*. Then $D_1 \cup D_2$ and $D_1 \cap D_2$ are also star domains with respect to z_*.

9. Which of the following domains are star-shaped?
    (a) $\{ z \in \mathbb{C}; \quad |z| < 1 \text{ and } |z+1| > \sqrt{2} \}$,
    (b) $\{ z \in \mathbb{C}; \quad |z| < 1 \text{ and } |z-2| > \sqrt{5} \}$,
    (c) $\{ z \in \mathbb{C}; \quad |z| < 2 \text{ and } |z+i| > 2 \}$.

    In each case determine the set of all star centers.

10. Show that the "sickle-shaped domain"
$$D = \{ z \in \mathbb{C}; \quad |z| < 1 , \ |z - 1/2| > 1/2 \}$$
    is an elementary domain.

11. Let $0 < r < R$ and f be the function
$$f : \dot{U}_R(0) \longrightarrow \mathbb{C} ,$$
$$z \longmapsto \frac{R+z}{(R-z)z} .$$

    Show that $f(z) = \dfrac{1}{z} + \dfrac{2}{R-z}$, and, by integrating along the curve α,
$$\alpha : [0, 2\pi] \longrightarrow \mathbb{C} , \quad \alpha(t) = r\exp(it) ,$$
    that
$$\frac{1}{2\pi} \int_0^{2\pi} \frac{R^2 - r^2}{R^2 - 2Rr\cos t + r^2}\, dt = 1 .$$

    Show in a similar manner:
$$\frac{1}{2\pi} \int_0^{2\pi} \frac{R\cos t}{R^2 - 2Rr\cos t + r^2}\, dt = \frac{r}{R^2 - r^2} , \quad \text{if } 0 \le r < R .$$

12. **Lemma on polynomial growth**
Let P be a nonconstant polynomial of degree n:

$$P(z) = a_n z^n + \cdots + a_0 , \quad a_\nu \in \mathbb{C} , \ 0 \leq \nu \leq n , \ n \geq 1 , \ a_n \neq 0 .$$

Then, for all $z \in \mathbb{C}$ with the property

$$|z| \geq \varrho := \max\left\{ 1, \frac{2}{|a_n|} \sum_{\nu=0}^{n-1} |a_\nu| \right\} ,$$

we have

$$\boxed{\frac{1}{2} |a_n| \, |z|^n \leq |P(z)| \leq \frac{3}{2} |a_n| \, |z|^n .}$$

Corollary. Any root of the polynomial P lies in the open disk with radius ρ centered at the origin.

13. **A proof of the Fundamental Theorem of Algebra**
Let P be a nonconstant polynomial of degree n,

$$P(z) = a_n z^n + \cdots + a_0 , \quad a_\nu \in \mathbb{C}, \ 0 \leq \nu \leq n, \ n \geq 1, \ a_n \neq 0 .$$

We have $P(z) = z(a_n z^{n-1} + \cdots + a_1) + a_0 = zQ(z) + a_0$.
Assumption: $P(z) \neq 0$ for all $z \in \mathbb{C}$.
Then for $z \neq 0$ we have

$$\frac{1}{z} = \frac{P(z)}{zP(z)} = \frac{zQ(z) + a_0}{zP(z)} = \frac{Q(z)}{P(z)} + \frac{a_0}{zP(z)} .$$

By integration along $\alpha(t) = R \exp(it)$, $0 \leq t \leq 2\pi$, $R > 0$, it follows that

$$2\pi i = \int_\alpha \frac{a_0}{zP(z)} \, dz .$$

By using the lemma on growth of polynomials, derive a contradiction (consider the limit $R \to \infty$).

14. Let $a \in \mathbb{R}$, $a > 0$. Consider the "rectangular path" α sketched in the figure.

$$\alpha = \alpha_1 \oplus \alpha_2 \oplus \alpha_3 \oplus \alpha_4 .$$

Since

$$f(z) = e^{-z^2/2}$$

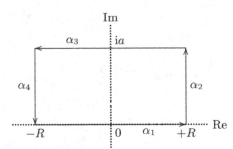

is analytic in $\mathbb{C}$, and $\mathbb{C}$ is star-shaped, it follows from the CAUCHY integral theorem for star domains that

$$0 = \int_\alpha f(z) \, dz = \int_{\alpha_1} f(z) \, dz + \int_{\alpha_2} f(z) \, dz + \int_{\alpha_3} f(z) \, dz + \int_{\alpha_4} f(z) \, dz .$$

Show:

$$\lim_{R \to \infty} \left| \int_{\alpha_2} f(z)\, dz \right| = \lim_{R \to \infty} \left| \int_{\alpha_4} f(z)\, dz \right| = 0$$

and deduce that

$$\int_{-\infty}^{\infty} e^{-\frac{1}{2}(x+ia)^2}\, dx = \int_{-\infty}^{\infty} e^{-x^2/2}\, dx \quad (= \sqrt{2\pi}) .$$

$$I(a) := \int_{-\infty}^{\infty} e^{-\frac{1}{2}(x+ia)^2} dx := \lim_{R \to \infty} \int_{-R}^{R} e^{-\frac{1}{2}(x+ia)^2} dx$$

is therefore independent of a and has the value $\sqrt{2\pi}$.

Corollary. (The FOURIER transform of $x \mapsto e^{-x^2/2}$)

$$\boxed{\int_0^{\infty} e^{-x^2/2} \cos(ax)\, dx = \frac{1}{2}\sqrt{2\pi}\, e^{-a^2/2} .}$$

15. Let $D \subset \mathbb{C}$ be a domain with the property

$$z \in D \implies -z \in D$$

and $f : D \to \mathbb{C}$ a continuous and even function ($f(z) = f(-z)$). Moreover, for some $r > 0$ let the closed disk $\overline{U}_r(0)$ be contained in D. Then

$$\int_{\alpha_r} f = 0 \quad \text{for} \quad \alpha_r(t) := r \exp(2\pi i t) , \ 0 \le t \le 1 .$$

16. **Continuous branches of the logarithm**

Lot $D \subset \mathbb{C}^{\bullet}$ be a domain which does not contain the origin. A continuous function $l : D \to \mathbb{C}$ with $\exp l(z) = z$ for all $z \in D$ is called *a continuous branch of the logarithm.*

Show:

(a) Any other continuous branch of the logarithm $\widetilde{l}$ has the form $\widetilde{l} = l + 2\pi i k$, $k \in \mathbb{Z}$.

(b) Any continuous branch of the logarithm l is in fact analytic, and $l'(z) = 1/z$.

(c) On D there exists a unique continuous branch of the logarithm only if the function $1/z$ has a primitive on D.

(d) Construct two domains D_1 and D_2 and continuous branches $l_1 : D_1 \to \mathbb{C}$, $l_2 : D_2 \to \mathbb{C}$ of the logarithm, such that their difference is not constant on $D_1 \cap D_2$.

17. **Fresnel Integrals**

Show:

$$\int_0^{\infty} \cos\left(t^2\right)\, dt = \int_0^{\infty} \sin\left(t^2\right)\, dt = \frac{1}{4}\sqrt{2\pi} .$$

Hint. Compare the function $f(z) := \exp(iz^2)$ on the real axis and on the first bisector. The value of the integral $\int_0^{\infty} \exp(-t^2)\, dt = \sqrt{\pi}/2$ can be used. Use also the inequality in Exercise 8, Sect. II.1.

II.3 The Cauchy Integral Formulas

The following lemma is a special case of the CAUCHY integral formula:

Lemma II.3.1 *One has*

$$\oint_\alpha \frac{d\zeta}{\zeta - a} = 2\pi i \ ,$$

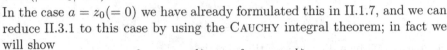

where integration is performed along the circle

$$\alpha(t) = z_0 + re^{it} \ , \qquad z_0 \in \mathbb{C} \ , \ 0 \le t \le 2\pi \ , \ r > 0 \ ,$$

and a is an arbitrary point in the interior of the disc ($|a - z_0| < r$).

In the case $a = z_0 (= 0)$ we have already formulated this in II.1.7, and we can reduce II.3.1 to this case by using the CAUCHY integral theorem; in fact we will show

$$\oint_{|\zeta - z_0| = r} \frac{d\zeta}{\zeta - a} = \oint_{|\zeta - a| = \varrho} \frac{d\zeta}{\zeta - a} \ ,$$

where $\varrho \le r - |z_0 - a|$.

Remark. We use a suggestive way of writing integrals along circles, which is self understanding.

Proof.

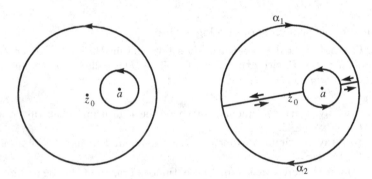

So it is claimed that the integrals along both of the circles drawn above agree. We shall limit ourselves to make the proof intuitively clear from the figure. It is easy but a little wearisome, to translate it into precise formulas. We introduce two additional curves α_1 and α_2 (see the above figure on the right and the next on the left). Slit the plane along the dashed lines, and get, in this way, a star domain in which the function $z \mapsto \frac{1}{z-a}$ is analytic. The integral "along the closed curve we have drawn", which is composed by a (small) circular arc, line segments and a (large) circular arc, vanishes by the CAUCHY integral theorem II.2.7 for star domains. The same argument holds for the figure reflected along the line connecting a and z_0, and the curve α_2 sketched on the right. Therefore

$$\int_{\alpha_1} \frac{1}{\zeta - a}\, d\zeta = 0 \text{ and } \int_{\alpha_2} \frac{1}{\zeta - a}\, d\zeta = 0 \ .$$

If one adds the two integrals, the contributions of the straight segments cancel, since the lines are traversed in reverse directions:

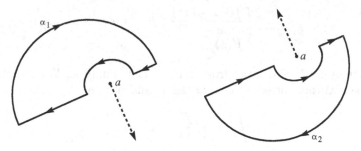

Therefore it follows that (taking into account the orientation!)

$$2\pi\mathrm{i} = \oint_{|\zeta - a| = \varrho} \frac{1}{\zeta - a}\, d\zeta = \oint_{|\zeta - z_0| = r} \frac{1}{\zeta - a}\, d\zeta \ .$$

□

From now on we shall use the notations

$$U_r(z_0) = \{\, z \in \mathbb{C}\, ;\quad |z - z_0| < r \,\}$$
$$\overline{U}_r(z_0) = \{\, z \in \mathbb{C}\, ;\quad |z - z_0| \leq r \,\}$$

for the respectively open and closed disks of radius $r > 0$ around $z_0 \in \mathbb{C}$.

Theorem II.3.2 (Cauchy Integral Formula, A.L. Cauchy, 1831) *Let*

$$f : D \longrightarrow \mathbb{C}\, ,\ D \subset \mathbb{C}\ open\, ,$$

be an analytic function. Assume that the closed disk $\overline{U}_r(z_0)$ is contained completely in D. Then for each point $z \in U_r(z_0)$

$$f(z) = \frac{1}{2\pi\mathrm{i}} \oint_\alpha \frac{f(\zeta)}{\zeta - z}\, d\zeta \ ,$$

where the integral is taken "around the circle α", i.e. along the closed curve

$$\alpha(t) = z_0 + r e^{\mathrm{i}t}\, ,\ 0 \leq t \leq 2\pi \ .$$

We emphasize that the point z needs not to be the center of the disk. It only has to lie in the interior of the disk!

Using the compactness of $\overline{U}_r(z_0)$ one can easily show that there exists an $R > r$ such that

$$D \supset U_R(z_0) \supset \overline{U}_r(z_0) \ .$$

We can thus assume that D is a disk. The function

$$g(w) := \begin{cases} \dfrac{f(w) - f(z)}{w - z} & \text{for } w \neq z \ , \\ f'(z) & \text{for } w = z \ , \end{cases}$$

is continuous in D and away from z is, in fact, analytic. We can therefore apply the CAUCHY integral theorem II.2.7$_1$ and obtain

$$\oint \frac{f(\zeta) - f(z)}{\zeta - z} \, d\zeta = 0 \ .$$

The assertion now follows from II.3.1. $\qquad\qquad\qquad\qquad\qquad\qquad\qquad$ □

In particular, the CAUCHY integral formula holds for $z = z_0$:

$$f(z_0) = \frac{1}{2\pi} \int_0^{2\pi} f\big(z_0 + r \exp(\mathrm{it})\big) \, dt$$

(this is the so-called *mean value equation*).

The essence of the Cauchy integral formula is that it computes the values of analytic functions in the interior of a disk from their values on the boundary.

From the LEIBNIZ rule one gets analogous formulas for the derivatives.

Lemma II.3.3 (Leibniz rule) *Let*

$$f : [a, b] \times D \longrightarrow \mathbb{C} \ , \quad D \subset \mathbb{C} \ \textit{open} \ ,$$

be a continuous function, which is analytic in D for any fixed $t \in [a, b]$. The derivative

$$\frac{\partial f}{\partial z} : [a, b] \times D \longrightarrow \mathbb{C}$$

is also assumed to be continuous. Then the function

$$g(z) := \int_a^b f(t, z) \, dt$$

is analytic in D, and

$$g'(z) = \int_a^b \frac{\partial f(t, z)}{\partial z} \, dt \ .$$

Proof. One can reduce II.3.3 to the analogous result for the real case, since complex differentiability can be expressed using partial derivatives (Theorem I.5.3). Thus one uses the real form of the LEIBNIZ criterion to verify the CAUCHY-RIEMANN equations and the formula for the derivative of g.

For the sake of completeness we shall formulate and prove the real form of the LEIBNIZ rule that we need.

Let $f : [a, b] \times [c, d] \longrightarrow \mathbb{R}$ be a continuous function. Suppose that the partial derivative

$$(t, x) \mapsto \frac{\partial}{\partial x} f(t, x)$$

exists and is continuous. Then

$$g(x) = \int_a^b f(t, x) \, dt$$

is also differentiable, and one has

$$g'(x) = \int_a^b \frac{\partial}{\partial x} f(t, x) \, dt .$$

Proof. We take the difference quotient at $x_0 \in D$:

$$\frac{g(x) - g(x_0)}{x - x_0} = \int_a^b \frac{f(t, x) - f(t, x_0)}{x - x_0} \, dt .$$

By the *mean value theorem of differential calculus*

$$\frac{f(t, x) - f(t, x_0)}{x - x_0} = \frac{\partial}{\partial x} f(t, \xi)$$

with a t-dependent point ξ between x_0 and x. By the *theorem of uniform continuity* (cf. Exercise 7 from I.3) for any given $\varepsilon > 0$ there exists a $\delta > 0$ with the property

$$\left| \frac{\partial}{\partial x} f(t_1, x_1) - \frac{\partial}{\partial x} f(t_2, x_2) \right| < \varepsilon \qquad \text{if } |x_1 - x_2| < \delta , \quad |t_1 - t_2| < \delta .$$

In particular,

$$\left| \frac{\partial}{\partial x} f(t, \xi) - \frac{\partial}{\partial x} f(t, x_0) \right| < \varepsilon \qquad \text{if } |x - x_0| < \delta .$$

It is decisive here that δ does not depend on t! Now we obtain

$$\left| \frac{g(x) - g(x_0)}{x - x_0} - \int_a^b \frac{\partial}{\partial x} f(t, x_0) \, dt \right| \leq \varepsilon (b - a) \qquad \text{if } |x - x_0| < \delta .$$

$\square$

Theorem II.3.4 (Generalized Cauchy Integral Formula) *With the assumptions and notation of II.3.2 we have: Every analytic function is arbitrarily often complex differentiable. Each derivative is again analytic. For $n \in \mathbb{N}_0$ and all z with $|z - z_0| < r$ we have*

$$f^{(n)}(z) = \frac{n!}{2\pi i} \oint_\alpha \frac{f(\zeta)}{(\zeta - z)^{n+1}} \, d\zeta \, ,$$

where $\alpha(t) = z_0 + r e^{it}$, $0 \le t \le 2\pi$.

The proof follows by induction on n with the help of II.3.2 and II.3.3.   □

For another proof see Exercise 10 in II.3.

Remark. Therefore it has also been proved that the assumptions of continuity of the derivative f', resp. of analyticity of f', we have previously made, were superfluous as they are automatically fulfilled. Moreover it follows that $u = \operatorname{Re} f$ and $v = \operatorname{Im} f$ are in fact $\mathcal{C}^\infty$-functions.

It was not necessary to use Lemma II.3.3 in its full generality for the proof of II.3.4. It would be possible just to check the required special case directly. Then one can get back II.3.3 from II.3.4 in full generality by using the FUBINI *theorem: If $f : [a,b] \times [c,d] \to \mathbb{C}$ is a continuous function, then*

$$\int_a^b \int_c^d f(x,y) \, dy \, dx = \int_c^d \int_a^b f(x,y) \, dx \, dy \, .$$

The following theorem gives a kind of partial converse to the CAUCHY integral theorem.

Theorem II.3.5 (Morera's Theorem, (G. Morera, 1886)) *Let $D \subset \mathbb{C}$ be open and*

$$f : D \longrightarrow \mathbb{C}$$

be continuous. For every triangular path $\langle z_1, z_2, z_3 \rangle$ whose triangle is entirely contained in D assume

$$\int_{\langle z_1, z_2, z_3 \rangle} f(\zeta) \, d\zeta = 0 \, .$$

Then f is analytic.

Proof. For each point $z_0 \in D$ there is an open neighborhood $U_\varepsilon(z_0) \subset D$. It is enough to show that f is analytic in $U_\varepsilon(z_0)$. For $z \in U_\varepsilon(z_0)$ let

$$F(z) := \int_{\sigma(z_0, z)} f(\zeta) \, d\zeta \, ,$$

where $\sigma(z_0, z)$ is the line segment connecting z_0 and z. As in II.2.4 (c) $\Rightarrow$ (a) one shows that F is a primitive of f in $U_\varepsilon(z_0)$, i. e. $F'(z) = f(z)$ for $z \in U_\varepsilon(z_0)$. In particular, f is analytic itself as the derivative of an analytic function.   □

Definition II.3.6 *An analytic function* $f : \mathbb{C} \to \mathbb{C}$ *is said to be* **entire**.

An entire function is thus an *analytic* function defined on the *entire complex plane* $\mathbb{C}$.

Examples. Polynomials $P : \mathbb{C} \to \mathbb{C}$, and $\exp, \cos, \sin : \mathbb{C} \to \mathbb{C}$ are entire functions.

Theorem II.3.7 (Liouville's Theorem, J. Liouville, 1847) *Every bounded entire function is constant.*
Equivalently: A nonconstant entire function cannot be bounded.

(In particular, for instance, cos cannot be bounded. In fact

$$\cos ix = \frac{e^x + e^{-x}}{2} \to \infty \quad \text{for} \quad x \to \infty .$$

LIOUVILLE actually only treated the special case of elliptic functions (cf. Chapter V and Exercise 7 in II.3).

Proof. We show $f'(z) = 0$ for every point $z \in \mathbb{C}$. From the CAUCHY integral formula

$$f'(z) = \frac{1}{2\pi i} \oint_{|\zeta - z| = r} \frac{f(\zeta)}{(\zeta - z)^2} \, d\zeta ,$$

which holds for *every* $r > 0$, it follows that

$$|f'(z)| \leq \frac{1}{2\pi} \underbrace{2\pi r}_{\substack{\text{arc} \\ \text{length}}} \frac{C}{r^2} = \frac{C}{r} .$$

The assertion can now be obtained by passing to the limit $r \to \infty$. □

From LIOUVILLE's Theorem follows easily:

Theorem II.3.8 (Fundamental Theorem of Algebra) *Each nonconstant complex polynomial has a root.*

Proof. Let

$$P(z) = a_0 + a_1 z + \cdots + a_n z^n , \qquad a_\nu \in \mathbb{C} , \ 0 \leq \nu \leq n , \ n \geq 1 , \ a_n \neq 0 .$$

be a polynomial of degree ≥ 1. Then

$$|P(z)| \to \infty \text{ for } |z| \to \infty$$

i.e. for each $C > 0$ there exists an $R > 0$ such that

$$|z| \geq R \implies |P(z)| \geq C ,$$

(Note:[2] One has $z^{-n} P(z) \to a_n$ for $|z| \to \infty$.) We assume that P has no complex root. Then $1/P$ is a bounded entire function and so $1/P$ is a constant by LIOUVILLE's theorem. □

---

[2] Cf. also Exercise 12 in II.2.

Corollary II.3.9 *Every polynomial*

$$P(z) = a_0 + a_1 z + \cdots + a_n z^n \,, \quad a_\nu \in \mathbb{C}, \ 0 \le \nu \le n \,,$$

of degree $n \ge 1$ can be written as a product of n linear factors and a constant $C \in \mathbb{C}^\bullet$

$$P(z) = C(z - \alpha_1) \cdots (z - \alpha_n) \,.$$

The numbers $\alpha_1, \ldots, \alpha_n \in \mathbb{C}$ are uniquely determined up to their order, and $C = a_n$.

Proof. If $n \ge 1$, there exists a zero α_1. We reorder the polynomial by powers of $(z - \alpha_1)$

$$P(z) = b_0 + b_1(z - \alpha_1) + \cdots \quad.$$

From $P(\alpha_1) = 0$ it follows that $b_0 = 0$ and therefore

$$P(z) = (z - \alpha_1)Q(z) \,, \quad \text{degree } Q = n - 1 \,.$$

The assertion then follows by induction on n. $\square$

If one collects equal α_ν, then one gets for P a formula

$$P(z) = C(z - \beta_1)^{\nu_1} \cdots (z - \beta_r)^{\nu_r}$$

with pairwise different $\beta_j \in \mathbb{C}$ and integers ν_j, for which we then have $\nu_1 + \cdots + \nu_r = n$.

We shall obtain other function-theoretic proofs of the fundamental theorem of algebra later (cf. also Exercise 13 in II.2 of this Chapter and application of the Residue Theorem III.6.3).

Exercises for II.3

We shall denote by $\alpha_{a;r}$ the curve whose image is the circle with center a and radius $r > 0$, i.e. with

$$\alpha_{a;r} : [0, 2\pi] \longrightarrow \mathbb{C}, \quad \alpha_{a,r}(t) = a + re^{it} \,.$$

1. Compute, using the CAUCHY integral theorem and the CAUCHY integral formula, the following integrals:

(a) $\displaystyle \int_{\alpha_{2;1}} \frac{z^7 + 1}{z^2(z^4 + 1)} \, dz \,,$ (b) $\displaystyle \int_{\alpha_{1;3/2}} \frac{z^7 + 1}{z^2(z^4 + 1)} \, dz \,,$

(c) $\displaystyle \int_{\alpha_{0;3}} \frac{e^{-z}}{(z + 2)^3} \, dz \,,$ (d) $\displaystyle \int_{\alpha_{0;3}} \frac{\cos \pi z}{z^2 - 1} \, dz \,,$

(e) $\displaystyle \int_{\alpha_{0;r}} \frac{\sin z}{z - b} \, dz \,,$ $(b \in \mathbb{C} \,, \ |b| \ne r) \,.$

2. Compute, using the CAUCHY integral theorem and the CAUCHY integral formula, the following integrals:

  (a) $\dfrac{1}{2\pi i}\displaystyle\int_{\alpha_{i;1}} \dfrac{e^z}{z^2+1}\,dz$,   (b) $\dfrac{1}{2\pi i}\displaystyle\int_{\alpha_{-i;1}} \dfrac{e^z}{z^2+1}\,dz$,

  (c) $\dfrac{1}{2\pi i}\displaystyle\int_{\alpha_{0;3}} \dfrac{e^z}{z^2+1}\,dz$,   (d) $\dfrac{1}{2\pi i}\displaystyle\int_{\alpha_{1+2i;5}} \dfrac{4z}{z^2+9}\,dz$.

3. Compute

  (a) $\displaystyle\int_{\alpha_{1;1}} \left(\dfrac{z}{z-1}\right)^n dz$, $n \in \mathbb{N}$,

  (b) $\displaystyle\int_{\alpha_{0;r}} \dfrac{1}{(z-a)^n(z-b)^m}\,dz$, $|a| < r < |b|$, $n, m \in \mathbb{N}$.

4. Let $\alpha = \alpha_1 \oplus \alpha_2$ be the curve sketched in the figure with $R > 1$ and

$$f(z) := \frac{1}{1+z^2} \ .$$

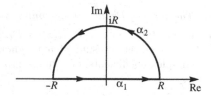

Show:

$$\int_\alpha f(z)\,dz = \int_{\alpha_1} f(z)\,dz + \int_{\alpha_2} f(z)\,dz = \pi$$

and

$$\lim_{R\to\infty} \left| \int_{\alpha_2} f(z)\,dz \right| = 0 \ .$$

Deduce that:

$$\int_{-\infty}^{\infty} \frac{1}{1+x^2}\,dx = \lim_{R\to\infty} \int_{-R}^{R} \frac{1}{1+x^2}\,dx = \pi \ .$$

These indefinite integrals could have been calculated more easily (arctan is a primitive!). However, this gives a first indication of how one can compute real integrals using complex methods. We shall return to this when applying the residue theorem cf. III.7).

5. Let α be the closed curve considered in Exercise 4 of II.1 ("figure eight"). Compute the integral

$$\int_\alpha \frac{1}{1-z^2}\,dz \ .$$

6. *Show:* If $f : \mathbb{C} \to \mathbb{C}$ is analytic and if there is a real number M such that for all $z \in \mathbb{C}$

$$\operatorname{Re} f(z) \leq M \ ,$$

then f is constant.

Hint. Consider $g := \exp \circ f$ and apply LIOUVILLE's theorem to g.

7. Let ω and ω' be complex numbers which are linearly independent over $\mathbb{R}$.
   Show: If $f : \mathbb{C} \to \mathbb{C}$ is analytic and

   $$f(z + \omega) = f(z) = f(z + \omega') \text{ for all } z \in \mathbb{C} ,$$

   then f is constant (J. LIOUVILLE, 1847).

8. **Gauss-Lucas Theorem** (C.F. GAUSS, 1816; F. LUCAS, 1879)

   Let P be a complex polynomial of degree $n \geq 1$, with n not necessarily different zeros $\zeta_1, \ldots, \zeta_n \in \mathbb{C}$. Show that for all $z \in \mathbb{C} \setminus \{\zeta_1, \ldots, \zeta_n\}$

   $$\frac{P'(z)}{P(z)} = \frac{1}{z - \zeta_1} + \frac{1}{z - \zeta_2} + \cdots + \frac{1}{z - \zeta_n} = \sum_{\nu=1}^{n} \frac{\overline{z - \zeta_\nu}}{|z - \zeta_\nu|^2} .$$

   Deduce from this the GAUSS-LUCAS theorem:
   For each zero ζ of P' there are n real numbers $\lambda_1, \ldots, \lambda_n$ with

   $$\lambda_1 \geq 0, \ldots, \lambda_n \geq 0, \quad \sum_{j=1}^{n} \lambda_j = 1 \text{ and } \zeta = \sum_{\nu=1}^{n} \lambda_\nu \zeta_\nu .$$

   Thus one can say: The zeros of P' lie in the "convex hull" of the zero set of P.

9. Show that every rational function R (i.e. $R(z) = P(z)/Q(z)$, P, Q polynomials, $Q \neq 0$) can be written as the sum of a polynomial and a finite linear combination, with complex coefficients, of "simple functions" of the form

   $$z \mapsto \frac{1}{(z - s)^n} , \quad n \in \mathbb{N} , \ s \in \mathbb{C} ,$$

   the so-called "partial fractions" (Partial fraction decomposition theorem), see also Chapter III, Appendix A to Sections III.4 and III.5, Proposition A.7).

   Deduce: If the coefficients of P and Q are real, then f has "a *real* partial fraction decomposition" (by putting together pairs of complex conjugate zeros, or rather by putting together the corresponding partial fractions (see also Exercise 10 in I.1).

10. A somewhat more direct proof of the generalized CAUCHY integral formula (Theorem II.3.4) is obtained with the following *Lemma:*
    Let $\alpha : [a, b] \to \mathbb{C}$ be a piecewise smooth curve and let $\varphi : \text{Image } \alpha \to \mathbb{C}$ be continuous. For $z \in D := \mathbb{C} \setminus \text{Image } \alpha$ and $m \in \mathbb{N}$ let

    $$F_m(z) := \frac{1}{2\pi \mathrm{i}} \int_\alpha \frac{\varphi(\zeta)}{(\zeta - z)^m} \, d\zeta .$$

    Then F_m is analytic in D and for all $z \in D$

    $$F'_m(z) = m \, F_{m+1}(z) .$$

    Prove this by direct estimate (without using the LEIBNIZ rule).

11. Let $D \subset \mathbb{C}$ be open, and $L \subset \mathbb{C}$ a line. If $f : D \to \mathbb{C}$ is a continuous function, which is analytic at all points $z \in D$, $z \notin L$, then f is analytic on the whole D.

12. **The Schwarz Reflection principle** (H.A. SCHWARZ, 1867)

    Let $D \neq \emptyset$ be a domain which is symmetric with respect to the real axis (i.e. $z \in D \implies \bar{z} \in D$). We consider the subsets

    $$
    \begin{aligned}
    D_+ &:= \{ z \in D \; ; \quad \text{Im } z > 0 \} \; , \\
    D_- &:= \{ z \in D \; ; \quad \text{Im } z < 0 \} \; , \\
    D_0 &:= \{ z \in D \; ; \quad \text{Im } z = 0 \} = D \cap \mathbb{R} \; .
    \end{aligned}
    $$

    If $f : D_+ \cup D_0 \to \mathbb{C}$ is continuous, $f \mid D_+$ analytic and $f(D_0) \subset \mathbb{R}$, then the function defined by

    $$
    \tilde{f}(z) := \begin{cases} f(z) & \text{for } z \in D_+ \cup D_0 \; , \\ \overline{f(\bar{z})} & \text{for } z \in D_+ \; , \end{cases}
    $$

    is analytic.

13. Let f be a continuous function on the compact interval $[a, b]$.

    Show: The function defined by

    $$
    F(z) = \int_a^b \exp(-zt) \, f(t) \, dt
    $$

    is analytic on the whole $\mathbb{C}$, and

    $$
    F'(z) = - \int_a^b \exp(-zt) t f(t) \, dt \; .
    $$

14. Let $D \subset \mathbb{C}$ be a domain and

    $$
    f : D \longrightarrow \mathbb{C}
    $$

    be an analytic function.

    Show: The function

    $$
    \varphi : D \times D \longrightarrow \mathbb{C}
    $$

    with

    $$
    \varphi(\zeta, z) := \begin{cases} \dfrac{f(\zeta) - f(z)}{\zeta - z} & \text{if } \zeta \neq z \; , \\ f'(\zeta) & \text{if } \zeta = z \; , \end{cases}
    $$

    is a continuous function of two variables.

    For each given $z \in D$ the function

    $$
    \zeta \longmapsto \varphi(\zeta, z)
    $$

    is analytic in D.

15. Determine all pairs (f, g) of entire functions with the property

    $$
    f^2 + g^2 = 1 \; .
    $$

    Result:

    $f = \cos \circ h$ and $g = \sin \circ h$, where h is an arbitrary entire function.

16. Let $f : \mathbb{C} \to \mathbb{C}$ be a non-constant, entire function. Then $f(\mathbb{C})$ is dense in $\mathbb{C}$.

III

Sequences and Series of Analytic Functions, the Residue Theorem

It is known from real analysis that *pointwise convergence* of a sequence of functions shows certain pathologies. For instance, the pointwise limit of a sequence of continuous functions is not necessarily continuous, and in general we cannot exchange limit processes, et cetera. Therefore we are led to introduce the notion of *uniform convergence,* which has better stability properties. For example, the limit of a uniformly convergent sequence of continuous functions is continuous. Another basic stability theorem holds for the (proper) integral:

A uniformly convergent sequence of integrable functions converges to an integrable function. The limit and integration can be exchanged.

However, differentiability in real analysis is not stable with respect to uniform convergence.

The corresponding stability theorems are more complicated and require additional conditions on the sequence of derivatives.

In function theory one introduces the concept of uniform convergence by analogy with real analysis. The stability of continuity and of the integral along curves can be obtained completely analogously to the real case, and in fact can be reduced to that case.

In contrast to the situation in real analysis, complex differentiability (on open subsets of the complex plane) is stable with respect to uniform convergence.

The reason is that the derivatives of an analytic function can be obtained through an integration process (CAUCHY's integral formula). Because of this stability, properties analogous to those for integration also hold for differentiation. In particular we have the WEIERSTRASS theorem which is characteristic for complex analysis:

A uniformly convergent sequence of analytic functions converges to an analytic function. Taking the limit and differentiation may be exchanged.

This has as a consequence the fact that proving the complex differentiability of a function defined as a limit is often much simpler than in the real case. We will see this already in the special case of power series.

For our purposes it is usually enough to require instead of uniform convergence just *local uniform convergence.* This implies uniform convergence on compact sets.

E. Freitag and R. Busam, *Complex Analysis,*
DOI: 10.1007/978-3-540-93983-2_III, © Springer-Verlag Berlin Heidelberg 2009

Local uniform convergence of series of functions is often proven using the WEIER-STRASS majorant test . Series to which this "test" can be applied are called normally convergent:

A series of functions is called normally convergent if each point of their domain of definition admits an open neighborhood and a majorant series of numbers which is valid in this neighborhood.

Examples of normally convergent series are *power series* in the interiors of their regions of convergence. There, in particular, they define analytic functions. Conversely we shall show that any function analytic in an open disk can be expanded into a power series there.

In particular, any analytic function can be locally expanded into a power series.

The key to this powerful expansion theorem is the CAUCHY integral formula.

However, this expansion theorem is just a special case of a general expansion theorem for functions analytic in *annuli* (ring domains)

$$\mathcal{A} = \{\ z \in \mathbb{C}\ ;\quad r < |z| < R\ \}\quad (0 \leq r < R \leq \infty)\ .$$

In such annuli the negative powers of z are also analytic. We shall show that any function analytic in an annulus can be expanded there in a so-called *Laurent series*

$$\sum_{n=-\infty}^{\infty} a_n z^n\ .$$

The case $r = 0$ is of special interest, for this is the case of an *isolated singularity*. We will describe a classification of singularities. The type of a singularity —*removable singularity, pole or essential singularity*— can be read off from both the LAURENT series and the behavior of the mapping.

Using the LAURENT expansion we shall prove the *Residue Theorem*, which can be used to compute integrals of analytic functions along closed curves avoiding the singularities of the function.

With these function-theoretic tools we can also get insight into the behavior of the mappings defined by analytic functions, and derive results that are unexpected from the point of view of real analysis. We shall, for instance, prove the *Open Mapping Theorem*:

The set of values of a nonconstant analytic function on a domain is also a domain.

In particular, the absolute value of such a function cannot attain a maximum (*maximum principle*).

We close the chapter with a small selection of applications of the residue theorem.

III.1 Uniform Approximation

A sequence of functions

$$f_0,\ f_1,\ f_2,\ \cdots : D \longrightarrow \mathbb{C}$$

defined on an arbitrary subset $D \subset \mathbb{C}$, is called *uniformly convergent* to the limit

$$f : D \longrightarrow \mathbb{C} \,,$$

if the following holds:

For each $\varepsilon > 0$ there exists a natural number N, such that

$$|f(z) - f_n(z)| < \varepsilon \text{ for all } n \geq N \text{ and all } z \in D \,.$$

In particular, N should not depend on z.

In this definition D can be an arbitrary set. We now assume that D is a subset of the complex plane, or more generally, a subset of $\mathbb{R}^p$.

The sequence (f_n) converges *locally uniformly* to f if for every point $a \in D$ there is a neighborhood U of a in $\mathbb{R}^p$ such that $f_n \mid U \cap D$ is uniformly convergent.

Using the HEINE-BOREL covering theorem it is easy to see that the sequence $(f_n \mid K)$ is uniformly convergent on any compact set K contained in D.

So one can say: A locally uniformly convergent sequence of functions f_n : $D \to \mathbb{C}$ is *compactly convergent*.

There is a converse of this if D is open, for then there exists for each point $a \in D$ a closed (and thus compact) disk with center a, which is contained in D.

The analogue of the following is well known in real analysis:

Remark III.1.1 *Let*

$$f_0, f_1, f_2, \cdots \,,$$

be a sequence of continuous functions which converges locally uniformly. Then its limit function is also continuous.

The proof follows as in the real case, see Exercise 1 in III.1.

For line integrals there is an analogous stability theorem.

Remark III.1.2 *Let*

$$f_0, f_1, f_2, \cdots : D \to \mathbb{C} \,, \quad D \subset \mathbb{C} \,,$$

be a sequence of continuous functions which converges locally uniformly to f. Then for any piecewise smooth curve $\alpha : [a, b] \to D$

$$\lim_{n \to \infty} \int_\alpha f_n(\zeta) \, d\zeta = \int_\alpha f(\zeta) \, d\zeta \,.$$

Proof: One has to use the fact that the image of α is compact, so that the sequence f_n is uniformly convergent on Image α. The assertion now immediately follows from the estimate

$$\left| \int_\alpha f_n - \int_\alpha f \right| \leq l(\alpha) \cdot \varepsilon \,,$$

if $|f_n(z) - f(z)| \leq \varepsilon$ for all $z \in$ Image α. Here $l(\alpha)$ is the arc length of the piecewise smooth curve α.   □

Theorem III.1.3 (K. Weierstrass, 1841) *Let*

$$f_0, f_1, f_2, \cdots : D \to \mathbb{C} , \quad D \subset \mathbb{C} \ open,$$

be a sequence of analytic functions which converges locally uniformly. Then the limit function f is analytic and the sequence of the derivatives (f_n') converges locally uniformly to f'.

Proof. The assertion follows immediately from the fact that complex differentiability can be characterized by a criterion involving integration (MORERA's Theorem, II.3.5), and the fact that our line integral is stable with respect to uniform convergence. The assertion about (f_n') results from the CAUCHY Integral Formula for f' resp. f_n' (cf. the proof of the Supplement to III.1.6). □

In this connection we should point out that the real analogue of III.1.3 is false. *By the Weierstrass Approximation Theorem, any continuous function*

$$f : [a, b] \longrightarrow \mathbb{R}$$

is actually the limit of a uniformly convergent sequence of polynomials!

However, in the real case there is also a stability theorem:

Let $f_0, f_1, f_2, \cdots : [a, b] \to \mathbb{R}$ be a sequence of continuously differentiable functions that converge pointwise to a function f. If the sequence (f_n') converges uniformly, then f is differentiable and $\lim_{n\to\infty} f_n'(x) = f'(x)$.

Theorem III.1.3 can naturally be rewritten for series:

A series of functions

$$f_0 + f_1 + f_2 + \cdots , \quad f_n : D \to \mathbb{C} , \quad D \ open \subset \mathbb{C} , \quad n \in \mathbb{N}_0 ,$$

*is called **(locally) uniformly convergent**, if the sequence (S_n) of the partial sums*

$$S_n := f_0 + f_1 + \cdots + f_n$$

is (locally) uniformly convergent.

Definition III.1.4 *A series $f_0 + f_1 + f_2 + \cdots$ of functions*

$$f_n : D \to \mathbb{C} , \quad D \subset \mathbb{C} , \quad n \in \mathbb{N}_0 ,$$

*is called **normally convergent** (in D), if for each point $a \in D$ there is a neighborhood U and a sequence $(M_n)_{n\geq 0}$ of non-negative real numbers, such that:*

$$|f_n(z)| \leq M_n \ for \ all \ z \in U \cap D , \ all \ n \in \mathbb{N}_0 ,$$

$$and \sum_{n=0}^{\infty} M_n \ converges.$$

Remark III.1.5 (Weierstrass majorant test) *A normally convergent series of functions converges **absolutely** and **locally uniformly**. A normally convergent series of functions can thus be **arbitrarily re-ordered** without changing its convergence or its limit.*

Theorem III.1.6 (K. Weierstrass, 1841) *Let*

$$f_0 + f_1 + f_2 + \cdots$$

be a normally convergent series of analytic functions on an open set $D \subset \mathbb{C}$. Then the limit function f is also analytic and

$$f' = f_0' + f_1' + f_2' + \cdots .$$

Supplement. *The series of derivatives is also normally convergent.*

Only the supplement remains to be proved. Let a be a point in D. Choose $\varepsilon > 0$ so small that the closed disk of radius 2ε is contained in D, and such that the series has a convergent majorant $\sum M_n$ in this closed disk. Then, for all z in the ε-neighborhood of a we have from the CAUCHY integral formula the estimate

$$|f_n'(z)| = \left| \frac{1}{2\pi i} \oint_{|\zeta - a| = 2\varepsilon} \frac{f_n(\zeta)}{(\zeta - z)^2} \, d\zeta \right| \leq 2\varepsilon^{-1} M_n .$$

□

An example of normal convergence.

Let $s \in \mathbb{C}$ and $s := \sigma + it$ with $\sigma, t \in \mathbb{R}$ (RIEMANN-LANDAU convention). For all $n \in \mathbb{N}$ the assignment

$$s \mapsto n^s := \exp(s \log n)$$

defines an analytic function in $\mathbb{C}$. One has $|n^s| = n^\sigma$. Then we have the following

Claim. *The series*

$$\sum_{n=1}^{\infty} \frac{1}{n^s}$$

converges absolutely and uniformly in every half-plane

$$\{ s \in \mathbb{C} ; \quad \mathrm{Re}\, s \geq 1 + \delta \} , \ \delta > 0 .$$

The convergence is normal in the half-plane

$$D := \{ s \in \mathbb{C} ; \quad \mathrm{Re}\, s > 1 \} .$$

*This defines a function ζ which is analytic in D, the so-called **Riemann Zeta Function**:*

$$\zeta(s) := \sum_{n=1}^{\infty} \frac{1}{n^s} , \quad Re\ s > 1 .$$

We shall discuss the properties of this function and its role in analytic number theory thoroughly in Chapter VII.

Proof of the assertion. For each $\delta > 0$ we have

$$\left| \frac{1}{n^s} \right| = \frac{1}{n^\sigma} \leq \frac{1}{n^{1+\delta}} \text{ for all } s \text{ with } \sigma \geq 1 + \delta .$$

$\square$

Exercises for III.1

1.   Prove Remark III.1.1: Let $D \subset \mathbb{C}$ and (f_n) be a sequence of continuous functions $f_n : D \to \mathbb{C}$ which converges locally uniformly in D. Then the limit function $f : D \to \mathbb{C}$ is also continuous.

2.   With the assumptions of Theorem III.1.3 show that for each $k \in \mathbb{N}$ the sequence $\left(f_n^{(k)} \right)$ of k-th derivatives converges locally uniformly to $f^{(k)}$.

3.   Let $D \subset \mathbb{C}$ be open and let (f_n) be a sequence of analytic functions $f_n : D \to \mathbb{C}$ with the property: For every closed disk $K \subset D$ there is a real number $M(K)$ such that $|f_n(z)| \leq M(K)$ for all $z \in K$ and all $n \in \mathbb{N}$.

      Show: The sequence (f'_n) has the analogous property.

4.   Show that the series

$$\sum_{\nu=1}^{\infty} \frac{z^{2\nu}}{1-z^\nu}$$

      converges normally in the unit disk $\mathbb{E} = \{\, z \in \mathbb{C} ; \quad |z| < 1 \,\}$.

5.   Show that the sequence

$$\sum_{\nu=1}^{\infty} \frac{(-1)^\nu}{z-\nu}$$

      converges locally uniformly, but not uniformly, in $D = \mathbb{C} - \mathbb{N}$.

6.   Show that the series

$$\sum_{\nu=1}^{\infty} \frac{1}{z^2 - (2\nu+1)z + \nu(\nu+1)}$$

      converges normally in $\mathbb{C} - \mathbb{N}_0$, and determine its limit.

7.   In which domain $D \subset \mathbb{C}$ defines the following series

$$\sum_{n=1}^{\infty} \frac{\sin(nz)}{2^n}$$

      an analytic function?

      (*Answer:* $D = \{\, z \in \mathbb{C} ; \quad |\operatorname{Im} z| < \log 2 \,\}$.)

      Is there any domain in which the series

$$\sum_{n=1}^{\infty} \frac{\sin(nz)}{n^2}$$

defines an analytic function?

8. Let f be a continuous function on the closed unit disk

$$\mathbb{E} := \{\, z \in \mathbb{C} \;;\quad |z| \le 1 \,\} \;,$$

such that $f|\mathbb{E}$ is analytic. Then $\oint_{|\zeta|=1} f(\zeta)\, d\zeta = 0$.

Hint. Consider, for $0 < r < 1$, the functions

$$f_r : \overline{U}_{1/r}(0) \longrightarrow \mathbb{C} \;,\quad z \longmapsto f(rz) \;.$$

III.2 Power Series

A power series (with center zero) is a series of the form

$$a_0 + a_1\, z + a_2\, z^2 + \ldots \quad ,$$

where the coefficients a_n, $n \in \mathbb{N}_0$, are given complex numbers, and z varies in $\mathbb{C}$.

Proposition III.2.1 (Convergence Theorem for Power Series) *For each power series*

$$a_0 + a_1\, z + a_2\, z^2 + \ldots$$

there exists a uniquely determined "number" $r \in [0, \infty] := [0, \infty[\,\cup\{\infty\}$ with the following properties:

(a) *The series converges normally in the open disk*

$$U_r(0) := \{\, z \in \mathbb{C} \;;\quad |z| < r \,\}.$$

(b) *The series diverges for each $z \in \mathbb{C}$ with $|z| > r$.*

Supplement. *The following holds:*

$$r = \sup\{\, t \ge 0 \;;\quad (a_n t^n) \text{ is a bounded sequence } \} \;, \qquad \text{and also}$$
$$r = \sup\{\, t \ge 0 \;;\quad (a_n t^n) \text{ is a null sequence } \} \;.$$

Proof (N.H. ABEL, 1826). Let r be one of the numbers defined in the *Supplement*. In the case $r = 0$ there is nothing to prove, so let r be > 0. It is clear that the series cannot converge for any z with $|z| > r$. So it is sufficient to show that for each ρ, $0 < \rho < r$, the given power series has a convergent majorant *not* depending on z. For this, we choose some ρ_1, $\rho < \rho_1 < r$. By definition of r as a supremum, the sequence $(a_n \rho_1^n)$ is (in both cases) bounded, let's say by the constant $M > 0$. Then we estimate for all z with $|z| \le \rho$:

$$\left|a_n z^n\right| = \left|a_n \rho_1^n \, \frac{z^n}{\rho_1^n}\right| \le M \cdot \left(\frac{\rho}{\rho_1}\right)^n .$$

The geometric series

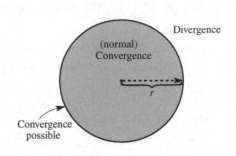

$$\sum_{n=0}^{\infty} (\rho/\rho_1)^n$$

is convergent, since we have $0 < \rho/\rho_1 < 1$.   □

Remark. *The quantity $r \in [0, \infty]$, uniquely determined by the properties in Proposition III.2.1, is called the* **convergence radius**. *The open disk $U_r(0)$, $r > 0$, is called the* **convergence disk** *of the given power series. In case of $r = \infty$ we set $U_r(0) = \mathbb{C}$, and in case $r = 0$ we have $U_r(0) = \emptyset$.*

The slightly more general case of a power series for an arbitrary center a,

$$\sum a_n \, (z - a)^n$$

can be reduced to the above case by means of the substitution $\zeta = z - a$.

Corollary III.2.1$_1$ *A power series with positive convergence radius $r > 0$ and center a gives rise to an analytic function $f : U_r(a) \to \mathbb{C}$ on the convergence disk $U_r(a)$. Its derivative is obtained by termwise differentiation.*

Explicitly, the function

$$f(z) := \sum_{n=0}^{\infty} a_n(z - a)^n$$

has in $U_r(a)$ the derivative

$$f'(z) = \sum_{n=1}^{\infty} n a_n(z - a)^{n-1} .$$

Remark. The convergence behavior of a power series with convergence radius r on the boundary of the convergence disk $\{\, z \in \mathbb{C} \;\; |z| = r \,\}$ is not touched by Proposition III.2.1. It may be different from case to case. The standard examples are (as in the real case) the following series:

(1) $\displaystyle\sum_{n=1}^{\infty} \frac{z^n}{n^2}$ with convergence radius $r = 1$. Because of the convergence of the

series $\displaystyle\sum_{n=1}^{\infty} \frac{1}{n^2}$, the power series converges for all z with $|z| \le 1$.

(2) The geometric series

$$\sum_{n=0}^{\infty} z^n$$

has the well known convergence radius $r = 1$, but it does *not* converge for any z with $|z| = 1$, for in such a case $(|z|^n)$ does not converge to zero.

(3) The *"logarithmic series"*

$$\sum_{n=1}^{\infty} (-1)^{n-1} \frac{z^n}{n} \quad (= \mathrm{Log}(1+z))$$

also has the convergence radius $r = 1$. At the boundary $|z| = 1$, it converges for $z = 1$ (LEIBNIZ' criterion), but diverges for $z = -1$ (harmonic series). There exist both convergence and divergence points of the power series at the boundary. By the way, -1 is the only divergence point! Prove this.

Theorem III.2.2 (Power series expansion, A.L. Cauchy, 1831) *Let the function*

$$f : D \longrightarrow \mathbb{C} , \ D \subset \mathbb{C} \ open ,$$

be analytic. We assume that the open disk $U_R(a)$ lies inside D. Then one has:

$$f(z) = \sum_{n=0}^{\infty} a_n \, (z-a)^n \ for \ all \ z \in U_R(a) , \quad where \ a_n := \frac{f^{(n)}(a)}{n!} , \ n \in \mathbb{N}_0 .$$

Especially, each analytic function can be *locally* expanded as a power series around each point of the domain of definition. Explicitly, for each $a \in D$ there exists a neighborhood $U(a)$ of a in D, and a power series $\sum_{n=0}^{\infty} a_n (z-a)^n$ which converges for all $z \in U(a)$, and represents the function f in $U(a)$. The convergence radius of the power series is thus $\geq R$.

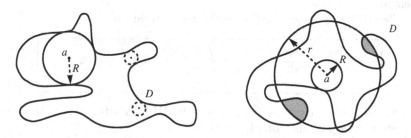

Supplement to III.2.2 *The involved coefficients have the integral representation*

$$a_n = \frac{1}{2\pi i} \oint_{|\zeta - a| = \rho} \frac{f(\zeta)}{(\zeta - a)^{n+1}} d\zeta \quad for \ 0 < \rho < R .$$

Remark. If f can be expanded as a power series in a neighborhood of a,

$$f(z) = \sum_{n=0}^{\infty} a_n (z-a)^n \ ,$$

then we necessarily have

$$a_n = \frac{f^{(n)}(a)}{n!} \qquad \left(= \frac{1}{2\pi i} \oint_{|\zeta - a| = \rho} \frac{f(\zeta)}{(\zeta - a)^{n+1}} d\zeta \quad \text{by II.3.4} \right) \ ,$$

because one obtains the higher derivatives of f by repeated termwise differentiation, Corollary III.2.1$_1$. The coefficients a_n are, as in the real case, the TAYLOR coefficients of f at a, and the power series introduced by f is the TAYLOR series at a.

Proof of III.2.2. Because of the uniqueness of the power series expansion, it is enough to exhibit for an arbitrary ρ, $0 < \rho < R$, an expansion in the smaller disk $|z - a| < \rho$. From the CAUCHY integral formula for disks (II.3.2) one has

$$f(z) = \frac{1}{2\pi i} \oint_{|\zeta - a| = \rho} \frac{f(\zeta)}{\zeta - z} \, d\zeta \quad \text{for } |z - a| < \rho \ .$$

The CAUCHY kernel $(z, \zeta) \to k(z, \zeta) := 1/(\zeta - z)$ can be expanded into a power series by means of the *geometric series*,

$$\frac{1}{\zeta - z} = \frac{1}{\zeta - a} \cdot \frac{1}{1 - \dfrac{z-a}{\zeta - a}} = \frac{1}{\zeta - a} \sum_{n=0}^{\infty} \left(\frac{z-a}{\zeta - a} \right)^n = \sum_{n=0}^{\infty} \frac{1}{(\zeta - a)^{n+1}} (z-a)^n \ .$$

(Also observe $q := \left| \dfrac{z-a}{\zeta - a} \right| < 1$.) After multiplication with $f(\zeta)$ and exchange of the line integral and the infinite sum, which is possible by III.1.2, we infer the claim. □

We collect: The coefficients a_n have the representation

$$a_n = \frac{f^{(n)}(a)}{n!} = \frac{1}{2\pi i} \oint_{|\zeta - a| = \rho} \frac{f(\zeta)}{\zeta - a} \, d\zeta \ , \quad n \in \mathbb{N}_0 \ .$$

The possibility of expansions into power series III.2.2 leads to a new *fundament of function theory* of one complex variable.

Analytic functions are exactly the functions, which are locally representable as a convergent power series.

Depending on the way whether one places the **complex differentiability**, or the **representability as power series** in the foreground, one deals with the CAUCHY-RIEMANN approach or with the WEIERSTRASS approach to complex analysis.

At our disposal now are different characterizations of the notion of an "analytic function". In the following theorem, which summarizes our already presented results, we collect the various equivalent characterizations of "being analytic". During the development of complex analysis there were different starting points, and this makes it understandable why even today different words such as "analytic", "regular", "holomorphic" and so on are still used. We prefer the term "analytic", or occasionally "holomorphic". The word "bianalytic" is not very nice. One should instead choose "biholomorphic" or "conformal".

Theorem III.2.3 *Let $D \subset \mathbb{C}$ be open. The following assertions are equivalent for a function $f : D \to \mathbb{C}$:*

(a) *f is analytic, i.e. complex differentiable at any point $z \in D$.*

(b) *f is totally differentiable in the sense of real analysis ($\mathbb{C} = \mathbb{R}^2$), and $u = \operatorname{Re} f$, $v = \operatorname{Im} f$ fulfill the Cauchy-Riemann differential equations:*

$$\frac{\partial u}{\partial x} = \frac{\partial v}{\partial y} \quad and \quad \frac{\partial u}{\partial y} = -\frac{\partial v}{\partial x} \ .$$

(c) *f is continuous, and for each triangle path $\langle z_1, z_2, z_3 \rangle$ whose convex hull Δ is contained in D one has:*

$$\int_{\langle z_1, z_2, z_3 \rangle} f(\zeta)\, d\zeta = 0 \qquad (Morera's\ condition) \ .$$

(d) *f locally admits primitives, i.e. for each point $a \in D$ there exists an open neighborhood $U(a) \subset D$, so that $f | U(a)$ admits a primitive.*

(e) *f is continuous, and for each disk $U_\rho(a)$ with $\overline{U}_\rho(a) \subset D$ it holds:*

$$f(z) = \frac{1}{2\pi i} \oint_{|\zeta - a| = \rho} \frac{f(\zeta)}{\zeta - z}\, d\zeta \qquad for \quad |z - a| < \rho \ .$$

(f) *f is locally representable as a convergent power series, i.e. for each point $a \in D$ there exists an open neighborhood of a in D, where f is representable as a power series.*

(g) *f is representable as a power series in each open disk fully contained in D.*

Here, the difference between real and complex analysis becomes clear! For a real interval $M \subset \mathbb{R}$ there exists "a lot of" functions $f : M \to \mathbb{R}$, which are 17 but not 18 times differentiable, or which are differentiable with a non-continuous derivative f'. Here is the standard example

$$f : \mathbb{R} \to \mathbb{R} \quad with \quad f(x) = \begin{cases} x^2 \sin(1/x) , & x \neq 0 , \\ 0 , & x = 0 . \end{cases}$$

In addition, CAUCHY's example (CAUCHY, 1823)

$$f : \mathbb{R} \to \mathbb{R} \quad \text{with} \quad f(x) = \begin{cases} \exp(-1/x^2) \,, & x \neq 0 \,, \\ 0 \,, & x = 0 \,, \end{cases}$$

shows the existence of a C^∞-function on $\mathbb{R}$, which is not represented by its TAYLOR series around 0 in any neighborhood of 0. The representation theorem III.2.2 has thus no direct counterpart in real analysis. The membership $f \in C^\infty(M)$ is a necessary condition for the power series expansion of f, but not a sufficient one. We will occur in the next paragraphs many other essential differences between C^∞-functions and analytic functions on an open set $D \subset \mathbb{C}$.

It is often inexpedient to compute the coefficients (or the convergence radius) from the equality

$$a_n = \frac{1}{2\pi i} \oint_{|\zeta - a| = \rho} \frac{f(\zeta)}{(\zeta - a)^{n+1}} \, d\zeta$$

or from the TAYLOR formula. It is better to use already known power series expansions.

Example. Let us expand the analytic function $f : \mathbb{C} \setminus \{\pm i\}$, $f(z) = 1/(1 + z^2)$, around the zero point.

The sum formula for the geometric series shows for $|z| < 1$

$$f(z) = \frac{1}{1 + z^2} = \sum_{n=0}^{\infty} (-1)^n \, z^{2n} \,.$$

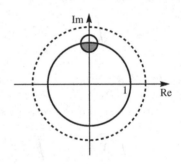

The radius of convergence r of this series is $r = 1$, as it follows from III.2.1. One can also determine it by pure complex analytic methods:

(1) r is at least 1, $r \geq 1$, which follows from III.2.2.
(2) r is at most 1, $r \leq 1$, for $f(z)$ is unbounded if z approaches i (compare with the figure).

The singularity in i, invisible in real analysis, is thus responsible for the fact that the radius cannot be bigger than 1. The convergence behavior of power series thus becomes more transparent in complex analysis.

Often, one uses such function theoretical arguments to determine the convergence radius of a power series. The formula in III.2.1 and analogous ones (e.g. Exercise 6 in III.2) are of minor importance in complex analysis.

The radius of convergence of the TAYLOR series of an analytic function can of course be bigger than the distance of the center to the boundary of the domain of definition. For instance, the principal value of the logarithm has in a point $a \in \mathbb{C}_-$ (slit plane, the negative part of the real axis is removed from $\mathbb{C}$) the expansion

$$\mathrm{Log}(a) + \sum_{\nu=1}^{\infty} \frac{(-1)^{\nu-1}}{a^\nu \nu} (z-a)^\nu ,$$

where the convergence radius is $|a|$. For a point $a \in \mathbb{C}_-$ with $\mathrm{Re}\, a < 0$ the distance to the boundary is strictly smaller than this radius.

Let us note that $\mathbb{C}_- \cap U_r(a)$ splits in two disjoint connected components. In the "upper" part of the disk the series represents the principal value of the logarithm, Log, but not also in the "lower" part, where it represents $\mathrm{Log} + 2\pi i$.

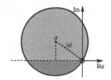

Computation Rules For Power Series

(A.L. CAUCHY, 1821; K. WEIERSTRASS, 1841)

1. *Identity of power series*
    If two power series

    $$\sum_{n=0}^{\infty} a_n z^n \quad \text{and} \quad \sum_{n=0}^{\infty} b_n z^n$$

    converge in *some* neighborhood of 0 and represent there the same function, then $a_n = b_n$ for all $n \in \mathbb{N}_0$.

2. *Cauchy's product formula*
    We assume that the radius of convergence for both series

    $$\sum_{n=0}^{\infty} a_n z^n \quad \text{and} \quad \sum_{n=0}^{\infty} b_n z^n$$

    is at least $R > 0$. Then one has

    $$\left(\sum_{n=0}^{\infty} a_n z^n \right) \cdot \left(\sum_{n=0}^{\infty} b_n z^n \right) = \sum_{n=0}^{\infty} c_n z^n \quad \text{for } |z| < R$$

    where

    $$c_n := \sum_{\nu=0}^{n} a_\nu b_{n-\nu} \quad \text{(compare with I.2.7)} .$$

3. *Multiplicative inversion of power series*
    Let $P(z) = a_0 + a_1 z + \cdots$ be a power series with positive convergence radius. We assume $a_0 \neq 0$. Then $P(z) \neq 0$ for all z in an open disk $|z| < r$. In this disk $Q := 1/P$ is analytic (see Theorem I.4.3), and thus

representable as a power series. (In the real case one cannot argue in this way!)

$$Q(z) = b_0 + b_1 z + \cdots \qquad \text{for } |z| < r .$$

From the equation $P(z) \cdot Q(z) = 1$ we get using 2.:

$$\sum_{\nu=0}^{n} a_\nu b_{n-\nu} = \begin{cases} 1 & \text{for } n = 0 , \\ 0 & \text{for } n > 0 . \end{cases}$$

This system of equations can be recursively solved:

$$\left. \begin{array}{lll} n = 0 : & a_0 b_0 = 1 \\ n = 1 : & a_0 b_1 + a_1 b_0 = 0 \\ n = 2 : & a_0 b_2 + a_1 b_1 + a_2 b_0 = 0 \\ & \vdots \end{array} \right\} \quad \text{which gives} \quad \begin{cases} b_0 = 1/a_0 , \\ b_1 , \\ b_2 , \\ \text{and so on.} \end{cases}$$

An example is inserted at the end of the computation rules.

4. *The double series theorem of Weierstrass*
    Let the power series

$$f_j(z) = \sum_{k=0}^{\infty} c_{jk}(z - a)^k , \qquad j \in \mathbb{N}_0 ,$$

be convergent in the disk $U_r(a)$, $r > 0$. We also assume the normal convergence of the series $\sum_{j=0}^{\infty} f_j$ in $U_r(a)$.
Then the limit function $F := \sum_{j=0}^{\infty} f_j$ is also analytic in $U_r(a)$, being represented as

$$F(z) = \sum_{k=0}^{\infty} \left(\sum_{j=0}^{\infty} c_{jk} \right)(z - a)^k.$$

Proof. By III.1.6 the limit function $F := \sum_{j=0}^{\infty} f_j$ is analytic in $U_r(a)$, too. It is represented by the TAYLOR series

$$F(z) = \sum_{k=0}^{\infty} b_k (z - a)^k \qquad \text{with} \qquad b_k = \frac{F^{(k)}(a)}{k!} \quad (k \in \mathbb{N}_0) .$$

On the other side, repeated use of III.1.6 gives $F^{(k)} = \sum_{j=0}^{\infty} f_j^{(k)}$, especially

$$\frac{F^{(k)}(a)}{k!} = \sum_{j=0}^{\infty} \frac{f_j^{(k)}(a)}{k!} = \sum_{j=0}^{\infty} c_{jk}$$

which gives rise to

$$b_k = \sum_{j=0}^{\infty} c_{jk} \qquad (k \in \mathbb{N}_0) .$$

□

Under the made assumptions, it is thus allowed to "sum up" infinitely many power series by reordering the sum. Subsequently, there will also be an example to 4.

5. *Rearrangement of power series*
    Let

$$P(z) = a_0 + a_1(z - a) + \cdots$$

be a power series with positive convergence radius r, and let b be an interior point in the convergence disk. Due to III.2.2, $P(z)$ must be representable as a power series in a neighborhood of b.

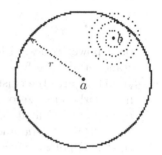

$$P(z) = b_0 + b_1(z-b) + b_2(z-b)^2 + \cdots \quad .$$

The coefficients fulfill the formula

$$b_n = \frac{P^{(n)}(b)}{n!} .$$

The convergence radius of the rearranged series is at least $r - |b - a|$.

The reader can now convince himself, that one obtains the same result if one uses the formula

$$(z - a)^n = (z - b + b - a)^n = \sum_{\nu=0}^{n} \binom{n}{\nu} (b - a)^{n-\nu} (z - b)^{\nu}$$

and naively rearranges in powers of $(z - b)$. So we obtained an exact argument for the "naive rearrangement".

6. *Substitution of a power series into another one*
    We restrict to the case

$$P(z) = a_0 + a_1 z + a_2 z^2 + \cdots ,$$
$$Q(z) = b_1 z + b_2 z^2 + \cdots ,$$

with $Q(0) = 0$. Then $P(Q(z))$ is well defined in a (small) neighborhood of $z = 0$, analytic and thus representable as a power series,

$$P(Q(z)) = c_0 + c_1 z + c_2 z^2 + \cdots \quad .$$

The coefficients c_n can be simply calculated:

$$c_0 = P\big(Q(0)\big) = P(0) = a_0 \ ,$$
$$c_1 = P'\big(Q(0)\big) \cdot Q'(0) = a_1 b_1 \ ,$$
$$c_2 = \frac{P''\big(Q(0)\big) \cdot Q'(0)^2 + P'\big(Q(0)\big) \cdot Q''(0)}{2} = a_2 b_1^2 + a_1 b_2 \ ,$$
$$\ldots$$

The "naive substitution" leads to the same result and can be justified in this way.

7.   *Inversion of power series with respect to composition*
Let

$$P(z) = \sum_{n=1}^{\infty} a_n z^n$$

be a power series with positive convergence radius (and without constant term!). We further assume $a_1 \neq 0$. The implicit function theorem (I.5.7) ensures the existence of an inverse function in a sufficiently small neighborhood of zero. This inverse is analytic, thus representable as a power series Q at 0. More precisely, there exists a number $\varepsilon > 0$, such that $P(Q(w)) = w$ and $Q(P(z)) = z$ for all $z, w \in U_\varepsilon(0)$. As in 6. the coefficients of the power series

$$Q(w) = \sum_{\nu=1}^{\infty} b_\nu w^\nu$$

can be computed recursively.

$$z = \sum_{\nu=1}^{\infty} b_\nu \left(\sum_{n=1}^{\infty} a_n z^n \right)^\nu = \sum_{\nu=1}^{\infty} \left(a_1^\nu b_\nu + R^{(\nu)}(a_1, \ldots, a_\nu, b_1, \ldots, b_{\nu-1}) \right) z^\nu \ .$$

Here, $R^{(\nu)}$ are polynomials in $a_1, \ldots, a_\nu$ and $b_1, \ldots, b_{\nu-1}$ obtained by iterated application of the CAUCHY multiplication formula:

$$1 = a_1 b_1 \ , \text{ thus } b_1 = \frac{1}{a_1} \ ,$$
$$0 = a_1^2 b_2 + a_2 b_1 \ ,$$
$$0 = a_1^3 b_3 + 2 a_1 a_2 b_2 + a_3 b_1 \ ,$$
$$\vdots$$
$$0 = a_1^\nu b_\nu + R^{(\nu)}(a_1, \ldots, a_\nu, b_1, \ldots, b_{\nu-1}) \ .$$

These formulas give conversely a proof for the local version of the implicit function theorem. One defines the coefficients b_n by this recursion scheme. A nontrivial point is then the convergence of the power series $Q(w) = \sum_{\nu=1}^{\infty} b_\nu w^\nu$. A direct (and thus also in the real case working) proof without using the power series expansion theorem was given by CAUCHY.

An example to 3. Multiplicative inversion of power series

Let

$$P(z) := \frac{\exp(z) - 1}{z} \qquad (z \neq 0) \,.$$

Then we have (for $z \neq 0$)

$$P(z) = \sum_{n=0}^{\infty} \frac{z^n}{(n+1)!} = 1 + \sum_{n=1}^{\infty} \frac{1}{(n+1)!} \, z^n =: \sum_{n=0}^{\infty} a_n \, z^n \,.$$

The right hand side is defined also at $z = 0$, where it takes the value 1; for this we also set $P(0) = 1$. Then, $Q = 1/P$ is analytic in an ε-neighborhood $U_\varepsilon(0)$ and has the power series expansion

$$Q(z) = b_0 + b_1 z + b_2 z^2 + \cdots \,.$$

The computation of the coefficients b_ν becomes simpler in the form

$$b_\nu = \frac{B_\nu}{\nu!}$$

i.e.

$$Q(z) = B_0 + \frac{B_1}{1!} z + \frac{B_2}{2!} z^2 + \ldots \,.$$

From $P(z)Q(z) = 1$ we infer

$$\sum_{\nu=0}^{n} \frac{1}{(\nu+1)!} \frac{B_{n-\nu}}{(n-\nu)!} = \begin{cases} 1, & \text{if } n = 0 \,, \\ 0, & \text{if } n > 0 \,. \end{cases}$$

We get $B_0 = 1$, and for $n \geq 1$ the equation

$$\frac{1}{1!} \frac{B_n}{n!} + \frac{1}{2!} \frac{B_{n-1}}{(n-1)!} + \cdots + \frac{1}{n!} \frac{B_1}{1!} + \frac{1}{(n+1)!} \frac{B_0}{0!} = 0 \,.$$

After multiplication with $(n+1)!$ we obtain the more transparent formula

$$\binom{1}{n+1} B_n + \binom{n+1}{2} B_{n-1} + \cdots + \binom{n+1}{n} B_1 + \binom{n+1}{n+1} B_0 = 0 \,. \qquad (*)$$

Hint. If we *formally replace* in the equation

$$(B+1)^{n+1} - B^{n+1} = 0 \qquad (n \geq 1) \qquad\qquad (**)$$

each occurrence of B^ν by B_ν (symbolically $B^\nu \mapsto B_\nu$), then $(**)$ leads to $(*)$. For instance

$$2B_1 + 1 = 0 \,,$$
$$3B_2 + 3B_1 + 1 = 0 \,,$$
$$4B_3 + 6B_2 + 4B_1 + 1 = 0 \,,$$
$$5B_4 + 10B_3 + 10B_2 + 5B_1 + 1 = 0 \,,$$

$$\vdots$$

with explicit solutions

$$B_1 = -\frac{1}{2} \ , \quad B_2 = \frac{1}{6} \ , \quad B_3 = 0 \ , \quad B_4 = -\frac{1}{30} \ .$$

The so called BERNOULLI numbers B_n (J. BERNOULLI, 1713) are rational numbers; they vanish for odd $n \geq 3$. The form of the first B_n should not lead to premature conclusions, one has for instance

$$B_{50} = \frac{495057205241079648212477525}{66} \quad \text{and}$$

$$B_{100} = -\frac{94598037819122125295227433069493721872702841533066936133385696204311395415197247711}{33330} \ .$$

The convergence radius for Q is finite, the CAUCHY-HADAMARD formula gives even $\limsup_{n \to \infty} |B_{2n}| = \infty$.

The denominators of the BERNOULLI numbers play an important role in many branches of mathematics. We will later come back to them. For the moment we list some other examples:

n	0	1	2	4	6	8	10	12	14
B_n	1	$-\frac{1}{2}$	$\frac{1}{6}$	$-\frac{1}{30}$	$\frac{1}{42}$	$-\frac{1}{30}$	$\frac{5}{66}$	$-\frac{691}{2730}$	$\frac{7}{6}$

n	16	20	30	40
B_n	$-\frac{3617}{510}$	$-\frac{174611}{330}$	$\frac{8615841276005}{14322}$	$-\frac{261082718496449122051}{13530}$

An example to 4. The double series theorem of Weierstrass

Let $D = \mathbb{E} = \{ z \in \mathbb{C} \ ; \quad |z| < 1 \ \}$, and for $z \in \mathbb{E}$ let

$$f_j(z) := \frac{z^j}{1 - z^j} \ , \quad j \in \mathbb{N} \ .$$

Then $\sum_{j=1}^{\infty} f_j$ is normally convergent in $\mathbb{E}$.

$$f_1(z) = \frac{z}{1 - z} = z + z^2 + z^3 + z^4 + z^5 + z^6 + z^7 + z^8 + \dots$$

$$f_2(z) = \frac{z^2}{1 - z^2} = z^2 + \quad z^4 + \quad z^6 + \quad z^8 + \dots$$

$$f_3(z) = \frac{z^3}{1 - z^3} = \quad z^3 + \quad z^6 + \quad \dots$$

$$f_4(z) = \frac{z^4}{1 - z^4} = \quad z^4 + \quad z^8 + \dots$$

$$\vdots$$

In the jth row of this scheme appears the term z^n, iff $n \in \mathbb{N}$ is a multiple of j. For $k \in \mathbb{N}$ let $d(k)$ be the number of (positive) divisors of k. (For a prime number p we have $d(p) = 2$.) Then 4. gives for $|z| < 1$

$$\sum_{j=1}^{\infty} \frac{z^j}{1-z^j} = \sum_{k=1}^{\infty} d(k) z^k \ ,$$

the so-called LAMBERT's series (J.H. LAMBERT, 1913).

An example to 5. Rearrangement of power series

Let us expand the function $f : \mathbb{C} \setminus \{1\} \to \mathbb{C}$, $z \to 1/(1-b)$, as a TAYLOR series at $i/2$ and find its radius of convergence. We have (first for arbitrary b with $|b| < 1$)

$$\frac{1}{1-z} = \frac{1}{1-b-(z-b)} = \frac{1}{1-b} \cdot \frac{1}{1 - \dfrac{z-b}{1-b}}$$

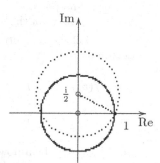

$$= \sum_{n=0}^{\infty} \frac{1}{(1-b)^{n+1}} (z-b)^n =: \tilde{f}(z) \ .$$

The radius of convergence of this series is $|1 - b|$, so especially for $b = i/2$ it is

$$\sqrt{1 + \frac{1}{4}} = \frac{\sqrt{5}}{2} \approx 1,118 > 1 \ .$$

The power series $1 + z + z^2 + \dots$ has the radius of convergence 1 and represents an analytic function a priori (by this formula) only for $|z| < 1$. After rearrangement, one obtains an analytic extension into a disk with bigger radius. The analytic extension is in the given example obvious, for we can use the sum formula

$$\frac{1}{1-z} = 1 + z + z^2 + \dots \quad .$$

This example shows that under certain circumstances rearrangement gives analytic continuations into larger domains.

Exercises for III.2

1. Find the convergence radius for each of the series:

    (a) $\displaystyle\sum_{n=0}^{\infty} n! \, z^n$, (b) $\displaystyle\sum_{n=0}^{\infty} \frac{z^n}{e^n}$,

    (c) $\displaystyle\sum_{n=1}^{\infty} \frac{n!}{n^n} z^n$, (d) $\displaystyle\sum_{n=1}^{\infty} a_n \, z^n$, $a_n := \begin{cases} a^n , & n \text{ even,} \\ b^n , & n \text{ odd,} \end{cases} \quad b > a > 0 \ .$

2. Show directly (without using theorem III.1.3):
    The power series $\sum_{n=0}^{\infty} c_n z^n$ and the termwise differentiated power series $\sum_{n=1}^{\infty} n \, c_n \, z^{n-1}$ have the same radius of convergence r. Moreover, for all $z \in U_r(0)$ one has $P'(z) = Q(z)$.

    Hint. For $z, b \in U_r(0)$

    $$P(z) - P(b) = \sum_{n=0}^{\infty} c_n (z^n - b^n) = (z - b) \sum_{n=1}^{\infty} c_n \varphi_n(z)$$

    with $\varphi_n(z) = z^{n-1} + z^{n-2} b + \dots + z b^{n-2} + b^{n-1} \ .$

3. Give examples of power series with finite radius of convergence $r \neq 0$, which have respectively one of the following properties:
    (a) the power series converges on the full boundary of the convergence disk,
    (b) the power series diverges on the full boundary of the convergence disk,
    (c) there are at least two convergence points and at least two divergence points on the boundary of the convergence disk.

4. A power series with positive radius of convergence $r < \infty$ converges *absolutely*
        either for all points
        or for no point
    on the boundary of the convergence domain. Give examples for these cases.

5. For the following functions f defined in a neighborhood of the point $a \in \mathbb{C}$ determine the TAYLOR series at a and the convergence radius:

    (a) $f(z) = \exp(z)$, $a = 1$,   (b) $f(z) = \dfrac{1}{z}$, $a = 1$,

    (c) $f(z) = \dfrac{1}{z^2 - 5z + 6}$, $a = 0$.   (d) $f(z) = \dfrac{1}{(z-1)(z-2)}$, $a = 0$,

6. Let $\sum_{n=0}^{\infty} a_n z^n$ be a power series with radius of convergence r.
    Show:
    (a) If $R := \lim\limits_{n \to \infty} |a_n|/|a_{n+1}|$ exists, then $r = R$.
    (b) If $\tilde{\rho} \lim\limits_{n \to \infty} \sqrt[n]{|a_n|} \in [0, \infty]$ exists, then $r = 1/\tilde{\rho}$. Here we formally use the conventions $1/0 = \infty$ and $1/\infty = 0$. ($r = 0$ for $\tilde{\rho} = \infty$, and $r = \infty$ for $\tilde{\rho} = 0$.)
    (c) If we set

    $$\rho := \varlimsup_{n \to \infty} \sqrt[n]{|a_n|} := \lim_{n \to \infty} \left(\sup \left\{ \sqrt[n]{|a_n|},\ \sqrt[n+1]{|a_{n+1}|},\ \sqrt[n+2]{|a_{n+1}|},\ \dots \right\} \right),$$

    then —following the same conventions as in (b)— one has:

    $$r = 1/\rho \qquad \text{(A.L. CAUCHY, 1821; J. HADAMARD, 1892)}$$

7. Let $f : D \to \mathbb{C}$ be an analytic function, defined on a domain $D \subset \mathbb{C}$, $a \in D$, and let $U_R(a)$ be the largest open disk inside D.
    Show:
    (a) If f is not bounded on $U_R(a)$, then R is the radius of convergence of the TAYLOR series of f in a.
    (b) Give an example with $r > R$, even in the case where there is no analytic continuation of f into a strictly larger domain.

8. Assume that the power series $P(z) = \sum_{n=0}^{\infty} a_n z^n$ has a positive radius of convergence, and that in the convergence disk the equality $P(z) = P(-z)$ holds. Then $a_n = 0$ for all odd n.

9. Determine in each case an entire function $f : \mathbb{C} \to \mathbb{C}$, which satisfies
        (a) $f(0) = 1$, $f'(z) = zf(z)$ for all $z \in \mathbb{C}$,
        (b) $f(0) = 1$, $f'(z) = z + 2f(z)$ for all $z \in \mathbb{C}$.

10. Determine the radius of convergence of the TAYLOR series of $1/\cos$ with center $a = 0$. The numbers E_{2n}, defined by the formula

$$\frac{1}{\cos z} = \sum_{n=0}^{\infty} \frac{E_{2n}}{(2n)!} z^{2n} \ ,$$

are called EULER numbers. Show that all E_{2n} are natural numbers, and compute $E_{2\nu}$ for $0 \le \nu \le 5$.

Result: $E_0 = 1 = E_2$, $E_4 = 5$, $E_6 = 61$, $E_8 = 1385$, $E_{10} = 50521$, $E_{12} = 2702765$.

11. Determine for the TAYLOR series with center 0 of $\tan := \sin/\cos$ the radius of convergence, and the first four coefficients.

12. Assume that the power series $P(z) = \sum_{n=0}^{\infty} c_n z^n$ has convergence radius r, $0 < r < \infty$. Let $D := U_r(0)$ be the corresponding convergence disk. A point $\rho \in \partial D := \{\ z \in \mathbb{C}\ ;\ |z| = r\ \}$ is called a *regular boundary point* for P, if there exists an ε-neighborhood $U = U_\varepsilon(\rho)$ and an analytic function g on U with $g|U \cap D = P|U \cap D$. A non-regular boundary point is called *singular*.

Show:

(a) There exists at least one singular boundary point for P.
(b) The series $1 + \sum_{n=1}^{\infty} z^{2^n}$ has convergence radius 1, and any boundary point is singular.

13. Determine an entire function $f : \mathbb{C} \to \mathbb{C}$ with

$$z^2 f''(z) + z f'(z) + z^2 f(z) = 0 \quad \text{for all } z \in \mathbb{C} \ .$$

Result: One solution is the BESSEL function of order 0,

$$f(z) := \mathcal{J}_0(z) := 1 + \sum_{n=1}^{\infty} \frac{(-1)^n}{(2 \cdot 4 \cdot 6 \cdots 2n)^2} z^{2n} \ .$$

14. Let the BESSEL function of order m ($m \in \mathbb{N}_0$) be defined by the formula

$$\mathcal{J}_m(z) = \sum_{n=0}^{\infty} \frac{(-1)^n (z/2)^{2n+m}}{n!(m+n)!} \ .$$

Show: Each $\mathcal{J}_m$ is an entire function.

15. Let $f(z) = \sum_{n=0}^{\infty} a_n z^n$ be a power series with convergence radius $r > 0$. Show that for each ρ with $0 < \rho < r$ the following inequality holds:

$$\sum_{n=0}^{\infty} |a_n|^2 \rho^{2n} \le M_f(\rho)^2 \quad \text{(GUTZMER's inequality, A. GUTZMER, 1888)} \quad (*)$$

Here, $M_f(\rho) := \sup\{\ |f(z)|\ ;\ |z| = \rho\ \}$. Derive from $(*)$ the CAUCHY estimate

$$|a_n| \le \frac{M_f(\rho)}{\rho^n} \ , \quad n \in \mathbb{N}_0 \ .$$

When does the equality in $(*)$ hold?

16. Let $f : \mathbb{C} \to \mathbb{C}$ be an entire function. Assume the existence of $m \in \mathbb{N}_0$, and of the positive constants M and R, such that for all z with $|z| \geq R$ the inequality $|f(z)| \leq M |z|^m$ is satisfied.

    Show: f is a polynomial of degree $\leq m$. What does this mean in the case $m = 0$?

17. Find all entire functions f with $f(f(z)) = z$ and $f(0) = 0$.

18. Fix $a, b, c \in \mathbb{C}$, $-c \notin \mathbb{N}_0$. The hypergeometric series

$$F(a, b, c; z) = \sum_{k=0}^{\infty} \frac{a(a+1)\cdots(a+k-1)b(b+1)\cdots(b+k-1)}{c(c+1)\cdots(c+k-1)} \frac{z^k}{k!}$$

converges for $|z| < 1$, and satisfies the differential equation

$$z(1-z)\,F''(z) + (c - (a+b+1)z)\,F'(z) - ab\,F(z) = 0 \ .$$

III.3 Mapping Properties of Analytic Functions

Let $D \subset \mathbb{C}$ be an open set. A subset $M \subset D$ is called *discrete in D*, iff there is no accumulation point of M in D.

Example: $M = \left\{ 1 \ , \ \dfrac{1}{2} \ , \ \dfrac{1}{3} \ , \ \dfrac{1}{4} \ , \ \dfrac{1}{5} \ , \ \dfrac{1}{6} \ , \ \cdots \right\}$.

This set is
     (a) discrete in $D = \mathbb{C}^\bullet$,
     (b) non-discrete in $D = \mathbb{C}$.

Being discrete is thus a relative notion.

Caution. The notion of a discrete subset is not consistently standardized in the literature.

Proposition III.3.1 *Let $f : D \to \mathbb{C}$ be an analytic function on a domain D, which is not identically zero. Then the set of zeros $N(f) := \{ z \in D ; \quad f(z) = 0 \}$ is discrete in D.*

Proof. We assume the contrary. Let a be an accumulation point of $N(f)$ in D. We consider the power series expansion of f around this point,

$$f(z) = \sum_{n=0}^{\infty} c_n \, (z-a)^n \ , \quad |z - a| < r \ .$$

Since a is an accumulation point of the zero set, we find in any neighborhood of a in D points $z \neq a$ with $f(z) = 0$. Since f is continuous, we get

$$c_0 = f(a) = 0 \ .$$

Now we can apply the same argument to the power series

$$\frac{f(z)}{z - a} := c_1 + c_2(z - a) + \cdots$$

to deduce $a_1 = 0$ and so on. All coefficients of the power series vanish, so $f(z) = 0$ for all z in a neighborhood of a. The set

$$U = \{ \ z \in D \ ; \quad z \ \text{is an accumulation point of } N(f) \ \}$$

is thus an open set! Trivially, the complement

$$V = \{ \ z \in D \ ; \quad z \ \text{is not an accumulation point of } N(f) \ \}$$

is also open. The function

$$g : D \longrightarrow \mathbb{R} \ ,$$

$$z \longmapsto g(z) := \begin{cases} 1 & \text{for } z \in U \ , \\ 0 & \text{for } z \in V \ , \end{cases}$$

is locally constant, because U, V are open sets. But D is connected, so g must be constant. Because U is not empty, we have $V = \emptyset$ and thus $f \equiv 0$. $\qquad\square$

Theorem III.3.2 (Identity Theorem for analytic functions) *Let $f, g :$ $D \to \mathbb{C}$ be two analytic functions, defined on a domain $D \neq \emptyset$. The following statements are equivalent:*

(a) $f = g$.

(b) *The "coincidence set"*

$$\{ \ z \in D \ ; \quad f(z) = g(z) \ \} \ .$$

has an accumulation point in D.

(c) *There exists a point $z_0 \in D$ with $f^{(n)}(z_0) = g^{(n)}(z_0)$ for all $n \in \mathbb{N}_0$.*

Proof. This is an application of III.3.1 with $f - g$ instead of f. $\qquad\square$

Corollary III.3.2$_1$ (Uniqueness of the analytic continuation)
*Let $D \subset \mathbb{C}$ be a domain, $M \subset D$ a subset with at least one accumulation point in D (for instance M open and non-empty) and let $f : M \to \mathbb{C}$ be a function. If there exists an **analytic** function $\widetilde{f} : D \to \mathbb{C}$ which extends f (i.e. $\widetilde{f}(z) = f(z)$ for all $z \in M$), then $\widetilde{f}$ is unique with this property.*

The Identity Theorem is so remarkable that we have to comment it.

(1) It states that the whole behavior of an analytic function in a domain $D \subset \mathbb{C}$ is already fully determined when its values are known on a "very small" subset of D, e.g. on a small path inside D.

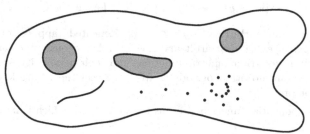

We can rephrase this:

Two analytic functions on D coincide, when they coincide on a small path, or on the elements $z_n \in D$, $z_n \neq a$, of a sequence (z_n) converging to $a \in D$. One can consider this as a massive solidarity between the values of an analytic function. In the real case, even if we restrict to $\mathcal{C}^\infty$ functions, the situation is completely different. A $\mathcal{C}^\infty$ function on an interval $M \subset \mathbb{R}$ can be changed on a subinterval $M_0 \subset M$, without losing the $\mathcal{C}^\infty$ smoothness on M.

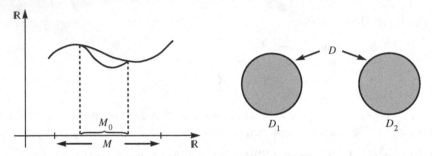

(2) In the assumption of the Identity Theorem it is essential that D is a domain, and thus especially connected. *Else*, if D would be of the form $D = D_1 \cup D_2$, $D_1 \neq \emptyset$, $D_2 \neq \emptyset$ and $D_1 \cap D_2 = \emptyset$, one can define functions $f, g : D \to \mathbb{C}$ by $f|D_1 = 1$ and $f \mid D_2 = 0$, and respectively $g = 0$. Though the restrictions $f \mid D_2 = g \mid D_2$ coincide, f and g do not coincide on D.

For the identity theorem it is essential that the coincidence set of f and g is *not discrete* in D as the following example shows: $D = \mathbb{C}^\bullet$, $f : D \to \mathbb{C}$ given by $z \mapsto \sin 1/z$, and $g : D \to \mathbb{C}$, $g = 0$. The coincidence set $\{z \in D; \quad f(z) = g(z)\}$ has the accumulation point 0, which does not lie in D but on its boundary.

(3) It becomes clear now that the real functions sin, cos, exp, cosh, sinh : $\mathbb{R} \to \mathbb{R}$ etc. can be extended to $\mathbb{C}$ *in only one way*.

If $D \subset \mathbb{C}$ is a domain with $D \cap \mathbb{R} \neq \emptyset$, and if $f, g : D \to \mathbb{C}$ are analytic functions with $f \mid D \cap \mathbb{R} = g \mid D \cap \mathbb{R}$, then we have $f(z) = g(z)$ for all $z \in D$.

Functional equations can be transferred from real to complex analysis. We want to illustrate this *permanence principle for functional equations* only in a few cases. From the functional equation of the real exponential function

$$\exp(x + y) = \exp(x) \exp(y) \,, \quad x, y \in \mathbb{R} \,,$$

we obtain by means of the Identity Theorem

$$\exp(z + y) = \exp(z) \exp(y) \qquad \text{for all } z \in \mathbb{C}$$

for any fixed (but arbitrary) $y \in \mathbb{R}$. Repeated application leads to $\exp(z + w) = \exp(z) \exp(w)$ for arbitrary $z, w \in \mathbb{C}$. Analogously one can follow the *Addition Theorems* for trigonometric functions and their periodicity from their real form. That there are no other periods as from the real theory are know is a fact which has to be proved.

The complex exponential function exp has the period $2\pi i$, which is invisible in real analysis.

The functional equation of the real logarithm $\log(xy) = \log x + \log y$ is true for complex x, y only in a restricted sense. For the principal value Log one has $\text{Log}(z_1 z_2) = \text{Log}(z_1) + \text{Log}(z_2)$, if $-\pi < \text{Arg}\, z_1 + \text{Arg}\, z_2 < \pi$ is required. (See also Exercise 22 in I.2.) Using the Identity Theorem we obtain a new proof for this.

(4) The real functions sin, cos and exp are "real analytic" functions. (An infinitely differentiable function $f : M \to \mathbb{R}$ defined on a non-degenerate interval M is called *real analytic* iff for any $a \in M$ it is represented by its TAYLOR series in some neighborhood of a.) We have:

Remark. *Let $M \subset \mathbb{R}$ be a non-degenerate interval. A function $f : M \to \mathbb{R}$ has an analytic continuation to a domain $D \subset \mathbb{C}$, $M \subset D$, iff f is real analytic.*

The condition is obviously necessary. For the converse one can argue as follows. For any $a \in M$ we can choose a positive number $\varepsilon(a)$ such that f is represented by its TAYLOR series in the interval of radius ε centered at a (intersected with M). We then define

$$D := \bigcup_{a \in M} U_{\varepsilon(a)}(a) .$$

Using the TAYLOR series we have for each $a \in M$ an analytic continuation into the disk $U_\varepsilon(a)$. The Identity Theorem ensures the coincidence of the local extensions in intersections of any two disks. They glue to a well defined function on D.

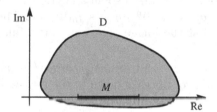

(5) The set of analytic functions on an open subset $D \subset \mathbb{C}$ is a commutative ring with unit (sum and product of analytic functions are analytic functions). We denote this ring by $\mathcal{O}(D)$. A direct consequence of the Identity Theorem ensures that $\mathcal{O}(D)$ has no zero divisors for a domain $D \subset \mathbb{C}$. Explicitly:

If the product of two analytic functions on a domain D vanishes identically, then one of the two functions vanishes identically on D.

Proof: Let $f, g \in \mathcal{O}(D)$ satisfy $fg = 0$. We show: $f = 0$ or $g = 0$. Equivalently, if $f \neq 0$ then $g = 0$.
From $f \neq 0$ follows the existence of an $a \in D$ with $f(a) \neq 0$. By continuity there exists a neighborhood $U \subset D$ of a with $f(z) \neq 0$ for all $z \in U$. The assumption $f(z)g(z) = 0$ for all $z \in D$ gives now $g(z) = 0$ for all $z \in U$. Hence $g \mid U$ is identically zero and by the Identity Theorem we have $g = 0 = $ zero function, which is the zero element in the ring $\mathcal{O}(D)$.   □

Conversely, if $\mathcal{O}(D)$ is an integral domain, then D is connected, and thus a domain.

The Identity Theorem couples thus an *algebraic statement* about the structure of the ring $\mathcal{O}(D)$ (its integrality) and the *topological* nature of D (its connectivity).

A further remarkable mapping property of analytic functions, which cannot be expected from the real theory, states:

Theorem III.3.3 (Open Mapping Theorem) *If f is a nonconstant analytic function on a domain $D \subset \mathbb{C}$, then its image $f(D)$ is open and arcwise connected, i.e. also a domain.*

Caution: The image of the real sine is $\sin(\mathbb{R}) = [-1, 1]$, which is not open in $\mathbb{R}$.

Proof. Let us fix an $a \in D$, and show that a full neighborhood of $b = f(a)$ is also contained in $f(D)$. Without loss of generality we can assume

$$a = b = f(a) = 0 \ .$$

We consider the power series expansion in a suitable neighborhood of 0,

$$f(z) = z^n (a_n + a_{n+1}z + \cdots) = z^n h(z) \ , \quad a_n \neq 0 \ , \quad n > 0 \ .$$

The function h defined by

$$h(z) = a_n + a_{n+1}z + \cdots$$

is analytic and without zero in a full disk $U_r(0)$, $r > 0$. Using II.2.9$_1$ we find an analytic n^{th} root of h, and thus also an analytic n^{th} root of f in this disk, $f(z) = f_0(z)^n$. Then $a_n = f_0'(0)^n$ implies $f_0'(0) \neq 0$. The Implicit Function Theorem I.5.7 shows that the image of f_0 contains a full neighborhood of 0. It remains to prove:

The function $z \mapsto z^n$ maps an arbitrary neighborhood of 0 onto a neighborhood of 0. (At this point the proof breaks down in the real case!)
One verifies this using polar coordinates

$$re^{i\varphi} \mapsto r^n e^{in\varphi} \ ,$$

the disk of radius r around 0 is mapped onto the disk of radius r^n.

$f(D)$ is also arcwise connected because of the continuity of f. The image $f(D)$ is thus a domain. $\square$

This proof clarifies the **local mapping behavior** of an analytic function.

*Each **non-constant** analytic function f with $f(0) = 0$ is in a small open neighborhood on 0 the composition of a conformal map with the nth power map. The angles in 0 are multiplied by n.*

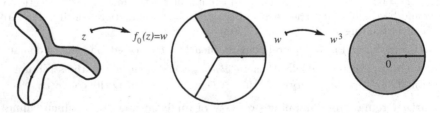

If f is especially injective in a neighborhood of a, the derivative f' of f is nonzero in a neighborhood of a.

As a simple *application* of the Open Mapping Theorem we get a result that follows also from the CAUCHY-RIEMANN differential equations I.5.5.

Corollary III.3.4 *If $D \subset \mathbb{C}$ is a domain, $f : D \to \mathbb{C}$ is analytic, and*

$$\operatorname{Re} f = const. \quad or \quad \operatorname{Im} f = const. \quad or \quad |f| = const. \ ,$$

then the function f is constant.

Proof. By assumption, for any $z \in D$ the value $f(z)$ is not an interior point of $f(D)$. □

Corollary III.3.5 (Maximum Principle) *If for an analytic function*

$$f : D \longrightarrow \mathbb{C} \ , \quad D \ a \ domain \ in \ \mathbb{C} \ ,$$

its modulus $|f| : D \to \mathbb{R}_+$ reaches its maximum on D, then f is constant. Here, we use the following terminology: $|f|$ has a maximum on D, iff there exists an $a \in D$ satisfying

$$|f(a)| \geq |f(z)| \ for \ all \ z \in D \ .$$

III.3.5₁ Supplement.

(a) *Because of the Identity Theorem, it is enough to assume the existence of a local maximum for $|f|$.*

(b) *If K is a compact subset of the domain D, and $f : D \to \mathbb{C}$ is analytic, then the restriction $f \mid K$ being a continuous function has a maximal modulus on K. By III.3.5 we can moreover affirm that the maximal modulus value is necessarily taken on the boundary of D.*

Proof of III.3.5: By the Open Mapping Theorem III.3.3, $f(a)$ is an *interior point* of $f(D)$ if f is non-constant. In each neighborhood of $f(a)$ there are then of course points $f(z)$, $z \in D$, with $|f(z)| > |f(a)|$. □

Directly from III.3.5 we get:

Corollary III.3.6 (Minimal Modulus Principle) *If $D \subset \mathbb{C}$ is a domain, and $f : D \to \mathbb{C}$ is analytic **and non-constant**, and if f has in $a \in D$ a (local) minimal modulus, then we necessarily have $f(a) = 0$.*

Proof. In case of $f(a) \neq 0$, the function $1/f$ would be well defined in a neighborhood of a and would take in a a maximal value of the modulus. □

From this we get another simple proof of the *Fundamental Theorem of Algebra*. Let P be a polynomial of degree $n \geq 1$. Because of $\lim_{|z| \to \infty} |P(z)| = \infty$, the modulus $|P|$ takes its minimal value on $\mathbb{C}$, and because of the Minimal Modulus Principle this shows the existence of a root of P.

An important *application* of III.3.5 is

Lemma III.3.7 (Schwarz' Lemma, H.A. Schwarz, 1869)   *We consider*
the unit disk $\mathbb{E} := \{\, z \in \mathbb{C} \; ; \; |z| < 1 \,\}$.
Let $f : \mathbb{E} \to \mathbb{C}$ *be an analytic function with* $|f(z)| \leq 1$ *for all* $z \in \mathbb{E}$, *and with*
zero as a fixed point, $f(0) = 0$. *Then one has for all* $z \in \mathbb{E}$

$$|f(z)| \leq |z| \qquad and \qquad |f'(0)| \leq 1 \ .$$

Proof (according to C. CARATHÉODORY, 1912). Let $f(z) = a_0 + a_1 z + \cdots$
be the TAYLOR series of f in 0. Because of $f(0) = 0$ we have $a_0 = 0$. Then
$f(z) = z g(z)$ with

$$g(z) = a_1 + a_2 z + a_3 z^2 + \cdots .$$

Clearly the function g is analytic in $\mathbb{E}$, and we have $f'(0) = a_1 = g(0)$. For
$r \in {]}0,1{[}$ we derive from the assumption $|f(z)| \leq 1$ with $z \in \mathbb{E}$ first

$$|g(z)| \leq \frac{1}{r} \text{ for all } z \in \mathbb{C} \text{ with } |z| = r \ (<1) \ .$$

From the Supplement to the Maximal Modulus Principle III.3.5$_1$ (b) then we
have even more,

$$|g(z)| \leq \frac{1}{r} \qquad \text{for all } z \in \mathbb{C} \text{ with } |z| \leq r \text{ and all } r \in {]}0,1{[} \ .$$

Passing to the limit with $r \to 1$ we get

$|g(z)| \leq 1$, and thus $|f(z)| \leq |z|$ for all $z \in \mathbb{E}$, and also $|g(0)| = |f'(0)| \leq 1$.

Let us now assume that there exists a point $a \in \mathbb{E}$, $a \neq 0$ with the property
$|f(a)| = |a|$. Then $|g|$ has in $\mathbb{E}$ a maximum, and is thus constant. It follows

$$f(z) = \zeta z$$

with a suitable constant ζ of modulus $|\zeta| = 1$. The same argument can be
applied in the case $|f'(0)| = 1$. □

As an important application of SCHWARZ's Lemma we determine all (globally)
conformal maps of $\mathbb{E}$ onto itself. If such a map f has the zero point as a fixed
point, then applying SCHWARZ' Lemma to f and its inverse f^{-1}

$$|f(z)| \leq |z| \qquad \text{and } |z| = \left|f^{-1}\big(f(z)\big)\right| \ \leq |f(z)|$$

we get $|f(z)| = |z|$, and then the following result:

Proposition III.3.8 *Let* $\varphi : \mathbb{E} \to \mathbb{E}$ *be a bijective map with zero as a fixed*
point, $\varphi(0) = 0$, *such that both* φ *and* φ^{-1} *are analytic. Then there exists a*
complex number ζ *with modulus 1 with*

$$\varphi(z) = \zeta z \ .$$

(φ is thus a rotation with center 0.)

One can rise the question, whether there are conformal maps $\mathbb{E} \to \mathbb{E}$, which don't fix 0.

Proposition III.3.9 *Fix $a \in \mathbb{E}$. Then the map $\varphi_a : \mathbb{E} \to \mathbb{E}$,*

$$\varphi_a(z) = \frac{z - a}{\overline{a}z - 1}$$

is a bijective map of the unit disk onto itself which satisfies

(a) $\varphi_a(a) = 0,$

(b) $\varphi_a(0) = a$ *and*

(c) $\varphi_a^{-1} = \varphi_a.$

Especially, φ_a is in both directions analytic.

Proof. The function is well defined on the unit disk, since the denominator has no zero in $\mathbb{E}$. We show first $|\varphi_a(z)| < 1$ (for $|z| < 1$). This is equivalent to $|z - a|^2 < |\overline{a}z - 1|^2 = |1 - \overline{a}z|^2$, and further to $(1 - |a|^2)(1 - |z|^2) > 0$, which is obviously true. A simple computation shows $\varphi_a(\varphi_a(z)) = z$. From this, φ is surjective and injective, and thus bijective. All other properties are clear. □

Theorem III.3.10 *Let $\varphi : \mathbb{E} \to \mathbb{E}$ be a conformal (i.e. bijective and in both directions analytic) map of the unit disk onto itself. Then there exists a complex number ζ of absolute value 1, and a point $a \in \mathbb{E}$ with the property:*

$$\varphi(z) = \zeta \, \frac{z - a}{\overline{a}z - 1} \; .$$

For the proof, let $a = \varphi^{-1}(0)$. The composition $\varphi \circ \varphi_a$ is a self map of the unit disk which fixes 0, thus a rotation. □

Remarks.

1. The set of all conformal (self) maps $D \to D$ of a domain $D \subset \mathbb{C}$ is a group with respect to the composition of maps. It is often denoted by $\mathrm{Aut}(D)$, the group of automorphisms of D. We have thus determined in III.3.10 $\mathrm{Aut}(\mathbb{E})$.

2. For the numerous applications of SCHWARZ's Lemma one should check the exercises for this and the next sections.

Exercises for III.3

1. Let (a_n) and (b_n) be two sequences of complex numbers. Two power series are defined by

$$P(z) := \sum a_n z^n \quad \text{and} \quad Q(z) := \sum b_n z^n \; .$$

Prove or refute: If the equation $P(z) = Q(z)$ has infinitely many solutions, then $P = Q$ and thus $a_n = b_n$ for all $n \in \mathbb{N}_0$.

2. Decide, whether there are analytic functions $f_j : \mathbb{E} \to \mathbb{C}, 1 \le j \le 4$, with

(a) $f_1\left(\dfrac{1}{2n}\right) = f_1\left(\dfrac{1}{2n-1}\right) = \dfrac{1}{n}$, $n \geq 1$.

(b) $f_2\left(\dfrac{1}{n}\right) = f_2\left(-\dfrac{1}{n}\right) = \dfrac{1}{n^2}$, $n \geq 1$.

(c) $f_3^{(n)}(0) = (n!)^2$, $n \geq 0$.

(d) $f_4^{(n)}(0) = \dfrac{n!}{n^2}$, $n \geq 0$.

3. Let $r > 0$, and $f : U_r(0) \to \mathbb{C}$ be analytic. For all $z \in U_r(0) \cap \mathbb{R}$ assume $f(z) \in \mathbb{R}$.

    Show: The TAYLOR coefficients of f with respect to the center $c = 0$ are real, and one has: $\overline{f(z)} = f(\bar{z})$.

4. Let $D \subset \mathbb{C}$ be an open set.

    Show: For a subset $M \subset D$ the following properties are equivalent:
    (a) M is discrete in D, i.e. no accumulation point of M lies in D.
    (b) For each $p \in M$ there exists an $\varepsilon > 0$, such that $U_\varepsilon(p) \cap M = \{p\}$, and M is closed in D (i.e. there exists a closed set $A \subset \mathbb{C}$ with $M = A \cap D$).
    (c) For each compact subset $K \subset D$ the intersection $M \cap K$ is finite.
    (d) M is *locally finite* in D, i.e. each point $z \in D$ has an ε-neighborhood $U_\varepsilon(z) \subset D$, such that $M \cap U_\varepsilon(z)$ is finite.

5. A discrete subset (see Exercise 4) is at most countable, i.e. either finite or countably infinite.

6. If the analytic function $f : D \to \mathbb{C}$ on the domain D is not constant zero, then the zeros of f are at most countable.

7. Let $f, g : \mathbb{C} \to \mathbb{C}$ be two analytic functions. We assume

    $$f\big(g(z)\big) = 0 \text{ for all } z \in \mathbb{C} .$$

    Show: If g is non-constant, then $f \equiv 0$.

8. Fix $R > 0$, and consider the closed disk $\overline{U}_R(0) := \{\, z \in \mathbb{C} \,;\, |z| \leq R \,\}$ and continuous functions $f, g : \overline{U}_R(0) \to \mathbb{C}$ which are analytic on the open disk $U_R(0)$, and such their absolute values coincide on its boundary:

    $$|f(z)| = |g(z)| \quad \text{for all } |z| = R .$$

    Show: If f and g have no zeros in $\overline{U}_R(0)$, then there exists a constant $\lambda \in \mathbb{C}$ with $|\lambda| = 1$ and $f = \lambda g$.

9. Let $f, g : \mathbb{E} \to \mathbb{E}$ be bijective analytic functions, which satisfy $f(0) = g(0)$ and $f'(0) = g'(0)$. Moreover, assume that f' and g' have no common zero.

    Show: $f(z) = g(z)$ for all $z \in \mathbb{E}$.

10. Determine the maximum of $|f|$ on $\overline{\mathbb{E}} := \{\, z \in \mathbb{C} \,;\, |z| \leq 1 \,\}$ for
    (a) $f(z) = \exp(z^2)$,
    (b) $f(z) = \dfrac{z+3}{z-3}$,
    (c) $f(z) = z^2 + z - 1$,
    (d) $f(z) = 3 - |z|^2$.
    In (d) the maximal modulus is attained in the (interior) point $a = 0$. Is there any contradiction to the maximum principle?

11. Let u be a non-constant harmonic function on a domain $D \subset \mathbb{R}^2$. Show that $u(D)$ is an open interval.

12. **Variant of the maximum principle for bounded domains.** If $D \subset \mathbb{C}$ is a bounded domain, and $f : \overline{D} \to \mathbb{C}$ is a continuous function on the closure of D which is analytic on D, then $|f|$ takes its maximum on the boundary of D.

Using the example of the strip

$$ S = \left\{ z \in \mathbb{C} ; \quad |\operatorname{Im} z| < \frac{\pi}{2} \right\} $$

and the function $f(z) = \exp\big(\exp(z)\big)$ show the necessity of the boundedness of D.

13. If $f : \mathbb{E} \to \mathbb{E}$ is an analytic map having two fixed points $a, b \in \mathbb{E}$, $a \neq b$, i.e. $f(a) = a$ and $f(b) = b$, then $f(z) = z$ for all $z \in \mathbb{E}$.

14. The image of a non-constant polynomial is closed, as one sees by studying its growth properties. Using this and the Open Mapping Theorem, give a new proof for the Fundamental Theorem of Algebra.

15. Let f be an analytic function on an open set containing the closed disk $\overline{U}_r(a)$. We assume $|f(a)| < |f(z)|$ for all z on the boundary of the disk. Then there exists a zero of f in the interior of the disk.

Using this, find a further proof of the Open Mapping Theorem.

16. A Maximality Property for $\operatorname{Aut}(\mathbb{E})$. (The **Schwarz-Pick Lemma**):
(a) *Show:* For each $\varphi \in \operatorname{Aut}(\mathbb{E})$ and each $z \in \mathbb{E}$, the following holds:

$$ \frac{|\varphi'(z)|}{1 - |\varphi(z)|^2} = \frac{1}{1 - |z|^2} \; . $$

(b) If $f : \mathbb{E} \to \mathbb{C}$ is a non-constant, analytic function with $|f(z)| \leq 1$ for all $z \in \mathbb{E}$, then we have either for all $z \in \mathbb{E}$ the strict inequality

$$ \frac{|f'(z)|}{1 - |f(z)|^2} < \frac{1}{1 - |z|^2}, $$

or $f \in \operatorname{Aut}(\mathbb{E})$ and the equality as in (a) holds.

17. Prove LIOUVILLE's Theorem (II.3.7) using SCHWARZ' Lemma.

III.4 Singularities of Analytic Functions

Functions like

$$ \frac{\sin z}{z} \, , \quad \frac{1}{z} \quad \text{and} \quad \exp \frac{1}{z} $$

are not defined at 0. But they are analytic in a *punctured neighborhood* $\overset{\bullet}{U}_r(0)$. Their behavior near 0 is quite different. They have different *singular behavior*. We will see that these three examples are characteristic.

Let $D \subset \mathbb{C}$ be open and $f : D \longrightarrow \mathbb{C}$ be an analytic function. Let a be a point, which is not in D, but has the property that for a suitable $r > 0$ the whole *punctured disk*

$$\overset{\bullet}{U}_r(a) := \{\, z \in \mathbb{C} \,;\quad 0 < |z - a| < r \,\}$$

is contained in D. We call a an **(isolated) singularity** of the function f. The set $D \cup \{a\} = D \cup U_r(a)$ is then open (as union of two open sets). Naturally, it may happen that a is not "really" a singularity, in other words it may be possible to extend f analytically into a. One calls a in this case a *removable singularity*.

Definition III.4.1 *A singularity a of an analytic function*

$$f : D \longrightarrow \mathbb{C} \,,\ D \subset \mathbb{C}\ \text{open},$$

*is called **removable**, iff f can be analytically extended to $D \cup \{a\}$, i.e. iff there exists an analytic function $\tilde{f} : D \cup \{a\} \to \mathbb{C}$ with $\tilde{f} \mid D = f$.*

We often write f instead of $\tilde{f}$ for the sake of simplicity, i.e. we define

$$f(a) := \lim_{z \to a} f(z).$$

From the removability the existence of this limit follows. This is the case in the example of the beginning of this section, $a = 0$ is a removable singularity for the function

$$f(z) = \frac{\sin z}{z} \,,\quad \text{and one has to set}\quad f(0) = \lim_{z \to 0} \frac{\sin z}{z} = 1 \,.$$

If a is a removable singularity, then f can be continuously extended to a. Especially, f is bounded in a neighborhood of a. The converse is also true.

Theorem III.4.2 (Riemann Removability Condition, B. Riemann, 1851) *A singularity a of an analytic function*

$$f : D \longrightarrow \mathbb{C} \,,\ D \subset \mathbb{C}\ \text{open},$$

is removable iff there exists a punctured neighborhood $\overset{\bullet}{U} := \overset{\bullet}{U}_r(a) \subset D$ of the point a where f is bounded.

Proof. We can assume without essential restrictions that a is the zero point. The function $h : U_r(0) \to \mathbb{C}$

$$h(z) = \begin{cases} z^2 f(z) \,, & z \neq 0 \\ 0 \,, & z = 0 \end{cases}$$

is differentiable in $\overset{\bullet}{U}_r(0)$. But h is also differentiable in $z = 0$, for we have

$$h'(0) = \lim_{z \to 0} \frac{h(z) - h(0)}{z - 0} = \lim_{z \to 0} z f(z) = 0 \,.$$

The function h is analytic, and thus representable as a power series:

$$h(z) = a_0 + a_1 z + a_2 z^2 + \cdots$$

Because of $h(0) = h'(0) = 0$ we have $a_0 = a_1 = 0$. We obtain for $z \neq 0$

$$f(z) = a_2 + a_3 z + a_4 z^2 + \cdots \quad .$$

The power series

$$a_2 + a_3 z + a_4 z^2 + \cdots$$

defines an analytic function $\tilde{f}$ in a neighborhood of zero (including zero). The function $\tilde{f}$ is the analytic continuation of f, which we looked for.   □

Definition III.4.3 *A singularity a of an analytic function*

$$f : D \longrightarrow \mathbb{C} , \ D \subset \mathbb{C} \ open,$$

*is called **non-essential**, iff there exists an integer $m \in \mathbb{Z}$, such that the function*

$$g(z) = (z - a)^m f(z)$$

has a removable singularity at a.

Removable singularities are of course non-essential ($m = 0$). A non-essential singularity which is not removable is called a *pole*.

If f has in a a non-essential singularity, then the function

$$g(z) = (z - a)^m f(z) \qquad \text{(with a suitable } m \in \mathbb{Z})$$

can be expanded as a power series near a,

$$. \quad g(z) = a_0 + a_1(z \quad a) + a_2(z - a)^2 + \cdots \quad .$$

If this power series does not vanish identically, then there exists a minimal integer $n \in \mathbb{N}_0$, such that $a_n \neq 0$:

$$g(z) = a_n(z - a)^n + a_{n+1}(z - a)^{n+1} + \cdots , \quad a_n \neq 0 .$$

Obviously, the function

$$h(z) = (z - a)^k f(z) , \quad k = m - n ,$$

has a removable singularity at $z = a$. We claim that k is the minimal integer with this property. If this is not the case, then the function

$$z \mapsto \frac{a_n}{z - a} + a_{n+1} + a_{n+2}(z - a) + \cdots$$

would have a removable singularity at a, and thus the same would happen for $z \mapsto (z - a)^{-1}$. This is not the case. We get:

Remark III.4.4 *Let a be a non-essential singularity of the analytic function*

$$f : D \longrightarrow \mathbb{C} \ , \ D \subset \mathbb{C} \ open.$$

If f does not vanish identically in some neighborhood of a, then there exists a minimal integer $k \in \mathbb{Z}$, such that

$$z \mapsto (z - a)^k f(z)$$

has a removable singularity at a.

Supplement. *One can characterize k also by the following two properties:*

(a) $h(z) = (z - a)^k f(z)$ *has a removable singularity at $z = a$.*
(b) $h(a) \neq 0$.

Definition III.4.5 *The negative $-k$ of the number k from III.4.4 is called the* **order** *of f in a.*

Notation. $\mathrm{ord}(f; a) := -k$.

Obviously we have:

(a) $\mathrm{ord}(f; a) \geq 0 \iff a$ is removable,
(b) $\mathrm{ord}(f; a) = 0 \iff a$ is removable and $f(a) \neq 0$,
(c) $\mathrm{ord}(f; a) < 0 \iff a$ is a pole.

In the last case $k = -\mathrm{ord}(f; a) \in \mathbb{N}$ is called the order of the pole a of f. A pole of order 1 is called *simple*.

Examples.

(1) $f(z) = (z - 1)^5 + 2(z - 1)^6 = (z - 1)^5 \left(1 + 2(z - 1)\right) = (z - 1)^5 h(z)$,
     and thus $\mathrm{ord}(f; 1) = 5$.

(2) $f(z) = \dfrac{1}{z^2} + \dfrac{1}{z} = z^{-2}(1 + z) = z^{-2} h(z)$,
     and thus $\mathrm{ord}(f; 0) = -2$. The function f has in 0 the order -2, the pole order is $+2$.

If $f : D \to \mathbb{C}$ vanishes in some suitable neighborhood of a (for a domain D we then have $f \equiv 0$), we complete our definition by setting

$$\mathrm{ord}(f; a) = \infty \ .$$

Remark III.4.6 *Let a be a non-essential singularity of the analytic functions*

$$f, g : D \longrightarrow \mathbb{C} \ , \quad D \subset \mathbb{C} \ open.$$

Then a is also a non-essential singularity of the functions

$$f \pm g \ , \qquad f \cdot g \qquad and \qquad \frac{f}{g} \ , \ if \ g(z) \neq 0 \ for \ all \ z \in D \setminus \{a\} \ ,$$

and we have

$$\mathrm{ord}(f \pm g; a) \geq \min\{ \mathrm{ord}(f;a) , \mathrm{ord}(g;a) \} ,$$
$$\mathrm{ord}(f \cdot g; a) = \mathrm{ord}(f;a) + \mathrm{ord}(g;a) ,$$
$$\mathrm{ord}\left(\frac{f}{g}; a\right) = \mathrm{ord}(f;a) - \mathrm{ord}(g;a) .$$

Here we use the usual convention

$$x + \infty = \infty + x = \infty \qquad \text{for all } x \in \mathbb{R} ,$$
$$\infty + \infty = \infty \qquad \text{and}$$
$$x < \infty \text{ for all } x \in \mathbb{R} .$$

The proof is simple, and we skip it.

Remark III.4.7 *Let*

$$f : D \longrightarrow \mathbb{C} , \ D \subset \mathbb{C} \ open,$$

be an analytic function, which has a pole at a. Then

$$\lim_{\substack{z \to a \\ z \in D}} |f(z)| = \infty .$$

In other words: For each $C > 0$ there exists a $\delta > 0$ with

$$|f(z)| \geq C \qquad if \ 0 < |z - a| < \delta , \ z \in D .$$

Proof. Let $k \in \mathbb{N}$ be the pole order of f in a. The function

$$h(z) = (z - a)^k f(z)$$

has then in $z = a$ a removable singularity, and we have $h(a) \neq 0$. Especially there exists a positive number $M > 0$ (for instance $M := |h(a)| / 2$), such that $|h(z)| \geq M > 0$ holds in a full neighborhood of a. This implies

$$|f(z)| \geq \frac{M}{|z - a|^k}$$

for all z in this neighborhood, excepting a. The claim now follows because k is positive. $\square$

Definition III.4.8 *A singularity of an analytic function*

$$f : D \longrightarrow \mathbb{C} , \ D \subset \mathbb{C} \ open,$$

*is called **essential** iff it is not non-essential.* [1]

---

[1] This definition is an impressive example of mathematical language arts.

Analytic functions have near essential singularities a completely different (rather "nervous") mapping behavior, namely:

Theorem III.4.9 (Casorati-Weierstrass, F. Casorati, 1868; K. Weierstrass, 1876) *Let a be an essential singularity of the analytic function*

$$f : D \longrightarrow \mathbb{C} \ , \ D \subset \mathbb{C} \ open.$$

*If $\overset{\bullet}{U} := \overset{\bullet}{U}_r(a)$ is an arbitrary punctured neighborhood of a, then the image $f(\overset{\bullet}{U} \cap D)$ is **dense** in $\mathbb{C}$, i.e. for any $b \in \mathbb{C}$ and any $\varepsilon > 0$ we have*

$$f(\overset{\bullet}{U} \cap D) \cap U_\varepsilon(b) \neq \emptyset \ .$$

Equivalently we have:

For any $b \in \mathbb{C}$ and any $\varepsilon > 0$ there exists $z \in \overset{\bullet}{U} \cap D$ with

$$|f(z) - b| < \varepsilon \ .$$

(So one can say: In an arbitrarily small punctured neighborhood of a, we can approach any given complex number with values of the function f arbitrarily close.)

Proof. We give an indirect proof, and assume that there exists a punctured neighborhood $\overset{\bullet}{U} := \overset{\bullet}{U}_r(a)$ such that $f(\overset{\bullet}{U} \cap D)$ is not dense in $\mathbb{C}$. Then there exists a $b \in \mathbb{C}$ and an $\varepsilon > 0$ with $|f(z) - b| \geq \varepsilon$ for all $z \in \overset{\bullet}{U} \cap D$. The function

$$g(z) := \frac{1}{f(z) - b}$$

is then bounded in a neighborhood of a. By RIEMANN Removability, g has in a a removable singularity.   □

We now see that by simple case by case considerations that the converse statements for III.4.7 and III.4.9 also hold. If, for instance, a is an isolated singularity of f satisfying $\lim_{z \to a} |f(z)| = \infty$, then it is neither removable, nor essential (the last using CASORATI-WEIERSTRASS), so it is a pole.

If, on the other side, the CASORATI-WEIERSTRASS property (from the assumption of IV.4.9) is fulfilled for f at a, then a is neither removable, nor a pole, else f could not approximate zero in the neighborhood of a because of $\lim_{z \to \infty} |f(z)| = \infty$. So a is an essential singularity of f.

We conclude:

Theorem III.4.10 (Classification of singularities by their mapping behavior) *Let $a \in \mathbb{C}$ be an isolated singularity of the analytic function*

$$f : D \longrightarrow \mathbb{C} \ , \ D \subset \mathbb{C} \ open.$$

The singularity a is

(1) **removable** $\Longleftrightarrow$ f is bounded in a suitable neighborhood of a,

(2) a **pole** $\Longleftrightarrow$ $\lim\limits_{z \to a} |f(z)| = \infty$,

(3) **essential** $\Longleftrightarrow$ in any (arbitrarily small) neighborhood of a the function f comes arbitrarily close to any given value.

The functions
  (1) $f_1 : \mathbb{C}^\bullet \longrightarrow \mathbb{C}$, with $f_1(z) = \sin(1/z)$, and
  (2) $f_2 : \mathbb{C}^\bullet \longrightarrow \mathbb{C}$, with $f_2(z) = \exp(1/z)$
both have at $a = 0$ an essential singularity. One can easily verify that

$$f_1\big(\overset{\bullet}{U}_r(0)\big) = \mathbb{C} , \quad \text{and}$$

$$f_2\big(\overset{\bullet}{U}_r(0)\big) = \mathbb{C}^\bullet , \quad \text{both for any } r > 0 .$$

These examples are typical. Namely one can prove:

Theorem. (So-called Big Theorem of Picard, E. Picard, 1879–80) *If a is an essential singularity of the analytic function $f : D \to \mathbb{C}$, then there are exactly two cases possible:*

If $\overset{\bullet}{U} \subset D$ is a punctured neighborhood of the point a than we have either

$$f(\overset{\bullet}{U}) = \mathbb{C} ,$$

or

$$f(\overset{\bullet}{U}) = \mathbb{C} \setminus \{c\} \quad \text{for a suitable } c .$$

Comparing with the Casorati-Weierstrass Theorem, the function f not only approaches each value arbitrarily close, it even takes each value with at most one exception.

The proof of this theorem is very intricate. In the second volume we will give a proof using the theory of RIEMANN surfaces. A direct proof can be found for instance in [ReS2].

We close this section with an example for the notions we have just introduced, and for an *application of the Cauchy Integral Formula.*

In *Fourier analysis* the following *Dirichlet integral*

$$\int_0^\infty \frac{\sin x}{x} \, dx \ \left(= \frac{1}{2} \int_{-\infty}^\infty \frac{\sin x}{x} \, dx \right)$$

plays an important role. At 0 it is harmless because of

$$\lim_{x \to 0} \frac{\sin x}{x} = 1 .$$

The integral is thus only at ∞ an improper integral. It is the standard example of a convergent integral, which is not absolutely convergent. The value of this integral can be computed using only methods of real analysis, using some special tricks. We want to compute the integral by function theoretical means, and claim

$$\boxed{\int_0^\infty \frac{\sin x}{x} \, dx = \frac{\pi}{2} \, .}$$

For the proof, we consider the analytic function

$$f : \mathbb{C}^\bullet \longrightarrow \mathbb{C} \, , \quad z \mapsto \frac{\exp(iz)}{z} \, ,$$

and integrate it along the following closed curve

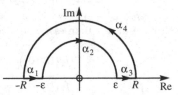

$$\alpha = \alpha_1 \oplus \alpha_2 \oplus \alpha_3 \oplus \alpha_4 \, .$$

If we slit the plane along its "negative imaginary axis", then the curve α runs in a star domain D, where f is analytic. The CAUCHY Integral Formula for star domains, II.2.7, then gives

$$0 = \int_\alpha f = \int_{\alpha_1} f + \int_{\alpha_2} f + \int_{\alpha_3} f + \int_{\alpha_4} f \, . \tag{$*$}$$

We consider each of the integrals:

(a)   We parametrize $\alpha_4(t) = R \exp(it)$, $0 \le t \le \pi$,

$$\int_{\alpha_4} f(\zeta) \, d\zeta = \int_0^\pi \frac{e^{iR\cos t} e^{-R\sin t}}{Re^{it}} \, iRe^{it} \, dt$$

and thus

$$\left| \int_{\alpha_4} f(\zeta) \, d\zeta \right| \le \int_0^\pi e^{-R\sin t} \, dt = 2 \int_0^{\pi/2} e^{-R\sin t} \, dt \, .$$

For $0 \le t \le \pi/2$ we can use the so-called *inequality of Jordan*

$$\frac{2t}{\pi} \le \sin t \ (\le t)$$

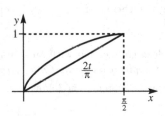

and thus

$$\left| \int_{\alpha_4} f(\zeta) \, d\zeta \right| \le 2 \int_0^{\pi/2} e^{-2Rt/\pi} \, dt = \frac{\pi}{R} \left(1 - e^{-R} \right) \, .$$

Taking the limit, we get

$$\lim_{R \to \infty} \int_{\alpha_4} f(\zeta) \, d\zeta = 0 \, .$$

(b) Putting together $\int_{\alpha_1}$ and $\int_{\alpha_3}$ we get:

$$\int_{\alpha_1} f(\zeta)\, d\zeta + \int_{\alpha_3} f(\zeta)\, d\zeta = \int_\varepsilon^R \frac{\exp(ix) - \exp(-ix)}{x}\, dx = 2i \int_\varepsilon^R \frac{\sin x}{x}\, dx \ .$$

(c) Finally

$$\int_{\alpha_2} \frac{\exp(i\zeta)}{\zeta}\, d\zeta = \int_{\alpha_2} \frac{1}{\zeta}\, d\zeta + \int_{\alpha_2} \frac{\exp(i\zeta) - 1}{\zeta}\, d\zeta = -\pi i + \int_{\alpha_2} \frac{\exp(i\zeta) - 1}{\zeta}\, d\zeta \ .$$

The function $z \to (e^{iz} - 1)/z$ has at $z = 0$ a removable singularity, and is thus bounded in a neighborhood of 0. This gives

$$\lim_{\varepsilon \to 0} \int_{\alpha_2} \frac{\exp(i\zeta) - 1}{\zeta}\, d\zeta = 0 \ .$$

Passing to the limits $\varepsilon \to 0$ and $R \to \infty$ we get from (*) and (a), (b) and (c)

$$0 = \lim_{R \to \infty} \left(\lim_{\varepsilon \to 0} \left(2i \int_\varepsilon^R \frac{\sin x}{x}\, dx \right) \right) - \pi i = 2i \lim_{R \to \infty} \int_0^R \frac{\sin x}{x}\, dx - \pi i \ ,$$

or

$$\frac{\pi}{2} = \lim_{R \to \infty} \int_0^R \frac{\sin x}{x}\, dx = \int_0^\infty \frac{\sin x}{x}\, dx \ . \qquad \square$$

This shows how —under certain circumstances— it is possible by function theoretical means to compute *real integrals*. We will come back to this, and systematically consider analogous integrals as applications of the *Residue Theorem* (compare with III.7).

Exercises for III.4

1. Let $D \subset \mathbb{C}$ be open and $f : D \setminus \{a\} \to \mathbb{C}$ an analytic function.

   Show:

   (a) The point a is a removable singularity of f, iff each one of the following conditions is satisfied:
      - (α) f is bounded in a punctured neighborhood of a.
      - (β) The limit $\lim_{z \to a} f(z)$ exists.
      - (γ) $\lim_{z \to a} (z - a) f(z) = 0$.

   (b) The point a is a simple pole of f, iff $\lim_{z \to a} (z - a) f(z)$ exists, and is $\neq 0$.

2. Let $f : \overset{\bullet}{U}_r(a) \to \mathbb{C}$ be analytic ($a \in \mathbb{C}$, $r > 0$). Show that the following properties are equivalent:

   (a) The point a is a pole of f of order $k \in \mathbb{N}$.

   (b) There exist an open neighborhood $U_\rho(a) \subset U_r(a)$ and an analytic function $h : U_\rho(a) \to \mathbb{C}$ such that $h(a) \neq 0$ and $f(z) = \dfrac{h(z)}{(z - a)^k}$ for all $z \in \overset{\bullet}{U}_\rho(a)$.

   (c) There exists an open neighborhood $U_\rho(a) \subset U_r(a)$ of a, and an analytic function $g : U_\rho(a) \to \mathbb{C}$ not vanishing in $\overset{\bullet}{U}_\rho(a)$, which has a zero of order k in a, such that $f = 1/g$ in $\overset{\bullet}{U}_\rho(a)$.

(d) There exist positive constants M_1 and M_2, such that we have for all z in a punctured neighborhood of a:

$$M_1 \, |z - a|^{-k} \, \leq \, |f(z)| \, \leq \, M_2 \, |z - a|^{-k} \; .$$

3. Prove the formulas in Remark III.4.6 for the order function ord.

4. Which of the following functions have a removable singularity at $a = 0$?

$$\text{(a)} \quad \frac{\exp(z)}{z^{17}} \; , \qquad \text{(b)} \quad \frac{(\exp(z) - 1)^2}{z^2} \; ,$$

$$\text{(c)} \quad \frac{z}{\exp(z) - 1} \; , \qquad \text{(d)} \quad \frac{\cos(z) - 1}{z^2} \; .$$

5. The functions defined by the following expressions have poles at $a = 0$. Find the orders of these poles.

$$\frac{\cos z}{z^2} \; , \qquad \frac{z^7 + 1}{z^7} \; , \qquad \frac{\exp(z) - 1}{z^4} \; .$$

6. If the singularity $a \in \mathbb{C}$ of the analytic functions is not removable, then the function $\exp \circ f$ has an essential singularity at a.

7. Prove the complex analogue of the rule of L'HOSPITAL:
Let $f, g : D \to \mathbb{C}$ be analytic functions, which have the same order k in $a \in D$. Then $h := f/g$ has in a a removable singularity, and the following formula holds:

$$\lim_{z \to a} \frac{f(z)}{g(z)} \; = \; \frac{f^{(k)}(a)}{g^{(k)}(a)} \; .$$

8. Let us consider the function

$$f(z) := \frac{(z - 1)^2 (z + 3)}{1 - \sin(\pi z / 2)} \; .$$

Find all singularities of f and determine for each one its type.

9. *Show:*

$$\int_0^\infty \frac{\sin^2 x}{x^2} \, dx = \frac{\pi}{2} \; .$$

10. *Show:*

$$\int_0^\infty \frac{\sin^4 x}{x^2} \, dx = \frac{\pi}{4} \; .$$

III.5 Laurent Decomposition

Let us fix numbers r, R with

$$0 \leq r < R \leq \infty$$

($r = 0$ and $R = \infty$ are allowed). We consider analytic functions in the *annulus*

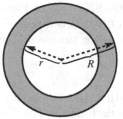

$$\mathcal{A} := \{\, z \in \mathbb{C} \,; \quad r < |z| < R \,\} \,.$$

Examples of such functions can be easily constructed. One can start with two analytic functions

$$g : U_R(0) \longrightarrow \mathbb{C} \,,$$
$$h : U_{1/r}(0) \longrightarrow \mathbb{C} \,.$$

Then the function $z \mapsto h(1/z)$ is analytic for $|z| > r$, and we can define

$$f(z) := g(z) + h(1/z) \qquad \text{for } r < |z| < R \,.$$

Indeed, each function which is analytic in an annulus can be decomposed in this way.

Theorem III.5.1 (Laurent decomposition, P.A. Laurent, 1843; K. Weierstrass, 1841, Nachlass, 1894) *Any function, which is analytic on an annulus*

$$\mathcal{A} = \{\, z \in \mathbb{C} \,; \quad r < |z| < R \,\} \,,$$

admits a decomposition as

$$f(z) = g(z) + h(1/z) \,. \tag{$*$}$$

Here,

$$g : U_R(0) \longrightarrow \mathbb{C} \ and$$
$$h : U_{1/r}(0) \longrightarrow \mathbb{C}$$

are analytic functions. If we further require $h(0) = 0$, then the above decomposition is unique.

After normalizing h by $h(0) = 0$, we call h the *principal part* of the Laurent decomposition $(*)$ for f.

Proof.
(1) *The uniqueness of the Laurent decomposition.* For this, we start with a *preliminary remark:*
Two analytic functions

$$f_\nu : D_\nu \longrightarrow \mathbb{C} \,, \quad D_\nu \subset \mathbb{C} \text{ open }, \nu = 1, 2 \,,$$

which coincide on $D_1 \cap D_2 \neq \emptyset$ can be uniquely merged to an analytic function $f : D_1 \cup D_2 \to \mathbb{C}$.

Since the difference of two LAURENT decompositions for the same function is a LAURENT decomposition for the zero function, it is enough to prove the uniqueness for this zero function. From the equation

$$g(z) + h(1/z) = 0$$

follows that the functions $z \mapsto g(z)$ and $z \mapsto -h(1/z)$ can be merged to an analytic function $H : \mathbb{C} \to \mathbb{C}$. By assumption, H is bounded. LIOUVILLE's Theorem shows that it is constant. Because of $\lim_{|z|\to\infty} H(z) = 0$ this constant is zero.

(2) *Existence of the Laurent decomposition.* We choose P, ϱ with the property

$$r < \varrho < P < R$$

and construct the LAURENT decomposition in the smaller annulus of all z with

$$\varrho < |z| < P .$$

Because of the uniqueness of the LAURENT decomposition, and because these smaller annuli exhaust the given annulus, we are done.

The claim will follow from the following auxiliary result, the CAUCHY Integral Formula for annuli, which is of independent interest for its own right.

Theorem III.5.1₁ *Let*

$$\mathcal{A} = \{ z \in \mathbb{C} ; \quad r < |z| < R \} \quad (0 \le r < R \le \infty)$$

be an annulus, and $G : \mathcal{A} \to \mathbb{C}$ an analytic function. If P and ϱ are chosen, such that

$$r < \varrho < P < R ,$$

then we have:

$$\oint_{|\zeta|=\varrho} G(\zeta)\, d\zeta = \oint_{|\zeta|=P} G(\zeta)\, d\zeta .$$

Proof. We reduce the proof to the CAUCHY Integral Formula for star domains, II.2.7, by finding suitable curves which lie in star domains of the shape of an annulus sector. There we can apply the CAUCHY Integral Formula, Sect. II.2. Adding the line integrals which appear in the following figure we obtain the claimed formula, because two integrals along two inverse curves cancel each other. We indicate this in the following figure:

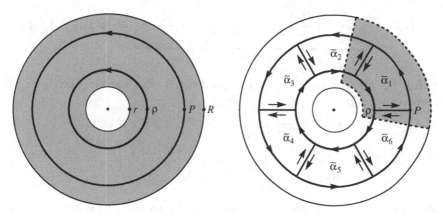

Because each $\tilde{\alpha}_\nu$ lies in a star domain, we have $\int_{\tilde{\alpha}_\nu} G(\zeta)\, d\zeta = 0$. After summing all integrals along the curves $\tilde{\alpha}_1, \ldots, \tilde{\alpha}_n$ in the right figure, the integrals over the radial segments cancel pairwise to give the conclusion.   □

We go back to the proof of III.5.1. Let $z \in \mathcal{A}$ be fixed. The function $G : \mathcal{A} \to \mathbb{C}$ with

$$G(\zeta) = \begin{cases} \dfrac{f(\zeta) - f(z)}{\zeta - z} & \zeta \neq z, \\ f'(\zeta) & \zeta = z, \end{cases}$$

is continuous in $\mathcal{A}$ and analytic in $\mathcal{A} \setminus \{z\}$. Using the power series expansion of f at z, or by the RIEMANN Removability Theorem, we see that G has at $\zeta = z$ a removable singularity. From the auxiliary result III.5.1$_1$ we deduce

$$\oint_{|\zeta|=\varrho} G(\zeta)\, d\zeta = \oint_{|\zeta|=P} G(\zeta)\, d\zeta,$$

and further

$$\oint_{|\zeta|=\varrho} \frac{f(\zeta)}{\zeta - z}\, d\zeta - f(z) \oint_{|\zeta|=\varrho} \frac{1}{\zeta - z}\, d\zeta$$
$$= \oint_{|\zeta|=P} \frac{f(\zeta)}{\zeta - z}\, d\zeta - f(z) \oint_{|\zeta|=P} \frac{1}{\zeta - z}\, d\zeta.$$

Let $z \in \mathcal{A}$ be chosen, such that $\varrho < |z| < P$ holds, i.e. z is an interior point of the smaller annulus. Then we have by II.3.1 on the one side

$$\oint_{|\zeta|=\varrho} \frac{1}{\zeta - z}\, d\zeta = 0 \quad \text{because of } |z| > \varrho,$$

and on the other side

$$\oint_{|\zeta|=P} \frac{1}{\zeta - z}\, d\zeta = 2\pi \mathrm{i} \quad \text{because of } |z| < P.$$

We get

$$f(z) \;=\; \underbrace{\frac{1}{2\pi i}\oint_{|\zeta|=P}\frac{f(\zeta)}{\zeta - z}\,d\zeta}_{g(z)} \;-\; \underbrace{\frac{1}{2\pi i}\oint_{|\zeta|=\varrho}\frac{f(\zeta)}{\zeta - z}\,d\zeta}_{-h(1/z)} \;=\; g(z) + h(1/z)\,,$$

which is the wanted LAURENT decomposition, it uses the functions g and h defined by

$$g(z) := \frac{1}{2\pi i}\oint_{|\zeta|=P}\frac{f(\zeta)}{\zeta - z}\,d\zeta\,, \qquad |z| < P\,,$$

and

$$h(z) := \frac{1}{2\pi i}\oint_{|\zeta|=\varrho}\frac{z f(\zeta)}{1 - \zeta z}\,d\zeta\,, \qquad |z| < \frac{1}{\rho}\,,$$

and using II.3.3 we see that g, h are analytic. Moreover $h(0) = 0$. $\qquad\square$

If we represent g and h as power series, then we get the so-called LAURENT series of f

$$g(z) = \sum_{n=0}^{\infty} a_n\, z^n \text{ for } |z| < R\,, \qquad h(z) = \sum_{n=1}^{\infty} b_n\, z^n \text{ for } |z| < \frac{1}{r}\,.$$

After setting $a_{-n} := b_n$ we obtain the LAURENT series in the form

$$f(z) = g(z) + h(1/z) = \sum_{n=-\infty}^{\infty} a_n\, z^n\,.$$

Convention: A series of the form

$$\sum_{n=-\infty}^{\infty} a_n \qquad \text{with } a_n \in \mathbb{C} \text{ for } n \in \mathbb{Z}\,,$$

is called *convergent* iff both series

$$\sum_{n=0}^{\infty} a_n \qquad \text{and} \qquad \sum_{n=1}^{\infty} a_{-n}$$

converge. As in the case of usual power series, one uses the same series symbol to refer to two different objects, on the one side the pair of the series $\sum_{n=0}^{\infty} a_n$ and $\sum_{n=1}^{\infty} a_{-n}$, and on the other side the number obtained as the sum of two numbers $\sum_{n=0}^{\infty} a_n + \sum_{n=1}^{\infty} a_{-n}$, assuming convergence.

In the same sense we use notions as *absolute, uniform* or *normal* convergence.

Corollary III.5.2 (Laurent expansion) *Let f be analytic in an annulus*

$$\mathcal{A} = \{\, z \in \mathbb{C}\,;\quad r < |z - a| < R\,\} \qquad (0 \le r < R \le \infty)\,.$$

Then one can represent f as a Laurent series which normally converges in this annulus,

$$f(z) = \sum_{n=-\infty}^{\infty} a_n \, (z-a)^n \qquad \text{for } z \in \mathcal{A} \, .$$

Supplement. *This Laurent expansion is unique, its coefficients are given by*

$$a_n = \frac{1}{2\pi i} \oint\limits_{|\zeta - a| = \varrho} \frac{f(\zeta) \, d\zeta}{(\zeta - a)^{n+1}} \, , \quad n \in \mathbb{Z} \, , \quad r < \varrho < R \, .$$

If we set $M_\varrho(f) := \sup \{ \, |f(\zeta)| \, ; \quad |\zeta - a| = \varrho \, \}$, then there hold the following **Cauchy estimates:**

$$|a_n| \le \frac{M_\varrho(f)}{\varrho^n} \, , \quad n \in \mathbb{Z} \, .$$

Using LAURENT series we can reformulate the classification of isolated singularities. If a is an isolated singularity of the analytic function f, then the restriction of f to

$$\dot{U}_r(a) = U_r(a) \setminus \{a\} \subset D \quad \text{is analytic for a suitable } r > 0 \, .$$

The punctured disk $\dot{U}_r(a)$ is an annulus, where we can expand f into a LAURENT series,

$$f(z) = \sum_{n=-\infty}^{\infty} a_n \, (z-a)^n \, .$$

The type of the singularity can now be read off from this expansion:

Remark III.5.3 (Classification of singularities) *The singularity a is*

(a) *removable, iff*
$$a_n - 0 \text{ for all } n < 0 \, ,$$

(b) *a pole, of order k ($\in \mathbb{N}$) iff*

$$a_{-k} \ne 0 \quad \text{and} \quad a_n = 0 \text{ for all } n < -k \, ,$$

(c) *essential, iff*

$$a_n \ne 0 \quad \text{for infinitely many } n < 0 \, .$$

The simple proof is left to the reader.   □

Remark. With the help of the LAURENT expansion one obtains a more transparent proof of the nontrivial direction in the RIEMANN Removability Theorem, compare with IV.2.

If the analytic function f is bounded in a suitable punctured neighborhood $\dot{U}_r(a)$ of the point a, then f has there a removable singularity.

Without restriction we can assume for this $a = 0$. Then the LAURENT decomposition looks like

$$f(z) = g(z) + h\left(\frac{1}{z}\right).$$

Here, h is an analytic function in the whole $\mathbb{C}$, and it is even bounded, because f is bounded near 0. By LIOUVILLE's Theorem h is constant.

One also can use the same strategy as in the proof of LIOUVILLE's Theorem, to get a direct proof, namely:

We expand f in $\overset{\bullet}{U}_\varepsilon(a)$ into a LAURENT series

$$f(z) = \sum_{n=-\infty}^{\infty} a_n\,(z-a)^n\,.$$

From the Supplement of III.5.2, we get

$$|a_n| \le \frac{M_\varrho(f)}{\varrho^n} \quad \text{for all } n \in \mathbb{Z} \text{ and } 0 < \varrho < \varepsilon\,.$$

We have to show $a_n = 0$ for all $n < 0$. From the assumtion we have $M_\varrho(f) \le M$ for a suitable $M > 0$, and hence

$$|a_n| \le \varrho^{-n} M \quad \text{for all } n \in \mathbb{Z}\,.$$

For $n < 0$, i.e. for $-n := k \ge 1$, we get

$$|a_{-k}| \le \varrho^k M \quad \text{for } k \in \mathbb{N}\,.$$

From $\lim_{\varrho \to 0} \varrho^k = 0$ follows

$$a_{-k} = 0 \quad \text{for all } k \in \mathbb{N}\,,$$

i.e. the principal part vanishes identically.   □

Examples.

(1) The function

$$f : \mathbb{C}^\bullet \longrightarrow \mathbb{C} \quad \text{with } f(z) = \frac{\sin z}{z}$$

has at $a = 0$ a removable singularity, because from

$$\sin z = z - \frac{z^3}{3!} + \frac{z^5}{5!} - \frac{z^7}{7!} \pm \cdots$$

we find for all $z \in \mathbb{C}^\bullet$

$$\frac{\sin z}{z} = 1 - \frac{z^2}{3!} + \frac{z^4}{5!} - \frac{z^6}{7!} \pm \cdots\,.$$

(2) The function

$$f(z) = \frac{z}{\exp z - 1} , \quad 0 < |z| < 2\pi ,$$

has at $a = 0$ a removable singularity. We have namely (see also Example 3 at the computation rules for power series in Sect. III.2)

$$f(z) = 1 + \sum_{n=1}^{\infty} \frac{B_n}{n!} z^n .$$

(3) The function

$$f(z) = \frac{\exp z}{z^3} \quad (z \neq 0)$$

has at $a = 0$ a pole of order 3, because we have:

$$f(z) = \frac{1 + z + \frac{z^2}{2!} + \frac{z^3}{3!} + \frac{z^4}{4!} + \frac{z^5}{5!} + \cdots}{z^3}$$

$$= \underbrace{\frac{1}{z^3} + \frac{1}{z^2} + \frac{1}{2!}\frac{1}{z}}_{h(1/z)} + \underbrace{\frac{1}{3!} + \frac{1}{4!} z + \frac{1}{5!} z^2 + \cdots}_{g(z)} .$$

(4) The function

$$f(z) = \exp\left(-\frac{1}{z^2}\right) \quad (z \neq 0)$$

has at $a = 0$ an essential singularity, because the LAURENT series is of the form

$$f(z) = 1 - \frac{1}{z^2} + \frac{1}{2!}\frac{1}{z^4} - \frac{1}{3!}\frac{1}{z^6} \pm \cdots = 1 + h(1/z) .$$

The principal part contains infinitely many coefficients $\neq 0$.

For the uniqueness in the Supplement to III.5.2 it is not enough to specify only the center a. One and the same function can have different LAURENT expansions in different annuli centered at a.

Example. We consider the analytic function given by the expression

$$f(z) := \frac{2}{z^2 - 4z + 3} ,$$

$f : \mathbb{C} \setminus \{1,3\} \to \mathbb{C}$. We want to expand f in three annuli centered at 0 into a LAURENT series,

$$0 < |z| < 1 , \quad 1 < |z| < 3 , \quad 3 < |z| .$$

The partial fraction decomposition of f is

$$f(z) = \frac{2}{z^2 - 4z + 3} = \frac{1}{1 - z} + \frac{1}{z - 3} .$$

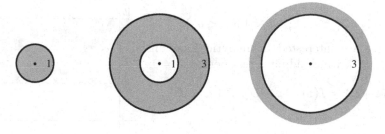

(a) For z with $0 < |z| < 1$ there holds

$$\frac{1}{1-z} = \sum_{n=0}^{\infty} z^n \quad \text{and} \quad \frac{1}{3-z} = \frac{1}{3}\left(\frac{1}{1-z/3}\right) = \frac{1}{3}\sum_{n=0}^{\infty}\left(\frac{z}{3}\right)^n .$$

That is why we have

$$f(z) = \frac{2}{z^2 - 4z + 3} = \sum_{n=0}^{\infty}\left(1 - \frac{1}{3^{n+1}}\right) z^n \quad \text{for } |z| < 1 .$$

The LAURENT series is in this case the power series expansion of f at 0.

(b) For $|z| > 1$ one has

$$\frac{1}{z-1} = \frac{1}{z}\left(\frac{1}{1-1/z}\right) = \sum_{n=0}^{\infty}\frac{1}{z^{n+1}}$$

and for $|z| < 3$

$$\frac{1}{3-z} = \sum_{n=0}^{\infty}\frac{z^n}{3^{n+1}} ,$$

so that putting this information together we get for $1 < |z| < 3$

$$f(z) = \frac{2}{z^2 - 4z + 3} = \underbrace{\sum_{n=1}^{\infty}\frac{-1}{z^n}}_{h(1/z)} + \underbrace{\sum_{n=0}^{\infty}\frac{-1}{3^{n+1}} z^n}_{g(z)} .$$

(c) For $|z| > 3$ there holds

$$\frac{1}{z-3} = \frac{1}{z(1-3/z)} = \sum_{n=0}^{\infty}\frac{3^n}{z^{n+1}}$$

and thus

$$f(z) = \frac{2}{z^2 - 4z + 3} = \sum_{n=1}^{\infty}(3^{n-1} - 1)\frac{1}{z^n} .$$

We conclude this section with an excursion about

Complex Fourier series

Let $]a, b[$ be an open interval in $\mathbb{R}$. We allow also $a = -\infty$ and $b = \infty$. So the interval can be a half-line or the whole real axis. We consider the horizontal strip

$$D = \{ z \in \mathbb{C} ;\quad a < \operatorname{Im} z < b \}$$

and we are interested in analytic functions $f : D \to \mathbb{C}$, which admit a real period $\omega \neq 0$,

$$f(z + \omega) = f(z) \quad (z \in D ,\ z + \omega \in D) .$$

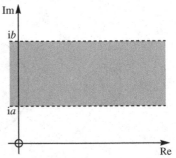

Then function $g(z) = f(\omega z)$ has the period 1. There is thus no loss of generality to start with a function with period 1, $f(z+1) = f(z)$.

We now consider the map

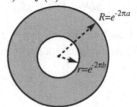

$$z \mapsto q := e^{2\pi i z} .$$

It maps D onto the annulus

$$\mathcal{A} = \{ \, q \in \mathbb{C} \, ; \quad r < |q| < R \, \} \, , \quad r := e^{-2\pi b} \, , \quad R := e^{-2\pi a} \, ,$$

where we use the conventions

$$e^{-2\pi a} = \infty \text{ if } a = -\infty$$

and

$$e^{-2\pi b} = 0 \text{ if } b = \infty \ .$$

As we know,

$$e^{2\pi i z} = e^{2\pi i z'} \iff z - z' \in \mathbb{Z} \ .$$

By

$$q \mapsto g(q) = g(e^{2\pi i z}) := f(z)$$

we thus define a function $g : \mathcal{A} \longrightarrow \mathbb{C}$, which is like f also analytic, because the map

$$D \longrightarrow \mathcal{A} \, , \quad z \mapsto e^{2\pi i z},$$

is locally conformal, since its derivative has no zero. Thus each point in D has an open neighborhood, which is conformally mapped onto an open neighborhood of the image point. The function g can be represented as a LAURENT series:

$$g(q) = \sum_{n=-\infty}^{\infty} a_n \, q^n \, ,$$

$$a_n = \frac{1}{2\pi i} \oint_{|\eta|=\varrho} \frac{g(\eta)}{\eta^{n+1}} \, d\eta = \int_0^1 \frac{g(\varrho e^{2\pi i x})}{\varrho^n e^{2\pi i n x}} \, dx \quad (r < \varrho < R) \ .$$

If we write

$$\varrho = e^{-2\pi y} \qquad (y \in \,]a, b[\,) \, ,$$

we get

$$a_n = \int_0^1 f(x + iy) e^{-2\pi i n(x+iy)} \, dx \ .$$

From this we are led to the following

Proposition III.5.4 *Let f be an analytic function in the strip*

$$D = \{\, z \in \mathbb{C} \;;\quad a < y < b \,\} \qquad (-\infty \le a < b \le \infty)\,,$$

which is 1-periodic, $f(z+1) = f(z)$ for all $z \in D$. Then one can represent f as a normally convergent complex Fourier series

$$f(z) = \sum_{n=-\infty}^{\infty} a_n\, e^{2\pi i n z}\,.$$

The coefficients a_n, $n \in \mathbb{Z}$, are the so-called Fourier coefficients. They are uniquely determined and can be given by the formula

$$a_n = \int_0^1 f(z) e^{-2\pi i n z}\, dx \qquad (z = x + iy)\,,$$

where y can be arbitrarily chosen in $]a, b[$.

Of course, one can recover from III.5.4 the LAURENT expansion.

Observation. One can also prove Proposition III.5.4 by using methods of real analysis, as follows:

For each fixed y the function $f(x+iy)$ is two times continuously differentiable in the variable x, and by a well-known result of real analysis it allows a FOURIER expansion

$$f(z) = \sum_{n=-\infty}^{\infty} a_n(y)\, e^{2\pi i n x}$$

with

$$a_n(y) = \int_0^1 f(z)\, e^{-2\pi i n x}\, dx$$

or also

$$a_n(y)\, e^{2\pi n y} = \int_0^1 f(z)\, e^{-2\pi i n z}\, dx\,.$$

We are done if we show that

$$a_n := a_n(y)\, e^{2\pi n y}$$

does not depend on y, because if this is the case then we have

$$f(z) = \sum_{n=-\infty}^{\infty} a_n\, e^{2\pi i n z}\,.$$

For this we show

$$\frac{d}{dy}\left(a_n(y) e^{2\pi n y} \right) = 0$$

or, equivalently,

$$a'_n(y) = -2\pi n a_n(y) \ .$$

For the proof we use the CAUCHY-RIEMANN differential equations

$$\frac{\partial f(z)}{\partial x} = -\mathrm{i}\frac{\partial f(z)}{\partial y} \ .$$

The integral formula for $a_n(y)$ together with the LEIBNIZ rule gives

$$a'_n(y) = \int_0^1 \frac{\partial f(z)}{\partial y} e^{-2\pi \mathrm{i} n x}\, dx = \int_0^1 \mathrm{i}\frac{\partial f(z)}{\partial x} e^{-2\pi \mathrm{i} n x}\, dx$$

and after partial integration we obtain the desired differential equation for $a_n(y)$. □

The power series expansion is a special case of the LAURENT expansion. We also get in this way a new proof for the local representability as power series for (twice continuously) differentiable complex functions. But one should have in mind that the real theory of FOURIER series is anything but trivial.

Exercises for III.5

1. Expand the function given by the formula $f(z) = z/(z^2+1)$ in

$$A = \{ \ z \in \mathbb{C} \ ; \quad 0 < |z-\mathrm{i}| < 2 \ \}$$

into a LAURENT series. What kind of singularity has f at $a = \mathrm{i}$?

2. Expand the function given by the formula $f(z) = \dfrac{1}{(z-1)(z-2)}$ into a LAURENT series in the annuli

$$A(a;r,R) := \{ \ z \in \mathbb{C} \ ; \quad r < |z-a| < R \ \}$$

for the following parameters:

$$(a;r,R) \in \{ \ (0;0,1), (0;1,2), (0;2,\infty), (1;0,1), (2;0,1) \ \}$$

3. Expand the function given by the formula $f(z) = \dfrac{1}{z(z-1)(z-2)}$ into a LAURENT series in each of the annuli $A(0;0,1)$, $A(0;1,2)$ and $A(0;2,\infty)$

4. Does the following "identity" contradict to the uniqueness of the LAURENT expansion

$$0 = \frac{1}{z-1} + \frac{1}{1-z} = \frac{1}{z}\cdot\frac{1}{1-1/z} + \frac{1}{1-z}$$

$$= \sum_{n=1}^{\infty}\frac{1}{z^n} + \sum_{n=0}^{\infty}z^n = \sum_{n=-\infty}^{\infty}z^n \ ?$$

5. Let us consider the recursively defined FIBONACCI sequence (f_n) with $f_0 = f_1 = 1$ and $f_n := f_{n-1} + f_{n-2}$ for $n \geq 2$.

Show:

(a) The power series $f(z) := \sum_{n=0}^{\infty} f_n z^n$ coincides with the rational function

$$z \mapsto \frac{1}{1 - z - z^2} \ .$$

(b) For all $n \in \mathbb{N}_0$ we have the following formula of BINET for the FIBONACCI numbers

$$f_n = \frac{1}{\sqrt{5}} \left(\frac{1 + \sqrt{5}}{2} \right)^{n+1} - \frac{1}{\sqrt{5}} \left(\frac{1 - \sqrt{5}}{2} \right)^{n+1} \ .$$

6. If f has at a a pole of order $m \in \mathbb{N}$, and if p is a polynomial of degree n, then the composition $g = p \circ f$ has at a a pole of order mn.

7. For $\nu \in \mathbb{Z}$, and $w \in \mathbb{C}$ let $\mathcal{J}_\nu(w)$ be the coefficient of z^ν in the LAURENT power series expansion of

$$f : \mathbb{C}^\bullet \longrightarrow \mathbb{C} \ , \quad f(z) := \exp\left(\frac{1}{2} \left(z - \frac{1}{z} \right) w \right) \ , \quad \text{i.e.}$$

$$f(z) = \sum_{\nu=-\infty}^{\infty} \mathcal{J}_\nu(w) z^\nu \ .$$

Show:

(a) $\mathcal{J}_\nu(-w) = \mathcal{J}_{-\nu}(w) = (-1)^\nu \mathcal{J}_\nu(w)$ for all $\nu \in \mathbb{Z}$ and all $w \in \mathbb{C}$.

(b) $\mathcal{J}_\nu(w) = \dfrac{1}{2\pi} \displaystyle\int_0^{2\pi} \cos(\nu t - w \sin t)\, dt = \dfrac{1}{\pi} \displaystyle\int_0^{\pi} \cos(\nu t - w \sin t)\, dt \ .$

(c) The functions $\mathcal{J}_\nu(w)$ are analytic at $\mathbb{C}$. Their TAYLOR expansions around the origin have for $\nu \ge 0$ the form

$$\mathcal{J}_\nu(w) = \sum_{\mu=0}^{\infty} \frac{(-1)^\mu \left(\frac{1}{2} w \right)^{2\mu+\nu}}{\mu! \, (\nu + \mu)!} \ .$$

(d) The functions $\mathcal{J}_\nu$ satisfy the BESSEL differential equation

$$w^2 f''(w) + w f'(w) + (w^2 - \nu^2) f(w) = 0 \ . \qquad\qquad (*)$$

The function $\mathcal{J}_\nu$ is called the BESSEL *function* of order ν (compare with page 123).

8. Show directly (without using more general propositions) that the function

$$f(z) := \exp \frac{1}{z}$$

takes in any punctured neighborhood $\overset{\bullet}{U}_r(0)$ any value $w \in \mathbb{C}^\bullet$ infinitely often!

9. Let $\mathcal{A}$ be the annulus

$$\mathcal{A} = \{ \ z \in \mathbb{C} ; \quad r < |z| < R \ \} \ , \quad 0 < r < R \ .$$

The function $f(z) = 1/z$ cannot be uniformly approximated in $\mathcal{A}$ by polynomials.

10. The function $z \to \cot \pi z$ has the period 1, and is analytic in both the upper and the lower half-plane. Determine the two corresponding FOURIER expansions.

A Appendix to III.4 and III.5

It seems to be natural to include also poles of analytic functions into their domain of definition, and to assign in each pole ∞ as function value. We thus set $\overline{\mathbb{C}} := \mathbb{C} \cup \{\infty\}$.

Definition A.1 *A map*

$$f : D \longrightarrow \overline{\mathbb{C}}, \quad D \subset \mathbb{C} \text{ open},$$

is called **meromorphic,** *iff the following conditions are satisfied:*

(a) *The set $S(f) = f^{-1}(\{\infty\})$ of places with value ∞ is discrete in D.*
(b) *The restriction f_0 of f from D to $D \setminus S(f)$,*

$$f_0 : D \setminus S(f) \longrightarrow \mathbb{C},$$

is analytic.

(c) *The points in $S(f)$ are poles of f_0.*

Addition of meromorphic functions

Let $f, g : D \to \overline{\mathbb{C}}$ be two meromorphic functions with S, T as pole sets. The function $f + g$ is analytic in the domain $D \setminus (S \cup T)$, and has in $S \cup T$ only non-essential singularities. (Some singularities can be removable.) In any case, one can extend $f+g$ from $D \setminus (S \cup T) \to \mathbb{C}$ to a meromorphic function $D \to \overline{\mathbb{C}}$. We denote this extension by $f + g$.

Analogously, one can define fg, f' and f/g, where the last function exists only in the case that the zeros of g lie discretely in D (or equivalently, $g \not\equiv 0$ on each connected component of D).

Remark A.2 *The set $\mathcal{M}(D)$ of all meromorphic functions on a domain D is a a field. In addition, for each $f \in \mathcal{M}(D)$ its derivative f' also belongs to $\mathcal{M}(D)$. The set $\mathcal{O}(D)$ of all analytic functions on D is a subring of $\mathcal{M}(D)$.*

Here, we committed a small inaccuracy, we have identified an analytic function $f : D \to \mathbb{C}$ with the corresponding meromorphic function $\tilde{f} = \iota \circ f : D \to \overline{\mathbb{C}}$, $\iota : \mathbb{C} \hookrightarrow \overline{\mathbb{C}}$ being the canonical inclusion.

Examples for meromorphic functions in whole $D = \mathbb{C}$ are the *rational functions*

$$R(z) = \frac{P(z)}{Q(z)} = \frac{a_n z^n + a_{n-1} z^{n-1} + \cdots + a_0}{b_m z^m + b_{m-1} z^{m-1} + \cdots + b_0},$$

where P and Q are polynomials and $m \in \mathbb{N}_0$, $b_m \neq 0$. The set of singularities of R is finite in this case, being contained in the zero set $N(R)$ of the denominator.

A typical example of a function with an infinite set of singularities is

$$\cot \pi z = \frac{\cos \pi z}{\sin \pi z} \, .$$

Because of

$$\sin \pi z \; = \; 0 \Longleftrightarrow z \in \mathbb{Z} \, ,$$

(and cos does not vanish on $\pi \mathbb{Z}$), the set of singularities of $\cot (\pi z)$ is $\mathbb{Z}$, the set of integers. If $f \in \mathcal{M}(D)$ and a is a pole of f, then knowing that the set $S(f)$ of poles is discrete in D, there exists a punctured neighborhood $\overset{\bullet}{U}(a)$ of a with $\overset{\bullet}{U}(a) \cap S(f) = \emptyset$. If k is the pole order of f in a, then for all $z \in \overset{\bullet}{U}(a)$ we have

$$f(z) \; = \; \frac{f_0(z)}{(z-a)^k}$$

with a suitable analytic function f_0 in $U(a)$, $f_0(a) \neq 0$.

Locally we can thus always represent a meromorphic function as quotient of two analytic functions. It is a non-trivial result that this is also globally possible, i.e. for *any* meromorphic $f \in \mathcal{M}(D)$ there exist analytic $g, h \in \mathcal{O}(D)$, $h \not\equiv 0$, which satisfy

$$f \; = \; \frac{g}{h} \, .$$

We will prove this result in case of $D = \mathbb{C}$ in Chapter IV using WEIERSTRASS products.

Algebraically, we can restate this result as follows: The quotient field

$$Q(\mathcal{O}(D)) = \left\{ \frac{g}{h} \; ; \quad g, h \in \mathcal{O}(D) \, , \; h \not\equiv 0 \right\}$$

of $\mathcal{O}(D)$ is the field $\mathcal{M}(D)$ of meromorphic functions.

Generalization.

We have already allowed ∞ as a value of meromorphic functions. Then why not also allow ∞ in the domain of definition? For this, we have to topologize the set $\overline{\mathbb{C}} = \mathbb{C} \cup \{\infty\}$.

Definition A.3 *A subset $D \subset \overline{\mathbb{C}}$ is **open** iff the following conditions are satisfied:*

(a) *$D \cap \mathbb{C}$ is open (in $\mathbb{C}$) .*
(b) *If $\infty \in D$, then there exists an $R > 0$ with*

$$D \supset \{\, z \in \mathbb{C} \, ; \quad |z| > R \,\} \, .$$

In the following we use the conventions:

$$\frac{1}{0} = \infty \, , \qquad \frac{1}{\infty} = 0 \, .$$

A set $D \subset \overline{\mathbb{C}}$ is obviously open iff the set $\{\, z \in \overline{\mathbb{C}} \, ; \quad z^{-1} \in D \,\}$ is open.

Definition A.4 *A function*

$$f : D \longrightarrow \overline{\mathbb{C}}, \quad D \subset \overline{\mathbb{C}} \ open ,$$

*is **meromorphic**, iff the following holds:*

(a) *f is meromorphic in $D \cap \mathbb{C}$.*
(b) *The function*

$$\widehat{f}(z) := f(1/z)$$

is meromorphic in the open set

$$\widehat{D} := \{ z \in \mathbb{C} ; \quad 1/z \in D \} .$$

If $\infty \notin D$, so D is a subset of $\mathbb{C}$, then this definition coincides with A.1. Else, if $\infty \in D$, the zero point lies in $\widehat{D}$ and (b) is a non-empty condition. It expresses the meromorphy of f at ∞. The behaviour of f "near" $z = \infty$ corresponds by definition to the behavior of $\widehat{f}$ "near" $z = 0$.

In this context, we use the *terminology:*

If $D \subset \overline{\mathbb{C}}$ is an open subset containing ∞, and if $f : D \setminus \{\infty\} \longrightarrow \mathbb{C}$ is an analytic function, then the singularity at ∞ of f is

(a) *removable,*
(b) *non-essential, respectively a pole of order $k \in \mathbb{N}$,*
(c) *essential,*

iff the function $\widehat{f} : \widehat{D} \setminus \{0\} \to \mathbb{C}$ has a corresponding property at 0.

The LAURENT expansion of f at ∞ is obtained from the LAURENT expansion of $\widehat{f}$ at 0 by substituting $1/z$ for z.

Examples.

(1) Let

$$p(z) = \sum_{\nu=0}^{n} a_\nu z^\nu , \quad a_\nu \in \mathbb{C} , \quad 0 \le \nu \le n ,$$

be a polynomial. What is the behavior of p at ∞? By definition, we have to investigate the behavior of $\widehat{p}$ near 0,

$$\widehat{p}(z) = p\left(\frac{1}{z}\right) = \frac{a_n}{z^n} + \frac{a_{n-1}}{z^{n-1}} + \cdots + a_0 .$$

If p is a non-constant polynomial, i.e. $n \ge 1$, $a_n \ne 0$, then $\widehat{p}$ has in $z = 0$ a pole of order n. By this, p has a pole of order n at ∞.

(2) By the exponential series

$$f(z) = \exp(z) = \sum_{n=0}^{\infty} \frac{z^n}{n!}$$

we introduce an entire function. To study its behaviour near ∞ we associate

$$\widehat{f}(z) = \exp\left(\frac{1}{z}\right) = \sum_{n=0}^{\infty} \frac{1}{n!\, z^n}.$$

The exponential function $f = \exp$ has thus at ∞ an essential singularity.

The entire functions are divided by their behavior at ∞ in two classes,

the *entire rational* functions (polynomials), which have a non-essential singularity at ∞, and

the *entire transcendental* functions, which are essentially singular at ∞.

In the case $D \subset \overline{\mathbb{C}}$, D may thus contain ∞, we also denote

by $\mathcal{M}(D)$ the set of all meromorphic functions in D, and

by $\mathcal{O}(D)$ the set of all analytic functions in D, i.e. by definition those meromorphic functions that don't take the value ∞.

An open set $D \subset \overline{\mathbb{C}}$ is called a **domain**, iff the intersection $D \cap \mathbb{C}$ is a domain in $\mathbb{C}$.

In analogy to A.2 we have

Remark A.5 *The set $\mathcal{M}(D)$ of all meromorphic functions on a domain $D \subset \overline{\mathbb{C}}$ is a field, which contains $\mathcal{O}(D)$ as a subring.*

Proposition A.6 *The meromorphic functions on the whole $\overline{\mathbb{C}}$ are exactly the rational functions.*

Proof. Consider a function $f \in \mathcal{M}(\overline{\mathbb{C}})$. The function f is analytic in a punctured neighborhood of ∞, i.e. in a region of the form

$$\{\, z \in \mathbb{C}\,;\quad |z| > C \,\}\,,\qquad \text{with a suitable } C \geq 0\,.$$

Therefore f has only finitely many poles. The principal part of f in such a pole s is of the form

$$h_s\left(\frac{1}{z-s}\right),\quad h_s \text{ polynomial}\,.$$

The function

$$g(z) = f(z) - \sum_{\substack{s \in \mathbb{C} \\ f(s) = \infty}} h_s\left(\frac{1}{z-s}\right)$$

is then analytic in the whole $\mathbb{C}$. By hypothesis, it has a non-essential singularity at ∞, so it is a polynomial. $\square$

We have not only proved A.6, but we also have

Proposition A.7 (Partial fraction decomposition) *Each rational function can be written as a sum of a polynomial and finitely many linear combinations of special rational functions (partial fractions) of the form*

$$z \mapsto (z-s)^{-n}\,,\quad n \in \mathbb{N}\,.$$

We further obtain:

Theorem A.8 (Variant of Liouville's Theorem) *Any analytic function f :* $\overline{\mathbb{C}} \to \mathbb{C}$ *is constant.*

One can argue as follows: $f \mid \mathbb{C}$ is a rational function without poles, i.e. a polynomial. Because it has no pole at ∞ it is constant. □

We will not make use of the following remark, but it is useful to notice that $\overline{\mathbb{C}}$ is a *compact topological space* with the topology given by A.3.
It is homeomorphic to the sphere

$$S^2 := \left\{ (w,t) \in \mathbb{C} \times \mathbb{R} \cong \mathbb{R}^3; \ |w|^2 + t^2 = 1 \right\},$$

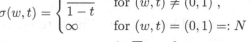

as one can see by using the stereographic projection $\sigma : S^2 \to \overline{\mathbb{C}} = \mathbb{C} \cup \{\infty\}$, which is defined by

$$\sigma(w,t) = \begin{cases} \dfrac{w}{1-t} & \text{for } (w,t) \neq (0,1) \ , \\ \infty & \text{for } (w,t) = (0,1) =: N \ . \end{cases}$$

The inverse map $\sigma^{-1} : \overline{\mathbb{C}} \to S^2$ is given by

$$\sigma^{-1}(z) = \begin{cases} \left(\dfrac{2z}{|z|^2+1} , \dfrac{|z|^2-1}{|z|^2+1} \right) & \text{for } z \neq \infty \ , \\ N & \text{for } z = \infty \ . \end{cases}$$

When one considers S^2 as a "model" for $\overline{\mathbb{C}}$, then one uses the terminology *Riemann sphere*.

The variant of LIOUVILLE's Theorem A.8 can be seen in a new light using the compactness of $\overline{\mathbb{C}}$. Each continuous function on $\overline{\mathbb{C}}$ with values in $\mathbb{C}$ has a maximum of its modulus. By the Maximum Principle, it is constant if analytic.

Möbius transformations

A rational function gives a *bijective* map $\overline{\mathbb{C}} \to \overline{\mathbb{C}}$ of the RIEMANN sphere onto itself, iff it is of the form

$$z \mapsto \frac{az+b}{cz+d} \ , \qquad a,b,c,d \in \mathbb{C} \ , \ ad - bc \neq 0 \ .$$

We call such transformations *linear fractional transformations*, or *Möbius transformations*. To each invertible matrix

$$M = \begin{pmatrix} a & b \\ c & d \end{pmatrix}$$

with complex entries we can assign the MÖBIUS transformation

$$Mz := \frac{az+b}{cz+d} .$$

The set of all invertible 2×2 matrices is the group $\mathrm{GL}(2,\mathbb{C})$. The set $\mathfrak{M}$ of all MÖBIUS transformations is also a group, the group law being the composition of functions.

Proposition A.9 *The map*

$$\mathrm{GL}(2,\mathbb{C}) \longrightarrow \mathfrak{M} ,$$

which associates to each invertible matrix M the corresponding Möbius transformation, is a homomorphism of groups. Two matrices define the same Möbius transformation iff they differ by a scalar factor $\neq 0$.

Corollary. *The inverse of the Möbius transformation given by the matrix M is*

$$M^{-1}z = \frac{dz-b}{-cz+a} .$$

In the exercises of this appendix one can find more about MÖBIUS transformations.

Exercises for the Appendix to III.4 and III.5

1. Let $D \subset \overline{\mathbb{C}}$ be a domain. The set $\mathcal{M}(D)$ of all meromorphic functions on D is a field.

2. The zero set of a non-vanishing meromorphic function one a domain is discrete.

3. Let ∞ be a singularity of an analytic function f. Classify the three possible types for this singularity by mapping properties of f.

4. Prove that the stereographic projection (see p.159)

$$\sigma : S^2 \longrightarrow \overline{\mathbb{C}}$$

is bijective by means of the mentioned formula for its inverse.

5. Let $f : \mathbb{C} \to \mathbb{C}$ be an entire function, which is injective. Show that f is of the form

$$f(z) = az + b , \quad a \neq 0 ,$$

and deduce that each such map is a conformal map from $\mathbb{C}$ onto itself. The group $\mathrm{Aut}(\mathbb{C})$ of conformal maps $\mathbb{C} \to \mathbb{C}$ consists exactly of the affine maps $z \mapsto az + b$, $a,b \in \mathbb{C}$, $a \neq 0$.

6. Find all entire functions f with $f(f(z)) = z$ for all $z \in \mathbb{C}$.

7. An *automorphism* of the RIEMANN sphere $\overline{\mathbb{C}}$ is a map $f : \overline{\mathbb{C}} \to \overline{\mathbb{C}}$ with the following properties
   (a) f is meromorphic, and
   (b) f is bijective.

   Show:
   (a) the inverse map f^{-1} is also meromorphic, and

(b)   each automorphism of $\overline{\mathbb{C}}$ is a MÖBIUS transformation, and conversely, i.e. $\text{Aut}(\overline{\mathbb{C}}) = \mathfrak{M}$.

8.   A non-identical MÖBIUS transformation has at least one, and at most two fixed points.

9.   Let a, b and c be three different points in the RIEMANN sphere $\overline{\mathbb{C}}$. Show the existence of exactly one MÖBIUS transformation M with the property:

$$Ma = 0, \quad Mb = 1, \quad Mc = \infty.$$

Hint. Consider

$$Mz := \frac{z-a}{z-c} : \frac{b-a}{b-c}.$$

Note: The expression on the right hand side of the above equation is called the *cross ratio* of the four complex numbers z, a, b and c, for short $\text{CR}(z,a,b,c)$.

10.   A subset of the RIEMANN sphere $\overline{\mathbb{C}}$ is called a *generalized circle,* iff it is
        either a circle in $\mathbb{C}$,
        or the union of a straight line in $\mathbb{C}$ (not necessarily passing through zero) with ∞.
      A map $\overline{\mathbb{C}} \to \overline{\mathbb{C}}$ is called *circle preserving,* iff it maps generalized circles to generalized circles.

      Show that the MÖBIUS transformations are circle preserving.

11.   For any two generalized circles, there exists a MÖBIUS transformation, mapping the first circle onto the second one.

12.   The following proposition is proved in linear algebra, by means of the JORDAN (normal) form.

      For any matrix $M \in \text{GL}(2,\mathbb{C})$, there exists a matrix $A \in \text{GL}(2,\mathbb{C})$ such that AMA^{-1} is either a diagonal matrix, or an upper triangular matrix with the same diagonal entries.

      Give a proof using function theory.

      Hint. Choose A conveniently, and by replacing M with AMA^{-1}, reduce to the case, where ∞ is a fixed point of M.

13.   For any matrix $M \in \text{SL}(2,\mathbb{C})$ of finite order, there exists a matrix $A \in \text{GL}(2,\mathbb{C})$ such that

$$AMA^{-1} = \begin{pmatrix} \zeta & 0 \\ 0 & \zeta^{-1} \end{pmatrix}$$

with a suitable root of unity $\zeta \in \mathbb{C}$.

III.6 The Residue Theorem

Preliminary observations about winding numbers

In II.2.8 we introduced the notion of an elementary domain.

A domain $D \subset \mathbb{C}$ is called an *elementary domain*, iff *any* analytic function $f : D \to \mathbb{C}$ possesses a global primitive in D. Equivalently, for *any* closed, piecewise smooth curve α in D, and any analytic function $f : D \to \mathbb{C}$ one has:

$$\int_\alpha f(\zeta)\, d\zeta = 0 \ .$$

A natural question in this context is the following one:

Let $D \subset \mathbb{C}$ be an *arbitrary* domain.

> How can we characterize all closed, piecewise smooth curves α in D, which satisfy $\int_\alpha f(\zeta)\, d\zeta = 0$ for any analytic function $f : D \to \mathbb{C}$?

In the Appendix B to Chapter IV we will see that this is the case exactly for those closed curves α in D which do not "surround" any point of the complement $\mathbb{C} \setminus D$. Especially, elementary domains are characterized by the property that the "interior of D" lies entirely in D. Intuitively this means that D has "no holes".

How is it possible to define rigorously the number of times a curve goes around a given point?
As a motivation for the forthcoming definition we consider an example:
For $k \in \mathbb{Z} \setminus \{0\}$, and $r > 0$, $z_0 \in \mathbb{C}$ let

$$\varepsilon_k(t) = z_0 + r \exp(2\pi i k t), \quad 0 \le t \le 1 \ ,$$

be the circle path, which circulates k times, with radius r and z_0,
Then we have:

$$\frac{1}{2\pi i} \int_{\varepsilon_k} \frac{1}{\zeta - z}\, d\zeta = \begin{cases} k & \text{for all } z \text{ with } |z - z_0| < r \ , \\ 0 & \text{for all } z \text{ with } |z - z_0| > r \ . \end{cases}$$

This example leads to:

Definition III.6.1 *Let α be a closed, piecewise smooth curve, whose image does not contain the point $z \in \mathbb{C}$. The **winding number** (or the **index**) of α with respect to the point z is defined by the formula:*

$$\boxed{\ \chi(\alpha; z) := \frac{1}{2\pi i} \int_\alpha \frac{1}{\zeta - z}\, d\zeta \ . \ }$$

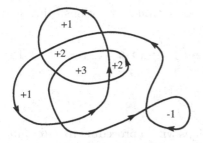 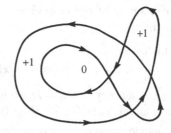

This definition is completely *ungeometrical*. For the moment we should be content that it respects geometric intuition, at least in the case of circle lines. The reader should realize that a rigorous definition supporting the intuition cannot be simple. One can prove (see also Exercise 3(e)) that the integral involved in the definition of the winding number for the curve α measures the *total variation of the argument of $\alpha(t)$* while t runs through the parameter interval of α.

In the appendices to Chapter IV we shall show that one can continuously deform any closed curve in the punctured plane into a circle line, which is k times surrounded for a suitable k in concordance with our geometrical intuition. Anyway, in the exercises to this section one can find and derive the main properties of the winding number, including its integrality. If α is a closed curve in an elementary domain D, then by the CAUCHY Integral Formula, the winding number of α with respect to any point of the complement of D is equal to zero. We will show in Appendix B of Chapter IV that the converse is also true. Intuitively, we see again that "*elementary domains are exactly the domains without holes*".

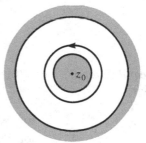

If $\mathcal{A}$ is an annulus, $r < |z| < R$, then the circle lines of radius ρ, $r < \varrho < R$, also surround points of the complement of the annulus, namely all z with $|z| \le r$.

Using the winding number, we can define what should be understood under "the interior" and "the exterior" of a curve.

If $\alpha : [a, b] \to \mathbb{C}$ is a piecewise smooth curve, then we define

$$\text{Int}\,(\alpha) := \{ \ z \in \mathbb{C} \setminus \text{Image } \alpha \ ; \quad \chi(\alpha; z) \ne 0 \ \} \ , \quad \text{the \textit{interior} of } \alpha \ ,$$
$$\text{Ext}\,(\alpha) := \{ \ z \in \mathbb{C} \setminus \text{Image } \alpha \ ; \quad \chi(\alpha; z) = 0 \ \} \ , \quad \text{the \textit{exterior} of } \alpha \ .$$

We always have

$$\mathbb{C} \setminus \text{Image } \alpha = \text{Int}\,(\alpha) \cup \text{Ext}\,(\alpha) \quad \text{(disjoint union)} \ .$$

In our example of the circle line $\alpha = \varepsilon_k$ the introduced notions coincide with our intuition:

$$\text{Int}\,(\alpha) = \{\, z \in \mathbb{C} \setminus \text{Image } \alpha \,;\quad \chi(\alpha; z) \neq 0 \,\} \quad = \{ z \in \mathbb{C}\,;\quad |z - z_0| < r \,\}\,,$$
$$\text{Ext}\,(\alpha) = \{\, z \in \mathbb{C} \setminus \text{Image } \alpha \,;\quad \chi(\alpha; z) = 0 \,\} \quad = \{\, z \in \mathbb{C}\,;\quad |z - z_0| > r \,\}\,.$$

For an elementary domain D we have:
If α is a closed curve in D, then $\text{Int}\,(\alpha) \subset D$.

We conclude these preliminary observations with a procedure to *determine the winding number* in concrete situations (e.g. in III.7.2). If we slit the complex plane along a half-line starting in $z \in \mathbb{C}$, then we obtain a star-shaped (and thus elementary) domain. The integral

$$\int_\alpha \frac{1}{\zeta - z}\, d\zeta$$

along an arbitrary curve $\alpha : [a, b] \to \mathbb{C}$, $z \notin \text{Image } \alpha$, depends only on the beginning point and the end point of α, as long as the curve does not intersect the half-line. This can be used to simplify a given curve without changing its winding number.

First Example. For both curves α and β in the figure we have

$$\int_\alpha \frac{1}{\zeta - z}\, d\zeta = \int_\beta \frac{1}{\zeta - z}\, d\zeta\,.$$

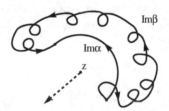

Second Example. Let r be > 0, and consider

$$\alpha(t) = \begin{cases} t & \text{for } -r \leq t \leq r\,, \\ re^{i(t-r)} & \text{for } r \leq t \leq r + \pi\,. \end{cases}$$

Then we have

$$\chi(\alpha; i) = \begin{cases} 1\,, & \text{if } r > 1\,, \\ 0\,, & \text{if } 0 < r < 1\,. \end{cases}$$

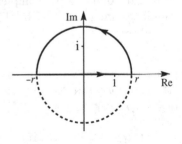

Instead of integrating along the interval from $-r$ to r, one can also integrate along the "lower half-circle" to obtain in totality an integral along the full circle.

Definition III.6.2 *Let* $f : \overset{\bullet}{U}_r(a) \to \mathbb{C}$ *be an analytic function, such that* a *is a singularity of* f, *and let*

$$f(z) = \sum_{n=-\infty}^{\infty} a_n(z-a)^n$$

be its Laurent series in $\overset{\bullet}{U}_r(a)$. *The coefficient* a_{-1} *in this expansion is called the **residue** of* f *at the point* a.

Notation. $\mathrm{Res}(f; a) := a_{-1}$.

Using the coefficient formula in III.5.2 we also can write

$$\mathrm{Res}(f; a) = \frac{1}{2\pi\mathrm{i}} \oint_{|\zeta-a|=\varrho} f(\zeta)\, d\zeta$$

for suitably small values of ϱ. At removable singularities the residue vanishes. But it can also vanish for true singularities. For instance, for $f_n(z) = z^n$, $n \in \mathbb{Z}$, we have $\mathrm{Res}(f_n; 0) = 0$ for $n \neq -1$ and $\mathrm{Res}(f_n; 0) = 1$ for $n = -1$.

We come now to the main result of this chapter.

Theorem III.6.3 (The Residue Theorem, A.L. Cauchy, 1826)   *Let* $D \subset \mathbb{C}$ *be an elementary domain, and* $z_1, \dots, z_k \in D$ *finitely many (pairwise distinct) points. Further, let* $f : D \setminus \{z_1, \dots, z_k\} \to \mathbb{C}$ *be an analytic function and* $\alpha : [a, b] \longrightarrow D \setminus \{z_1, \dots, z_k\}$ *a closed, piecewise smooth curve. Then the following formula holds:*

The Residue Formula

$$\int_\alpha f(\zeta)\, d\zeta = 2\pi\mathrm{i} \sum_{j=1}^{k} \mathrm{Res}(f; z_j)\, \chi(\alpha; z_j) .$$

Proof. We expand f around each of its singularities z_j into a LAURENT series,

$$f(z) = \sum_{n=-\infty}^{\infty} a_n^{(j)}(z - z_j)^n , \quad 1 \leq j \leq k .$$

By definition, $a_{-1}^{(j)} = \mathrm{Res}(f; z_j)$, $1 \leq j \leq k$. Because each principal part

$$h_j\left(\frac{1}{z - z_j}\right) := \sum_{n=-1}^{-\infty} a_n^{(j)}(z - z_j)^n$$

defines an analytic function in $\mathbb{C} \setminus \{z_j\}$, the auxiliary function

$$g(z) = f(z) - \sum_{j=1}^{k} h_j\left(\frac{1}{z - z_j}\right)$$

has at all z_j, $1 \le j \le k$, removable singularities. It can be thus analytically extended to the whole D. Because D is an elementary domain, it follows

$$
\begin{aligned}
0 = \int_\alpha g(\zeta)\, d\zeta &= \int_\alpha \left(f(\zeta) - \sum_{j=1}^{k} h_j \left(\frac{1}{\zeta - z_j} \right) \right) d\zeta \\
&= \int_\alpha f(\zeta)\, d\zeta - \sum_{j=1}^{k} \int_\alpha h_j \left(\frac{1}{\zeta - z_j} \right) d\zeta \\
&= \int_\alpha f(\zeta)\, d\zeta - \sum_{j=1}^{k} \int_\alpha \sum_{n=-1}^{-\infty} a_n^{(j)} (\zeta - z_j)^n \, d\zeta \\
&= \int_\alpha f(\zeta)\, d\zeta - \sum_{j=1}^{k} \sum_{n=-1}^{-\infty} a_n^{(j)} \int_\alpha (\zeta - z_j)^n \, d\zeta \\
&= \int_\alpha f(\zeta)\, d\zeta - \sum_{j=1}^{k} a_{-1}^{(j)} \int_\alpha \frac{1}{\zeta - z_j}\, d\zeta \\
&= \int_\alpha f(\zeta)\, d\zeta - 2\pi i \sum_{j=1}^{k} \mathrm{Res}(f; z_j)\, \chi(\alpha; z_j) ,
\end{aligned}
$$

by definition of residue and winding number. We obtain

$$
\int_\alpha f(\zeta)\, d\zeta = 2\pi i \sum_{j=1}^{k} \mathrm{Res}(f; z_j)\, \chi(\alpha; z_j) . \qquad \square
$$

Remarks.

(1) In the residue formula of Theorem III.6.3 there is a contribution only from those points z_j with $\chi(\alpha; z_j) \ne 0$, i.e. points that are surrounded by α, i.e. points in the interior of α, $z_j \in \mathrm{Int}\,(\alpha)$.

(2) If f can be analytically extended into the points $z_1, \ldots, z_k$, then we have

$$
\int_\alpha f(\zeta)\, d\zeta = 0 .
$$

Hence the Residue Theorem is a generalization of the CAUCHY Integral Formula for elementary domains.

(3) If f is analytic in D (elementary domain), then for any $z \in D$ the function

$$
h : D \setminus \{z\} \longrightarrow \mathbb{C} , \quad \zeta \longmapsto \frac{f(\zeta)}{\zeta - z} ,
$$

is analytic, and we have $\mathrm{Res}(h; z) = f(z)$. The Residue Formula now gives

$$
\frac{1}{2\pi i} \int_\alpha h(\zeta)\, d\zeta = \frac{1}{2\pi i} \int_\alpha \frac{f(\zeta)}{\zeta - z}\, d\zeta = \mathrm{Res}(h; z)\, \chi(\alpha; z) = f(z)\, \chi(\alpha; z) .
$$

This gives a generalization or the CAUCHY Integral Formula for arbitrary curves,

$$\chi(\alpha; z)f(z) = \frac{1}{2\pi i} \int_\alpha \frac{f(\zeta)}{\zeta - z}\, d\zeta \ .$$

Before we consider applications of the Residue Theorem, we list some useful methods for the computation of residues at non-essential singular points.

Remark III.6.4 *Let $D \subset \mathbb{C}$ be a domain, $a \in D$ a point in D, and $f, g : D \setminus \{a\} \to \mathbb{C}$ analytic functions with non-essential singularities at a. Then we have*

(1) *If $\operatorname{ord}(f; a)$ is ≥ -1, then*

$$\operatorname{Res}(f; a) = \lim_{z \to a}(z - a)f(z) \ .$$

More general, if the point a is a pole of order k, then

$$\operatorname{Res}(f; a) = \frac{\widetilde{f}^{(k-1)}(a)}{(k - 1)!} \qquad with \ \widetilde{f}(z) = (z - a)^k f(z) \ .$$

(2) *If $\operatorname{ord}(f; a) \geq 0$, and $\operatorname{ord}(g; a) = 1$, then*

$$\operatorname{Res}(f/g; a) = \frac{f(a)}{g'(a)} \ .$$

(3) *If $f \not\equiv 0$, then on has for all $a \in D$*

$$\operatorname{Res}(f'/f; a) = \operatorname{ord}(f; a) \ .$$

(4) *If g is analytic, then*

$$\operatorname{Res}\left(g\, \frac{f'}{f} \ ; \ a \right) = g(a)\operatorname{ord}(f; a) \ .$$

For the proof, use the LAURENT series of f, g. $\square$

Examples.

(1) The function

$$h(z) = \frac{\exp(iz)}{z^2 + 1}$$

has at $a = i$ a pole of first order.
From III.6.4 (1), using $z^2 + 1 = (z - i)(z + i)$ it follows

$$\operatorname{Res}(h; i) = \lim_{z \to i}(z - i)h(z) = -\frac{i}{2e} \ .$$

The same result follows using III.6.4 (2) with $f(z) = \exp(iz)$ and $g(z) = z^2 + 1$,

$$\operatorname{Res}(h; i) = \frac{f(i)}{g'(i)} = \frac{\exp(-1)}{2i} = -\frac{i}{2e} \ .$$

(2) The function
$$h(z) = \pi \frac{\cos(\pi z)}{\sin(\pi z)}$$

has at $k \in \mathbb{Z}$ poles of order one, and we compute
$$\text{Res}(h; k) = \pi \frac{\cos(\pi k)}{\pi \cos(\pi k)} = 1 \ .$$

(3) The function
$$f(z) = \frac{1}{(z^2 + 1)^3}$$

has at $z = i$ a pole of order 3. From III.6.4 (1) we get
$$\text{Res}(f; i) = \frac{\tilde{f}^{(2)}(i)}{2!} \qquad \text{with } \tilde{f}(z) = \frac{1}{(z + i)^3} \ ,$$

hence
$$\text{Res}(f; i) = -\frac{3i}{16} \ .$$

Exercises for III.6

1.   For the functions defined by the following expressions compute all residues in all singular points.

(a) $\dfrac{1 - \cos z}{z^2}$,   (b) $\dfrac{z^3}{(1 + z)^3}$,   (c) $\dfrac{1}{(z^2 + 1)^3}$,

(d) $\dfrac{1}{(z^2 + 1)(z - 1)^2}$,   (e) $\dfrac{\exp(z)}{(z - 1)^2}$,   (f) $z \exp\left(\dfrac{1}{1 - z}\right)$,

(g) $\dfrac{1}{(z^2 + 1)(z - i)^3}$,   (h) $\dfrac{1}{\exp(z) + 1}$,   (i) $\dfrac{1}{\sin \pi z}$.

2.   Let $D \subset \mathbb{C}$ be a domain, $\alpha : [0, 1] \rightarrow D$ a smooth closed curve, and $a \notin \text{Image } \alpha$. *Show:* The *winding number*
$$\chi(\alpha; a) = \frac{1}{2\pi i} \int_\alpha \frac{1}{\zeta - a} d\zeta$$

is always an integer.

Hint. Define for $t \in [0, 1]$
$$G(t) := \int_0^t \frac{\alpha'(s)}{\alpha(s) - a} ds \quad \text{and} \quad F(t) := (\alpha(t) - a) \exp(-G(t)) \ ,$$

then compute $F'(t)$, and finally show $\alpha(t) - a = (\alpha(0) - a) \exp G(t)$ for all $t \in [0, 1]$.

3. *Computation rules for the winding number*

   (a) If α is a closed curve in $\mathbb{C}$, then the function

   $$\mathbb{C} \setminus \text{Image } \alpha \longrightarrow \mathbb{C} , \quad z \longmapsto \chi(\alpha; z) ,$$

   is locally constant.

   (b) If α and β are two composable curves then we have

   $$\chi(\alpha \oplus \beta; z) = \chi(\alpha; z) + \chi(\beta; z) ,$$

   for all allowed z (not lying in the image of α or β). Especially,

   $$\chi(\alpha^-; z) = -\chi(\alpha; z) .$$

   (c) The interior of a closed curve is always bounded, and its exterior is always unbounded (and thus non-empty).

   (d) If a curve α is runs in an open disk, then the exterior of α contains the complement of the disk.

   (e) If $\alpha : [0, 1] \to \mathbb{C}$ is a curve, and a is a point in the complement of the image of α, then there exists a partition $0 = a_0 < a_1 < \cdots < a_n = 1$, and elementary domains (even open disks) $D_1, \ldots, D_n$, which don't contain a, and such that $\alpha[a_{\nu-1}, a_\nu] \subset D_\nu$, $1 \le \nu \le n$. Because in each D_ν there exists a continuous branch of the logarithm, we obtain another proof of the integrality of the winding number for a closed curve α, $\alpha(0) = \alpha(1)$.

4. Assume that f has at ∞ an isolated singularity. We define

   $$\text{Res}(f; \infty) := -\text{Res}(\tilde{f}; 0) , \quad \text{where we set}$$
   $$\tilde{f}(z) := \frac{1}{z^2} \hat{f}(z) = \frac{1}{z^2} f\left(\frac{1}{z}\right) .$$

   The factor z^{-2} is natural, as it will become transparent from the following computation rules, especially Exercise 5.

   (a) *Show:*

   $$\text{Res}(f; \infty) = -\frac{1}{2\pi i} \oint_{\alpha_R} f(\zeta) \, d\zeta ,$$

   where $\alpha_R(t) = R \exp(it)$, $t \in [0, 2\pi]$, and R is chosen big enough to ensure that f is analytic in the complement of the closed disk centered at 0 with radius R.

   (b) The function

   $$f(z) = \begin{cases} 1/z , & \text{if } z \ne \infty , \\ 0 , & \text{if } z = \infty , \end{cases}$$

   has at ∞ a removable singularity, but $\text{Res}(f; \infty) = -1$ (not zero!).

   It seems that ∞ plays a special role. The special role disappears if one defines the notion of the "residue" for the *differential* $f(z) \, dz$ instead of the function f. The notion of residue of f is related to the differential $f(z) \, dz$, and the notion of order of f to the function f itself.

5. Let $f : \overline{\mathbb{C}} \to \overline{\mathbb{C}}$ be a rational function.

    Show:
    $$\sum_{p \in \overline{\mathbb{C}}} \mathrm{Res}(f; p) = 0.$$

6. Compute the following integrals:

    (a) $I := \oint_{|\zeta|=2} \dfrac{1}{(\zeta - 3)(\zeta^{13} - 1)} \, d\zeta$,   (b) $I := \oint_{|\zeta|=10} \dfrac{\zeta^3}{\zeta^4 - 1} \, d\zeta$.

7. If f has at $a \in \mathbb{C}$ a pole of order 1, and if g is analytic in an open neighborhood of a, then
    $$\mathrm{Res}(fg; a) = g(a) \, \mathrm{Res}(f; a) .$$

8. The residue of an analytic function f at a singularity $a \in \mathbb{C}$ is the uniquely determined complex number c, such that the function
    $$f(z) - \frac{c}{z - a}$$
    admits a primitive in a punctured neighborhood of the point a.

9. Let f be analytic in $\overset{\bullet}{U}_r(0) := U_r(0) \setminus \{0\}$, $r > 0$.

    Show: $\mathrm{Res}(f'; 0) = 0$.

10. Let $\varphi : D \to \widetilde{D}$ be a conformal map between two domains in the complex plane, α a curve in D, and $\widetilde{\alpha} = \varphi \circ \alpha$ its image in $\widetilde{D}$. Then we have for any continuous function $f : \widetilde{D} \to \mathbb{C}$ the "substitution rule"
    $$\int_{\widetilde{\alpha}} f(\eta) \, d\eta = \int_{\alpha} f(\varphi(\zeta)) \varphi'(\zeta) \, d\zeta .$$

    Deduce from this the *transformation formula for residues*:
    $$\mathrm{Res}(f; \varphi(a)) = \mathrm{Res}((f \circ \varphi)\varphi'; a) ,$$

    where f is an analytic function on $\widetilde{D} \setminus \{\varphi(a)\}$. This property also covers the invariance of the winding number with respect to conformal transformations.

III.7 Applications of the Residue Theorem

From the richness of the applications of the Residue Theorem we only choose a few. First, we consider applications inside the theory of functions of a complex variable, as for instance the existence of an integral that counts zeros and poles of a given analytic function. Another important application is the calculus of integrals. The Residue Theorem offers an instrument to compute many integrals, including real integrals. Finally, we apply the Residue Theorem to summation of series, and especially compute the partial fraction decomposition of the cotangent function. For other applications of the Residue Theorem, e.g. computations of GAUSS sums, the reader will find some exercises at the end of this section.

Function theoretical consequences of the Residue Theorem

We start with a result that connects the number of zeros and the number of poles of an analytic function defined on an elementary domain. From III.6.4, (3) we immediately obtain

Proposition III.7.1 *Let $D \subset \mathbb{C}$ be an elementary domain, e.g. a star domain. Let f be a meromorphic function on D with the zeros $a_1, \ldots, a_n \in D$, and the poles $b_1, \ldots, b_m \in D$. Then, for any closed piecewise smooth curve α in D, which avoids in its image all zeros and poles, one has:*

$$\frac{1}{2\pi i} \int_\alpha \frac{f'}{f}(\zeta)\, d\zeta = \sum_{\mu=1}^{n} \operatorname{ord}(f; a_\mu) \chi(\alpha; a_\mu) + \sum_{\nu=1}^{m} \operatorname{ord}(f; b_\nu) \chi(\alpha; b_\nu) \ .$$

An application of III.7.1 is the following result of A. HURWITZ:

Theorem III.7.2 (A. Hurwitz, 1889) *Let f_0, f_1, f_2, $\cdots : D \to \mathbb{C}$ be a sequence of analytic functions that converges locally uniformly to (the analytic function) $f : D \to \mathbb{C}$. Assume that none of the functions f_n has a zero in D.*

Then f is either identically zero, or it has no zero in D.

Proof. We assume the contrary, i.e. that $f \not\equiv 0$ but there exists a point a with $f(a) = 0$, and derive a contradiction. We can choose an $\varepsilon > 0$ small enough such that the disk centered at a of radius 2ε is contained in D, and there are no zeros of f in this disk, excepting a. One easily sees that the sequence f_n'/f_n converges locally uniformly in $U_{2\varepsilon}(a) \setminus \{a\}$ to f'/f, which gives

$$0 = \frac{1}{2\pi i} \oint_{\partial U_\varepsilon(a)} \frac{f_n'}{f_n} \ \to \ \frac{1}{2\pi i} \oint_{\partial U_\varepsilon(a)} \frac{f'}{f} \ ,$$

in contradiction to the assumption $f(a) = 0$. $\square$

Corollary III.7.3 *Let $D \subset \mathbb{C}$ be a domain, and f_0, f_1, f_2, $\ldots$ a sequence of **injective** analytic functions $f_n : D \to \mathbb{C}$, which converges locally uniformly to the (analytic) function $f : D \to \mathbb{C}$. Then f is either constant, or injective.*

Proof. We assume f to be non-constant, and pick an arbitrary $a \in D$. Because of the injectivity, each function $z \to f_n(z) - f_n(a)$ does not vanish in $D \setminus \{a\}$. By III.7.2, this is also the case for the limit function

$$z \mapsto f(z) - f(a) \ ,$$

and thus $f(z) \neq f(a)$ for all $z \in D \setminus \{a\}$. $\square$

Variants of Proposition III.7.1 and further applications

Proposition III.7.4 (Special case of III.7.1) *Using the notations of III.7.1 we define*

$$N(0) := \sum_{\mu=1}^{n} \mathrm{ord}(f; a_\mu) \quad = \; total \; number \; of \; zeros \; of \; f \; ,$$

$$N(\infty) := -\sum_{\nu=1}^{m} \mathrm{ord}(f; b_\nu) \quad = \; total \; number \; of \; poles \; of \; f \; ,$$

in both cases counting multiplicities. We assume that the curve α surrounds all zeros and poles with index 1. Then we have:

$$\boxed{\begin{array}{c} \textbf{\textit{Number of zeros and poles}} \\[2mm] \dfrac{1}{2\pi\mathrm{i}} \displaystyle\int_\alpha \frac{f'}{f}(\zeta)\,d\zeta = N(0) - N(\infty) \; . \end{array}}$$

If f has no poles, we get a formula for the number of zeros. Using this formula, it is often numerically possible to decide whether an analytic function has zeros in a given domain.

Remark III.7.5 (Argument Principle) *Let $f : D \to \mathbb{C}$ be an analytic function, α a closed curve in the domain D, such that f does not vanish on its image. Then we have:*

$$\boxed{\; \dfrac{1}{2\pi\mathrm{i}} \displaystyle\int_\alpha \frac{f'(\zeta)}{f(\zeta)}\,d\zeta = \chi(f \circ \alpha; 0) \; . \;}$$

The result is an integer. (See also Exercise 2 in III.6 or the Consequence to A.10.) Under the assumption of III.7.1 and III.7.4, and using the same notations, this number equals $N(0)$.

The number of zeros of f, counted with multiplicities, is thus equal to the winding number of the image curve $f \circ \alpha$ around 0. The correctness of III.7.5 follows from III.7.4 and the substitution rule for integrals.   □

During the proof of the Open Mapping Theorem, III.3.3, we showed that in a domain D containing 0, any non-constant analytic function f with $f(0) = 0$ can be written in a small open neighborhood of 0 as the composition of a conformal map with the nth power map. From this we derive the following result, which can also be proven by means of the Residue Theorem.

Theorem III.7.6 *Let $f : D \to \mathbb{C}$ be a non-constant analytic function in a domain $D \subset \mathbb{C}$. Let $a \in D$ be fixed, and $b := f(a)$. Let $n \in \mathbb{N}$ be the order of $f(z) - b$ at $z = a$. Then there exist open neighborhoods $U \subset D$ of a, and $V \subset \mathbb{C}$ of b, such that any $w \in V$, $w \neq b$ has exactly n pre-images $z_1, \ldots, z_n \in U$. We thus have $f(z_j) = w$ for $1 \le j \le n$. Moreover, the order of $f(z) - w$ at any point z_j is exactly 1.*

Proof using the Argument Principle. We choose an ε-neighborhood of a, whose closure is contained in D. We can choose ε small enough, such that f does not take the value b on its boundary, and such that the derivative $f'(z)$ has no zero in

$0 < |z - a| \leq \varepsilon$. We choose $U = U_\varepsilon(a)$ and $V = V_\delta(b)$, where $\delta > 0$ is taken small enough to ensure

$$V \cap f(\partial U) = \emptyset \ .$$

This is possible, since the image of ∂U is compact. Hence the complement of $f(\partial U)$ is open and contains b. Let us pick $w \in V$. The number of places in U where f takes the value w is equal to the winding number $\chi(f \circ \alpha; w)$ by the Argument Principle. Here α is the curve,

$$\alpha(t) = a + \varepsilon e^{2\pi i t} \ , \quad 0 \leq t \leq 1 \ .$$

This winding number depends continuously on w, and takes only integral values. It is thus constant ($= n$). That the order of $f(z) - w_j$ at any point z_j is one follows finally from the condition imposed on f'. □

Theorem III.7.6 contains another proof of the Open Mapping Theorem. In the case $n = 1$ it implies the injectivity of f on U, wich gives a new proof of the Theorem of Inverse Functions.

Corollary III.7.6₁ Let $f : D \to \mathbb{C}$ be an analytic function on an open subset $D \subset \mathbb{C}$, and let $a \in D$. The function f is injective on a small open neighborhood $a \in U \subset D$ iff $f'(a) \neq 0$. In this case f maps a small open neighborhood of a conformally onto an open neighborhood of $f(a)$.

This is the promised proof for the local version of the Open Mapping Theorem, which does not involve the real Implicit Function Theorem.

Theorem III.7.7 (Rouché's Theorem, E. Rouché, 1862) *Let f, g be analytic functions defined on an elementary domain D, and let α be a closed curve in D, which surrounds each point in its interior $\mathrm{Int}(\alpha)$ exactly once. For simplicity, we assume that f and $f + g$ have only finitely many zeros in D. (This condition is superfluous, see also Chapter IV, Appendix B).*

Assumption: $|g(\zeta)| < |f(\zeta)|$ *for* $\zeta \in$ Image α .

Then the functions $f, f+g$ have no zeros on the image of α, and the functions f and $f + g$ have in the interior of α the same number of zeros, counting multiplicities.

This result ensures the invariance of the number of zeros of an analytic function for "small deformations".

Proof of III.7.7. We consider the family of functions

$$h_s(z) = f(z) + sg(z) \ , \quad 0 \leq s \leq 1 \ ,$$

which connects f ($= h_0$) with $f + g$ ($= h_1$). It is clear that these functions have no zeros on the image of α. The integral which "counts the zeros" depends then continuously on the parameter s, and is an integer by III.7.5, hence constant. □

If we are also interested in the position of the zeros of a given analytic function, and not only just their number, then there is the following generalization of III.7.1:

Proposition III.7.8 *Let $D \subset \mathbb{C}$ be an elementary domain, and let f be a mero-morphic function in D with zeros in $a_1, \ldots, a_n$ and poles in $b_1, \ldots, b_m \in D$. Let*

$$g : D \longrightarrow \mathbb{C}$$

be an analytic function. Then for any closed, piecewise smooth curve $\alpha : [a, b] \to D$, which avoids the zeros and poles of f, the following formula is true:

$$\frac{1}{2\pi i} \int_\alpha \frac{f'}{f} g = \sum_{\mu=1}^{n} \operatorname{ord}(f; a_\mu) \, \chi(\alpha; a_\mu) \, g(a_\mu) + \sum_{\nu=1}^{m} \operatorname{ord}(f; b_\nu) \, \chi(\alpha; b_\nu) \, g(b_\nu) \, .$$

For instance, if we know that f has exactly one simple zero, than we can find its position by choosing $g(z) = z$.

Examples and applications

(1) Using III.7.4, we discover further proofs of the *Fundamental Theorem of Algebra*. For instance, we can argue as follows:

Because of $\lim_{|z| \to \infty} |P(z)| = \infty$ there exists an $R > 0$, such that P has no roots z with $|z| \geq R$. The number of all roots of P is

$$N(0) = \frac{1}{2\pi i} \oint_{|\zeta|=R} \frac{P'(\zeta)}{P(\zeta)} \, d\zeta \, .$$

The function P'/P hat at ∞ a simple zero. The LAURENT series in ∞ has the form

$$\frac{n}{z} + \frac{c_2}{z^2} + \frac{c_3}{z^3} + \frac{c_4}{z^4} + \cdots \quad (n = \deg P) \, .$$

This gives

$$N(0) = n = \deg P \, .$$

A polynomial of degree n has thus exactly n roots, counting multiplicities.

A slightly different proof uses the Theorem of ROUCHÉ, applied to the functions

$$f(z) = a_n z^n \quad \text{and} \quad g(z) = P(z) - f(z) \, ,$$

$P(z) = a_n z^n + \cdots + a_0$ being here the given polynomial of degree $n > 0$.

The Theorem of ROUCHÉ can be used to solve equations, especially one gets more information about the *position of solutions*; it is somehow possible to "separate" them. As an illustration, we give two examples:

(2) We consider the polynomial $P(z) = z^4 + 6z + 3$.

Let f, g be the polynomials

$$f(z) = z^4 \, , \quad \text{and} \quad g(z) = 6z + 3 \, ,$$

then we estimate for $|z| = 2$:

$$|g(z)| \leq 6\,|z| + 3 = 15 < 16 = |f(z)| \, .$$

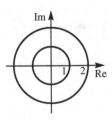

The functions f and $f + g = P$ have the same number of zeros in the disk $|z| < 2$ by III.7.7, and because f has at 0 its only zero of order four, we get that P has in the disk $|z| < 2$ also exactly four zeros.

We apply once more the same idea for the new decomposition $P = f_1 + g_1$ with $f_1(z) := 6z$ and $g_1(z) := z^4 + 3$. For $|z| = 1$ we have:

$$|g_1(z)| = |z^4 + 3| \leq |z|^4 + 3 = 1 + 3 = 4 < 6 = |6z| = |f_1(z)| .$$

The Theorem of ROUCHÉ claims then that f_1 and $P = f_1 + g_1$ have the same numbers of zeros in the unity disk $U_1(0) = \mathbb{E}$, namely one as for f_1. We now have the following information about the zeros of P: Among the four roots of P exactly one, say a, lies in the unit disk $\mathbb{E}$, and the other three lie in the annulus $1 < |z| < 2$. The precise position of a in $\mathbb{E}$ can now numerically be "determined" by evaluating the integral

$$\oint_{|\zeta|=1} \zeta \frac{4\zeta^3 + 6}{\zeta^4 + 6\zeta + 3} \, d\zeta ,$$

which gives approximately $a \approx -0.5113996194\ldots$.

(3) Let $\alpha \in \mathbb{C}$ be a complex number of modulus $|\alpha| > e = \exp(1)$. We claim that the equation

$$\alpha z \exp(z) = 1 \qquad (\text{ i.e. } \alpha z - \exp(-z) = 0) \tag{$*$}$$

has exactly one solution in $\mathbb{E}$.

In addition: If $\alpha > 0$ is real and positive, then the solution is also real and positive.

Here, it is natural to introduce the auxiliary functions

$$f(z) = \alpha z \quad \text{and} \quad g(z) = \exp(-z) .$$

There is exactly one zero for f at $z_0 = 0$, and for $|z| = 1$ we can estimate

$$|g(z)| = |\exp(-z)| = \exp(-\operatorname{Re}(z)) \leq e < |\alpha| = |f(z)| .$$

ROUCHÉ's Theorem ensures that $f + g$ has in $\mathbb{E}$ exactly one zero, i.e. the equation $(*)$ has exactly one solution in $\mathbb{E}$.

The additional remark follows from the Mean Value Theorem of real analysis.

Computation of integrals using the residue theorem

If $a \in \mathbb{C}$ is an isolated singularity of an analytic function f, then f is in a punctured r-neighborhood $\overset{\bullet}{U}_r(a)$ representable as a LAURENT series,

$$\sum_{n=-\infty}^{\infty} a_n (z - a)^n ,$$

and we have by III.5.2

$$\operatorname{Res}(f;a) = a_{-1} = \frac{1}{2\pi i} \oint_{\partial U_\varrho(a)} f(\zeta)\, d\zeta\ , \quad 0 < \varrho < r\ .$$

If we have other methods at our disposal to compute the residue $\operatorname{Res}(f;a)$, then the residue formula gives us the possibility to compute integrals. We restrict to three types:

Type I. Integrals of the form

$$\int_0^{2\pi} R(\cos t, \sin t)\, dt\ ,$$

where R is a complex rational function, which we write as a quotient of two polynomials P, Q in the variables x, y,

$$R(x,y) = \frac{P(x,y)}{Q(x,y)}\ .$$

We require $Q(x,y) \neq 0$ for all $x, y \in \mathbb{R}$ with $x^2 + y^2 = 1$. Such integrals can be computed by suitable substitutions (e.g. $u = \tan(t/2)$), which reduce the given integral to the integral of a rational function, which can be computed by means of the partial fraction decomposition.

It is often much simpler to apply the Residue Theorem, by interpreting the given integral as an integral along a suitable closed curve.

Proposition III.7.9 *Let P and Q be polynomials of two variables x, y, and assume $Q(x,y) \neq 0$ for all $(x,y) \in \mathbb{R}^2$ with $x^2 + y^2 = 1$. Then*

$$\int_0^{2\pi} \frac{P(\cos t, \sin t)}{Q(\cos t, \sin t)}\, dt\ =\ 2\pi i \sum_{a \in \mathbb{E}} \operatorname{Res}(f;a)\ ,$$

where $\mathbb{E}$ is the unit disk, and f is the rational function

$$f(z) = \frac{1}{iz}\, \frac{P\left(\frac{1}{2}\left(z + \frac{1}{z}\right),\ \frac{1}{2i}\left(z - \frac{1}{z}\right) \right)}{Q\left(\frac{1}{2}\left(z + \frac{1}{z}\right),\ \frac{1}{2i}\left(z - \frac{1}{z}\right) \right)}\ .$$

Proof. Because of

$$\cos t = \frac{\exp(it) + \exp(-it)}{2}\ , \quad \sin t = \frac{\exp(it) - \exp(-it)}{2i}\ ,$$

the rational function f has no poles on the unit circle. For all $a \in \mathbb{E}$ the winding number of this unit circle with respect to a is one. The Residue Formula in III.6.3 gives

$$2\pi i \sum_{a \in E} \operatorname{Res}(f; a) = \oint_{|\zeta|=1} f(\zeta)\, d\zeta = \int_0^{2\pi} f(e^{it})\, i e^{it}\, dt = \int_0^{2\pi} \frac{P(\cos t, \sin t)}{Q(\cos t, \sin t)}\, dt \ .$$

□

Examples.

(1) One has for all $a \in E$

$$\int_0^{2\pi} \frac{1}{1 - 2a \cos t + a^2}\, dt = \frac{2\pi}{1 - a^2} \ .$$

For $a = 0$ this is evident. Else, we associate by III.7.9 the rational function

$$f(z) = \frac{1}{iz \left(1 + a^2 - az - (a/z)\right)} = \frac{i/a}{(z - a)(z - 1/a)} \ .$$

There is exactly one pole of f in E, located at a. It is a simple pole, so we can use III.6.4 (1), to compute

$$\operatorname{Res}(f; a) = \lim_{\substack{z \to a \\ z \neq a}} (z - a) f(z) = \frac{i}{a^2 - 1} \ .$$

As claimed, we get

$$\int_0^{2\pi} \frac{1}{1 - 2a \cos t + a^2}\, dt = 2\pi i \frac{i}{a^2 - 1} = \frac{2\pi}{1 - a^2} \ .$$

(2) Analogously, we get for $a, b \in \mathbb{R}$ with $a > b > 0$

$$\int_0^{2\pi} \frac{1}{(a + b \cos t)^2}\, dt = \frac{2\pi a}{\sqrt{(a^2 - b^2)^3}} \ .$$

Further examples can be found in the exercises.

Type II. Improper convergent integrals of the type

$$\int_{-\infty}^{\infty} f(x)\, dx \ .$$

Remark. We take the notion of improper integral as granted, see also VI.1. To apply the Residue Theorem, we reformulate this integral as the limit

$$\lim_{R \to \infty} \int_{-R}^{R} f(x)\, dx \ ,$$

the so-called CAUCHY *principal value*. From the existence of this limit with "correlated" integration limits one cannot in general deduce the convergence of

$\int_{-\infty}^{\infty} f(x)\, dx$. Its existence, i.e. its convergence, is equivalent with the existence of both separated limits

$$\int_{0}^{\infty} f(x)\, dx := \lim_{R_1 \to \infty} \int_{0}^{R_1} f(x)\, dx \quad \text{and} \quad \int_{-\infty}^{0} f(x)\, dx := \lim_{R_2 \to \infty} \int_{-R_2}^{0} f(x)\, dx$$

(see also Sect. VI.1). The existence of the improper integral $\int_{-\infty}^{\infty} f(x)\, dx$ implies the existence of the CAUCHY principal value, both having the same value. For an *even* or a *non-negative* function f one can also conversely deduce from the existence of the CAUCHY principal value the existence of the improper integral, both having the same value.

The computation of improper integrals is based on the following idea. Let $D \subset \mathbb{C}$ be an elementary domain containing the closed upper half-plane

$$\overline{\mathbb{H}} = \{\, z \in \mathbb{C} \; ; \quad \operatorname{Im} z \geq 0 \,\}\, .$$

Let $a_1, \dots, a_k \in \mathbb{H}$ be pairwise distinct points in the (open) upper half-plane, and let

$$f : D \setminus \{a_1, \dots, a_k\} \longrightarrow \mathbb{C}$$

be an analytic function. We choose $r > 0$ large enough, such that

$$r > |a_\nu| \qquad \text{for } 1 \leq \nu \leq k\, .$$

We consider then the curve α, which is sketched in the figure. It is built from of the line segment $[-r, r]$ and the half-circle α_r from r to $-r$. The Residue Formula gives (because of $\chi(\alpha; a_\nu) = 1$)

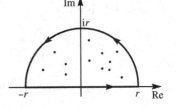

$$\int_{-r}^{r} f(x)\, dx + \int_{\alpha_r} f(z)\, dz = \int_{\alpha} f(z)\, dz = 2\pi i \sum_{\nu=1}^{k} \operatorname{Res}(f; a_\nu)\, .$$

If one can knows

$$\lim_{r \to \infty} \int_{\alpha_r} f(z)\, dz = 0\, ,$$

then

$$\boxed{\lim_{r \to \infty} \int_{-r}^{r} f(x)\, dx = 2\pi i \sum_{\nu=1}^{k} \operatorname{Res}(f; a_\nu)\, .}$$

If we independently know that $\int_{-\infty}^{\infty} f(x)\, dx$ exists, then

$$\boxed{\int_{-\infty}^{\infty} f(x)\, dx = 2\pi i \sum_{\nu=1}^{k} \operatorname{Res}(f; a_\nu)\, .}$$

Let P and Q be two polynomials, such that their degrees satisfy

$$\deg Q \geq 2 + \deg P \ .$$

We assume that Q has no real roots. The rational function

$$f(z) = \frac{P(z)}{Q(z)}$$

defined on the domain

$$\operatorname{Im} z > -\varepsilon \ , \quad \varepsilon > 0 \text{ sufficiently small} \ ,$$

trivially satisfies the assumption $\lim\limits_{r \to \infty} \int_{\alpha_r} f(z)\, dz = 0$. We then have

Proposition III.7.10 *Let* P, Q *be two polynomials with* $\deg Q \geq \deg P + 2$. *Assume that* Q *has no real roots. Let* $a_1, \ldots, a_k \in \mathbb{H}$ *be the complete set of poles in the upper half-plane of* $f = P/Q$. *Then*

$$\boxed{\int_{-\infty}^{\infty} f(x)\, dx = 2\pi i \sum_{\nu=1}^{k} \operatorname{Res}(f; a_\nu) \ .}$$

Examples.

(1) We compute the integral

$$I = \int_0^\infty \frac{1}{1 + t^6}\, dt = \frac{1}{2} \int_{-\infty}^\infty \frac{1}{1 + t^6}\, dt \ .$$

The zeros of $Q(z) = z^6 + 1$ in $\mathbb{H}$ are

$$a_1 = \exp\left(\frac{\pi}{6} i\right) \ , \quad a_2 = \exp\left(\frac{\pi}{2} i\right) \ , \quad \text{and} \quad a_3 = \exp\left(\frac{5\pi}{6} i\right) \ .$$

By III.6.4, (2), we have

$$\operatorname{Res}\left(\frac{1}{Q} ; a_\nu\right) = \frac{1}{6 a_\nu^5} = -\frac{a_\nu}{6} \ .$$

This gives

$$I = \frac{1}{2} \int_{-\infty}^\infty \frac{1}{1 + t^6}\, dt = -\frac{\pi i}{6}\left(\exp\left(\frac{\pi}{6} i\right) + \exp\left(\frac{\pi}{2} i\right) + \exp\left(\frac{5\pi}{6} i\right)\right)$$

$$= \frac{\pi}{6}\left(2 \sin\frac{\pi}{6} + 1\right) = \frac{\pi}{3} \ .$$

(2) We show

$$\int_{-\infty}^\infty \frac{1}{(t^2 + 1)^n}\, dt = \frac{\pi}{2^{2n-2}} \cdot \frac{(2n - 2)!}{\left((n-1)!\right)^2} \qquad (n \in \mathbb{N}) \ ,$$

especially

$$\int_{-\infty}^{\infty} \frac{1}{(t^2+1)}\, dt = \pi\,, \quad \int_{-\infty}^{\infty} \frac{1}{(t^2+1)^2}\, dt = \frac{\pi}{2}\,, \quad \int_{-\infty}^{\infty} \frac{1}{(t^2+1)^3}\, dt = \frac{3\pi}{8}\,.$$

The meromorphic function $f(z) = 1/(z^2+1)^n$ has in $\mathbb{H}$ its only pole at $z_0 = i$. The LAURENT series of f at this pole is obtained using the geometric series, or III.6.4 (1), and one shows

$$\mathrm{Res}(f;i) = \frac{1}{i}\binom{2n-2}{n-1}\frac{1}{2^{2n-1}} = \frac{1}{2^{2n-1}i}\cdot\frac{(2n-2)!}{((n-1)!)^2}\,.$$

(3) Let $k, n \in \mathbb{Z}$, $0 \le k < n$. Then

$$\boxed{\int_{-\infty}^{\infty} \frac{t^{2k}}{1+t^{2n}}\, dt = \frac{\pi}{n\sin\big((2k+1)\pi/2n\big)}\,.}$$

The roots of $Q(z) = 1+z^{2n}$ in $\mathbb{H}$ are

$$a_\nu = \exp\left(\frac{(2\nu+1)\pi i}{2n}\right)\,, \quad 0 \le \nu < n\,.$$

The derivative Q' is $\neq 0$ in all these places, hence all a_ν are simple roots. By III.6.4 we have for the function

$$R = f/g\,, \quad f(z) = z^{2k} \text{ and } g(z) = 1+z^{2n} \quad \text{the residues}$$

$$\mathrm{Res}(R;a_\nu) = \frac{1}{2n}a_\nu^{2k-2n+1} = -\frac{1}{2n}a_\nu^{2k+1}\,.$$

From the functional equation of the exponential function we further have

$$\sum_{\nu=0}^{n-1} a_\nu^{2k+1} = \sum_{\nu=0}^{n-1}\exp\left(\frac{\pi i}{2n}(2\nu+1)(2k+1)\right)$$

$$= \exp\left(\frac{(2k+1)\pi i}{2n}\right)\sum_{\nu=0}^{n-1}\exp\left(\frac{\pi i(2k+1)\nu}{n}\right)$$

$$= \exp\left(\frac{(2k+1)\pi i}{2n}\right)\cdot\frac{1-\exp((2k+1)\pi i\ )}{1-\exp((2k+1)\pi i/n)}$$

$$= \frac{i}{\sin((2k+1)\pi/2n)}\,.$$

Apply now III.7.10.

The following proposition can be seen as a generalization of III.7.10.

Proposition III.7.11 *Let P and Q be polynomials, and let $\alpha \geq 0$. Assume that the polynomial Q has no roots on the real line, and also that the degree inequality holds*

$\deg Q \geq 2 + \deg P$ *in case of* $\alpha = 0$, *and*

$\deg Q \geq 1 + \deg P$.

Let $a_1, \ldots, a_k$ be all roots of Q in the upper half-plane. Then

$$\int_{-\infty}^{\infty} \frac{P(t)}{Q(t)} \exp(i\alpha t)\, dt = 2\pi i \sum_{\nu=1}^{k} \mathrm{Res}(f; a_\nu) \ .$$

The meromorphic function f in the R.H.S. is the integrand in the L.H.S.,

$$f(z) = \frac{P(z)}{Q(z)} \exp(i\alpha z) \ .$$

Proof: We have already considered the case $\alpha = 0$. The sharp degree inequality $\deg Q \geq 2 + \deg P$ always ensures the absolute convergence of the integral, we have to use only the simple estimate $|\exp(i\alpha t)| \leq 1$.

In the case $\alpha > 0$ it is enough to demand $\deg Q \geq 1 + \deg P$. This can be seen as follows:

We first choose $R > 1$, such that all roots of Q lie in the disk $U_R(0)$. For an arbitrary $r > R$ we consider the closed polygonal path with vertices in $-r, r, r + ir, -r + ir$. It contains in its interior the half-disk of radius R, and hence also all roots of Q that lie in $\mathbb{H}$. The contributions of $\mathrm{Im}\, z \leq \sqrt{R}$ and $\mathrm{Im}\, z \geq \sqrt{R}$ are estimated separately.

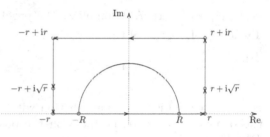

(1) The standard estimate for integrals gives immediately

$$\left| \int_{\pm r}^{\pm r + i\sqrt{r}} \right| \leq C \frac{\sqrt{r}}{r}$$

with a suitable constant C. This expression goes to zero for $r \to \infty$.

(2) For $\mathrm{Im}\, z \geq \sqrt{r}$ we have $|\exp(i\alpha z)| \leq e^{-\alpha\sqrt{r}}$. This expression converges to zero for $r \to \infty$ quicker than any rational function.

From (1) and (2) we obtain

$$\int_{-\infty}^{\infty} = \lim_{r \to \infty} \int_{-r}^{r} = \lim_{r \to \infty} \int$$

which is $2\pi i$ times the sum of residues.

This proof shows III.7.11 first for the CAUCHY principal value of the integral. But this integral is also convergent (but not absolutely) as an improper integral. To show this, one can for instance repeat the argument for a rectangle which is not symmetric with respect to the imaginary axis.

Example.

We show for $a > 0$

$$\int_0^\infty \frac{\cos t}{t^2 + a^2}\, dt = \frac{\pi}{2a} e^{-a}\,.$$

We obviously have

$$\int_0^\infty \frac{\cos t}{t^2 + a^2}\, dt = \frac{1}{2} \mathrm{Re}\left(\int_{-\infty}^\infty \frac{\exp(it)}{t^2 + a^2}\, dt \right)\,.$$

The function $f(z) = \dfrac{\exp(iz)}{z^2 + a^2}$ has only a simple pole in the upper half-plane, namely ia. Using III.6.4, we get

$$\mathrm{Res}(f; ia) = \frac{e^{-a}}{2ai}\,,$$

and Proposition III.7.11 completes the proof.

Type III. Integrals of the form

$$\int_0^\infty x^{\lambda - 1} R(x)\, dx\,, \qquad \lambda \in \mathbb{R}\,,\ \lambda > 0\,,\ \lambda \notin \mathbb{Z}\,.$$

Here, $R = P/Q$ is a rational function with polynomials P and Q, such that the denominator Q has no roots on the positive real axis $\mathbb{R}_+$. Also assume $R(0) \neq 0$, and

$$\lim_{x \to \infty} x^\lambda\, |R(x)| = 0$$

which is equivalent to $\deg Q > \lambda + \deg P$. We then consider in the slit plane $\mathbb{C}_+ := \mathbb{C} \setminus \mathbb{R}_{\geq 0}$ the function

$$f(z) = (-z)^{\lambda - 1} R(z) \qquad \text{for } z \in \mathbb{C}_+ := \mathbb{C} \setminus \mathbb{R}_{\geq 0}\,.$$

Here $(-z)^{\lambda - 1} := \exp\big((\lambda - 1) \mathrm{Log}(-z) \big)$, which uses the *principal branch of the logarithm* $\mathrm{Log} : \mathbb{C}_- \to \mathbb{C}$. From $z \in \mathbb{C}_+$ we get $-z \in \mathbb{C}_-$. The function f is hence analytic in $\mathbb{C}_+$.

Proposition III.7.12 *Using the above assumptions we have*

$$\boxed{\int_0^\infty x^{\lambda - 1} R(x)\, dx = \frac{\pi}{\sin(\lambda \pi)} \sum_{a \in \mathbb{C}_+} \mathrm{Res}(f; a)\,.}$$

Sketch of the proof. The function f is meromorphic in $\mathbb{C}_+$. We consider then the closed curve $\alpha := \alpha_1 \oplus \alpha_2 \oplus \alpha_3 \oplus \alpha_4$, where the curves α_j, $1 \leq j \leq 4$, are up to a translation of the parameter intervals (such that can be composed)

$$\alpha_1(t) := \exp(\mathrm{i}\varphi)\, t\ ,\qquad \frac{1}{r} \leq t \leq r\ ,$$

$$\alpha_2(t) := r\exp(\mathrm{i}t)\ ,\qquad \varphi \leq t \leq 2\pi - \varphi\ ,$$

$$\alpha_3(t) := -\exp(-\mathrm{i}\varphi)\, t\ ,\qquad -r \leq t \leq -\frac{1}{r}\ ,$$

$$\alpha_4(t) := \frac{1}{r}\exp\left(\,\mathrm{i}(2\pi - t)\,\right)\ ,\quad \varphi \leq t \leq 2\pi - \varphi\ .$$

We of course assume $r > 1$ and $0 < \varphi < \pi$.

Because $\mathbb{C}_+$ is an elementary domain, we can find a sufficiently large $r > 1$ with

$$\int_\alpha f(z)\, dz = \int_{\alpha_1} f(z)\, dz + \int_{\alpha_2} f(z)\, dz + \int_{\alpha_3} f(z)\, dz + \int_{\alpha_4} f(z)\, dz$$

$$= 2\pi\mathrm{i} \sum_{a\in\mathbb{C}_+} \mathrm{Res}(f; a)\ . \tag{$*$}$$

For this r, we take the limit $\varphi \to 0$. Because of the definition of $(-z)^{\lambda-1}$, the integrals on α_1 and α_3 converge to

$$\exp\left(-(\lambda-1)\pi\mathrm{i}\right) \int_{1/r}^r x^{\lambda-1} R(x)\, dx\ ,$$

and respectively to

$$-\exp\left(\ (\lambda-1)\pi\mathrm{i}\right) \int_{1/r}^r x^{\lambda-1} R(x)\, dx\ .$$

On the other side, the other two integrals do not contribute to the result, because one easily can show

$$\lim_{r\to\infty} \int_{\alpha_2} f(z)\, dz = \lim_{r\to\infty} \int_{\alpha_4} f(z)\, dz = 0$$

uniformly in φ. $\qquad\square$

Example. $\displaystyle \int_0^\infty \frac{x^{\lambda-1}}{1+x}\, dx = \frac{\pi}{\sin(\lambda\pi)}\ .$

The partial fraction decomposition of the cotangent

As another application, we deduce the *partial fraction decomposition of the cotangent*

$$\cot \pi z := \frac{\cos \pi z}{\sin \pi z}\ ,\qquad z \in \mathbb{C} \setminus \mathbb{Z}\ .$$

Proposition III.7.13 *For all $z \in \mathbb{C} - \mathbb{Z}$*

$$\pi \cot \pi z = \frac{1}{z} + \sum_{\substack{n \in \mathbb{Z} \\ n \neq 0}} \left[\frac{1}{z-n} + \frac{1}{n} \right]$$

$$\left(= \frac{1}{z} + \sum_{n=1}^{\infty} \left\{ \frac{1}{z-n} + \frac{1}{z+n} \right\} = \frac{1}{z} + \sum_{n=1}^{\infty} \frac{2z}{z^2 - n^2} \right) .$$

The involved series are absolutely (even normally) convergent.

We recall the definition

$$\sum_{n \neq 0} a_n := \sum_{n=1}^{\infty} a_n + \sum_{n=1}^{\infty} a_{-n} .$$

Proof. The absolute convergence follows from

$$\frac{1}{z-n} + \frac{1}{n} = \frac{z}{(z-n)n} ,$$

and from the convergence of the series $1 + 1/4 + 1/9 + \dots$. By the way, the absolute convergence of a series $\sum_{n \neq 0} a_n$, $n \in \mathbb{Z}$, implies by standard arguments of real analyis

$$\sum_{n \neq 0} a_n = \lim_{N \to \infty} \sum_{n \in S_N} a_n ,$$

where $S_1, S_2, S_3, \dots$ is an increasing sequence of sets exhausting $\mathbb{Z} \setminus \{0\}$, i.e.

$$S_1 \subset S_2 \subset S_3 \subset \cdots \quad \text{and} \quad \mathbb{Z} \setminus \{0\} = S_1 \cup S_2 \cup S_3 \cup \cdots$$

To prove the partial fraction decomposition formula, we introduce for a fixed $z \in \mathbb{C} \setminus \mathbb{Z}$ the function

$$f(w) = \frac{z}{w(z-w)} \pi \cot \pi w .$$

Its singularities are

$$w = z \text{ and } w = n \in \mathbb{Z} .$$

All singularities are simple poles, excepting $w = 0$ which is a pole of second order. The residues in the simple poles are obviously

$$-\pi \cot \pi z \text{ and } \frac{z}{n(z-n)} \text{ for } n \neq 0 .$$

A short computation gives for the residue in 0 the value $\frac{1}{z}$. All involved residues are exactly the summands in the partial fraction decomposition III.7.13!

We now integrate the function f along the boundary ∂Q_N of the square Q_N with vertices in $(\pm 1 \pm i)\left(N + \frac{1}{2}\right)$. Its edges are parallel to the axes, and their length is $2N + 1$, $N \in \mathbb{N}$, and we moreover assume $N > |z|$.

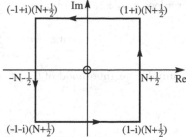

On the integration path there are no singularities of f. We obtain

$$\frac{1}{2\pi i} \int_{\partial Q_N} f(\zeta)\, d\zeta = -\pi \cot \pi z + \frac{1}{z} + \sum_{0 < |n| \leq N} \frac{z}{n(z - n)} \,.$$

It remains to show that the integral in the L.H.S. converges to zero for $N \to \infty$. For this is enough to show that $\pi \cot \pi w$ is bounded on ∂Q_N, because of the estimate

$$\left| \int_{\partial Q_N} f(\zeta)\, d\zeta \right| \leq \text{const} \cdot 4(2N + 1) \frac{|z|}{\left(N + \frac{1}{2}\right)\left(N + \frac{1}{2} - |z|\right)} \,.$$

In the region $|y| \geq 1$, $y = \operatorname{Im} z$, we have

$$|\cot \pi z| \leq \frac{1 + \exp(-2\pi |y|)}{1 - \exp(-2\pi |y|)} \leq \frac{1 + \exp(-2\pi)}{1 - \exp(-2\pi)} \,.$$

The boundedness of

$$\pi \cot \pi \left(\pm \left(N + \frac{1}{2}\right) + iy \right) \text{ for } |y| \leq 1 \,,$$

follows from the periodicity of the cotangent. $\qquad\qquad\square$

In Section III.2 we have introduced the BERNOULLI numbers B_n by the TAYLOR series

$$g(z) := \frac{z}{\exp(z) - 1} = B_0 + B_1 z + \sum_{k=1}^{\infty} \frac{B_{2k}}{(2k)!} z^{2k}$$

and we found $B_0 = 1$ and $B_1 = -\dfrac{1}{2}$.

Using the BERNOULLI numbers, there is a direct connection to the TAYLOR expansion of $\pi z \cot \pi z$ at the zero point, and then further to the special values of the RIEMANN zeta function at natural numbers $2, 4, \ldots$. First, we have by definition

$$z \cot z = iz\, \frac{\exp(iz) + \exp(-iz)}{\exp(iz) - \exp(-iz)} = iz\, \frac{\exp(2iz) + 1}{\exp(2iz) - 1}$$

$$= \frac{2iz}{\exp(2iz) - 1} + iz = g(2iz) + iz$$

$$= iz + 1 - \frac{2iz}{2} + \sum_{k=1}^{\infty} \frac{B_{2k}}{(2k)!} (2iz)^{2k} = 1 + \sum_{k=1}^{\infty} (-1)^k \frac{2^{2k}}{(2k)!} B_{2k}\, z^{2k} \,.$$

We replace z by πz, and obtain

$$\pi z \cot \pi z = 1 + \sum_{k=1}^{\infty} (-1)^k \frac{2^{2k}}{(2k)!} \pi^{2k} B_{2k} \, z^{2k} \tag{$*$}$$

in a suitable neighborhood of 0. On the other side, (compare with III.7.13)

$$\pi z \cot \pi z = 1 + \sum_{n=1}^{\infty} \frac{2z^2}{z^2 - n^2} \, .$$

Using the geometric series we see

$$\frac{1}{z^2 - n^2} = -\frac{1}{n^2} \sum_{k=0}^{\infty} \left(\frac{z^2}{n^2} \right)^k ,$$

hence

$$\pi z \cot \pi z = 1 + 2z^2 \sum_{n=1}^{\infty} \left(-\frac{1}{n^2} \sum_{k=0}^{\infty} \left(\frac{z^2}{n^2} \right)^k \right) .$$

We exchange the order of summation, to finally obtain

$$\pi z \cot \pi z = 1 - 2 \sum_{k=1}^{\infty} \left(\sum_{n=1}^{\infty} \frac{1}{n^{2k}} \right) z^{2k} \, .$$

Comparing with $(*)$ we can isolate the values $\zeta(2k)$.

Proposition III.7.14 (L. Euler, 1737) *The values of the Riemann zeta function at the even natural numbers are given by the Euler formula:*

$$\boxed{\zeta(2k) := \sum_{n=1}^{\infty} \frac{1}{n^{2k}} = \frac{(-1)^{k+1}(2\pi)^{2k}}{2(2k)!} B_{2k} \, , \quad k \in \mathbb{N} \, .}$$

Examples.

Using the values of B_{2k} listed at the end of Sect. III.2 we get

$$\zeta(2) = \frac{\pi^2}{6}, \ \zeta(4) = \frac{\pi^4}{90}, \ \zeta(6) = \frac{\pi^6}{945}, \ \zeta(8) = \frac{\pi^8}{9450} \text{ and } \zeta(10) = \frac{\pi^{10}}{3^5 \cdot 5 \cdot 7 \cdot 11} \, .$$

The values $\zeta(2n+1)$, $n \in \mathbb{N}$, are less understood. However, we know that $\zeta(3)$ is irrational, [Apé], (R. Apéry, 1978).

Exercises for III.7

1. Find the number of solutions of each of the following equations in the given domains:

$$2z^4 - 5z + 2 = 0 \qquad \text{in } \{ \, z \in \mathbb{C} \, ; \quad |z| > 1 \, \} \, ,$$
$$z^7 - 5z^4 + iz^2 - 2 = 0 \qquad \text{in } \{ \, z \in \mathbb{C} \, ; \quad |z| < 1 \, \} \, ,$$
$$z^5 + iz^3 - 4z + i = 0 \qquad \text{in } \{ \, z \in \mathbb{C} \, ; \quad 1 < |z| < 2 \, \} \, .$$

2. The polynomial $P(z) = z^4 - 5z + 1$ has
    (a) one root a with $|a| < \frac{1}{4}$.
    (b) and the other three roots in the annulus $\frac{3}{2} < |z| < \frac{15}{8}$.

3. Let $\lambda > 1$. Show that the equation $\exp(-z) + z = \lambda$ has in the right open half-plane $\{ z \in \mathbb{C} \; ; \; \operatorname{Re} z > 0 \}$ exactly one solution, which is real.

4. For $n \in \mathbb{N}_0$ define
    $$e_n(z) = \sum_{\nu=0}^{n} \frac{z^\nu}{\nu!} \,.$$
    For a given $R > 0$ there exists an n_0, such that for all $n \geq n_0$ the function e_n has no zero in $U_R(0)$.

5. Let f be analytic in an open set D containing the closed unit disk $\overline{\mathbb{E}} = \{ z \in \mathbb{C} \; ; \; |z| \leq 1 \}$. Assume $|f(z)| < 1$ for $|z| = 1$.
    For any $n \in \mathbb{N}$ the equation $f(z) = z^n$ has exactly n solutions in $\mathbb{E}$. Especially, f has exactly one fixed point in $\mathbb{E}$.

6. Let $f : D \to \mathbb{C}$ be an injective analytic function on a domain $D \subset \mathbb{C}$. Let $\overline{U}_\rho(a) \subset D$ be a closed disk in D. For $w \in f(U_\rho(a))$ prove the following explicit formula for the inverse function f^{-1} of f,
    $$f^{-1}(w) = \frac{1}{2\pi i} \int_{|\zeta - a| = \rho} \frac{\zeta f'(\zeta)}{f(\zeta) - w} \, d\zeta \,.$$

7. Let $a_1, \ldots, a_l \in \mathbb{C}$ be pairwise different non-integral numbers, Let f be an analytic function in $\mathbb{C} \setminus \{a_1, \ldots, a_l\}$, such that $|z^2 f(z)|$ is bounded outside a suitable compact set. We set
    $$g(z) := \pi \cot(\pi z) f(z) \quad \text{and} \quad h(z) := \frac{\pi}{\sin \pi z} f(z)$$
    Show:
    $$\lim_{N \to \infty} \sum_{n=-N}^{N} f(n) = -\sum_{j=1}^{l} \operatorname{Res}(g; a_j) \,,$$
    $$\lim_{N \to \infty} \sum_{n=-N}^{N} (-1)^n f(n) = -\sum_{j=1}^{l} \operatorname{Res}(h; a_j) \,.$$

8. Using exercise 7, show
    $$\sum_{n=1}^{\infty} \frac{1}{n^2} = \frac{\pi^2}{6} \quad \text{and} \quad \sum_{n=1}^{\infty} (-1)^{n+1} \frac{1}{n^2} = \frac{\pi^2}{12} \,.$$

9. Compute the integrals:
    $$\int_0^{2\pi} \frac{\cos 3t}{5 - 4\cos t} \, dt \,, \quad \int_0^{\pi} \frac{1}{(a + \cos t)^2} \, dt \,, \quad a \in \mathbb{R} \,, \ a > 1 \,.$$

10. *Show:*
    $$\int_0^{2\pi} \frac{\sin 3t}{5 - 3\cos t} \, dt = 0 \,, \quad \int_0^{2\pi} \frac{1}{(5 - 3\sin t)^2} \, dt = \frac{5\pi}{32} \,.$$

11. *Show:*

(a) $\displaystyle\int_{-\infty}^{\infty} \frac{1}{x^4+1}\, dx = \frac{\pi}{\sqrt{2}}$, (b) $\displaystyle\int_{0}^{\infty} \frac{x}{x^4+1}\, dx = \frac{\pi}{4}$,

(c) $\displaystyle\int_{0}^{\infty} \frac{x^2}{x^6+1}\, dx = \frac{\pi}{6}$, (d) $\displaystyle\int_{0}^{\infty} \frac{1}{x^4+x^2+1}\, dx = \frac{\pi}{2\sqrt{3}}$.

12. *Show:*

(a) $\displaystyle\int_{-\infty}^{\infty} \frac{x^2}{(x^2+a^2)^2}\, dx = \frac{\pi}{2a}$, $(a>0)$

(b) $\displaystyle\int_{-\infty}^{\infty} \frac{dx}{(x^2+4x+5)^2} = \frac{\pi}{2}$,

(c) $\displaystyle\int_{0}^{\infty} \frac{dx}{(x^2+a^2)(x^2+b^2)} = \frac{\pi}{2ab(a+b)}$, $(a,b>0)$.

13. *Show:*

(a) $\displaystyle\int_{-\infty}^{\infty} \frac{\cos x}{(x^2+1)^2}\, dx = \frac{\pi}{e}$,

(b) $\displaystyle\int_{-\infty}^{\infty} \frac{\cos x}{(x^2+a^2)(x^2+b^2)}\, dx = \frac{\pi}{a^2-b^2}\left(\frac{e^{-b}}{b}-\frac{e^{-a}}{a}\right)$, $(a,b>0,\ a\neq b)$,

(c) $\displaystyle\int_{0}^{\infty} \frac{\cos 2\pi x}{x^4+x^2+1}\, dx = \frac{-\pi}{2\sqrt{3}}e^{-\pi\sqrt{3}}$.

14. *Show:*

$$\int_{0}^{\infty} \frac{dx}{1+x^5} = \frac{\pi\sqrt{10}\sqrt{5+\sqrt{5}}}{25} \approx 1.069\,896\ldots \quad .$$

Hint. Let ζ be a primitive root of unity of order 5. The integrand takes on the half-lines $\{\,t\ ;\ t\geq 0\,\}$ and $\{\,t\zeta\ ;\ t\geq 0\,\}$ the same values. Compare the integrals along these half-lines.

Generalize the exponent 5 to an arbitrary odd natural exponent, and compute the integral.

15. *Show:*

$$\int_{0}^{\infty} \frac{\log^2 x}{1+x^2}\, dx = \frac{\pi^3}{8} , \qquad \int_{0}^{\infty} \frac{\log x}{1+x^2}\, dx = 0 .$$

16. *Show:*

$$\int_{0}^{\infty} \frac{x\sin x}{x^2+1}\, dx = \frac{\pi}{2e} .$$

17. Give a proof for the following formula

$$\boxed{\ \int_{-\infty}^{\infty} e^{-t^2}\, dt = \sqrt{\pi}\ ,}$$

by integrating

$$f(z) = \frac{\exp(-z^2)}{1+\exp(-2az)} \qquad \text{with } a := e^{\pi i/4}\sqrt{\pi}$$

along the parallelogram with vertices in $-R$, $-R + a$, $R + a$ and R, and by taking the limit $R \to \infty$. Also use the identity

$$f(z) - f(z + a) = \exp(-z^2) .$$

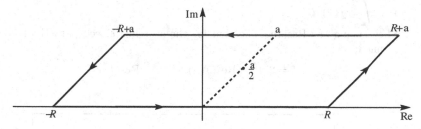

18. (MORDELL's trick) Compute in two different ways

$$\int_\alpha \frac{\exp(\, 2\pi i\, z^2/n \,)}{\exp(\, 2\pi i\, z \,) - 1}\, dz \, ,$$

in order to obtain for the GAUSS sum $G_n := \sum_{k=0}^{n-1} \exp \dfrac{2\pi i k^2}{n}$, $n \in \mathbb{N}$, the following explicit formula:

$$G_n = \frac{1 + (-i)^n}{1 + (-i)}\, \sqrt{n} \, .$$

The integral is taken along one of the curves $\alpha = \alpha(R)$ sketched in the figure.

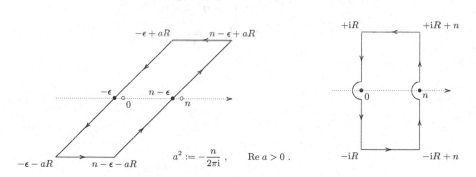

Especially we have $G_1 = 1$, $G_2 = 0$, $G_3 = i\sqrt{3}$, $G_4 = 2(1 + i)$, ...

19. Assume that the polynomials P and Q, and the number α fulfill the same properties as listed in Proposition III.7.11, excepting the fact that we allow more generally **simple** poles on the real axis $x_1 < x_2 < \cdots < x_p$ for P/Q. We consider the function $f(z) = \dfrac{P(z)}{Q(z)} \exp(i\alpha z)$, $\alpha > 0$, and the following integral for sufficiently large values of $r > 0$, and sufficiently small values of $\epsilon > 0$:

$$I(r, \epsilon) := \left(\int_{-r}^{x_1 - \epsilon} + \int_{x_1 + \epsilon}^{x_2 - \epsilon} + \cdots + \int_{x_{p-1} + \epsilon}^{x_p - \epsilon} + \int_{x_p + \epsilon}^{r} \right) f(x)\, dx \, .$$

Then the limit
$$I := \lim_{\substack{r \to \infty \\ \epsilon \to 0}} I(r, \epsilon)$$
is called the CAUCHY principal value of the integral, sometimes denoted by
P.V. $\int_{-\infty}^{\infty} f(x)\, dx$.

Show, using the Residue Theorem for the sketched closed path the following formula:
$$I = 2\pi i \sum_{a \in \mathbb{H}} \mathrm{Res}(f; a) + \pi i \sum_{j=1}^{p} \mathrm{Res}(f; x_j) \ .$$

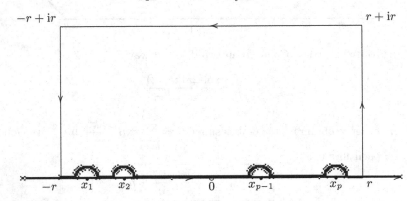

Examples.

(a) $\qquad$ P.V. $\int_{-\infty}^{\infty} \dfrac{1}{(x - i)^2 (x - 1)}\, dx = \dfrac{\pi}{2}$,

(b) $\qquad$ P.V. $\int_{-\infty}^{\infty} \dfrac{1}{x(x^2 - 1)}\, dx = 0$.

IV

Construction of Analytic Functions

In this (central) chapter, we are concerned with the *construction of analytic functions*. We will meet three different construction principles.

(1) We first study in detail a classical function using methods of the function theory of one complex variable, namely the **Gamma function**.

(2) We treat the theorems of WEIERSTRASS and MITTAG-LEFFLER for the construction of analytic functions with prescribed zeros and respectively poles with specified principle parts.

(3) We prove the RIEMANN Mapping Theorem, which claims that an elementary domain $D \neq \mathbb{C}$ can be conformally mapped onto the unit disk $\mathbb{E}$. In this context we will once more review the CAUCHY Integral Theorem, prove more general variants of it, and gain different topological characterizations of elementary domains as regions "without holes".

The zero set and the pole set of a given analytic function $f \neq 0$ are *discrete* subsets of the domain of definition of f. The following question naturally arises:

> Let S be a discrete subset of $D \subset \mathbb{C}$. Let us fix for each $s \in S$ a natural number $m(s)$. Does there exist any analytic function $f : D \to \mathbb{C}$, whose zero set $N(f)$ is exactly S, and such that for any $s \in S = N(f)$ we have $\operatorname{ord}(f; s) = m(s)$?

The answer is always yes, but we will give a proof only in the case $D = \mathbb{C}$. As a corollary, we obtain the existence of a meromorphic function with prescribed (discretely chosen) zeros and poles of given orders. Another proposition claims that for fixed prescribed poles, and fixed prescribed corresponding principal parts, there exists a meromorphic function with exactly this singularity behavior. (But the control on the zero set gets lost.) The solutions to both problems are closely connected with the names WEIERSTRASS and MITTAG-LEFFLER (the WEIERSTRASS Factorization Theorem and the MITTAG-LEFFLER Theorem). This way, we obtain new interesting examples of analytic and meromorphic functions, which are important for further applications. Moreover we also discover new alternative formulas for already known functions, and new relations between them.

Both principles of construction are already encountered in the example of the Gamma function, which is the first we want to study in this chapter.

E. Freitag and R. Busam, *Complex Analysis*,
DOI: 10.1007/978-3-540-93983-2_IV, © Springer-Verlag Berlin Heidelberg 2009

IV.1 The Gamma Function

We introduce the Gamma function as an EULERian integral of the second kind (L. EULER, 1729/30):

$$\Gamma(z) = \int_0^\infty t^{z-1} e^{-t} \, dt \ ,$$

with $t^{z-1} := e^{(z-1)\log t}$, $\log t \in \mathbb{R}$, Re $(z) > 0$.

Name and notation go back to A.M. LEGENDRE (1811).

At the beginning, we have to make some comments about improper integrals.

Remark. Let $S \subset \mathbb{C}$ be an *unbounded* set, let $l \in \mathbb{C}$, and let $f : S \to \mathbb{C}$ be a function. The terminology

$$f(s) \to l \quad (s \to \infty) \quad \text{or also} \quad \lim_{s\to\infty} f(s) = l$$

has the following meaning:

For any $\varepsilon > 0$ there exists a constant $C > 0$ with

$$|f(s) - l| < \varepsilon \quad \text{for all } s \in S \subset \mathbb{C} \text{ with } |s| > C \ .$$

In the special case of $S = \mathbb{N}$, we obtain the notion of a convergent sequence $(f(n))$. The usual rules for the manipulation of limits remain valid. We don't need to reformulate or prove them, because of

$$\lim_{s\to\infty} f(s) = \lim_{\varepsilon\to 0} f(1/\varepsilon) \ .$$

A continuous function

$$f : [a, b[\longrightarrow \mathbb{C} \ , \quad a < b \le \infty \ (b = \infty \text{ is allowed}) \ ,$$

*is called **improperly integrable**, iff the limit*

$$\int_a^b f(x) \, dx := \lim_{t\to b} \int_a^t f(x) \, dx$$

exists.

The function f is called *absolutely integrable*, iff the function $|f|$ is integrable. The absolute integrability implies integrability. More precisely:

*The continuous function $f : [a, b[\to \mathbb{C}$ is **(improperly) absolutely integrable**, iff there exists a constant $C \ge 0$ with the property*

$$\int_a^t |f(x)| \, dx \le C \text{ for all } t \in [a, b[\ .$$

This is a direct simple generalization of the corresponding proposition (in real analysis) for real valued functions, after splitting a complex function into the real and imaginary parts. We do not prove the result in real analysis, but mention that it also uses a splitting, namely into the positive and negative part, in order to reduce the assertion to non-negative functions. Then one can use a monotonicity criterion.

In full analogy, one introduces the notion of improper integrability for left open intervals, and continuous functions

$$f :]a, b] \longrightarrow \mathbb{C} , \quad -\infty \le a < b ,$$

and finally for (left and right) open intervals:

A continuous function

$$f :]a, b[\longrightarrow \mathbb{C} , \quad -\infty \le a < b \le \infty ,$$

*is called **improperly integrable**, iff for some (or any) $c \in]a, b[$ the restrictions of f to $]a, c]$, and to $[c, b[$ are both improperly integrable.*

It is clear that this condition and the definition

$$\int_a^b f(x) \, dx := \int_a^c f(x) \, dx + \int_c^b f(x) \, dx$$

are independent of c.

Proposition IV.1.1 *The Gamma integral*

$$\Gamma(z) := \int_0^\infty t^{z-1} e^{-t} \, dt$$

converges absolutely in the half-plane $\mathrm{Re}\, z > 0$, where it represents an analytic function. The derivatives of the Gamma function are f given by

$$\Gamma^{(k)}(z) = \int_0^\infty t^{z-1} (\log t)^k e^{-t} \, dt \qquad (k \in \mathbb{N}_0) .$$

Proof. We split the Γ-integral into two integrals as

$$\Gamma(z) = \int_0^1 t^{z-1} e^{-t} \, dt + \int_1^\infty t^{z-1} e^{-t} \, dt$$

and use

$$\left| t^{z-1} e^{-t} \right| = t^{x-1} e^{-t} \qquad (x = \mathrm{Re}\, z).$$

For any $x_0 > 0$ there exists a number $C > 0$ with the property

$$t^{x-1} \le C e^{t/2} \text{ for all } x \text{ with } 0 < x \le x_0 , \text{ and for all } t \ge 1 .$$

This ensures the absolute convergence of

$$\int_1^\infty t^{z-1} e^{-t}\, dt$$

for all $z \in \mathbb{C}$.

For the convergence of the first integral we use the estimate

$$\left| t^{z-1} e^{-t} \right| < t^{x-1} \quad \text{for} \ \ t > 0$$

and the existence of

$$\int_0^1 \frac{1}{t^s}\, dt \qquad (s < 1)\,.$$

Moreover, this estimate shows that the sequence of functions

$$f_n(z) := \int_{1/n}^n t^{z-1} e^{-t}\, dt$$

converges locally uniformly to Γ. Hence Γ is an analytic function. (The same argument shows that the integral from 1 (instead of 0) to ∞ is an entire function.) The formulas for the derivatives of Γ are obtained by applying the LEIBNIZ rule to the functions f_n, see also II.3.3, followed by passing to the limit $n \to \infty$. $\qquad\square$

Obviously

$$\Gamma(1) = \int_0^\infty e^{-t}\, dt = -e^{-t}\Big|_0^\infty = 1\,.$$

By partial integration ($u(t) = t^z$, $v'(t) = e^{-t}$), we obtain the *functional equation*

$$\Gamma(z+1) = z\,\Gamma(z) \quad \text{for} \ \ \operatorname{Re} z > 0\,.$$

Especially, for $n \in \mathbb{N}_0$

$$\boxed{\ \Gamma(n+1) = n!\,. \ }$$

The Γ-function "interpolates" the factorials.

An iterated application of the functional equation gives

$$\Gamma(z) = \frac{\Gamma(z+n+1)}{z\cdot(z+1)\cdots(z+n)}\,.$$

The R.H.S. has a bigger domain of definition than the L.H.S.! It gives thus an analytic continuation to the set of all $z \in \mathbb{C}$ which satisfy

$$\operatorname{Re} z > -(n+1) \ \text{and} \ z \neq 0\,,\ -1\,,\ -2\,,\ \dots\,,\ -n\,.$$

All these analytic continuations (obtained for various n), are unique by III.3.2, so they glue to a function that we also denote by Γ.

We conclude by collecting all previous properties of the Gamma function:

Proposition IV.1.2 *The Γ-function can be uniquely extended as an analytic function to $\mathbb{C} \setminus S$, where S is the set*

$$S := \{\, 0,\, -1,\, -2,\, -3,\, \ldots \,\} \,,$$

and it satisfies for $z \in \mathbb{C} \setminus S$ the functional equation

$$\boxed{\Gamma(z+1) = z\,\Gamma(z)\,.}$$

The elements of S are all simple poles with corresponding residues given by

$$\boxed{\operatorname{Res}(\Gamma; -n) = \frac{(-1)^n}{n!}\,.}$$

The Γ-function is a meromorphic function in $\mathbb{C}$ with pole set S.

Proof. It remains to compute the residues:

$$\operatorname{Res}(\Gamma; -n) = \lim_{z \to -n} (z+n)\,\Gamma(z) = \frac{\Gamma(1)}{(-n)(-n+1)\cdots(-1)} = \frac{(-1)^n}{n!}\,. \qquad \square$$

Because of the estimate

$$|\Gamma(z)| \le \Gamma(x) \ \text{ for } x > 0 \ \ (x = \operatorname{Re} z)\,,$$

the Γ-function is bounded in any closed vertical strip of the form

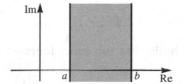

$$0 < a \le x \le b\,.$$

These properties of the Γ-function already determine it:

Proposition IV.1.3 (Characterization of the Γ-function, H. Wielandt, 1939) *Let $D \subset \mathbb{C}$ be a domain containing the vertical strip $V \subset \mathbb{C}$*

$$V := \{\, z = x + iy \,;\ x, y \in \mathbb{R}\,,\ 1 \le x < 2 \,\}\,.$$

Let $f : D \to \mathbb{C}$ be an analytic function with the following properties:

(1) *f is bounded in this vertical strip.*

(2) *The following functional equation is satisfied:*

$$f(z+1) = z\,f(z) \ \text{ for } z,\ z+1 \in D\,.$$

Then

$$f(z) = f(1)\,\Gamma(z) \ \text{ for } z \in D\,.$$

Proof. Using the functional equation, we are able —in full analogy with the argument for the analytic continuation of Γ— to extend f to an analytic function $\mathbb{C} \setminus S$, also denoted by f. Once more we have to avoid the points

$$z \in S = \{\, 0\,,\ -1\,,\ -2\,,\ -3,\ \ldots \,\}\,,$$

and the extension fulfills the functional equation ($z \in \mathbb{C} \setminus S$)

$$f(z+1) = z\,f(z) \ .$$

At the points of S it has simple poles, or removable singularities, and

$$\operatorname{Res}(f; -n) = \frac{(-1)^n}{n!} f(1) \ .$$

The function $h(z) := f(z) - f(1)\,\Gamma(z)$ has thus in S removable singularities, hence it is an *entire* function. The function h is bounded in the vertical strip V. One can use the functional equation to enlarge the strip to the left, and prove the boundedness of h in strips of the form

$$a \le x \le b < 2,$$

first under the additional condition $|\operatorname{Im}\ z| \ge 1$. But the set $|\operatorname{Im}\ z| \le 1$, $a \le \operatorname{Re} z \le b$ is compact.

Unfortunately this is not enough to apply LIOUVILLE's Theorem. But one can use a trick: From the functional equation

$$h(z+1) = z\,h(z)\ , \quad (h(z) = f(z) - f(1)\,\Gamma(z))\ ,$$

we obtain that the *entire* function $H : \mathbb{C} \to \mathbb{C}$, defined by

$$H(z) := h(z)\,h(1-z)$$

is periodic up to the sign:

$$H(z+1) = -H(z).$$

The function H is bounded in the vertical strip $\{\ z \in \mathbb{C}\ ;\ 0 \le \operatorname{Re} z < 1\ \}$ since this strip is invariant under $z \mapsto 1 - z$. Therefore, H is bounded in $\mathbb{C}$. Now LIOUVILLE's Theorem implies that $H(z)$ is constant. Because of $h(1) = 0$ this constant is zero. It follows $h = 0$. □

We would like now to develop a product formula for the Γ-function. First we have to establish some fundamental facts about *infinite products*. We will reduce the study of infinite products to the study of infinite series, by using the complex logarithm. In principle we want to define

$$\prod_{n=1}^{\infty} b_n := \exp \sum_{n=1}^{\infty} \log b_n \ .$$

But this has to be done with caution. First, some factors could vanish, and secondly because of the problematic of the complex logarithm. We will assume from the beginning that the sequence b_n converges to 1, ("just as" the sequence of the summands of a convergent series converges to 0). We then write

$$b_n = 1 + a_n \ ,$$

and (a_n) is a null sequence. Then there exists a natural number N with

$$|a_n| < 1 \quad \text{for} \quad n > N \ .$$

We can now set

$$\prod_{n=1}^{\infty} b_n := \prod_{n=1}^{N} b_n \cdot \exp \left(\sum_{n=N+1}^{\infty} \text{Log}(1 + a_n) \right) \ ,$$

where Log is the principal value of the logarithm. It is given in the domain $(|z| < 1)$ by the series

$$\text{Log}(1 + z) = - \sum_{n=1}^{\infty} (-1)^n \frac{z^n}{n} \ .$$

We call the infinite product (absolutely) convergent, if the corresponding series is *absolutely* convergent. (We do not want to deal with non-absolutely convergent products.) For sufficiently small values of $|z|$, e.g. $|z| \leq 1/2$, we have the inequalities

$$\frac{1}{2} |z| \ \leq \ |\text{Log}(1 + z)| \ \leq \ 2 |z| \ .$$

The absolute convergence of the series of logarithms is thus equivalent to the convergence of the series

$$\sum_{n=1}^{\infty} |a_n| \ .$$

Conversely, this condition implies that the sequence (b_n) converges to 1. It has the advantage that there is no need of N. We finally define:

Definition IV.1.4 *The infinite product*

$$(1 + a_1)(1 + a_2)(1 + a_3) \cdots$$

converges **absolutely,** *iff the series*

$$|a_1| + |a_2| + |a_3| + \cdots$$

converges.

The preliminary discussion has led us to the following

Lemma IV.1.5 *If the series*

$$a_1 + a_2 + a_3 + \cdots$$

is absolutely convergent, then there exists a natural number N with $|a_\nu| < 1$ for $\nu > N$, and such that we have

(a) $\displaystyle\sum_{n=N+1}^{\infty} \mathrm{Log}(1+a_n)$ *converges absolutely.*

(b) $\displaystyle\lim_{n\to\infty} \prod_{\nu=1}^{n}(1+a_\nu) = (1+a_1)\cdots(1+a_N)\exp\left(\sum_{n=N+1}^{\infty}\mathrm{Log}(1+a_n)\right).$

The limit in (b) is independent of the chosen N. We call it the *value* of the infinite product, and denote it by

$$\prod_{n=1}^{\infty}(1+a_n)\ .$$

From IV.1.5 (b) we also deduce:

Remark IV.1.6 *The value of the absolutely convergent (infinite) product*

$$(1+a_1)(1+a_2)(1+a_3)\ldots$$

is not equal to zero, iff all factors $(1+a_n)$ are not equal to zero.

Caution: The product

$$\lim_{n\to\infty}\prod_{\nu=1}^{n}\frac{1}{\nu} = \lim_{n\to\infty}\frac{1}{n!} = 0$$

is not an absolutely convergent infinite product in our sense since its factors don't approach 1.

Remark IV.1.7 *Let*

$$f_1 + f_2 + f_3 + \cdots$$

be a normally convergent series of analytic functions defined on an open set $D \subset \mathbb{C}$. Then the infinite product

$$(1+f_1)(1+f_2)(1+f_3)\cdots$$

defines an analytic function $F : D \to \mathbb{C}$.

In addition: *The zero set $N(F)$ of F is the union of the zero sets of the functions $1 + f_n(z)$, $n \in \mathbb{N}$. If F is not identically zero, then for $z \notin N(F)$*

$$\frac{F'(z)}{F(z)} = \sum_{n=1}^{\infty}\frac{f_n'(z)}{1+f_n(z)}\ ,$$

and the above series converges normally in the complement of $N(F)$.

Terminology. The infinite product $(1+f_1)(1+f_2)(1+f_3)\cdots$ is called **normally convergent**, iff the corresponding series $f_1 + f_2 + f_3 + \cdots$ is normally convergent.

Proof. In any given compact set we have $|f_n(z)| \leq 1/2$ for all but finitely many n. Lemma IV.1.5 then shows the normal convergence of the series underlying our infinite product. The additional result follows from WEIERSTRASS' Theorem after termwise differentiation.   □

After this excursion to infinite products, we come back to the Γ-function. The function $1/\Gamma$ has zeros in

$$z = 0, -1, -2, -3, \ldots \quad .$$

One may speculate that it is related to the infinite product

$$\left(1 + \frac{z}{1}\right)\left(1 + \frac{z}{2}\right)\left(1 + \frac{z}{3}\right) \cdots \quad ,$$

but this doesn't converge absolutely. But we have

Lemma IV.1.8 *The series*

$$\sum_{n=1}^{\infty} \left[\left(1 + \frac{z}{n}\right) \cdot e^{-\frac{z}{n}} - 1\right]$$

converges normally in the entire complex plane $\mathbb{C}$.

Corollary. *The infinite product*

$$H(z) := \prod_{n=1}^{\infty} \left(1 + \frac{z}{n}\right) \cdot e^{-\frac{z}{n}}$$

defines an entire function H with the property

$$H(z) = 0 \iff -z \in \mathbb{N} .$$

Proof. Considering the TAYLOR expansion we get

$$(1 + w)e^{-w} - 1 = -\frac{w^2}{2} + \text{higher order terms} .$$

Hence, for any compact set $K \subset \mathbb{C}$ there exists a constant $C = C_K$ with the property

$$|(1 + w)e^{-w} - 1| \leq C |w|^2 \quad \text{for all } w \in K .$$

The series in IV.1.8 is thus up to a constant factor dominated by

$$\sum_{n=1}^{\infty} \frac{1}{n^2} .$$

□

There is one further important way to rewrite the infinite product $H(z)$, which goes back to GAUSS (1811) but was already known to EULER (1729, 1776). From real analysis we recall the well-known existence of the limit

$$\gamma := \lim_{n\to\infty} \left(1 + \frac{1}{2} + \cdots + \frac{1}{n} - \log n \right) (\approx 0.577\,215\,664\,901\,532\,860\,606\,512\ldots),$$

the EULER-MASCHERONI Constant (see also Exercise 3 in Sect. IV.1).

Lemma IV.1.9 *Let*

$$G_n(z) = ze^{-z\log n} \prod_{\nu=1}^{n} \left(1 + \frac{z}{\nu} \right) .$$

Then

$$\lim_{n\to\infty} G_n(z) = ze^{\gamma z} H(z) .$$

Corollary. *The function*

$$G(z) = \lim_{n\to\infty} G_n(z)$$

is analytic in the entire $\mathbb{C}$*, and has simple zeros exactly in the elements of the set*

$$S = \{\, 0, -1, -2, \ldots \,\} .$$

Proof. We rewrite

$$G_n(z) = ze^{z(1+\cdots+1/n-\log n)} \prod_{\nu=1}^{n} \left(1 + \frac{z}{\nu} \right) e^{-z/\nu} . \qquad \square$$

Proposition IV.1.10 (Gauss' Product Representation, C.F. Gauss, 1811-12) *For all* $z \in \mathbb{C}$

$$\boxed{\frac{1}{\Gamma(z)} = G(z) = \lim_{n\to\infty} \frac{n^{-z}}{n!}\, z(z+1)\cdots(z+n) .}$$

Corollary. *The Gamma function has no zeros.*

Proof. We check the characterizing properties of the Gamma function, IV.1.3, for the function $1/G$. First, we observe that $1/G$ is analytic in a domain containing the vertical strip $1 \leq x < 2$, $x := \operatorname{Re} z$.

(1) $1/G(z)$ *is bounded in this vertical strip.*

This is because

$$\left| n^{-z} \right| = n^{-x} ,$$

and

$$|z + \nu| \geq x + \nu .$$

(2) *Functional equation.*

A trivial computation shows

$$zG_n(z+1) = \frac{z+n+1}{n} G_n(z) .$$

(3) *Normalization.*

$$G_n(1) = 1 + \frac{1}{n} \text{ for all } n .$$ □

As already pointed out during the proof of IV.1.3 (for the function h), it is useful to associate to Γ the function

$$f(z) := \Gamma(z)\,\Gamma(1-z) .$$

It is periodic up to sign,

$$f(z+1) = -f(z) ,$$

having thus the period 2. It has simple poles at all integers, and the corresponding residues are

$$\text{Res}(f; -n) = \lim_{z \to -n} (z+n)\,\Gamma(z)\,\Gamma(1-z) = (-1)^n .$$

The same properties are are shared by the function

$$\frac{\pi}{\sin \pi z} .$$

Proposition IV.1.11 (L. Euler, 1749) *For all $z \in \mathbb{C} - \mathbb{Z}$ we have:*

> **Completion Formula**
>
> $$\Gamma(z)\,\Gamma(1-z) = \frac{\pi}{\sin \pi z} .$$

First Consequence. $\Gamma(1/2) = \sqrt{\pi}$, *and in general we have:*

$$\Gamma\left(n + \frac{1}{2}\right) = \sqrt{\pi} \prod_{k=0}^{n-1} \left(k + \frac{1}{2}\right) , n \in \mathbb{N}_0 .$$

Second Consequence.

$$\frac{\sin \pi z}{\pi} = z \prod_{n=1}^{\infty} \left(1 - \frac{z^2}{n^2}\right) \quad \text{(absolutely convergent product) .}$$

Proof. The function

$$h(z) := \Gamma(z)\,\Gamma(1-z) - \frac{\pi}{\sin \pi z}$$

is bounded on the set

$$0 \le x \le 1 , \quad |y| \ge 1 ,$$

and has at $z = n$, for $n \in \mathbb{Z}$, a removable singularity. It is thus an entire function. The rectangle

$$0 \le x \le 1 \,, \quad |y| \le 1 \,,$$

is compact, therefore, h is bounded in the strip $0 \le x \le 1$. Being periodic up to sign, h is bounded in $\mathbb{C}$. By the LIOUVILLE Theorem, h is a constant. The constant has to be zero because of

$$h(z) = -h(-z) \,.$$

The first consequence is trivial. The second consequence follows from the infinite product representations of $1/\Gamma(z)$.   □

Remark. From the product expansion for the sine, we get a new proof for the partial fraction decomposition of the cotangent, III.7.13, just using

$$\frac{\sin' z}{\sin z} = \cot z$$

to conclude by the additional part of Remark IV.1.7,

$$\pi \cot \pi z = \frac{1}{z} + \sum_{\substack{n \in \mathbb{Z} \\ n \neq 0}} \left[\frac{1}{z-n} + \frac{1}{n} \right] \,.$$

Proposition IV.1.12 (The Legendre Relation, A.M. Legendre, 1811)

> **Duplication Formula**
> $$\Gamma\left(\frac{z}{2}\right) \Gamma\left(\frac{z+1}{2}\right) = \frac{\sqrt{\pi}}{2^{z-1}} \Gamma(z) \,.$$

Proof. The function

$$f(z) = 2^{z-1} \Gamma\left(\frac{z}{2}\right) \Gamma\left(\frac{z+1}{2}\right)$$

has the characterizing properties of the Gamma function. The normalizing constant

$$f(1) = \sqrt{\pi}$$

is determined by IV.1.11.   ⊓

To conclude this section, we generalize the classical STIRLING Formula (J. STIRLING, 1730),

$$1 \le \frac{n!}{\sqrt{2\pi}\, n^{n+\frac{1}{2}} e^{-n}} \le e^{\frac{1}{12n}}$$

from the factorials to the Gamma function, and prove it using complex analysis.

We denote by $\operatorname{Log} z$ the principal branch of the logarithm on the slit plane $\mathbb{C}_-$. The function

$$z^{z-\frac{1}{2}} := e^{(z-\frac{1}{2})\operatorname{Log} z}$$

is analytic on this domain.

We now look for an analytic function H, defined on the slit plane $\mathbb{C}_-$, such that the associated function

$$h(z) = z^{z-\frac{1}{2}}e^{-z}e^{H(z)}$$

possesses the characterizing properties of the Gamma function, i.e. we want to prove the generalization of STIRLING's formula from an *identity*

$$\Gamma(z) = A \cdot h(z), \quad A \in \mathbb{C}.$$

The functional equation $h(z+1) = z\, h(z)$ is of course satisfied, when we have

$$H(z) - H(z+1) = H_0(z)$$

with

$$H_0(z) = \left(z + \frac{1}{2}\right)\left[\operatorname{Log}(z+1) - \operatorname{Log} z\right] - 1 .$$

The natural candidate of such a function is given by the series

$$H(z) := \sum_{n=0}^{\infty} H_0(z+n) ,$$

provided it converges (GUDERMANN's series, C. GUDERMANN, 1845).

Lemma IV.1.13 *We have*

$$|H_0(z)| \le \frac{1}{2}\left|\frac{1}{2z+1}\right|^2 \ \textit{for } z \in \mathbb{C}_- , \ \left|z + \frac{1}{2}\right| > 1 .$$

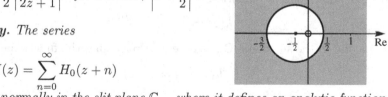

Corollary. *The series*

$$H(z) = \sum_{n=0}^{\infty} H_0(z+n)$$

converges normally in the slit plane $\mathbb{C}_-$, where it defines an analytic function. In any angular domain

$$W_\delta := \left\{\, z = |z|\, e^{i\varphi} \, ; \, -\pi+\delta \le \varphi \le \pi-\delta \,\right\}$$

with $0 < \delta \le \pi$ we have

$$\lim_{\substack{z \to \infty \\ z \in W_\delta}} H(z) = 0 .$$

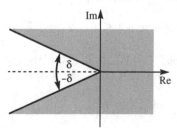

Proof. In the region Re $z > 1$ we have the identity

$$H_0(z) = \left(z + \frac{1}{2}\right)\left[\text{Log}(1+z) - \text{Log}\,z\right] - 1 = \frac{1}{2w}\,\text{Log}\,\frac{1+w}{1-w} - 1$$

with $\quad w := \dfrac{1}{2z+1}$.

It is an identity of analytic functions, for in the given region, the numbers z, $1+z$, $1+1/z$ and w are not negatively real. It is thus sufficient to prove it only for real positive numbers, but this is clear.

In the given region we have $|w| < 1$, and a simple computation leads to the TAYLOR expansion

$$H_0(z) = \frac{w^2}{3} + \frac{w^4}{5} + \frac{w^6}{7} + \cdots \quad .$$

By means of the geometric series we get for $|w| \leq 1/2$ the estimate

$$|H_0(z)| \leq \frac{4}{9}|w|^2 \leq \frac{1}{2}|w|^2 \; .$$

Proof of the Corollary. The normal convergence of the series defining $H(z)$ in the slit plane is an obvious direct consequence of the Lemma IV.1.13. We still have to show that $H(z)$ converges to 0 in angular domains. In any angular domain W_δ we have the estimate

$$|H_0(z+n)| \leq \frac{C(\delta)}{n^2} \; , \quad n \geq N(\delta) \; ,$$

where $C(\delta)$ and $N(\delta)$ depend only on δ. For a given $\varepsilon > 0$ we can then enlarge $N(\delta)$ to also ensure $|\sum_{n>N(\delta)} \frac{C(\delta)}{n^2}| < \varepsilon$.

Each of the finitely many summands corresponding to $n \leq N(\delta)$ converges to 0. We obtain, as claimed, that $H(z)$ converges to 0 in W_δ.   □

Now we know that the function

$$h(z) := z^{z-\frac{1}{2}} e^{-z} e^{H(z)}$$

is analytic in the slit plane $\mathbb{C}_-$, where it also satisfies the functional equation

$$h(z+1) = z\,h(z) \; .$$

Claim. The function h is bounded in the vertical strip

$$\{\, x + iy \; ; \; 1 \leq x \leq 2,\, y \in \mathbb{R} \,\} :$$

Proof. (1) We first show that the function

$$z^{z-\frac{1}{2}} = \exp\left(\left(z - \frac{1}{2}\right)\text{Log}\,z\right)$$

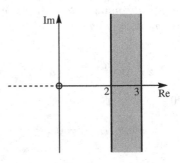

is bounded in the vertical strip

$$\{\, (x,y)\,;\quad a \le x \le b\,,\; y \in \mathbb{R}\,\}\,,\quad 0 < a < b\,.$$

For this we show that the expression

$$\operatorname{Re}\left(\left(z - \frac{1}{2}\right)\operatorname{Log} z\right) = \left(x - \frac{1}{2}\right)\operatorname{Log}|z| - y\operatorname{Arg} z$$

is bounded from above. For $y = \operatorname{Im} z \to \pm\infty$ we have correspondingly

$$\operatorname{Arg} z \to \pm\frac{\pi}{2}\,,$$

which easily implies

$$\operatorname{Re}\left(\left(z - \frac{1}{2}\right)\operatorname{Log} z\right) \to -\infty \qquad \text{for } |y| \to \infty \text{ because of}$$

$$\left(\frac{|y|}{\log|z|}\right)^{-1} \to 0 \qquad \text{for } |y| \to \infty\,.$$

(2) From IV.1.13 we see that $H(z)$, and hence also $\exp\bigl(H(z)\bigr)$ is bounded in the strip $1 \le x \le 2$.

We have checked the characterizing properties of the Gamma function, and obtain $\Gamma(z) = A \cdot h(z)$. The normalizing factor can be determined by means of LEGENDRE's Relation IV.1.12:

$$\sqrt{\pi} = A\left(1 + \frac{1}{n}\right)^{n/2} \cdot 2^{-\frac{1}{2}} \cdot \exp\left(-\frac{1}{2} + H\left(\frac{n}{2}\right) + H\left(\frac{n+1}{2}\right) - H(n)\right)\,.$$

For $x \to \infty$, the convergence of $H(x) \to 0$, and of

$$\left(1 + \frac{1}{x}\right)^{x/2} \to \sqrt{e}\,,$$

lead to $A = \sqrt{2\pi}$.   □

Proposition IV.1.14 (The Stirling Formula) *Let H be the function*

$$H(z) = \sum_{n=0}^{\infty}\left(\left(z + n + \frac{1}{2}\right)\cdot\operatorname{Log}\left(1 + \frac{1}{z+n}\right) - 1\right)\,.$$

Then for all $z \in \mathbb{C}_-$

$$\boxed{\Gamma(z) = \sqrt{2\pi}\; z^{z - \frac{1}{2}}\, e^{-z}\, e^{H(z)}\,.}$$

In any angular domain $W(\delta)$, $0 < \delta \le \pi$, we have $H(z) \to 0$ for $z \to \infty$.

The ordinary Stirling Formula

From the estimates for $H_0(z)$ we obtain for any positive real x

$$0 < H_0(x) < \frac{1}{12x(x+1)} = \frac{1}{12}\left(\frac{1}{x} - \frac{1}{x+1}\right),$$

hence

$$0 < H(x) < \frac{1}{12}\sum_{n=0}^{\infty}\left(\frac{1}{x+n} - \frac{1}{x+n+1}\right) = \frac{1}{12x},$$

and thus finally

$$H(x) = \frac{\theta}{12x} \text{ with } 0 < \theta = \theta(x) < 1 \text{ for } x > 0.$$

From $n! = n\,\Gamma(n)$ follows

$$n! = \sqrt{2\pi n}\,\left(\frac{n}{e}\right)^n e^{\frac{\theta(n)}{12n}}, \quad 0 < \theta(n) < 1.$$

This is the ordinary STIRLING Formula for $n!$.

Exercises for IV.1

1.   Which of the following products are absolutely convergent? Find the corresponding values, when they exist.

   (a) $\displaystyle\prod_{\nu=2}^{\infty}\left(1 - \frac{1}{\nu}\right),$ (b) $\displaystyle\prod_{\nu=2}^{\infty}\left(1 - \frac{1}{\nu^2}\right),$

   (c) $\displaystyle\prod_{\nu=2}^{\infty}\left(1 - \frac{2}{\nu(\nu+1)}\right),$ (d) $\displaystyle\prod_{\nu=2}^{\infty}\left(1 - \frac{2}{\nu^3+1}\right).$

2.   The product $\prod_{\nu=0}^{\infty}\left(1 + z^{2^{\nu}}\right)$ is absolutely convergent, *iff* $|z| < 1$. If this is the case, then

$$\prod_{\nu=0}^{\infty}\left(1 + z^{2^{\nu}}\right) = \frac{1}{1-z}.$$

3.   Show that the sequence (γ_n) defined by

$$\gamma_n := 1 + \frac{1}{2} + \frac{1}{3} + \cdots + \frac{1}{n} - \log n$$

is (strictly) decreasing, and bounded from below by 0. Hence the following limit exists:

$$\gamma := \lim_{n\to\infty} \gamma_n \approx 0.\,577\,215\,664\,901\,532\,860\,606\,512\,090\,082\,402\,431\,042\,159\ldots$$

(*The Euler-Mascheroni Constant*).

4. Show that the EULER Product Formula for $1/\Gamma$ can be obtained from GAUSS' representation of Γ, and conversely. For this, recall:

$$\Gamma(z) = \lim_{n\to\infty} \frac{n!\, n^z}{z(z+1)\dots(z+n)} \qquad \text{(C.F. GAUSS)}$$

$$\frac{1}{\Gamma(z)} = z e^{\gamma z} \prod_{n=1}^{\infty} \left(1 + \frac{z}{n}\right) e^{-z/n} \qquad \text{(L. EULER)}$$

5. Show for $z \in \mathbb{C} \setminus S$, $S := \{\, 0\,,\ -1\,,\ -2\,,\ -3\,,\ \dots\,\}$

$$\lim_{n\to\infty} \frac{\Gamma(z+n)}{n^z \Gamma(n)} = 1\,.$$

6. **A further characterization of Γ** : Let $f : \mathbb{C} \setminus S \to \mathbb{C}$ (S as in Exercise 5) be an analytic function with the following properties:

$$\text{(a)} \quad f(z+1) = z\, f(z) \quad \text{and} \quad \text{(b)} \quad \lim_{n\to\infty} \frac{f(z+n)}{n^z\, f(n)} = 1\,.$$

Then $f(z) = f(1)\Gamma(z)$ for all $z \in \mathbb{C} - S$.

7. *Show:*

$$\Gamma\left(\frac{1}{6}\right) = 2^{-1/3} \left(\frac{3}{\pi}\right)^{1/2} \Gamma\left(\frac{1}{3}\right)^2\,.$$

8. *Show:*

$$|\Gamma(iy)|^2 = \frac{\pi}{y \sinh \pi y}\,, \qquad \left|\Gamma\left(\frac{1}{2} + iy\right)\right|^2 = \frac{\pi}{\cosh \pi y}\,.$$

9. An alternative proof of the *Duplication Formula*. The expressions $\Gamma(z)\Gamma\left(z + \frac{1}{2}\right)$ and $\Gamma(2z)$ define two meromorphic functions with the same poles, which are simple. Hence there exists an analytic entire function $g : \mathbb{C} \to \mathbb{C}$ with

$$\Gamma(z)\Gamma\left(z + \frac{1}{2}\right) = \exp(g(z))\, \Gamma(2z)\,.$$

Show that g is a polynomial of degree ≤ 1, and deduce

$$\Gamma(z)\Gamma\left(z + \frac{1}{2}\right) = 2^{1-2z} \sqrt{\pi}\, \Gamma(2z)\,.$$

10. A characterization of Γ using the *Duplication Formula*. Let $f : \mathbb{C} \to \overline{\mathbb{C}}$ be a meromorphic function, and assume $f(x) > 0$ for all $x > 0$. Also assume

$$f(z+1) = z\, f(z) \quad \text{and} \quad \sqrt{\pi} f(2z) = 2^{2z-1} f(z) f\left(z + \frac{1}{2}\right)\,.$$

Then $f(z) = \Gamma(z)$ for all $z \in \mathbb{C}$. For the proof, use the following auxiliary result:

If $g : \mathbb{C} \to \mathbb{C}$ is an analytic function which satisfies $g(z+1) = g(z)$, $g(2z) = g(z)g\left(z + \frac{1}{2}\right)$ for all $z \in \mathbb{C}$, and $g(x) > 0$ for all $x > 0$, then $g(z) = ae^{bz}$ with suitable constants a and b.

11. The GAUSS Multiplication Formula. For any $p \in \mathbb{N}$

$$\Gamma\left(\frac{z}{p}\right)\Gamma\left(\frac{z+1}{p}\right)\cdots\Gamma\left(\frac{z+p-1}{p}\right) = (2\pi)^{\frac{p-1}{2}}\, p^{\frac{1}{2}-z}\, \Gamma(z) .$$

Hint. Prove for

$$f(z) := (2\pi)^{\frac{1-p}{2}} p^{z-\frac{1}{2}} \Gamma\left(\frac{z}{p}\right)\Gamma\left(\frac{z+1}{p}\right)\cdots\Gamma\left(\frac{z+p-1}{p}\right)$$

the characterizing properties of Γ.

12. The EULER beta function. Let $D \subset \mathbb{C}$ be the half-plane Re $z > 0$.
For $z, w \in D$ let

$$B(z,w) := \int_0^1 t^{z-1}(1-t)^{w-1}\, dt .$$

B is called EULER's beta function. (Following A.M. LEGENDRE, 1811, it is
Euler's integral of the first kind.)
Show:
(a) B is continuous (as a function of two variables).
(b) For any fixed $w \in D$ the map $D \to \mathbb{C}$, $z \mapsto B(z,w)$, is analytic.
     For any fixed $z \in D$ the map $D \to \mathbb{C}$, $w \mapsto B(z,w)$, is analytic.
(c) For all $z, w \in D$

$$B(z+1,w) = \frac{z}{z+w} \cdot B(z,w) , \qquad B(1,w) = \frac{1}{w} .$$

(d) The function

$$f(z) := \frac{B(z,w)\Gamma(z+w)}{\Gamma(w)}$$

has the characterizing properties of Γ. We thus have for Re $z > 0$ and
Re $w > 0$:

$$\boxed{\; B(z,w) = \frac{\Gamma(z)\Gamma(w)}{\Gamma(z+w)} \;}$$

We can thus reduce the study of the beta function to the study of the
Gamma function.
(e) $B(z,w) = \displaystyle\int_0^\infty \frac{t^{z-1}}{(1+t)^{z+w}}\, dt .$
(f) $B(z,w) = 2\displaystyle\int_0^{\pi/2} (\sin\varphi)^{2z-1}(\cos\varphi)^{2w-1}\, d\varphi .$

13. If μ_n is the volume of the n-dimensional unit ball in $\mathbb{R}^n$, then

$$\mu_n = 2\mu_{n-1}\int_0^1 \left(1-t^2\right)^{\frac{n-1}{2}}\, dt = \frac{\pi^{n/2}}{\Gamma\left(\frac{n}{2}+1\right)} .$$

14. The GAUSS ψ-function is defined by $\psi(z) := \Gamma'(z)/\Gamma(z) .$
Show:

(a) ψ is meromorphic in $\mathbb{C}$ with simple poles in $S := \{\, -n\; ;\quad n \in \mathbb{N}_0 \,\}$ and
     Res $(\psi; -n) = -1$.

(b) $\psi(1) = -\gamma$. (γ is the EULER-MASCHERONI Constant).

(c) $\psi(z+1) - \psi(z) = \dfrac{1}{z}$.

(d) $\psi(1-z) - \psi(z) = \pi \cot \pi z$.

(e) $\psi(z) = -\gamma - \dfrac{1}{z} - \displaystyle\sum_{\nu=1}^{\infty} \left(\dfrac{1}{z+\nu} - \dfrac{1}{\nu} \right)$.

(f) $\psi'(z) = \displaystyle\sum_{\nu=0}^{\infty} \dfrac{1}{(z+\nu)^2}$, where the series normally converges in $\mathbb{C}$.

(g) For any positive x

$$(\log \Gamma)''(x) = \sum_{\nu=0}^{\infty} \frac{1}{(x+\nu)^2} > 0 ,$$

the real Γ-function is thus logarithmic convex.

15. **The Bohr-Mollerup Theorem** (H. BOHR, J. MOLLERUP, 1922). Let $f : \mathbb{R}_+^{\bullet} \to \mathbb{R}_+^{\bullet}$ be a function with the following properties:
(a) $f(x+1) = xf(x)$ for all $x > 0$ and   (b) $\log f$ is convex.
Then $f(x) = f(1)\Gamma(x)$ for all $x > 0$.

16. For $\alpha \in \mathbb{C}$, and $n \in \mathbb{N}$ let

$$\binom{\alpha}{n} := \frac{\alpha(\alpha-1)\cdots(\alpha-n-1)}{n!} , \qquad \binom{\alpha}{0} := 1 .$$

Show that for all $\alpha \in \mathbb{C} \setminus \mathbb{N}_0$

$$\binom{\alpha}{n} = \frac{(-1)^n \, \Gamma(n-\alpha)}{\Gamma(-\alpha)\Gamma(n+1)} \sim \frac{(-1)^n}{\Gamma(-\alpha)} n^{-\alpha-1} \quad \text{for } n \to \infty ,$$

i.e. the quotient of the expressions on both sides converges to 1 for $n \to \infty$.

17. *The Hankel Integral Representation* for $\dfrac{1}{\Gamma}$ (H. HANKEL, 1864). For any $z \in \mathbb{C}$

$$\frac{1}{\Gamma(z)} = \frac{1}{2\pi i} \int_{\gamma_{r,\varepsilon}} w^{-z} \exp(w) \, dw ,$$

where $\gamma_{r,\varepsilon}$ is the HANKEL contour sketched in the figure. (This path is called in German *"uneigentlicher Schleifenweg"*.)

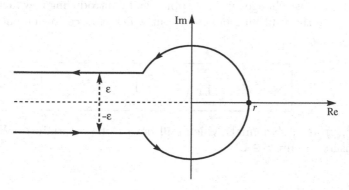

IV.2 The Weierstrass Product Formula

We consider the following *problem:*

Let there be given a domain $D \subset \mathbb{C}$, and a discrete subset S in D. For any $s \in S$ let us fix a natural number $m_s (\geq 1)$.

Does there exist an analytic function $f : D \to \mathbb{C}$ with the following properties:

(a) $f(z) = 0 \iff z \in S$, and
(b) $\mathrm{ord}(f; s) = m_s$ for each $s \in S$?

One can indeed construct such functions using WEIERSTRASS *products*. We will restrict for simplicity to the case $D = \mathbb{C}$.

Because the closed disks in $\mathbb{C}$ are compact, there exist only finitely many $s \in S$ with $|s| \leq N$. We can thus count the elements of the set S, and we can sort them to have increasing absolute values

$$S = \{ s_1, s_2, \dots \} ,$$
$$|s_1| \leq |s_2| \leq |s_3| \leq \cdots .$$

If S is a finite set, a solution to our problem is:

$$\prod_{s \in S} (z - s)^{m_s} .$$

For an infinite S, this product will in general not converge. We can and will assume, that the zero point is *not* contained in S, because we can multiply, a posteriori, with z^m.

The advantage is that we can consider the infinite product

$$\prod_{n-1}^{\infty} \left(1 - \frac{z}{s_n} \right)^{m_n} , \quad m_n := m_{s_n} ,$$

which has some better chances to converge.

Sometimes, this product converges normally, e.g. for $s_n = n^2$, $m_n = 1$, but this does not always happen, e.g. for $s_n = n$ and $m_n = 1$. Following WEIERSTRASS we change our first approach, by introducing new factors which don't change the vanishing behavior, but which enforce convergence.

New approach.

$$\boxed{ f(z) := \prod_{n=1}^{\infty} \left(1 - \frac{z}{s_n} \right)^{m_n} e^{P_n(z)} . }$$

Here $P_n(z)$ are polynomials which still have to be determined. We must at least ensure for any $z \in \mathbb{C}$

$$\lim_{n \to \infty} \left(1 - \frac{z}{s_n} \right)^{m_n} e^{P_n(z)} = 1 \ .$$

In this connection we remark:

There exists an analytic function A_n on the open disk $U_{|s_n|}(0)$ with

$$\left(1 - \frac{z}{s_n} \right)^{m_n} e^{A_n(z)} = 1 \ \text{for any} \ z \in U_{|s_n|}(0) \ , \qquad \text{and}$$

$$A_n(0) = 0 \ .$$

The existence of A_n follows directly from II.2.9. Obviously, A_n is unique, and we will give an explicit formula a little later.

The power series expansion of A_n in the disk $U_{|s_n|}(0)$ converges uniformly in any compact set $K \subset U_{|s_n|}(0)$. If we truncate this power series at some suitable order, we obtain a polynomial P_n with the property

$$\left| 1 - \left(1 - \frac{z}{s_n} \right)^{m_n} e^{P_n(z)} \right| \leq \frac{1}{n^2} \quad \text{for all} \ z \ \text{with} \ |z| \leq \frac{1}{2} |s_n| \ .$$

From the convergence of the series $1 + \frac{1}{4} + \frac{1}{9} + \cdots$ follows:

The series

$$\sum_{n=1}^{\infty} \left| 1 - \left(1 - \frac{z}{s_n} \right)^{m_n} e^{P_n(z)} \right|$$

is normally convergent, because in any compact disk $|z| \leq R$ it is dominated by the series $\sum \frac{1}{n^2}$, with the exception of the finitely many terms with $\frac{1}{2} |s_n| \leq R$.

This discussion has led to the following Theorem:

Theorem IV.2.1 (The Weierstrass Product Theorem, first variant, K. Weierstrass, 1876)
Consider a discrete subset $S \subset \mathbb{C}$ and a map

$$m : S \longrightarrow \mathbb{N} \ , \qquad s \mapsto m_s \ .$$

Then there exists an analytic function

$$f : \mathbb{C} \longrightarrow \mathbb{C}$$

with the following properties:

(a) $S = N(f) := \left\{ \ z \in \mathbb{C} \ ; \ \ f(z) = 0 \ \right\}$, *and*
(b) $m_s = \text{ord}(f; s)$ *for all* $s \in S$.

f has thus its zeros exactly at the points $s \in S$, and in any $s \in S$ the order is the prescribed value m_s. By construction, f has the form of a (finite or infinite) product, from which one can read off the position and order of the zeros of f. We use the terminology:

f is a solution of the **zero distribution** $\{\ (s, m_s)\ ;\ s \in S\ \}$.

Together with f, any other function of the form

$$F(z) := \exp\bigl(h(z)\bigr)\, f(z)$$

for entire h is also a solution of the same zero distribution.

Conversely, for any other solution for the same vanishing distribution F the quotient $g := F/f$ is a non-vanishing entire function, and II.2.9 ensures the existence of an entire h with $\exp h = g = F/f$.

An important application of the WEIERSTRASS Product Theorem is

Proposition IV.2.2 *A meromorphic function $\mathbb{C} \to \overline{\mathbb{C}}$ is representable as a quotient of two entire functions. In other words, the field $\mathcal{M}(\mathbb{C})$ of all mero-morphic functions in $\mathbb{C}$ is the quotient field of the integral domain $\mathcal{O}(\mathbb{C})$ of all entire functions.*

Proof. Let $f \in \mathcal{M}(\mathbb{C})$, $f \not\equiv 0$, and let $S := S(f)$ be the set of poles of f. Then S lies discretely in $\mathbb{C}$. We set $m_s := -\operatorname{ord}(f; s)$, the pole order of f for any $s \in S$. Using these data as zero distribution, there exists an entire function h with $N(h) = S$, and $\operatorname{ord}(h; s) = m_s$. The meromorphic function $g := fh$ on $\mathbb{C}$ has then only removable singularities, so it is an analytic function on $\mathbb{C}$. We have $f = g/h$. By the way, in this representation the functions g and h have no common zero.   □

Practical construction of Weierstrass products

Our existence proof gives us polynomials P_n, which often have very large degrees. An improvement is obtained by refining our argument. First of all, we determine the power series A_n. One simply checks,

$$A_n(z) = m_n \left(\frac{z}{s_n} + \frac{1}{2}\left(\frac{z}{s_n}\right)^2 + \frac{1}{3}\left(\frac{z}{s_n}\right)^3 + \cdots \right).$$

The polynomial P_n is obtained by truncation at a suitable position,

$$P_n(z) = m_n \sum_{\nu=1}^{k_n} \frac{1}{\nu}\left(\frac{z}{s_n}\right)^\nu \qquad \text{with a suitable } k_n \in \mathbb{N}.$$

We introduce the so-called WEIERSTRASS elementary factors E_k,

$$E_0(z) := (1 - z), \quad E_k(z) := (1 - z)\exp\left(z + \frac{z^2}{2} + \cdots + \frac{z^k}{k}\right), \quad k \in \mathbb{N}.$$

The infinite product takes the form

$$\prod_{n=1}^{\infty} \left(E_{k_n}\!\left(\frac{z}{s_n}\right) \right)^{m_n}.$$

For this infinite product, a so-called WEIERSTRASS product, we work out an improved proof for the convergence, which gives more precise conditions for the choice of the degrees of the polynomials P_n. This proof is based on the following two Lemmas:

Lemma IV.2.3 *Let $m > 0$, and $k \geq 0$ be two integers. Under the assumption*

$$2\,|z| \leq 1 \qquad and \qquad 2m\,|z|^{k+1} \leq 1$$

we have

$$|E_k(z)^m - 1| \leq 4m\,|z|^{k+1} \ .$$

The elementary proof is left to the reader. □

Lemma IV.2.4 *Let $(s_n)_{n \geq 1}$ be a sequence of complex numbers not equal to 0 such that*

$$\lim_{n \to \infty} |s_n| = \infty \ .$$

Let $(m_n)_{n \geq 1}$ be an arbitrary sequence of natural numbers. Then there exists a sequence $(k_n)_{n \geq 1}$ of non-negative integers, such that the series

$$\sum_{n=1}^{\infty} m_n \left| \frac{z}{s_n} \right|^{k_n + 1}$$

converges for all z in $\mathbb{C}$. For instance, the choice $k_n \geq m_n + n$ is possible.

Proof of IV.2.4. Let us fix $z \in \mathbb{C}$. Because of $\lim_{n \to \infty} |s_n| = \infty$ there exists an $n_0 \in \mathbb{N}$, such that for all $n \geq n_0$

$$\left| \frac{z}{s_n} \right| \leq \frac{1}{2} \ .$$

Hence for $n \geq n_0$

$$m_n \left| \frac{z}{s_n} \right|^{k_n + 1} \leq m_n \left(\frac{1}{2} \right)^{n + m_n} < \left(\frac{1}{2} \right)^n \ . \qquad\qquad □$$

We obtain a second form of the WEIERSTRASS Product Theorem:

Theorem IV.2.5 (The Weierstrass Product Theorem, second variant) *With the choice of the sequence (k_n) as in IV.2.4, the Weierstrass product*

$$\prod_{n=1}^{\infty} \left(E_{k_n}\!\left(\frac{z}{s_n} \right) \right)^{m_n}$$

converges normally in $\mathbb{C}$, and it defines an analytic function $f : \mathbb{C} \to \mathbb{C}$, whose zeros are located exactly at the points $s_1, s_2, s_3, \ldots$ with the prescribed orders $m_1, m_2, m_3, \ldots$.

The function $f_0(z) := z^{m_0} f(z)$ has an additional zero of order m_0 at the origin.

Starting from the convergence of

$$\sum_{n=1}^{\infty} m_n \left| \frac{z}{s_n} \right|^{k_n + 1}$$

for all $z \in \mathbb{C}$, we only have to show the (normal) convergence of

$$\prod_{n=1}^{\infty} \left(E_{k_n}\left(\frac{z}{s_n}\right) \right)^{m_n} .$$

Equivalently, we have to show the (normal) convergence of

$$\sum_{n=1}^{\infty} \left(E_{k_n}\left(\frac{z}{s_n}\right)^{m_n} - 1 \right) .$$

Let $R > 0$ be arbitrary. We then choose N sufficiently large, such that for any $n \geq N$

$$\frac{R}{|s_n|} \leq \frac{1}{2} .$$

The summands of a convergent series tend to zero, hence after enlarging N if necessary, we have

$$2m_n \left(\frac{R}{|s_n|} \right)^{k_n+1} \leq 1 \text{ for } n \geq N .$$

The estimate of Lemma IV.2.3 gives for any $n \geq N$, and any z with $|z| \leq R$

$$\left| E_{k_n}\left(\frac{z}{s_n}\right)^{m_n} - 1 \right| \leq 4m_n \left(\frac{|z|}{|s_n|} \right)^{k_n+1} \leq 4m_n \left(\frac{R}{|s_n|} \right)^{k_n+1} .$$

The claimed normal convergence follows now using Lemma IV.2.4.   □

Examples for the Weierstrass Product Theorem

In the following we use the more comfortable, refined version IV.2.5 of the Product Theorem. One could avoid this, because in each particular example the convergence of the involved series could be proven directly.

1. We search for an entire function f, which has simple zeros at the squares of integers ≥ 0. Because $\sum_{n=1}^{\infty} |z \cdot n^{-2}|$ converges for all $z \in \mathbb{C}$, we can choose $k_n = 0$ for all $n \in \mathbb{N}$. A solution is

$$f(z) := z \cdot \prod_{n=1}^{\infty} \left(1 - \frac{z}{n^2} \right) .$$

2. We search for an entire function f, which has simple zeros in $\mathbb{Z}$. For this we order the integers:

$$s_0 = 0 , \; s_1 = 1 , \; s_2 = -1 , \; \dots , \; s_{2n-1} = n , \; s_{2n} = -n , \; \dots .$$

The WEIERSTRASS Product Theorem gives the solution

$$f(z) := z \cdot \prod_{n=1}^{\infty} \left(1 - \frac{z}{s_n} \right) e^{z/s_n} ,$$

because the series

$$\sum_{n=1}^{\infty} \left| \frac{z}{s_n} \right|^2 = |z|^2 \cdot \sum_{n=1}^{\infty} \frac{1}{|s_n|^2}$$

converges for any $z \in \mathbb{C}$. One has:

$$f(z) = z \lim_{N \to \infty} \prod_{n=1}^{2N} \left(1 - \frac{z}{s_n}\right) e^{z/s_n}$$

$$= z \lim_{N \to \infty} \prod_{n=1}^{N} \left(\left(1 - \frac{z}{n}\right) e^{z/n} \right) \left(\left(1 + \frac{z}{n}\right) e^{-z/n} \right)$$

$$= z \lim_{N \to \infty} \prod_{n=1}^{N} \left(1 - \frac{z^2}{n^2}\right)$$

$$= z \prod_{n=1}^{\infty} \left(1 - \frac{z^2}{n^2}\right) .$$

The last infinite product converges absolutely!

Another solution to the same problem is $\sin \pi z$. The logarithmic derivatives of both solutions coincide because of the partial fraction decomposition of the cotangent function, III.7.13:

$$\pi \frac{\cos \pi z}{\sin \pi z} = \pi \cot \pi z = \frac{1}{z} + \sum_{n=1}^{\infty} \frac{2z}{z^2 - n^2} .$$

The two solutions differ by a constant factor. After dividing $\sin \pi z$ by z, and passing to the limit $z \to 0$, we find for this constant the value π.

We have thus found a new proof (compared with Sect. IV.1) for the product expansion $\sin \pi z$ as an infinite product,

$$\boxed{\sin \pi z = \pi z \prod_{n=1}^{\infty} \left(1 - \frac{z^2}{n^2}\right) .}$$

3. Let ω_1, $\omega_2 \in \mathbb{C}$ be two complex numbers, which are linearly independent over $\mathbb{R}$, i.e. they don't lie on the same line through the origin. We call

$$L := L(\omega_1, \omega_2) := \mathbb{Z}\omega_1 + \mathbb{Z}\omega_2$$

the *lattice* spanned by ω_1 and ω_2. We now are looking for an entire function $\sigma : \mathbb{C} \to \mathbb{C}$, which has simple zeros exactly at the lattice points. For $k \in \mathbb{N}$ let

$$L_k := \{ t_1\omega_1 + t_2\omega_2 ; \quad t_1, t_2 \in \mathbb{Z} , \ \max \{|t_1|, |t_2|\} = k \} .$$

This set contains $8k$ elements, and $L = \bigcup_{k=0}^{\infty} L_k$. Corresponding to this decomposition of L we order its elements as follows:

$s_0 = 0$, $s_1 = \omega_1$, $s_2 = \omega_1 + \omega_2$, $s_3 = \omega_2$, $s_4 = -\omega_1 + \omega_2$, $s_5 = -\omega_1$, $s_6 = -\omega_1 - \omega_2$, $s_7 = -\omega_2$, $s_8 = \omega_1 - \omega_2$, $s_9 = 2\omega_1$, $s_{10} = 2\omega_1 + \omega_2$, ...

The sequence of the absolute values of (s_n) is not monoton increasing, but we have $\lim |s_n| = \infty$.

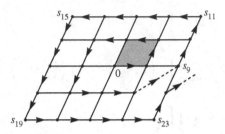

Lemma IV.2.6 *For any $z \in \mathbb{C}$ the following series converges:*

$$\sum_{n=1}^{\infty} \left| \frac{z}{s_n} \right|^3 .$$

Proof. We can find a constant d, such that $|\omega| \geq kd$ for all $\omega \in L_k$. Using the estimate

$$\sum_{n=1}^{\infty} \left| \frac{z}{s_n} \right|^3 = \sum_{k=1}^{\infty} \sum_{s_n \in L_k} \left| \frac{z}{s_n} \right|^3 \leq \sum_{k=1}^{\infty} 8k \left(\frac{|z|}{kd} \right)^3 = \frac{8|z|^3}{d^3} \sum_{k=1}^{\infty} \frac{1}{k^2} < \infty$$

we are done.   □

We take $k_n = 2$ for all $n \in \mathbb{N}$, and find

$$\sigma(z) := \sigma(z; L) := z \cdot \prod_{n=1}^{\infty} \left\{ \left(1 - \frac{z}{s_n} \right) \cdot \exp\left(\frac{z}{s_n} + \frac{1}{2} \left(\frac{z}{s_n} \right)^2 \right) \right\} ,$$

an entire function with the requested properties.

Because of the absolute convergence of this product, we can write

$$\sigma(z; L) := z \cdot \prod_{\substack{w \in L \\ w \neq 0}} \left\{ \left(1 - \frac{z}{\omega} \right) \cdot \exp\left(\frac{z}{\omega} + \frac{1}{2} \left(\frac{z}{\omega} \right)^2 \right) \right\} .$$

The function σ is called the WEIERSTRASS σ-function for the lattice L (WEIERSTRASS, 1862/63). The logarithmic derivative of the σ-function

$$\zeta(z) := \zeta(z; L) := \frac{\sigma'(z)}{\sigma(z)} = \frac{1}{z} + \sum_{\substack{w \in L \\ w \neq 0}} \left\{ \frac{1}{z - \omega} + \frac{1}{\omega} + \frac{z}{\omega^2} \right\}$$

is called the WEIERSTRASS ζ-function (of the lattice L). The negative derivative

$$-\zeta'(z) = -\zeta'(z;L) =: \wp(z;L)$$

is the WEIERSTRASS $\wp$-*function* (of the lattice L), explicitly:

$$\wp(z;L) = \frac{1}{z^2} + \sum_{\substack{\omega \in L \\ \omega \neq 0}} \left\{ \frac{1}{(z-\omega)^2} - \frac{1}{\omega^2} \right\}.$$

This function plays a fundamental role in the theory of *elliptic functions*, Chapter V. One can interpret the WEIERSTRASS $\wp$-function as a MITTAG-LEFFLER partial fraction series. Such series will be treated in the next section.

Exercises for IV.2

1. Show for the WEIERSTRASS elementary factors E_k the properties:

   (a) $E_k'(z) = -z^k \exp \left(z + \frac{z^2}{2} + \cdots + \frac{z^k}{k} \right)$.

   (b) If $E_k(z) = \sum_{\nu=0}^{\infty} a_\nu z^\nu$ is the TAYLOR series of E_k in zero, then $a_0 = 1$, $a_1 = a_2 = \cdots = a_k = 0$, and $a_\nu \leq 0$ for $\nu > k$.

   (c) For $|z| \leq 1$ one has $|E_k(z) - 1| \leq |z|^{k+1}$.

2. The WALLIS Product Formula (J. WALLIS, 1655).

   $$\frac{\pi}{2} = \lim_{n \to \infty} \prod_{\nu=1}^{n} \frac{4\nu^2}{4\nu^2 - 1}.$$

   Hint. Use the product formula for $\sin \pi z$.

3. *Show:*

   (a) $\cos \pi z = \prod_{n=1}^{\infty} \left(1 - \frac{4z^2}{(2n-1)^2} \right) = \prod_{n=-\infty}^{\infty} \left(1 - \frac{2z}{2n-1} \right) e^{\frac{2z}{2n-1}}$,

   (b) $\cos \frac{\pi z}{4} - \sin \frac{\pi z}{4} = \prod_{n=1}^{\infty} \left(1 + \frac{(-1)^n}{2n-1} z \right)$.

4. Let $f : \mathbb{C} \to \overline{\mathbb{C}}$ be a meromorphic function, such that all its poles are simple with integral residues. Then there exists a meromorphic function $h : \mathbb{C} \longrightarrow \overline{\mathbb{C}}$ with $f(z) = h'(z)/h(z)$.

5. Let R be a commutative ring with unit element 1. The set

   $$R^\bullet := \{ r \in R \; ; \; rs = 1 \text{ for a suitable } s \in R \}$$

   is the group of units of R. An element $r \in R \setminus \{0\}$ is called *irreducible*, iff it is not a unit, and the relation $r = ab$ implies either $a \in R^\bullet$ or $b \in R^\bullet$. A *prime* element $p \in R \setminus \{0\}$ is characterized by the property

   $$p \notin R^\bullet \text{ and } (p \mid ab \implies p \mid a \text{ or } p \mid b),$$

   where "$\mid$" denotes the divisibility relation. Moreover, if the ring R as above has no zero divisors, then it is called an *integral domain*. If any element $r \in R$,

$r \neq 0$, has a product decomposition (unique up to a possible permutation of the factors) of the form

$$r = \varepsilon p_1 p_2 \cdots p_m \; , \quad \varepsilon \in R^{\bullet} \; ,$$

with finitely many *prime* factors $p_1, p_2, \ldots, p_m$, then R is called *factorial*, or a *UFD* (unique factorization domain).

An *ideal* in a ring R is an additive subgroup $\mathbf{a}$ of R with the property

$$a \in \mathbf{a} \; , \; r \in R \implies ra \in \mathbf{a} \; .$$

An ideal is called *finitely generated*, iff there are finitely many elements $a_1, \ldots, a_n \in R$ with

$$\mathbf{a} = \left\{ \sum_{\nu=1}^{n} r_\nu a_\nu \; ; \quad r_\nu \in R \right\} \; .$$

Finally, an ideal $\mathbf{a}$ is said to be *principal*, iff (it is finitely generated and) in the above representation we can choose $n = 1$.

Let $R = \mathcal{O}(\mathbb{C})$ be the ring of analytic functions in $\mathbb{C}$.

(a) Let $\mathbf{a}$ be the set of all entire functions f with the following property. There exists a natural number m, such that f vanishes at all points of $m\mathbb{Z} = \{ 0, \pm m, \pm 2m, \ldots \}$. Show that $\mathbf{a}$ is not finitely generated.

(b) Which are the irreducible elements in $\mathcal{O}(\mathbb{C})$? Which are the prime elements in $\mathcal{O}(\mathbb{C})$?

(c) Which are the invertible elements (i.e. the units) in $\mathcal{O}(\mathbb{C})$?

(d) $\mathcal{O}(\mathbb{C})$ is not a UFD, i.e. there exist elements $\neq 0$ in $\mathcal{O}(\mathbb{C})$ which cannot be wrtiten as product of finitely many prime elements.

(e) Any finitely generated ideal in $\mathcal{O}(\mathbb{C})$ is principal.

Hint. This Exercise is not obvious. It is enough to show that any two entire functions f, g with no common zeros already generate the unit ideal. This means that there exist $A, B \in \mathcal{O}(\mathbb{C})$ with $Af + Bg = 1$.

Take $A = (1 + hg)/f$ for suitable $h \in \mathcal{O}(\mathbb{C})$.

For the proof, it can be used that for any discrete subset $S \subset \mathbb{C}$, and for any function $h_0 : S \to \mathbb{C}$ there exists an entire function $h : \mathbb{C} \to \mathbb{C}$ which equals h_0 on S. In fact, more is true, one can even prescribe for any $s \in S$ finitely many TAYLOR coefficients (see also Exercise 5 in the next section).

IV.3 The Mittag-Leffler Partial Fraction Decomposition

If we replace in IV.2.5 the function f by $1/f$, then we obtain meromorphic functions with prescribed poles of prescribed orders. But we can do more, for each pole we can prescribe the principal part of the LAURENT series.

Theorem IV.3.1 (The Mittag-Leffler Partial Fraction Decomposition, M.G. Mittag-Leffler, 1877) *Let $S \subset \mathbb{C}$ be a discrete subset. For any point $s \in S$ let us fix an entire function*

$$h_s : \mathbb{C} \longrightarrow \mathbb{C} , \quad h_s(0) = 0 .$$

Then there exists an analytic function

$$f : \mathbb{C} \setminus S \longrightarrow \mathbb{C} ,$$

with the principal part in each $s \in S$ encoded by h_s, i.e.

$$f(z) - h_s\left(\frac{1}{z - s}\right) , \quad s \in S ,$$

has a removable singularity at $z = s$.

If the set S is finite, then we can immediately write down a solution of the problem,

$$f(z) = \sum_{s \in S} h_s\left(\frac{1}{z - s}\right) .$$

But we cannot expect convergence for infinite S. Similar as in the case of the WEIERSTRASS product, we can introduce new summands that enforce convergence. So let S be an infinite, discrete set in $\mathbb{C}$. We order the elements of S, such their absolute values increase,

$$S = \{ s_0, s_1, s_2, \ldots \} , \quad |s_0| \le |s_1| \le |s_2| \le \cdots .$$

Each of the functions

$$z \mapsto h_n\left(\frac{1}{z - s_n}\right) , \quad h_n := h_{s_n} , \quad n \in \mathbb{N} ,$$

is analytic in the disk

$$|z| < |s_n| ,$$

and hence representable as a power series. After a suitable truncation we find polynomials P_n with the following property:

The series

$$\sum_{n=N}^{\infty} \left[h_n\left(\frac{1}{z - s_n}\right) - P_n(z) \right]$$

converges normally in the region $|z| < |s_N|$.

For instance, we can choose P_n such that

$$\left| h_n\left(\frac{1}{z - s_n}\right) - P_n(z) \right| \le \frac{1}{n^2} \quad \text{for} \quad |z| \le \frac{1}{2} |s_n| .$$

So we obtain that

$$f(z) := h_0\left(\frac{1}{z - s_0}\right) + \sum_{n=1}^{\infty}\left[h_n\left(\frac{1}{z - s_n}\right) - P_n(z)\right]$$

defines an analytic function in the domain $\mathbb{C}\setminus S$ having the prescribed singular behavior in S. Any series which is obtained by this procedure will be called a MITTAG-LEFFLER partial fraction series. We say that

f is a solution of the given principal part distribution.

If f is a solution of a principal part distribution, then so is also

$$f_0 := f + g\,, \quad \text{with } g \text{ an arbitrary entire function.}$$

This is the *general* solution of the principal part distribution, because for two solutions f_0 and f for the same principal part distribution in the difference $f_0 - f =: g$ the principal parts cancel and we get an entire function. Conversely, adding an entire function does not change singularities and corresponding principal parts.

Examples.

1. *Partial fraction series for* $\dfrac{\pi}{\sin \pi z}$.

This function has singularities located in $S = \mathbb{Z}$. All of them are simple poles with principal part

$$\frac{(-1)^n}{z - n}\,.$$

The power series expansion of this function is

$$\frac{(-1)^n}{z - n} = \frac{(-1)^{n+1}}{n}\cdot\left(1 + \frac{z}{n} + \frac{z^2}{n^2} + \cdots\right)\,.$$

We truncate at the first position. In the region $|z| \le r,\, r > 0$, we have

$$\left|\frac{(-1)^n}{z - n} - \frac{(-1)^{n+1}}{n}\right| \le \frac{2r}{n^2} \text{ for } n \ge 2r\,.$$

The series

$$h(z) = \frac{1}{z} + \sum_{n\ne 0}\left[\frac{(-1)^n}{z - n} + \frac{(-1)^n}{n}\right]$$

is then a MITTAG-LEFFLER partial fraction series. Grouping the terms for n and $-n$ together, we obtain

$$h(z) = \frac{1}{z} + \sum_{n=1}^{\infty}(-1)^n\left[\frac{1}{z - n} + \frac{1}{z + n}\right]\,.$$

Claim. We have the identity

$$\frac{\pi}{\sin \pi z} = \frac{1}{z} + \sum_{n=1}^{\infty} (-1)^n \left[\frac{1}{z-n} + \frac{1}{z+n} \right] .$$

Proof. Once more, we use the partial fraction decomposition of the cotangent, III.7.13, and the relation

$$\frac{1}{\sin z} = \cot \frac{z}{2} - \cot z .$$

A direct proof by means of LIOUVILLE's Theorem can also be given. We leave this as a non-trivial exercise to the reader.

2. The Γ-function has the singularities in $S = \{-n;\ n \in \mathbb{N}_0\}$. All of them are simple poles, and the residue in $z \in S$ is $\text{Res}(\Gamma; -n) = \frac{(-1)^n}{n!}$. The principal parts are thus

$$h_n \left(\frac{1}{z+n} \right) = \frac{(-1)^n}{n!} \frac{1}{z+n} .$$

The function

$$g(z) := \Gamma(z) - \sum_{n=0}^{\infty} \frac{(-1)^n}{n!} \frac{1}{z+n}$$

is entire, for the convergence is ensured by the factor $n!$ in the denominator. We can find another expression for g, in fact

$$g(z) = \int_1^{\infty} t^{z-1} e^{-t}\, dt .$$

This can be seen by considering the difference

$$\int_0^1 t^{z-1} e^{-t}\, dt = \Gamma(z) - \int_1^{\infty} t^{z-1} e^{-t}\, dt \quad (\text{Re } z > 0) ,$$

then expanding e^{-t} as a power series around zero, and finally exchanging the integral and the sum. By the way, this is a new proof of the analytic continuation of Γ to $\mathbb{C} \setminus S$.

For the Γ-function we conclude the decomposition (E.F. PRYM, 1876)

$$\Gamma(z) = \sum_{n=0}^{\infty} \frac{(-1)^n}{n!} \frac{1}{z+n} + \int_1^{\infty} t^{z-1} e^{-t}\, dt .$$

3. We come back to the WEIERSTRASS $\wp$-function, Sect. IV.2, Example 3. We are searching for a function which has in all lattice points poles of second order with residue 0, and principal part

$$h_n \left(\frac{1}{z - s_n} \right) = \frac{1}{(z - s_n)^2} .$$

For $n \geq 1$ we once more consider the power series expansion

$$h_n\left(\frac{1}{z-s_n}\right) = \frac{1}{s_n^2}\,\frac{1}{(1-z/s_n)^2} = \frac{1}{s_n^2} + 2\cdot\frac{z}{s_n^3} + 3\cdot\frac{z^2}{s_n^4} + \cdots,$$

and it is enough to truncate at the first position, $P_n(z) := 1/s_n^2$. Then we have:

$$h_n\left(\frac{1}{z-s_n}\right) - P_n(z) = \frac{1}{(z-s_n)^2} - \frac{1}{s_n^2} = \frac{2zs_n - z^2}{s_n^2(z-s_n)^2}.$$

For the proof of convergence we choose an arbitrary $R > 0$. For all but finitely many n one has $|s_n| > 2R$, and then for $|z| \leq R$ we can estimate

$$\left|h_n\left(\frac{1}{z-s_n}\right) - P_n(z)\right| \leq \frac{R(2|s_n| + R)}{|s_n|^2(|s_n| - R)^2} < \frac{3R|s_n|}{|s_n|^2\left(\frac{1}{2}|s_n|\right)^2} = \frac{12R}{|s_n|^3}.$$

Recall the convergence of the series $\sum|s_n|^{-3}$, IV.2.6. So a solution to the given principal part distribution is

$$\wp(z; L) := \frac{1}{z^2} + \sum_{n=1}^{\infty}\left\{\frac{1}{(z-s_n)^2} - \frac{1}{s_n^2}\right\}.$$

Because of the absolute convergence we can write

$$\wp(z; L) := \frac{1}{z^2} + \sum_{\substack{w\in L \\ w\neq 0}}\left\{\frac{1}{(z-w)^2} - \frac{1}{w^2}\right\}.$$

We conclude: The WEIERSTRASS $\wp$-function for the lattice L is a meromorphic function with the singularities at the lattice points. Each $w \in L$ is a pole of second order, the principal part being

$$h\left(\frac{1}{z-w}\right) = \frac{1}{(z-w)^2}.$$

Especially, all residues are zero. In Chapter V we will extensively study this function.

Exercises for IV.3

1. Prove the WEIERSTRASS Product Theorem by means of MITTAG-LEFFLER's Theorem, by first considering the principal part distribution

$$\left\{\frac{m_n}{z-s_n} \; ; \; n\in\mathbb{N}\right\},$$

and then observing that f is a solution for the zero distribution $\{(s_n, m_n); n\in\mathbb{N}\}$, iff $\frac{f'}{f}$ is a solution for the principal part distribution $\left\{\frac{m_n}{z-s_n} \; ; \; n\in\mathbb{N}\right\}$.

2.   Using the relation

$$\cot \frac{z}{2} - \tan \frac{z}{2} = 2 \cot z$$

and the partial fraction decomposition of the cotangent, prove

$$\pi \tan(\pi z) = 8z \sum_{n=0}^{\infty} \frac{1}{(2n+1)^2 - 4z^2} \ .$$

3.   *Show:*

$$\frac{\pi}{\cos \pi z} = 4 \sum_{n=0}^{\infty} \frac{(-1)^n (2n+1)}{(2n+1)^2 - 4z^2} \ ,$$

and derive from this

$$\frac{\pi}{4} = \sum_{n=0}^{\infty} (-1)^n \frac{1}{2n+1} = 1 - \frac{1}{3} + \frac{1}{5} - \frac{1}{7} + \cdots \ .$$

4.   Find a meromorphic function f in $\mathbb{C}$ which has simple poles in

$$S = \{ \ \sqrt{n} \ ; \quad n \in \mathbb{N} \ \}$$

with corresponding residues $\mathrm{Res}(f; \sqrt{n}) = \sqrt{n}$, and is analytic in $\mathbb{C} \setminus S$.

5.   Prove the following refinement of the MITTAG-LEFFLER Theorem:

Theorem. (Mittag-Leffler) Let $S \subset \mathbb{C}$ be a discrete subset. Then one can construct an analytic function $f : \mathbb{C} \setminus S \to \mathbb{C}$ which has at any $s \in S$ not only given principal parts but also finitely many LAURENT coefficients for non-negative indices.

In german this is called the *Anschmiegungssatz*. ("Anschmiegen" means "to cling".)

Hint. Consider a suitable product of a partial fraction series with a WEIER-STRASS product.

IV.4 The Riemann Mapping Theorem

The RIEMANN Mapping Theorem claims that any elementary domain, which is not $\mathbb{C}$, is conform equivalent to the unit disk $\mathbb{E}$. There is a more general form of the RIEMANN Mapping Theorem, usually called the **Uniformization Theorem** (P. KOEBE, H. Poincaré, 1907). This result asserts the following:

Any simply connected Riemann surface is conform equivalent to exactly one of the following three Riemann surfaces:
  (a) *the Riemann sphere* $\overline{\mathbb{C}} \cong \mathbb{P}^1(\mathbb{C})$,
  (b) *the plane* $\mathbb{C}$,
  (c) *the unit disk* $\mathbb{E}$.

We will prove in this volume a weaker version of this theorem, the "RIEMANN Mapping Theorem". In the second volume, we will give a proof for the general uniformization theorem.

In Section I.5 we have already introduced the notion of a *conformal map*, and studied some related elementary geometric properties. Let us once more give the definition of a *(globally) conformal map* between two open sets $D, D' \subset \mathbb{C}$.

Definition IV.4.1 *A map*

$$\varphi : D \longrightarrow D'$$

*between two open sets D, D' in the complex plane is called **conformal**, iff the following conditions are satisfied:*

(a) φ *is bijective*,
(b) φ *is analytic*,
(c) φ^{-1} *is analytic*.

Instead of (c), we can request that the derivative of φ has no zero. Remarkably, this third condition is automatically satisfied:

Remark IV.4.2 *In IV.4.1 the condition (c) follows from (a) and (b).*

Proof. From the Open Mapping Theorem III.3.3 we know that $\varphi(D)$ is open. The map φ^{-1} is hence continuous. (The inverse image of an open set $U \subset D$ by φ^{-1} is exactly its image under φ, and hence open.)
By the Implicit Function Theorem I.5.7, φ^{-1} is analytic in the complement of the set of all $w = \varphi(z) \in D'$ with $\varphi'(z) = 0$. The set of these exceptional places is the image under φ of a discrete set in D, hence discrete in D'. Now the claim follows from RIEMANN's Removability Theorem, III.4.2.   □

Two domains D and D' are called *conform equivalent*, iff there exists a conformal map $\varphi : D \to D'$. The conformal equivalence is surely an equivalence relation on the set of all domains in $\mathbb{C}$. Recall once again (II.2.12):

Remark IV.4.3 *Any domain which is conform equivalent with an elementary domain is itself elementary.*

Remark IV.4.4 *The two elementary domains*

$$\mathbb{C} \quad and \quad \mathbb{E} = \{\, z \in \mathbb{C} \ ; \ |z| < 1 \,\}$$

are not conform equivalent.

Proof. LIOUVILLE's Theorem implies that any analytic function $\varphi : \mathbb{C} \to \mathbb{E}$ is constant, since $\mathbb{E}$ is bounded.   □

However, $\mathbb{C}$ and $\mathbb{E}$ are *topologically equivalent* (homeomorphic), explicit topological maps between them are

$$\mathbb{C} \longrightarrow \mathbb{E}, \qquad z \longmapsto \frac{z}{1 + |z|} \ ,$$

$$\mathbb{E} \longrightarrow \mathbb{C}, \qquad w \longmapsto \frac{w}{1 - |w|} \ .$$

This example shows that the *necessary* condition of topological equivalence is not sufficient for the *conformal* equivalence.

A *main problem* of the theory of conformal maps is given by the following question:

(1) When do two domains $D, D' \subset \mathbb{C}$ belong to the same equivalence class?
(2) How many different maps realize the conformal equivalence for two domains in the same class?

The second question is equivalent to the determination of the group of conformal automorphisms (self maps) of a fixed domain D of a given class. It is easy to that

$$\text{Aut}(D) := \{ \, \varphi : D \to D \; ; \quad \varphi \text{ is conformal} \, \}$$

is a group with respect to the composition of maps.

Why is it enough to study $\text{Aut}(D)$? Because for two conformal maps $\varphi, \psi : D \to D'$ the map $\theta = \psi^{-1}\varphi : D \to D$ lies in $\text{Aut}(D)$.

The first question leads us to the search of a full list of *"standard domains"*, such that

(1) any domain is conform equivalent to a standard domain, and
(2) any two different standard domains are not conform equivalent.

We will restrict here to elementary domains. (The general case is more complicated, see for instance [Co2], [Sp] or [Ga].)

The standard elementary domains are the complex plane and the unit disk ($\mathbb{C}$ and $\mathbb{E}$).

Theorem IV.4.5 (The Riemann Mapping Theorem, B. Riemann, 1851) *Any elementary domain $D \subset \mathbb{C}$, which is not $\mathbb{C}$, is conform equivalent to the unit disk $\mathbb{E}$.*

The proof will cover seven steps. Before we start the proof, let us give an overview of the main parts.

In a first main part (steps 1 and 2), we map the given elementary domain $D \neq \mathbb{C}$ conformally onto an elementary domain $D^* \subset \mathbb{E}$ containing 0. .

In a second main part (step 3), we consider the set $\mathcal{M}$ of all injective, analytic functions φ mapping D^* to $\mathbb{E}$, and fixing 0. If there exists a map $\varphi \in \mathcal{M}$ with maximal $|\varphi'(0)|$, then one can show that φ is surjective, i.e. conformal as a map $D^* \to \mathbb{E}$. This reduction to the solution of an extremal problem goes back to L. FÉJER and F. RIESZ (1922).

Finally, in a third main part (steps 4 to 7) we solve this extremal problem.

Proof of the Riemann mapping theorem

Step 1. For any elementary domain $D \subset \mathbb{C}$, $D \neq \mathbb{C}$, there exists a conform equivalent domain $D_1 \subset \mathbb{C}$ containing a full disk in its complement $\mathbb{C} \setminus D_1$.

By assumption there exists a point $b \in \mathbb{C}$, $b \notin D$. The function $f(z) = z - b$ is then analytic in the domain D, and has no zero there. It possesses hence an analytic square root g (II.2.9₁)

$$g : D \longrightarrow \mathbb{C} , \quad g^2(z) = z - b .$$

Obviously, g is injective,

$$g(z_1) = g(z_2) \implies g^2(z_1) = g^2(z_2) \implies z_1 = z_2 ,$$

hence it defines a conformal map onto a domain $D_1 = g(D)$. The argument for the injectivity shows more, namely $g(z_1) = -g(z_2)$ also implies $z_1 = z_2$. In other words:

If a non-zero point w lies in D_1, then $-w$ does not lie in D_1.

Since D_1 is open and non-empty, there exists a disk fully contained in D_1 and not containing 0. Then the disc, which is obtained by reflection at the origin, is contained in the complement of D_1.

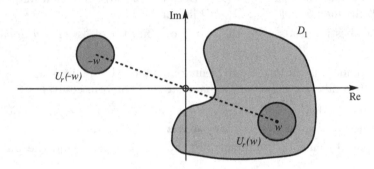

This first step can be illustrated using the example of the slit plane $\mathbb{C}_-$. We choose $b = 0$, and let g be the principal branch of the square root. Then g maps conformally $\mathbb{C}_-$ onto the right half-plane.

Step 2. For any elementary domain $D \subset \mathbb{C}$, $D \neq \mathbb{C}$, there exists a conform equivalent domain D_2 such that

$$0 \in D_2 \subset \mathbb{E} .$$

By the first step, we can assume that a full disk $U_r(a)$ is contained in the complement of D. (We can replace D by D_1.) The mapping

$$z \longmapsto \frac{1}{z - a}$$

maps then D conformally onto a *bounded* domain D_1', because of

$$z \in D \implies |z - a| > r \implies \frac{1}{|z - a|} < \frac{1}{r} .$$

After a suitable translation $z \mapsto z + \alpha$ we obtain a conform equivalent domain containing 0, that after a suitable contraction is moreover contained in the unit disk $\mathbb{E}$.

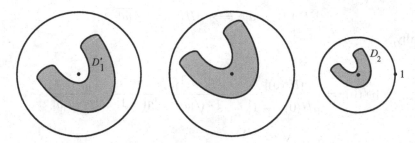

We are now in position to start with the main part of the proof. For a better understanding, we insert an auxiliary result.

Step 3.

Lemma IV.4.6 *Let D be an elementary domain, $0 \in D \subset \mathbb{E}$. If D is strictly contained in $\mathbb{E}$, then there exists an injective analytic function $\psi : D \to \mathbb{E}$ with the following properties:*

(a) $\psi(0) = 0$, *and*
(b) $|\psi'(0)| > 1$.

The corresponding assertion is false for $D = \mathbb{E}$ by the SCHWARZ Lemma, III.3.7.

Proof of IV.4.6. Let us choose a point $a \in \mathbb{E}$, $a \notin D$. By III.3.9, the mapping

$$ h(z) = \frac{z - a}{\bar{a}z - 1} $$

maps the unit disk conformally onto itself. The function h has no zeros in the elementary domain D, and hence by II.2.9$_1$ there exists an analytic square root

$$ H : D \longrightarrow \mathbb{C} \quad \text{with} \quad H(z)^2 = h(z) . $$

Then H still maps D injectively onto a subset of the unit disk $\mathbb{E}$. A further application of III.3.9 shows that also the function

$$ \psi(z) = \frac{H(z) - H(0)}{\overline{H(0)}H(z) - 1} $$

maps D injectively into $\mathbb{E}$. Obviously, $\psi(0) = 0$. We have to compute the derivative. A simple computation leads to

$$ \psi'(0) = \frac{H'(0)}{|H(0)|^2 - 1} . $$

We have

$$ H^2(z) = \frac{z - a}{\bar{a}z - 1} \quad \Longrightarrow \quad 2H(0) \cdot H'(0) = |a|^2 - 1 . $$

Furthermore,

$$|H(0)|^2 = |a| \implies |H(0)| = \sqrt{|a|} \ .$$

Finally,

$$|\psi'(0)| = \frac{|H'(0)|}{\left||H(0)|^2 - 1\right|} = \frac{\left||a|^2 - 1\right|}{2 \cdot \sqrt{|a|}} \cdot \frac{1}{\left||a| - 1\right|} = \frac{|a| + 1}{2 \cdot \sqrt{|a|}}$$

$$= 1 + \frac{(\sqrt{|a|} - 1)^2}{2 \cdot \sqrt{|a|}} > 1 \ .$$

$\square$

A direct consequence of the Lemma is:

*Let D be an elementary domain, $0 \in D \subset \mathbb{E}$. We **assume** that there exists among all injective, analytic maps $\varphi : D \to \mathbb{E}$ with $\varphi(0) = 0$ one with a maximal value of $|\varphi'(0)|$. Such a φ is surjective. Especially, D and $\mathbb{E}$ are conform equivalent.*

This is because for any non-surjective map $\varphi : D \to \mathbb{E}$ with $\varphi(0) = 0$, applying Lemma IV.4.6 on $\varphi(D)$, we can find an injective, analytic

$$\psi : \varphi(D) \to \mathbb{E} \ , \quad \psi(0) = 0 \ ,$$

with $|\psi'(0)| > 1$. Then

$$|(\psi \circ \varphi)'(0)| > |\varphi'(0)| \ ,$$

Hence maximality implies surjectivity. $\square$

We have thus reduced the RIEMANN Mapping Theorem to an *extremal value problem*.

Let D be a bounded domain, which contains 0. Does there exist among all injective analytic $\varphi : D \to \mathbb{E}$, $\varphi(0) = 0$, a map with a maximal value of $|\varphi'(0)|$?

In the remaining steps we show that the answer to this extremal value problem is always positive. There is no need to assume that D is an elementary domain.

Step 4. Let D be a bounded domain, $0 \in D$. We denote by $\mathcal{M}$ the non-empty set of all injective analytic functions

$$\varphi : D \longrightarrow \mathbb{E} \ , \quad \varphi(0) = 0 \ ,$$

and let

$$M := \sup\{ \ |\varphi'(0)| \ ; \ \varphi \in \mathcal{M} \ \} \ , \quad \text{where } M = \infty \text{ is also allowed} \ .$$

We choose a sequence $\varphi_1, \varphi_2, \varphi_3, \dots$ of functions from $\mathcal{M}$, such that

$$|\varphi_n'(0)| \to M \text{ for } n \to \infty \ .$$

(M can be ∞. Then for any $C > 0$ one has $|\varphi_n'(0)| > C$ for almost all n.)

Main problem. We will show:

(1) The sequence (φ_n) has a locally uniformly convergent subsequence.
(2) The limit φ of this subsequence is also injective.
(3) $\varphi(D) \subset \mathbb{E}$.

Then the limit φ is an injective analytic function with the property $|\varphi'(0)| = M$. Especially, $0 < M < \infty$. At this point we will be done with the proof.

Step 5. The sequence (φ_n) admits a locally uniformly convergent subsequence. This is a consequence of MONTEL's Theorem, that will be treated in the following. First we give two preliminary results.

Lemma IV.4.7 *Let us fix $D \subset \mathbb{C}$ open, $K \subset D$ compact, and $C > 0$. For any $\varepsilon > 0$ there exists a $\delta = \delta(D, C, K) > 0$ with the following property:*

If $f : D \to \mathbb{C}$ is an analytic function, which is bounded on D by C, i.e. $|f(z)| \leq C$ for all $z \in D$, then for all $a, z \in K$:

$$|f(z) - f(a)| < \varepsilon , \quad \text{if} \quad |z - a| < \delta .$$

Observation. In case of $K = \{a\}$, the Lemma can be reformulated in a standard terminology as:

*The set $\mathcal{F}$ of all analytic functions $f : D \to \mathbb{C}$ with $|f(z)| \leq C$ for all $z \in D$ is **equicontinuous** at a.*

Because a varies in the Lemma in a compact set, we can speak of a *locally uniform equicontinuity.*

Proof of IV.4.7. We first assume that K is a compact disk, i.e. there exist z_0 and $r > 0$ with

$$K := \overline{U}_r(z_0) = \{ z \in \mathbb{C} ; \quad |z - z_0| \leq r \} \subset D .$$

We moreover assume that the closed disk of doubled radius $\overline{U}_{2r}(z_0)$ is still contained in D. The CAUCHY Integral Formula II.3.2 gives then for any $z, a \in K$:

$$
\begin{aligned}
|f(z) - f(a)| &= \left| \frac{1}{2\pi i} \oint_{|\zeta - z_0| = 2r} \left(\frac{f(\zeta)}{\zeta - z} - \frac{f(\zeta)}{\zeta - a} \right) d\zeta \right| \\
&= \frac{|z - a|}{2\pi} \left| \oint_{|\zeta - z_0| = 2r} \frac{f(\zeta)}{(\zeta - z)(\zeta - a)} d\zeta \right| \\
&\leq \frac{|z - a|}{2\pi} \cdot 4\pi r \cdot \frac{C}{r^2} = \frac{2C}{r} |z - a| .
\end{aligned}
$$

For a given $\varepsilon > 0$ we then pick a $\delta > 0$ with

$$\delta < \min\left\{ r , \frac{r}{2C}\varepsilon \right\} ,$$

so that we have $|f(z) - f(a)| < \varepsilon$ for all $a, z \in K$ with $|z - a| < \delta$.

Let now $K \subset D$ be an arbitrary compact set. Then there exists a number $r > 0$ with the following property:

For any a in K, the closed disk $U_r(a)$ of radius r centered at a is fully contained in D.

The above number r is sometimes called a LEBESGUE's Number for the compact set $K \subset D$. The existence of this number is a standard application of the notion of compactness. For any $a \in K$ there exists an $r(a) > 0$, such that the disk of doubled radius $2r(a)$ is contained in D. Applying compactness we find finitely many $a_1, \ldots, a_n \in K$ with $K \subset U_{r(a_1)}(a_1) \cup \cdots \cup U_{r(a_n)}(a_n)$. The minimum of the numbers $r_{a_1}, \ldots, r_{a_n}$ is then a LEBESGUE Number.

From the existence of a LEBESGUE number easily follows that K can be covered by finitely many disks $U_r(a)$, $a \in K$ with $\overline{U}_{2r}(a) \subset D$. The Lemma has now been reduced to the preceding special case:

Lemma IV.4.8 *Let*

$$f_1, \ f_2, \ f_3, \cdots : D \longrightarrow \mathbb{C} \ , \quad D \subset \mathbb{C} \ open \ ,$$

be a sequence of analytic functions which is bounded (i.e. $|f_n(z)| \leq C$ for all $z \in D$ and $n \in \mathbb{N}$). If the sequence (f_n) converges pointwise in a dense subset $S \subset D$, then it converges locally uniformly in D.

Proof of IV.4.8. We show that (f_n) is a locally uniform CAUCHY sequence, i.e.

For any compact set $K \subset D$, and any $\varepsilon > 0$ there exists a natural number $N > 0$, such that for all $m, n \geq N$, and all $z \in K$

$$|f_m(z) - f_n(z)| < \varepsilon \ .$$

It is enough to restrict to closed disks K. In this case K is the closure of its interior, and $K \cap S$ is dense in K.

It is simple to show, and well-known in the real analysis, that any locally uniform CAUCHY sequence is locally uniformly convergent.

Let $\varepsilon > 0$ be arbitrary. We then choose the number δ as in Lemma IV.4.7. The compactness of K implies the existence of finitely many points $a_1, \ldots, a_l \in S \cap K$ with

$$K \subset \bigcup_{j=1}^{l} U_\delta(a_j) \ .$$

(For this, one chooses a sufficiently small LEBESGUE Number $r > 0$, and covers K with disks $U_{r/2}(a)$, $a \in K$. Clearly, K is then covered by the disks $U_{3r/4}(a)$, $a \in S \cap K$.) Now let z be an arbitrary point in K. Then there exists a point a_j with the property $|z - a_j| < \delta$. The triangle inequality gives

$$|f_m(z) - f_n(z)| \leq |f_m(z) - f_m(a_j)| + |f_m(a_j) - f_n(a_j)| + |f_n(z) - f_n(a_j)| \ .$$

The first and third terms are by the Lemma IV.4.7 smaller than ε, the middle term is also controlled by ε for sufficiently large m, n, more exactly for all $m, n \geq N$ with a suitable N which works for all of the finitely many points a_j.   □

Theorem IV.4.9 (Montel's Theorem, P. Montel, 1912) *Let*

$$f_1, f_2, f_3, \cdots : D \longrightarrow \mathbb{C}, \quad D \subset \mathbb{C} \text{ open},$$

be a bounded sequence of analytic functions, i.e. there exists a constant $C > 0$ with the property $|f_n(z)| \leq C$ for all $z \in D$ and all $n \in \mathbb{N}$.
Then there exists a subsequence $f_{\nu 1}, f_{\nu 2}, f_{\nu 3}, \ldots$, which converges locally uniformly.

Proof. Let

$$S \subset D, \quad S = \{s_1, s_2, s_3, \ldots\}$$

be a countable, dense subset of D. Such a subset exists, for instance $S = \{ z = x + iy \in D ; \ x \in \mathbb{Q}, \ y \in \mathbb{Q} \}$. By the BOLZANO-WEIERSTRASS Theorem there exists a subsequence of $f_1, f_2, f_3, \ldots$, which converges at the point $s_1 \in S$. Let us denote this subsequence by

$$f_{11}, \ f_{12}, \ f_{13}, \ \cdots \ .$$

The same argument gives a subsequence of this subsequence, which converges also at the point s_2. Inductively, we construct subsequences

$$f_{n1}, \ f_{n2}, \ f_{n3}, \ \cdots \ ,$$
$$\text{converging in } s_1, s_2, \ldots, s_n \ ,$$

of already constructed subsequences $f_{j1}, f_{j2}, f_{j3}, \ldots, j < n$.
The diagonal trick gives us the (sub)sequence

$$f_{11}, f_{22}, f_{33}, \cdots$$

which converges at all $s \in S$. The claim follows now by IV.4.8.   □

Step 6. φ is injective. Recall that we constructed φ as a locally uniform limit of functions in the family $\mathcal{M}$ of all injective, analytic functions $D \to \mathbb{E}$ with 0 as a fixed point. This limit solves our extremal problem, if it also belongs to the same family. We have to prove the injectivity of φ. It follows from HURWITZ's Theorem, III.7.2:

Let $f_1, f_2, f_3, \ldots$ be a sequence of injective, analytic functions on a domain $D \subset \mathbb{C}$, which converges locally uniformly. The limit is then either constant, or injective.

So we have to exclude the case of a constant limit. But for all functions of the non-empty class $\mathcal{M}$, the derivative at zero does not vanish, so by the extremal property also $\varphi'(0)$ does not vanish.

Step 7 (the final step). $\varphi(D) \subset \mathbb{E}$.

We only know, at this point, that the image of φ is contained in the *closure* of $\mathbb{E}$. But by the Maximum Principle, φ would be a constant, if there were a point of $\partial \mathbb{E}$ in its image,

This concludes the proof of the RIEMANN Mapping Theorem IV.4.5.   □

Exercises for IV.4

1.   Let $D = \{\, z \in \mathbb{C} \;;\;\; |z| > 1 \,\}$. Can there exist any conformal map from D onto the punctured plane $\mathbb{C}^{\bullet}$?

2.   The two annuli

$$r_\nu < |z| < R_\nu \quad (0 \le r_\nu < R_\nu < \infty,\; 1 \le \nu \le 2)$$

are conform equivalent, if the ratios R_ν/r_ν, $\nu = 1, 2$ are equal. (The converse is also true, as we will see in the second volume by means of the theory of RIEMANN surfaces.)

3.   The map

$$f : \mathbb{E} \longrightarrow \mathbb{C}_- \,, \quad z \longmapsto \left(\frac{1-z}{1+z}\right)^2 ,$$

is conformal.

4.   The map

$$\varphi(z) = \frac{(1+z^2)^2 - i(1-z^2)^2}{(1+z^2)^2 + i(1-z^2)^2}$$

defines a conformal map of $D := \{\, z = re^{i\varphi} \;;\;\; 0 < \varphi \le \frac{\pi}{2} \,,\; 0 < r < 1 \,\}$ onto the unit disk $\mathbb{E}$.

5.   Determine the image of

$$D = \{\, z \in \mathbb{C} \;;\;\; |\operatorname{Re} z|\,|\operatorname{Im} z| > 1 \,,\; 0 < \operatorname{Re} z,\, \operatorname{Im} z \,\}$$

under the map $\varphi(z) = z^2$.

6.   Let $D, D^* \subset \mathbb{C}$ be conform equivalent domains. Show that the groups of (conformal) automorphism $\operatorname{Aut}(D)$ and $\operatorname{Aut}(D^*)$ are isomorphic.

7.   Prove the following uniqueness result (H. POINCARÉ, 1884): If $D \subset \mathbb{C}$ is an elementary domain which is not $\mathbb{C}$, and if $z_0 \in D$ is a fixed point in D, then there exist exactly one conformal map

$$\varphi : D \longrightarrow \mathbb{E} \quad \text{with } \varphi(z_0) = 0 \text{ and } \varphi'(z_0) > 0 \,.$$

8.   If $D = \{\, z \in \mathbb{E} \;;\;\; \operatorname{Re} z > 0 \,\}$, and $z_0 := \sqrt{2} - 1$, then the map

$$\varphi(z) = -\frac{z^2 + 2z - 1}{z^2 - 2z - 1}$$

is the unique conformal map $\varphi : D \to \mathbb{E}$ with $\varphi(z_0) = 0$ and $\varphi'(z_0) > 0$ as in Exercise 7.

Also show that φ can be extended to a homeomorphism $\overline{D} \to \overline{\mathbb{E}}$.

9. Let $D \subset \mathbb{C}$ be an elementary domain, and let $f : D \to \mathbb{E}$ be a conformal map. If (z_n) is a sequence in D with $\lim_{n \to \infty} z_n = r \in \partial D$, then the sequence $(|f(z_n)|)$ converges to 1. Give an example of a sequence (z_n) converging to a boundary point of D, such that the image sequence $(f(z_n))$ by a conformal map $f : D \to \mathbb{E}$ does not converge to a boundary point of $\mathbb{E}$.

10. Let
$$D = \{ z \in \mathbb{C} ; \quad \text{Im } z > 0 \} \setminus \{ z = iy ; \quad 0 \leq y \leq 1 \} .$$

    (a) Map D conformally onto the upper half-plane $\mathbb{H}$.
    (b) Map D conformally onto $\mathbb{E}$.

11. Any conformal map $f : \mathbb{H} \to \mathbb{E}$ is of the form

$$z \longmapsto e^{i\varphi} \frac{z - \lambda}{z - \bar{\lambda}} \quad \text{with } \lambda \in \mathbb{H} , \ \varphi \in \mathbb{R} .$$

In the special case $\varphi = 0$, $\lambda = i$, we obtain the so-called CAYLEY map.

A Appendix : The Homotopical Version of the Cauchy Integral Theorem

We would like to show that the notion of an "elementary domain" is of purely topological nature. In other words, if

$$\varphi : D \longrightarrow D' \quad \text{is a homeomorphism,} \quad D, \ D' \subset \mathbb{C} \text{ domains },$$

and if D is *elementary*, then D' is elementary too.

In this framework, it is not convenient to consider only piecewise smooth curves. We shortly prove that one can also integrate *analytic* functions along arbitrary continuous curves.

Lemma A.1 *Let*

$$\alpha : [a, b] \longrightarrow D , \quad D \subset \mathbb{C} \ open,$$

be a (continuous) curve. Then there exists a partition

$$a = a_0 < a_1 < \cdots < a_n = b$$

and an $r > 0$ with the property $U_r\big(\alpha(a_\nu)\big) \subset D$ and

$$\alpha([a_\nu, a_{\nu+1}]) \subset U_r\big(\alpha(a_\nu)\big) \cap U_r\big(\alpha(a_{\nu+1})\big) \subset D \ \text{for } 0 \leq \nu < n .$$

Supplement. *Let $f : D \to \mathbb{C}$ be an analytic function. The number*

$$\sum_{\nu=0}^{n-1} \int_{\alpha(a_\nu)}^{\alpha(a_{\nu+1})} f(\zeta)\, d\zeta$$

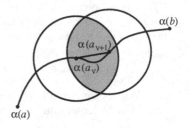

does not depend on the choice of the particular partition of $[a, b]$. (The integral is taken along the line segment connecting $\alpha(a_\nu), \alpha(a_{\nu+1})$ in $\mathbb{C}$.) If α is piecewise smooth, then the above sum coincides with the integral $\int_\alpha f(\zeta)\, d\zeta$ of f along the curve α.

We *extend our definition* of a line integral to continuous curves by the above Lemma.

The proof of Lemma A.1 immediately follows from the following facts (A) and (B):

(A) Existence of the LEBESGUE Number, Sect. IV.4. If $K \subset D$ is a compact subset of the open set $D \subset \mathbb{R}^n$, then there exists an $\varepsilon > 0$, such that

$$x \in K \implies U_\varepsilon(x) \subset D \, .$$

(B) *The Uniform Continuity Theorem:* For any $\varepsilon > 0$, there exists a $\delta > 0$ such that
$$x, y \in [a, b] \text{ and } |x - y| < \delta \implies |\alpha(x) - \alpha(y)| < \varepsilon \, . \qquad \square$$

We now consider continuous maps

$$H : Q \longrightarrow D \, , \quad D \subset \mathbb{C} \text{ open} \, ,$$

of the square

$$Q = \{\, z \in \mathbb{C} \, ; \quad 0 \le x, y \le 1 \,\} = [0, 1] \times [0, 1]$$

into an open set $D \subset \mathbb{C}$.

The image of the boundary ∂Q of Q can be interpreted as a closed curve:

$$\alpha = \alpha_1 \oplus \alpha_2 \oplus \alpha_3 \oplus \alpha_4 \, ,$$
$$\alpha_1(t) = H(t, 0) \qquad \text{for } 0 \le t \le 1 \, ,$$
$$\alpha_2(t) = H(1, t - 1) \quad \text{for } 1 \le t \le 2 \, ,$$
$$\alpha_3(t) = H(3 - t, 1) \quad \text{for } 2 \le t \le 3 \, ,$$
$$\alpha_4(t) = H(0, 4 - t) \quad \text{for } 3 \le t \le 4 \, .$$

We will denote this curve simply by $H|\partial Q$.

This curve is surely closed, but *not* every closed curve can be realized in this way. Intuitively, only those curves can be realized which can be "filled up" in D.

Proposition A.2 *Let*

$$H : Q \longrightarrow D , \quad D \subset \mathbb{C} \ \text{open} ,$$

be a continuous map, and let $f : D \to \mathbb{C}$ *be an analytic function. Then*

$$\int_{H|\partial Q} f(\zeta) \, d\zeta = 0 .$$

Proof. Let n be a natural number. We split Q as a union of n^2 squares

$$Q_{\mu\nu} = \left\{ z \in Q ; \ \frac{\mu}{n} \leq x \leq \frac{\mu+1}{n} , \ \frac{\nu}{n} \leq y \leq \frac{\nu+1}{n} \right\} \quad (0 \leq \mu, \nu \leq n-1) .$$

Because $H(Q) \subset D$ is a compact set, there exists after the choice of a sufficiently large n, for any (μ, ν) an open disk $U_{\mu\nu}$ with

$$H(Q_{\mu\nu}) \subset U_{\mu\nu} \subset D .$$

By the CAUCHY Integral Theorem we have

$$\int_{H|\partial Q_{\mu\nu}} f(\zeta) \, d\zeta = 0$$

and hence

$$\int_{H|\partial Q} f(\zeta) \, d\zeta = \sum_{0 \leq \mu, \nu \leq n-1} \int_{H|\partial Q_{\mu\nu}} f(\zeta) \, d\zeta = 0 . \qquad \square$$

Definition A.3 *Two curves*

$$\alpha, \beta : [0, 1] \longrightarrow D , \quad D \subset \mathbb{C} \ \text{open} ,$$

with the same starting and end points, $\alpha(0) = \beta(0)$ *and* $\alpha(1) = \beta(1)$*, are called* **homotopic** *in D (with fixed end points), iff there exists a continuous map (a so-called homotopy) $H : Q \to D$ with the following properties:*

(a) $\alpha(t) = H(t, 0)$,

(b) $\beta(t) = H(t, 1)$,

(c) $\alpha(0) = \beta(0) = H(0, s)$ *and*
$\alpha(1) = \beta(1) = H(1, s)$ *for*
$0 \leq s \leq 1$.

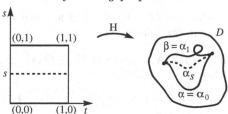

α and β have the same starting and end points, and this happens also for any curve $\alpha_s : [0, 1] \to D$,

$$\alpha_s(t) := H(t, s) \ ,$$

and we have

$$\alpha_0 = \alpha \text{ and } \alpha_1 = \beta \ .$$

Intuitively, the family (α_s) is a continuous deformation of α into β inside the domain D, such that the initial and end points are fixed during deformation.

Remark A.4 *In a convex domain $D \subset \mathbb{C}$, any two curves α and β with the same initial and end points are homotopic.*

Proof. The following homotopy deforms α into β:

$$H(t, s) = \alpha(t) + s(\beta(t) - \alpha(t)) \ . \qquad \qquad \square$$

From A.2 we infer:

Theorem A.5 (Homotopical version of the Cauchy Integral Theorem)
*Let $D \subset \mathbb{C}$ be open,, and let α, β be two curves which are homotopic in D. Then for **any** analytic function $f : D \to \mathbb{C}$*

$$\int_\alpha f = \int_\beta f \ .$$

Proof. Let $H : Q = [0, 1] \times [0, 1] \to D$ be a homotopy between α and β. Using A.2 we have

$$\int_{H|\partial Q} f(\zeta) \, d\zeta = 0$$

for any analytic function $f : D \to \mathbb{C}$. We have

$$0 = \int_{H|\partial Q} f(\zeta) \, d\zeta = \int_\alpha f + \int_{\beta^-} f = \int_\alpha f - \int_\beta f \ ,$$

hence

$$\int_\alpha f = \int_\beta f \ . \qquad \qquad \square$$

Definition A.6 *Let $\alpha : [0, 1] \to D$ be a closed curve in D, $\alpha(0) = \alpha(1) = z_0$. Then α is called **null-homotopic** in D, iff α is homotopic to the constant curve $\beta(t) := z_0$.*

*A domain $D \subset \mathbb{C}$ is called **simply connected**, iff any closed curve in D is null-homotopic in D.*

Remark A.7 *If $\alpha : [0, 1] \to D$ is a closed curve in D which is null-homotopic, then*

$$\int_\alpha f = 0$$

for any analytic function $f : D \to \mathbb{C}$.

Corollary. *If $D \subset \mathbb{C}$ is a simply connected domain, then*

$$\int_\alpha f = 0$$

*for **any** closed curve α in D and **any** analytic function $f : D \to \mathbb{C}$.*

Any simply connected domain is thus an elementary domain.

The converse of the last statement is also true:

Proposition A.8 *For a domain $D \subset \mathbb{C}$ the following propositions are equivalent:*

(a) *D is an elementary domain.*
(b) *D is simply connected.*

Proof. We have already seen, that (b) implies (a).
The implication (a) $\Rightarrow$ (b) follows by the following remarks:
(1) Being simply connected is a topological property, i.e. if $\varphi : D \to D'$ is a homeomorphism between two domains $D, D' \subset \mathbb{C}$, then D is simply connected iff D' is simply connected. (Any homotopy $H : [0,1] \times [0,1] \to D$ between two curves in D can be transported to a homotopy $H' := \varphi \circ H$ between the corresponding curves in D'.)
(2) The unit disk $\mathbb{E}$ and the complex plane $\mathbb{C}$ are simply connected (A.4).
The proof is now a corollary of the RIEMANN Mapping Theorem. $\square$

As a byproduct, we get the following deep topological result for the plane:

Proposition A.9 *Any two simply connected domains in the complex plane are homeomorphic (i.e. topologically equivalent).*

Proof. Any conformal map is automatically a homeomorphism. By the RIEMANN Mapping Theorem, it is enough to observe that the two standard elementary domains $\mathbb{C}$ and $\mathbb{E}$ are topologically equivalent. $\square$

As a final result in this appendix, we prove

Proposition A.10 *Any closed curve α in the domain $D := \mathbb{C}^\bullet$ with $\alpha(0) = \alpha(1) = 1$ is homotopically equivalent to a circle path of index $k \in \mathbb{Z}$ (i.e. "running k times around the zero point").*

The definition of the winding number by means of an integral is thus *a posteriori* completely justified. Especially, we have

Consequence. *The winding number, as defined by III.6.1, is always an integer.*

The proof uses the exponential function

$$\exp : \mathbb{C} \longrightarrow \mathbb{C}^\bullet \, ,$$

and is based on the following

Claim. *Let* $\alpha : [0,1] \to \mathbb{C}^{\bullet}$ *be a closed curve,* $\alpha(0) = \alpha(1) = 1$. *Then there exists a (uniquely determined) curve*

$$\tilde{\alpha} : [0,1] \longrightarrow \mathbb{C}$$

with the following properties:

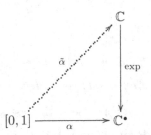

(a) $\tilde{\alpha}(0) = 0$,
(b) $\exp \circ \tilde{\alpha} = \alpha$.

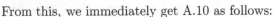

From this, we immediately get A.10 as follows:

From $\alpha(1) = 1$ we have $\tilde{\alpha}(1) = 2\pi i k$ with a suitable $k \in \mathbb{Z}$. Since $\mathbb{C}$ is convex there exists a homotopy $\tilde{H}$ between $\tilde{\alpha}$ and the line segment σ joining 0 and $2\pi i k$. Composing with exp, we get a homotopy

$$H := \exp \circ \tilde{H}$$

between α and $\exp \circ \sigma$.

Let us use the notation $\exp \circ \sigma =: \varepsilon_k$, the k-times passed circle line. Hence α and ε_k are homotopic, and in particular they have the same winding number, namely k. Up to the Claim, we are done.   □

Proof of the claim. We first show:

For any point $a \in \mathbb{C}^{\bullet}$ there exists an open neighborhood $V = V(a)$, such that the inverse image of V under exp *splits into a disjoint union of open sets U_n,*

$$\exp^{-1}(V) = \bigcup_{n \in \mathbb{Z}} U_n \quad (\textit{disjoint union}) ,$$

where exp *maps each U_n topologically onto V.*

When a does not lie on the negative real axis, then we can take V to be the slit plane $\mathbb{C}_-$. The inverse image of $\mathbb{C}_-$ under exp splits as a disjoint union of parallel horizontal strips of height 2π, as already observed in I.5.9.

If a lies on the negative real axis, then we slit the plane along the positive real axis. The inverse image of V also splits as a disjoint union of parallel horizontal strips of height 2π.

Using HEINE-BOREL's Theorem, we deduce the existence of a partition

$$0 = a_0 < a_1 < \cdots < a_m = 1 ,$$

such that each piece of the curve

$$\alpha\big([a_\nu, a_{\nu+1}]\big), \quad 0 \le \nu < m ,$$

is contained in an open set V as above.

We can then lift the whole curve, piece by piece, into the horizontal strips. □

B Appendix : A Homological Version of the Cauchy Integral Theorem

In connection with the CAUCHY Integral Theorem for star domains, we were led to the following question:

(1) For which *domains* $D \subset \mathbb{C}$ do we have

$$\int_\alpha f = 0$$

for *any* analytic function $f : D \to \mathbb{C}$, and *any* closed curve α in D?

By definition, we called such domains elementary, and in appendix A we saw that the *elementary* domains are exactly the simply connected domains in topology. In this appendix we will focus on another characterization of elementary domains. In a more general context, we will be concerned with the following question:

(2) Let $D \subset \mathbb{C}$ be an *arbitrary domain*. How can we characterize those closed curves α in D, for which

$$\int_\alpha f = 0 \qquad \text{for any analytic function } f : D \longrightarrow \mathbb{C} ? \qquad (*)$$

The answer will be that α satisfies $(*)$ iff the interior of α is contained in D.

Definition B.1 *A closed curve in a domain D is called* **null-homologous** *in D, iff its interior*

$$\text{Int}(\alpha) := \{ \, z \in \mathbb{C} \setminus \text{Image} \, (\alpha) \, ; \quad \chi(\alpha; z) \neq 0 \, \}$$

is contained in D.

Remark B.2 *Any null-homotopic curve in D is also null-homologous.*

Proof. For $D = \mathbb{C}$ this is clear. Else we can find an $a \in \mathbb{C} \setminus D$ in the complement of D. The function

$$f(z) = \frac{1}{z - a}$$

is analytic in D, so that its integral along α vanishes, i.e. a lies in the exterior of α. $\square$

The converse is not true! A counterexample is given (without proof) in the figure for $D := \mathbb{C} \setminus \{-1, 1\}$.

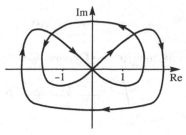

Now we come to the main result of this appendix. It represents a *global version* of the CAUCHY Integral Theorem. For its proof, we reproduce the surprisingly simple argumentation of J.D. DIXON, [Dix]. This source has become standard in textbooks (see also [La1], [Ru] or [McG]).

Theorem B.3 *Let α be a closed curve in an open set $D \subset \mathbb{C}$. Then the following properties are equivalent:*

(1) *For any analytic function $f : D \to \mathbb{C}$ we have*

$$\boxed{\begin{array}{c} \textbf{\textit{The General Cauchy Integral Formula}} \\[6pt] f(z)\,\chi(\alpha;z) = \dfrac{1}{2\pi\mathrm{i}} \displaystyle\int_\alpha \dfrac{f(\zeta)}{\zeta - z}\,d\zeta \ . \end{array}}$$

    for all $z \in D \setminus \mathrm{Image}\,(\alpha)$.

(2) $\int_\alpha f = 0$ *for any analytic function $f : D \to \mathbb{C}$.*

(3) α *is null-homologous in D.*

Proof. $(1) \Rightarrow (2)$.
Let us pick a point $a \in D \setminus \mathrm{Image}\,\alpha$ (it exists since D is non-compact), and consider

$$g : D \longrightarrow \mathbb{C} \quad \text{with} \quad g(z) := (z - a)f(z) \ .$$

Then g is analytic in D, $g(a) = 0$, and the General CAUCHY Integral Formula for g (instead of f) gives

$$\int_\alpha f(\zeta)\,d\zeta = \int_\alpha \frac{g(\zeta)}{\zeta - a}\,d\zeta = 2\pi\mathrm{i}g(a)\,\chi(\alpha;a) = 0 \ .$$

$(2) \Rightarrow (3)$.
For $a \in \mathbb{C} \setminus D$ the function $D \to \mathbb{C}$, $z \mapsto 1/(z - a)$, is analytic in D. Then (2) implies $\dfrac{1}{2\pi\mathrm{i}} \displaystyle\int_\alpha \frac{1}{z - a}\,dz = 0$.

$(3) \Rightarrow (1)$.
Let α be a closed, null-homologous curve in D. Let $f : D \to \mathbb{C}$ be an analytic function. From the definition of the winding number,

$$\chi(\alpha;z) = \frac{1}{2\pi\mathrm{i}} \int_\alpha \frac{1}{\zeta - z}\,d\zeta$$

we can rewrite the General CAUCHY Integral Formula (1) as

$$\int_\alpha \frac{f(\zeta) - f(z)}{\zeta - z}\,d\zeta = 0 \qquad \text{for all } z \in D \setminus \mathrm{Image}\,\alpha \ .$$

The idea is now to show that the integral in the L.H.S. is an analytic function $G : D \setminus \text{Image } \alpha \to \mathbb{C}$ in the variable z, which can be extended to an *entire* function $F : \mathbb{C} \to \mathbb{C}$. Then we show that F is bounded, and apply LIOUVILLE's Theorem. The proof will also give the value 0 for the (constant) function F.

We start to study the quotient $\quad \frac{f(\zeta) - f(z)}{\zeta - z} \quad$ considered as a function of ζ and z. First, we need some Lemmas.

Lemma B.3$_1$ *Let $D \subset \mathbb{C}$ be open. Let $f : D \to \mathbb{C}$ be an analytic function. The map*

$$\varphi : D \times D \longrightarrow \mathbb{C} \; ,$$

$$(\zeta, z) \longmapsto \begin{cases} \dfrac{f(\zeta) - f(z)}{\zeta - z} & \textit{if } \zeta \neq z \; , \\ f'(z) & \textit{if } \zeta = z \; , \end{cases}$$

is continuous as a function of the total variable $(\zeta, z) \in D \times D$.

Proof (See also Exercise 14 in II.3). We have to show continuity only in the points (ζ_0, z_0) of the diagonal, i.e. $\zeta_0 = z_0$. We choose $\delta > 0$, such that the disk $U = U_\delta(a)$ of radius δ centered at $a = \zeta_0 = z_0$ is contained in D. The Main Theorem of differential and integral calculus gives

$$\frac{f(\zeta) - f(z)}{\zeta - z} = \int_0^1 f'\big((1 - t)z + t\zeta\big) \, dt \; ,$$

for two different values ζ and z in U. Then

$$\varphi(\zeta, z) - \varphi(a, a) = \int_0^1 \Big[f'(\sigma(t)) - f'(a) \Big] \, dt \qquad \text{where } \sigma(t) := (1 - t)z + t\zeta \; .$$

The last equality also holds for $\zeta = z$. The Lemma easily follows from the continuity of f'.

The function $\varphi(\zeta, \cdot) : D \to \mathbb{C}$, $z \to \varphi(\zeta, z)$, is for each fixed value of ζ an analytic function in the variable z. (The analyticity in $z = \zeta$ follows from RIEMANN's Removability Theorem, or already from II.2.7$_1$.) The LEIBNIZ rule, II.3.3, then implies:

Lemma B.3$_2$ *The function $G : D \to \mathbb{C}$,*

$$G(z) = \int_\alpha \varphi(\zeta, z) \, d\zeta$$

is analytic in D. □

Lemma B.3$_3$ *There exists an entire function*

$$F : \mathbb{C} \longrightarrow \mathbb{C} \quad \textit{with} \quad F|D = G \; .$$

Proof. We now use essentially the assumption $\text{Int}(\alpha) \subset D$. Let

$$\text{Ext}(\alpha) = \{ \, z \in \mathbb{C} \setminus \text{Image } \alpha \; ; \quad \chi(\alpha; z) = 0 \, \}$$

be the exterior of α. Then $\text{Ext}(\alpha)$ is open, and

$$D \cup \text{Ext}(\alpha) = \mathbb{C} \, .$$

For $z \in D \cap \text{Ext}(\alpha)$ we use $\chi(\alpha; z) = 0$ to conclude

$$G(z) = \int_\alpha \frac{f(\zeta) - f(z)}{\zeta - z} \, d\zeta = \int_\alpha \frac{f(\zeta)}{\zeta - z} \, d\zeta - 2\pi i f(z) \chi(\alpha; z) = \int_\alpha \frac{f(\zeta)}{\zeta - z} \, d\zeta \, .$$

The function

$$H : \mathbb{C} \setminus \text{Image } \alpha \longrightarrow \mathbb{C} \, ,$$

$$H(z) = \int_\alpha \frac{f(\zeta)}{\zeta - z} \, d\zeta \, ,$$

is an analytic function (once more by LEIBNIZ' rule), and in particular H is also analytic in $\text{Ext}(\alpha)$. Because H coincides in the intersection $D \cap \text{Ext}(\alpha)$ with G, we obtain an entire function

$$F : \mathbb{C} \longrightarrow \mathbb{C} \quad \text{with} \quad F(z) = \begin{cases} G(z) \, , & \text{if } z \in D \, , \\ H(z) \, , & \text{if } z \in \text{Ext}(\alpha) \, , \end{cases}$$

hence an analytic extension of G to the whole $\mathbb{C}$.   □

Lemma B.3$_4$ *F is identically zero.*

Proof. The image of our curve α is bounded. It lies in a disk $U_R(0)$ of radius, say $R > 0$. Then the complement of this disk is in the exterior of α,

$$\{ \, z \in \mathbb{C} \; ; \quad |z| > R \, \} \subset \text{Ext}(\alpha) \, .$$

We then choose a piecewise smooth curve β, which runs inside $D \cap U_R(0)$ and is homotopic to α. This is possible by A.1, taking β to be a polygonal curve. We can and do replace α by β in the definition of F. Then we apply the standard estimate for integrals for $|z| > R$. Then $F(z) = H(z)$, and hence

$$|F(z)| = |H(z)| = \left| \int_\beta \frac{f(\zeta)}{\zeta - z} \, d\zeta \right| \leq \frac{C}{|z| - R}$$

with a suitable constant C. Hence F is bounded on $\mathbb{C}$. It is also entire, hence constant by LIOUVILLE's Theorem. The estimate for $|z| > R$ also gives

$$\lim_{|z| \to \infty} |F(z)| = 0 \, ,$$

so F is constant zero, $F \equiv 0$.   □

We finally come back to the proof of Theorem B.3. From $F \equiv 0$, we have in particular $G \equiv 0$, i.e. we have for any $z \in D \setminus \text{Image } \alpha$:

$$\int_\alpha \frac{f(\zeta) - f(z)}{\zeta - z} \, d\zeta = \int_\alpha \varphi(\zeta, z) \, d\zeta = G(z) = 0 \; .$$

Using the definition of the winding number, we get

$$f(z)\chi(\alpha; z) = \frac{1}{2\pi i} \int_\alpha \frac{f(\zeta)}{\zeta - z} \, d\zeta \; .$$

This finishes the proof of Theorem B.3.   □

Definition B.4 *Two closed curves α, β in a domain D are called homologous in D, iff for any point a in the complement of D the two winding numbers are equal,*

$$\chi(\alpha; a) = \chi(\beta; a) \; .$$

The base points of the two curves α, β can be different. Theorem B.3 is often formulated in the following form:

Corollary B.5 *Let $D \subset \mathbb{C}$ be a domain, and let α, β be two homologous closed curves in D. Then for any analytic function $f : D \to \mathbb{C}$*

$$\int_\alpha f = \int_\beta f \; .$$

Proof. We connect the initial points of α and β in D by some suitable curve σ, and consider the curve

$$\gamma := (\alpha^-)_0 \oplus \sigma_0 \oplus \beta_0 \oplus (\sigma^-)_0.$$

The lower index 0 reflects the fact that we possibly have to shift the parameter intervals such that $\oplus$ can be defined. Then γ is a closed, null-homologous curve, and we apply B.3, (2) $\Rightarrow$ (1).   □

Using the homological version of the CAUCHY Integral Theorem, we also can state a generalized version of the Residue Theorem (compare with III.6.3).

Theorem B.6 *Let $D \subset \mathbb{C}$ be open, and let $S \subset D$ be discrete in D. We also consider an analytic function $f : D \setminus S \to \mathbb{C}$, and let α be a closed curve in $D \setminus S$, with its interior contained in D, $\text{Int}(\alpha) \subset D$. (So α is null-homologous in D.) Then*

$$\int_\alpha f(\zeta) \, d\zeta = 2\pi i \sum_{s \in S} \text{Res}(f; s) \, \chi(\alpha; s) \; .$$

Proof. The set

$$\text{Int}(\alpha) \cup \text{Image}(\alpha)$$

is bounded and closed, hence compact. There exist hence only finitely many points $s \in S$ in the interior of α. The sum in the Theorem is a finite sum, thus well defined. We can now use Theorem B.3, and proceed as in the proof of the Residue Theorem.   □

C Appendix : Characterizations of Elementary Domains

We have already seen in Appendix A, that the notion of an *elementary domain* is actually a topological notion, A.8:

The elementary domains $D \subset \mathbb{C}$ are exactly the simply connected domains.

Using the homological version of the CAUCHY Integral Theorem, we obtain further characterizations. In the following theorem, we list a series of equivalent properties for the fact that a domain $D \subset \mathbb{C}$ is simply connected. Some authors consider this theorem to be an *aesthetic peak* of the classical complex analysis. However, its practical use should not be overestimated.

Theorem C.1 *For a domain $D \subset \mathbb{C}$, the following properties are equivalent:*

Function theoretical characterizations

(1) *Any analytic function in the domain D admits primitives, i.e. D is an elementary domain.*

(2) *For any analytic function in D, and any closed curve α in D, we have*

$$\int_\alpha f = 0 \,,$$

i.e. the general version of the Cauchy Integral Theorem is valid in D.

(3) *For any analytic function $f : D \to \mathbb{C}$, and any closed curve α in D, and any $z \in D \setminus \text{Image } \alpha$*

$$f(z)\chi(\alpha; z) = \frac{1}{2\pi i} \int_\alpha \frac{f(\zeta)}{\zeta - z} \, d\zeta \,,$$

i.e. the general version of the Cauchy Integral Formula is valid in D.

(4) *Any analytic function $f : D \to \mathbb{C}$, which has no zero in D, possesses an analytic logarithm in D, i.e. there exists an analytic function $l : D \to \mathbb{C}$ with $f = \exp \circ l$.*

(5) *Any analytic function $f : D \to \mathbb{C}$, which has no zero in D, possesses an analytic square root in D.*

(6) *D is either $\mathbb{C}$, or conform equivalent to the unit disk $\mathbb{E}$.*

A potential theoretic characterization

(7) *Any harmonic function $D \to \mathbb{R}$ is the real part of an analytic function $D \to \mathbb{C}$.*

Geometric characterizations

(8) *D is simply connected, i.e. any closed curve in D is null-homotopic in D (this is sometimes called "homotopically simply connected").*

(9) *The interior of any closed curve in D is contained in D (this is sometimes called "homologically simply connected").*

(10) *D is homeomorphic to the unit disk* $\mathbb{E} = \{\, z \in \mathbb{C} ;\ |z| < 1 \,\}$.

(11) *The complement of D in the Riemann sphere is connected, i.e. any locally constant* $h : \overline{\mathbb{C}} \setminus D \to \mathbb{C}$ *is constant.*

One can reformulate (11), if one doesn't to want use $\overline{\mathbb{C}}$, also in the following way:

(12) *If* $\mathbb{C} \setminus D = K \cup A$, *is a disjoint decomposition of* $\mathbb{C} \setminus D$ *into two sets, with K compact, and A closed,* $(K \cap A = \emptyset,)$ *then* $K = \emptyset$.

If one knows the notion of a connected component, one reformulate (12) as follows:

(13) $\mathbb{C} \setminus D$ *has no bounded connected component.*

Proof. All function theoretical properties are equivalent, as already proven. The most intricate step was the proof of the RIEMANN Mapping Theorem. In this proof, we used the assumption "elementary domain", only in the form: *Any non-vanishing analytic function on an elementary domain admits a square root.* This gives (5) ⇒ (6).

We have also shown, that the function theoretical characterization (1) implies the potential theoretical property (7). Conversely, from the potential theoretical property it follows the existence of a logarithm for an analytic function f, because $\log|f|$ is harmonic.

We also know that the geometrical properties (8)–(10) and the function theoretical properties are equivalent. It remains to show that property (12) characterizes the simple connectedness.

(We do not intend to make use of the notion of a *connected component*, or the *topology of the Riemann sphere*. Who knows them is invited to fill in the details for the equivalence of (11)–(13). However, for this restriction we have to pay a price, that we have to work with the rather clumsy condition (12) instead of the handy conditions (11) and (13).)

We show now (12) ⇒ (9). Let α be a closed curve in D. We decompose the complement of D in two disjoint sets:

$$K := \{\, a \in \mathbb{C} \setminus D ;\ \chi(\alpha;a) \neq 0 \,\},$$
$$A := \{\, a \in \mathbb{C} \setminus D ;\ \chi(\alpha;a) = 0 \,\}.$$

Both sets are closed, as inverse images of the closed sets $\{0\}$, $\mathbb{Z} \setminus \{0\}$ under a continuous map. The winding number with respect to α vanishes for all points in the complement of a disk containing (the image of) α, hence K is bounded, hence compact. By (12), $K = \emptyset$, hence $\text{Int}(\alpha) \subset D$.

For the converse (12) $\Leftarrow$ (9), we give an indirect proof. We assume that we have a disjoint decomposition $\mathbb{C} \setminus D = A \cup K$, into a closed set A and a compact set K. Then

$$D \cup K = \mathbb{C} \setminus A = D \cup (\mathbb{C} \setminus A) \,.$$

The set $U = D \cup K$ is thus open! (Think of K as a "hole" in D.) To finish the proof of this implication, and thus of the whole theorem, we need the following

Lemma C.2 *Let $U \subset \mathbb{C}$ be open, and let $K \subset U$ be a non-empty compact subset. Then the set $D := U \setminus K$ is not simply connected.*

This intuitively clear assertion once more confirms our feeling that simply connected domains are "domains without holes", and gives a rigorous meaning of it.

Proof of C.2. We exhibit a closed curve α in D, and a point a in K, such that the corresponding winding number $\chi(\alpha; a)$ is not zero. Then the interior of α is not contained in D.

An exact proof is tricky, but the idea is simple. We pave K by a net of squares, and form α from edges.

Pavement construction. Let n be a natural number. We consider the (finite) set of all squares indexed by integers

$$Q_{\mu\nu} := \left\{ z = x + \mathrm{i}y \;;\; \frac{\mu}{n} \le x \le \frac{\mu+1}{n} \,,\; \frac{\nu}{n} \le y \le \frac{\nu+1}{n} \right\},$$

which intersect K. A simple compactness argument shows that for sufficiently large n all these squares are contained in U. Let Q be the finite union of these squares. After enlarging n if necessary, we can assume that K is contained in the interior of Q. We will now construct (finitely many) closed curves $\alpha_1, \dots, \alpha_k$, such that the union of their images is ∂Q. This construction will also give

$$\sum_{j=1}^{k} \int_{\alpha_j} \frac{d\zeta}{\zeta - a} = 2\pi\mathrm{i}$$

for all a in the interior of Q. In particular, any point of K is surrounded by at least one of the constructed curves. During the construction, we will have to manage combinatorial difficulties. For instance, we have to avoid that some boundary edge is passed twice.

Construction of the boundary curves (according to Leutbecher [Le]). The boundary of Q is a finite union of some edges of the squares $(Q_{\mu\nu})$. In the following we call these edges *boundary edges*. Each boundary edge is the boundary of exactly one square of the finite set. We call this the touching square. We want to orient the boundary edges (this simply means an ordering of its two vertices). Each boundary edge is oriented in such a way such that the touching square is to the left. The heart of the construction consists in uniquely associating to any oriented boundary edge s another oriented boundary edge s' as its "successor'. The end point of s is a vertex of four squares of the net, we started with. There are 4 possible configurations depending on the fact which of the four belong to Q. The following figure shows these 4 cases and the way how we select the successor s'.

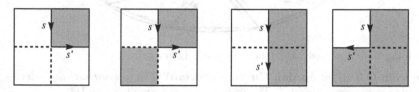

From the figure we extract the following combinatorial information:

If two oriented boundary edges have the same successor, then they coincide.

Let us construct α_1. We pick some oriented boundary edge s_0. We consider the chain of successors, $s_1 = s_0'$, $s_2 = s_1'$, ... There are only finitely many edges, so this chain has repetitions. Let m be minimal with $s_m \in \{s_0, \ldots, s_{m-1}\}$. From the combinatorial information we easily infer $s_m = s_0$. Thus the chain gives us an oriented polygonal path, denoted by α_1. If α_1 does not exhaust the boundary of Q, we repeat the argument for some new oriented boundary edge, as we did for s_0, to obtain α_2, *et cetera*.

Let now q_0 be one of the squares $Q_{\mu\nu}$ in the pavement, and let $a \in q_0$ be an arbitrary inner point. Then

$$\sum_{j=1}^{k} \int_{\alpha_j} \frac{d\zeta}{\zeta - a} = \sum_{q} \int_{\partial q} \frac{d\zeta}{\zeta - a},$$

where the sum in the R.H.S. runs through all squares q of the pavement. This is because the integrals over non-boundary edges cancel pairwise. Here the boundaries of the squares are oriented in the usual way (the squares are to the left). The integrals in the L.H.S. are either $2\pi i$ if $q = q_0$, or zero else. The L.H.S. becomes

$$\sum_{j=1}^{k} \int_{\alpha_j} \frac{d\zeta}{\zeta - a} = 2\pi i.$$

By continuity, this formula is true for all a in the interior of Q.   □

Jordan curves

A closed curve is called a JORDAN curve, iff it has no double point excepting the starting and end points. In the old function theory books, the CAUCHY Integral Theorem was usually proven only for JORDAN curves, which enough for practical use. It seems to be natural to approximate a JORDAN curve α by a polygonal path, decompose the polygon into (sufficiently small) triangles, and then split an integral along α as a sum of integrals along closed triangle curves.

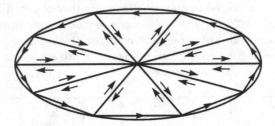

For doing that rigorously, the following result is needed:

Theorem C.3 (The Jordan Curve Theorem) *The interior and the exterior of a Jordan curve are connected. The interior is even simply connected.*

Unfortunately, this intuitively clear statement is rather deep. Even if we take this Theorem as granted, performing the above program is intricate, as demonstrated in the text book of DINGHAS, which is else an excellent source.

The reduction to JORDAN curves brings no simplifications, but unnecessary complications.

Exercises for Appendices A,B,C

1. Show the claimed invariance of an integral (of analytic functions, along continuous curves) with respect to taking subdivisions in A.1.

2. **A method for computing the winding number**
    Let $\alpha : [0,1] \to \mathbb{C}^\bullet$ be a closed curve which intersects the real axis $\operatorname{Im} z = 0$ in finitely many points, $\alpha(t_1), \alpha(t_2), \ldots, \alpha(t_N)$ corresponding to the arguments $t_1 < t_2 < \ldots < t_N$. Let $\alpha(t) = \xi(t) + i\eta(t)$, $\xi(t), \eta(t) \in \mathbb{R}$, be the usual decomposition in real and imaginary parts. Assume that η changes sign when passing t_ν $(1 \le \nu \le N)$. Without loss of generality, we can assume $[0,1] = [t_1, t_N]$, hence $\alpha(t_1) = \alpha(t_N)$, and also $\alpha(t) \neq 0$ for all $t \in [0,1]$. We extend α periodically to $\mathbb{R}$, with period $1 = t_N - t_1$. The points $t_1, \ldots, t_N$ fall into four different disjoint classes $M_1, \ldots, M_4$:

    M_1 $\xi(t_\nu) > 0$, and $\eta(t)$ changes sign from $t_\nu - \varepsilon$ to $t_\nu + \varepsilon$, from $-$ to $+$.
    M_2 $\xi(t_\nu) > 0$, and $\eta(t)$ changes sign from $t_\nu - \varepsilon$ to $t_\nu + \varepsilon$, from $+$ to $-$.
    M_3 $\xi(t_\nu) < 0$, and $\eta(t)$ changes sign from $t_\nu - \varepsilon$ to $t_\nu + \varepsilon$, from $+$ to $-$.
    M_4 $\xi(t_\nu) < 0$, and $\eta(t)$ changes sign from $t_\nu - \varepsilon$ to $t_\nu + \varepsilon$, from $-$ to $+$.

    We set for $1 \le \nu \le N$

    $$
    \delta_\nu = \begin{cases} +1 , & \text{if } t_\nu \in M_1 \cup M_3 , \\[2mm] -1 , & \text{if } t_\nu \in M_2 \cup M_4 . \end{cases}
    $$

    Then

    $$
    \chi(\alpha; 0) = \frac{1}{2} \sum_{\nu=1}^{N-1} \delta_\nu .
    $$

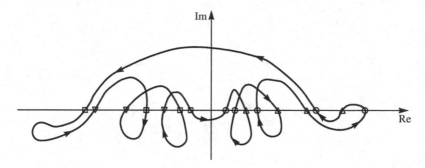

3. A domain $D \subset \mathbb{C}$ is simply connected, *iff* any two curves α and β running in D, which have the same starting point, and the same end point, are homotopic in D.

4. If $\alpha = (\alpha_1, \ldots, \alpha_n)$ is a system of closed curves α_ν, and Image $\alpha_\nu \subset D$ ($D \subset \mathbb{C}$, D domain), then we define for $a \notin \bigcup_{\nu=1}^{n}$ Image α_ν:

$$\chi(\alpha; a) := \sum_{\nu=1}^{n} \chi(\alpha_\nu; a) .$$

Two such systems α and β are called homologous, iff $\chi(\alpha; z) = \chi(\beta; z)$ for all $z \in \mathbb{C} \setminus D$.

Show: If $f : D \to \mathbb{C}$ is analytic, and if α and β are two homologous systems (of closed curves) in D, then

$$\int_\alpha f = \int_\beta f \quad \left(:= \sum \int_{\beta_\nu} f(\zeta)\, d\zeta \right) .$$

(Compare with Corollary B.5).

5. Recall the function G from Lemma B.3$_2$. Show that G is analytic, by checking its continuity and using MORERA's Theorem.

6. Give more detailed arguments for the equivalences in the proof of C.1.

V

Elliptic Functions

Historically, the starting point of the theory of elliptic functions were the *elliptic integrals,* named in this way because of their direct connection to computing arc lengths of ellipses. Already in 1718 (G.C. FAGNANO), a very special elliptic integral was extensively investigated,

$$E(x) := \int_0^x \frac{dt}{\sqrt{1-t^4}} \ .$$

It represents in the interval $]0, 1[$ a strictly increasing (continuous) function. So we can consider its inverse function f. A result of N.H. ABEL (1827) affirms that f has a meromorphic continuation into the entire $\mathbb{C}$. In addition to an obvious real period, ABEL discovered a hidden complex period. So the function f turned out to be a *doubly periodic function.* Nowadays, a meromorphic function in the plane with two independent periods is also called *elliptic.* Many results that were already know for the elliptic integral, as for instance the famous EULER Addition Theorem for elliptic integrals, appeared to be surprisingly simple corollaries of properties of elliptic functions. This motivated K. WEIERSTRASS to turn the tables. In his lectures in the winter term 1862/1863 he gave a purely function theoretical introduction to the theory of elliptic functions. In the center of this new setup, there is a special function, the $\wp$-function. It satisfies a differential equation which immediately shows the inverse function of $\wp$ to be an elliptic integral. The theory of elliptic integrals was thus derived as a byproduct of the theory of elliptic functions.

We already came across the WEIERSTRASS $\wp$-function as an example of a MITTAG-LEFFLER series, but no periodicity aspects were discussed there. We will see that all elliptic functions can be constructed from the $\wp$-function.

The historically older approach to the theory of elliptic functions (ABEL, 1827/1828, JACOBI, 1828) was not leading to the $\wp$-function, but rather to so-called theta functions. In connection with the ABEL Theorem describing the possible zeros and poles of elliptic functions, we will also touch lightly this approach at the end of Sect. V.6.

Functions with two independent periods ω_1 and ω_2 can be considered as functions on the factor group $\mathbb{C}/L$, $L = \mathbb{Z}\omega_1 + \mathbb{Z}\omega_2$. This quotient can be geometrically realized by gluing opposite sides of the parallelogram

$$\{ t_1\omega_1 + t_2\omega_2 \ ; \quad 0 \leq t_1, t_2 \leq 1 \} \ .$$

E. Freitag and R. Busam, *Complex Analysis,*
DOI: 10.1007/978-3-540-93983-2_V, © Springer-Verlag Berlin Heidelberg 2009

This identification of corresponding points on parallel sides gives rise to a *torus*. Two tori are always *topologically equivalent*. But they are only conform equivalent, iff the corresponding lattices can be obtained from each other by rotation and dilation. We call the two lattices *equivalent* if this is the case. The study of equivalence classes of lattices leads to the theory of *modular functions*. We will introduce them at the end of this chapter, and systematically investigate them in the next one.

V.1 Liouville's Theorems

Let us recall the notion of a *meromorphic* function on an open subset $D \subset \mathbb{C}$. (See also the appendix A to III.4 and III.5 at page 155.) Such a function is a map

$$f : D \longrightarrow \overline{\mathbb{C}} = \mathbb{C} \cup \{\infty\}$$

with the following properties:

(a) The set of points with value ∞

$$S = f^{-1}(\infty) = \{\, a \in D ; \quad f(a) = \infty \,\}$$

is discrete in D, i.e. it has no accumulation point in D.

(b) The restriction

$$f_0 : D \setminus S \longrightarrow \mathbb{C} ,$$
$$f_0(z) = f(z) \ \text{ for } \ z \in D , \quad z \notin S ,$$

is analytic.

(c) The points with value ∞ are poles of f_0.

We recall the definition of the sum of two meromorphic functions f and g. First, we can consider the restricted analytic function (also denoted by) $f + g$,

$$(f + g)(z) := f(z) + g(z) \text{ on } \mathbb{C} \setminus (S \cup T) , \quad S = f^{-1}(\infty) , \quad T = g^{-1}(\infty) .$$

It has in $S \cup T$ only non-essential (possibly removable) singularities, and we set

$$(f + g)(a) := \lim_{z \to a} (f(z) + g(z)) \qquad \text{if it exists in } \mathbb{C} ,$$
$$:= \infty \qquad\qquad\qquad \text{else, i.e. if } a \text{ is a pole of } f + g .$$

We thus obtain a meromorphic function

$$f + g : D \longrightarrow \overline{\mathbb{C}} .$$

Analogously, we define the product $f \cdot g$ and the quotient f/g, where in the last case the zero set of the denominator g must be discrete. If D is a domain, this means that g is not identically zero. We obtain

The set of meromorphic functions on a domain $D \subset \mathbb{C}$ is a field.

Elliptic functions are doubly periodic meromorphic functions on $\mathbb{C}$.

Definition V.1.1 *A subset $L \subset \mathbb{C}$ is a **lattice**[1], iff there exist two "vectors" ω_1 and ω_2 in $\mathbb{C}$, which are linearly independent over $\mathbb{R}$, and generate L as an abelian group, i.e.*

$$L = \mathbb{Z}\omega_1 + \mathbb{Z}\omega_2 = \{m\omega_1 + n\omega_2; \quad m, n \in \mathbb{Z}\}.$$

(*Observation.* Two complex numbers are $\mathbb{R}$-linearly independent, iff none of them is zero and their quotient is non-real.)

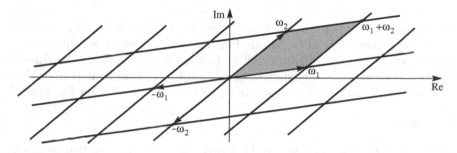

Definition V.1.2 *An elliptic function for the lattice L is a meromorphic function*

$$f : \mathbb{C} \longrightarrow \overline{\mathbb{C}} = \mathbb{C} \cup \{\infty\}$$

with the property

$$f(z + \omega) = f(z) \quad for \ \ \omega \in L \ and \ z \in \mathbb{C} \ .$$

It is enough to require the ω-periodicity only for two generators ω_1 and ω_2 of L:

$$f(z + \omega_1) = f(z + \omega_2) = f(z) \quad \text{for all} \ \ z \in \mathbb{C} \ .$$

Because of this, elliptic functions are also called *doubly periodic*.

The set $\mathcal{P}$ of poles of an elliptic function is "periodic", too,

$$a \in \mathcal{P} \implies a + \omega \in \mathcal{P} \ \text{for} \ \ \omega \in L \ .$$

The same also holds for the set of zeros.

J. LIOUVILLE proved in 1847 in his lectures the following three fundamental results on elliptic functions:

Theorem V.1.3 (The First Liouville Theorem, J. Liouville, 1847) *Any elliptic function without poles is constant.*

---

[1] This is an *ad-hoc* definition. We will replace it in the appendix by an invariant version.

Proof. The set

$$\mathcal{F} = \mathcal{F}(\omega_1, \omega_2) = \{\, t_1\omega_1 + t_2\omega_2\,;\quad 0 \le t_1, t_2 \le 1\,\}$$

is called a *fundamental region* for a lattice L, or also a *fundamental parallelogram* with respect to the basis ω_1, ω_2 (of L over $\mathbb{Z}$, or of $\mathbb{C}$ over $\mathbb{R}$).

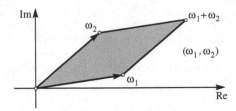

Obviously, for any point $z \in \mathbb{C}$ there exists a lattice point $\omega \in L$, such that $z - \omega \in \mathcal{F}$. Any value of an elliptic function is hence also taken in a fundamental region $\mathcal{F}$. But $\mathcal{F}$ is bounded and closed in $\mathbb{C}$, hence, any continuous function defined on $\mathcal{F}$ is bounded. An elliptic function without poles is thus bounded on $\mathcal{F}$, and hence also on the entire $\mathbb{C}$, so it is a constant function. □

The torus

Let f be an elliptic function for the lattice L. If z and w are two points in $\mathbb{C}$, such that their difference lies in L, then we have $f(z) = f(w)$. It is thus natural to introduce the *factor group* $\mathbb{C}/L$. The elements of this factor group are equivalence classes with respect to the equivalence relation

$$z \equiv w \quad \mathrm{mod}\ L \iff z - w \in L\,.$$

We denote the equivalence class (the orbit) of z by $[z]$, hence

$$[z] = \{\, w \in \mathbb{C}\,;\quad w - z \in L\,\} = z + L\,.$$

The addition in $\mathbb{C}$ induces an addition in $\mathbb{C}/L$

$$[z] + [w] := [z + w]\,.$$

It does not depend on the choice of the representatives of z and w. Hence we get on $\mathbb{C}/L$ a *structure as an abelian group*.

If f is an elliptic function for the lattice L, then there exists a uniquely determined map

$$\widehat{f} : \mathbb{C}/L \longrightarrow \overline{\mathbb{C}}\,,$$

such that the diagram

$$
\begin{array}{ccc}
z \quad \in & \mathbb{C} \xrightarrow{\ f\ } & \overline{\mathbb{C}} \\
\downarrow & \downarrow \quad \diagup \widehat{f} & \\
[z] \quad \in & \mathbb{C}/L &
\end{array}
$$

commutes. The map $\widehat{f}$ is well defined, i.e. the definition

$$\widehat{f}([z]) := f(z)$$

does not depend on the choice of the representative z.

By a slight abuse of notation we will write f instead of $\widehat{f}$ in the following. There will be no danger of confusion. An elliptic function $f : \mathbb{C} \to \overline{\mathbb{C}}$ can thus be considered as a function on the torus $\mathbb{C}/L$, that we also denote by $f : \mathbb{C}/L \to \overline{\mathbb{C}}$. This aspect will turn out to be useful.

Geometric picture of the torus

As mentioned in the proof of the First LIOUVILLE Theorem, each point in $\mathbb{C}/L$ has a representative in the fundamental region

$$\mathcal{F} = \{ \, z = t_1\omega_1 + t_2\omega_2 \, ; \quad 0 \leq t_1, t_2 \leq 1 \, \} \, .$$

Two points $z, w \in \mathcal{F}$ induce the same point in $\mathbb{C}/L$, iff they coincide or they both lie at the boundary of $\mathcal{F}$ at corresponding "opposite" places. We obtain a *geometrical model* of $\mathbb{C}/L$ by gluing opposite edges of a fundamental parallelogram:

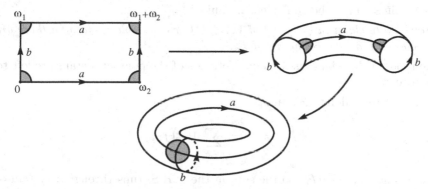

In the second volume we will obtain a more geometrical meaning of $\mathbb{C}/L$ as a compact RIEMANN surface, for the moment we use it just for visualization. But it is useful to have in mind that an elliptic function "lives" on a torus.

Knowing the First LIOUVILLE Theorem it is natural to study the poles of an elliptic function.

As already mentioned, if z is a pole of an elliptic function, then the whole orbit $[z]$ consists of poles. The translations $z \mapsto z + \omega$, $\omega \in L$, leave an elliptic function invariant, so the residues of f in z and $z + \omega$ coincide,

$$\mathrm{Res}(f; z) = \mathrm{Res}(f; z + \omega) \, .$$

This allows us to introduce (independently of the choice of the representative z)

$$\mathrm{Res}(f; [z]) := \mathrm{Res}(f; z) \, .$$

Theorem V.1.4 (The Second Liouville Theorem) *An elliptic function has only finitely many poles modulo L (i.e. on the torus $\mathbb{C}/L$), and the sum of their residues vanishes:*

$$\sum_z \operatorname{Res}(f; z) = 0 .$$

Here, the sum is taken over a system of representatives modulo L for all poles of f.

Proof. The pole set $\mathcal{P}$ of an elliptic function is discrete, its intersection with a compact set, e.g. the fundamental region $\mathcal{F}$, is hence finite. So there are only finitely many poles modulo L. We compute the residue sum by integrating along the boundary of a translated fundamental parallelogram, the translation being necessary to avoid possible poles that may lie on the boundary $\partial\mathcal{F}$ of $\mathcal{F}$. So let

$$\mathcal{F}_a = a + \mathcal{F} = \{\, a + z \; ; \quad z \in \mathcal{F} \,\}$$

be the fundamental parallelogram obtained by a-translation. It has, like $\mathcal{F}$, the property that any orbit has representatives in $\mathcal{F}_a$. When two different points of $\mathcal{F}_a$ are equivalent, then they are boundary points. Two different inner points cannot be equivalent modulo L.

Intermediate claim in the proof of V.1.4. After a suitable choice of a there are no poles of f on the boundary of $\mathcal{F}_a$.

The proof of this claim uses the discreteness of the pole set, details are left to the reader.

We integrate f along $\partial\mathcal{F}_a$ and get

$$\int_{\partial\mathcal{F}_a} f = 2\pi \mathrm{i} \sum_{z \in \mathcal{F}_a} \operatorname{Res}(f; z) .$$

There is no pole in $\partial\mathcal{F}_a$, so the sum in the R.H.S. runs through a system of representatives modulo L. To finish the proof we observe the vanishing of the integral in the L.H.S. This is because the integrals along opposite edges of the parallelogram $\mathcal{F}_a$ cancel, for instance

$$\int_a^{a+\omega_1} f(\zeta)\, d\zeta = \int_{a+\omega_2}^{a+\omega_1+\omega_2} f(\zeta)\, d\zeta = -\int_{a+\omega_1+\omega_2}^{a+\omega_2} f(\zeta)\, d\zeta . \qquad \square$$

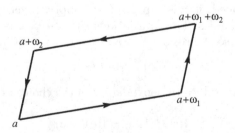

For an important consequence of Theorem V.1.4, we need the following

Definition V.1.5 *The **order** of an elliptic function is the number of all its poles on the torus* $\mathbb{C}/L$, *counting multiplicities (i.e. each pole is counted as many times as its pole order),*

$$\mathrm{Ord}(f) = -\sum_a \mathrm{ord}(f; a) \ .$$

In the above sum, a is taken in a system of representatives modulo L for the poles of f.

(The minus sign was introduced to have $\mathrm{Ord}(f) \geq 0$, because $\mathrm{ord}(f, a) < 0$ if a is a pole of f.) The First LIOUVILLE Theorem can be restated as

$$\mathrm{Ord}(f) = 0 \iff f \text{ is constant.}$$

A direct consequence of the Second LIOUVILLE Theorem V.1.4 is the fact that the pole set modulo L cannot consist of only one pole (mod L) which is *simple*. (The residue of a simple pole is always $\neq 0$.) We summarize:

Proposition V.1.6 *There exists no elliptic function of order* 1.

Let us now also study the zero set of an elliptic function. Analogously to the definition of the pole order, we can also consider the zero order of an elliptic function f which does not vanish. It is the number of the zeros of f, counted with multiplicities. Equivalently, the zero order of f is the pole order of the function $1/f$. We further generalize this notion.

Let f be a non-constant elliptic function, and let $b \in \mathbb{C}$ be fixed. Then

$$g(z) = f(z) - b$$

is also an elliptic function. We use the notation

b-$\mathrm{Ord}\ f :=$ number of zeros of g in $\mathbb{C}/L$, counted with multiplicities. Additionally we set

$$\infty\text{-}\mathrm{Ord}\ f := \mathrm{Ord}\ f \ .$$

Theorem V.1.7 (The Third Liouville Theorem) *The b-order of a non-constant elliptic function f is independent of $b \in \overline{\mathbb{C}}$,*

$$\mathrm{Ord}\ f = b\text{-}\mathrm{Ord}\ f \qquad \text{for any } b \in \overline{\mathbb{C}} \ .$$

In particular, f has modulo L as many zeros as poles.

Proof. Together with f, the derivative f' is also an elliptic function:

$$f(z + \omega) = f(z) \implies f'(z + \omega) = f'(z) \text{ for } \omega \in L \ .$$

Since f is non-constant and elliptic, the same is also true for

$$g(z) = \frac{f'(z)}{f(z)} \ .$$

We apply the Second LIOUVILLE Theorem V.1.4 to g. The point a is a pole of g, iff it is either a pole or a zero of f, and

$$\mathrm{Res}(g;a) = \mathrm{ord}(f;a) \quad \begin{cases} < 0\,, & \text{for a pole } a \text{ of } f\,, \\ > 0\,, & \text{for a zero } a \text{ of } f\,, \end{cases}$$

as can be seen by using the LAURENT series of f, see also III.6.4 (3). □

Let f be a non-constant elliptic function. A point $b \in \overline{\mathbb{C}}$ is called a *ramification point* (with respect to f), iff there exists an $a \in \mathbb{C}$ such that $\mathrm{ord}(f(z) - b; a) \geq 2$. In case of $b = \infty$ then a is a pole of order at least two.

The Third LIOUVILLE Theorem implies:

Remark V.1.8 *Let f be a non-constant elliptic function having order N, considered as a function on the torus,*

$$f : \mathbb{C}/L \longrightarrow \overline{\mathbb{C}} \ .$$

Then there are only finitely many ramification points $b \in \overline{\mathbb{C}}$, and hence also only finitely many pre-images $[a]$ in $\mathbb{C}/L$ of ramification points b, $f(a) = b$. For the number $\#f^{-1}(z)$ of pre-images of an arbitrary $z \in \overline{\mathbb{C}}$ we have

$$0 < \#f^{-1}(z) \quad \begin{cases} < N\,, & \text{if } z \text{ is a ramification point of } f\,, \\ = N\,, & else\,. \end{cases}$$

We make a further assertion about the position of the ramification points:

A power series converging in a suitable neighborhood of a,

$$a_0 + a_1(z - a) + a_2(z - a)^2 + \cdots$$

has a zero in $z = a$, iff $a_0 = 0$. This zero is multiple iff also $a_1 = 0$, i.e. iff f' also vanishes at a. From this simple fact follows:

Remark V.1.9 *Let*

$$f : \mathbb{C}/L \longrightarrow \overline{\mathbb{C}}$$

be an elliptic function, and let $b \in \mathbb{C}$ be a finite point $(b \neq \infty)$. The point b is a ramification point exactly if there exists a pre-image

$$a \in \mathbb{C}\,, \quad f(a) = b$$

at which the derivative of f vanishes.

Supplement. ∞ *is a ramification point of $f \not\equiv 0$, iff 0 is a ramification point of $1/f$.*

A Appendix to the Definition of the Period Lattice

Let $L \subset \mathbb{R}^n$ be an additive subgroup, i.e.

$$a, b \in L \implies a \pm b \in L .$$

If L is *discrete*, then one can show the existence of k linearly independent vectors $\omega_1, \ldots, \omega_k \in \mathbb{R}^n$, $0 \le k \le n$, with the property

$$L = \mathbb{Z}\omega_1 + \cdots + \mathbb{Z}\omega_k .$$

In the second volume we will prove this *structure theorem* in connection with ABELian functions. The group L is thus isomorphic to $\mathbb{Z}^k$. The number k is the *rank* of L. In case of $k = n$ the discrete additive group L is called a *lattice*. In case $n = 2$ there are three possibilities for the rank, and thus three possibilities for discrete subgroups of $\mathbb{R}^2$:

(1) $L = \{0\}$, $k = 0$,
(2) $L = \mathbb{Z}\omega_1$, $\omega_1 \in \mathbb{C}$, $\omega_1 \ne 0$, (cyclic group) $k = 1$,
(3) $L = \mathbb{Z}\omega_1 + \mathbb{Z}\omega_2$, $\omega_1, \omega_2 \in \mathbb{C}$, and ($\omega_1$ and ω_2) are
    $\mathbb{R}$-linearly independent, $k = 2$.

Let $f : \mathbb{C} \to \overline{\mathbb{C}}$ be a non-constant meromorphic function. The set of its periods

$$L_f := \{\, \omega \in \mathbb{C}; \quad f(z + \omega) = f(z) \text{ for all } z \in \mathbb{C} \,\}$$

is a *discrete subgroup* of $\mathbb{C}$ by the identity principle. There are three possibilities:

(1) $L_f = \{0\}$ (f has no non-trivial periods).
(2) L_f is cyclic (f is simply periodic).
(3) L_f is a lattice (f is an elliptic function).

In case $n = 2$ the proof of the *structure theorem* is very simple, see also Exercise 3 in this section.

Exercises for V.1

1.   Let $\mathcal{F}$ be a fundamental parallelogram of the lattice L. Show

$$\mathbb{C} = \bigcup_{\omega \in L} (\omega + \mathcal{F}) .$$

2.   Let $f : \mathbb{C} \to \overline{\mathbb{C}}$ be a non-constant meromorphic function. The set of its periods

$$L_f := \{\, \omega \in \mathbb{C}; \quad f(z + \omega) = f(z) \text{ for all } z \in \mathbb{C} \,\}$$

is a discrete subgroup of $\mathbb{C}$.

3. Prove the *structure theorem for discrete subgroups* $L \subset \mathbb{C}$.

   Hint. If $L \neq \{0\}$, then there exists a period $\omega_1 \neq 0$ in L of minimal absolute value. Then

   $$L \cap \mathbb{R}\omega_1 = \mathbb{Z}\omega_1 \ .$$

   If L lies in the real line $\mathbb{R}\omega_1$ generated by ω_1, then the structure theorem easily follows. Else, there exists an ω_2 in L, which does not lie in $\mathbb{R}\omega_1$, having minimal absolute value with this property. Show then $L = \mathbb{Z}\omega_1 + \mathbb{Z}\omega_2$.

   From the structure theorem we can prove:

   If $L \subset \mathbb{C}$ is a discrete subgroup which contains a lattice, then it is itself a lattice. In particular, any group L' which sits between two lattices L and L'', $L \subset L' \subset L''$, is also a lattice.

4. The number of minimal vectors (i.e. non-zero vectors of minimal modulus) in a lattice L is 2, 4 or 6. Give also explicit examples for each case.

5. Let f and g be elliptic functions for the same lattice.

   (a) If f and g have the same poles, and for each pole respectively the same principal parts, then f and g differ by an additive constant.

   (b) If f and g have the same pole set and the same zero set, and if for any pole or zero the corresponding multiplicities coincide, then f and g differ by a multiplicative constant.

6. Two lattices $L = \mathbb{Z}\omega_1 + \mathbb{Z}\omega_2$ and $L' = \mathbb{Z}\omega_1' + \mathbb{Z}\omega_2'$ coincide iff there exists a matrix with integral entries $\begin{pmatrix} a & b \\ c & d \end{pmatrix}$ and determinant ± 1 with the property

   $$\begin{pmatrix} \omega_1' \\ \omega_2' \end{pmatrix} = \begin{pmatrix} a & b \\ c & d \end{pmatrix} \begin{pmatrix} \omega_1 \\ \omega_2 \end{pmatrix} \ .$$

7. Let

   $$\mathcal{F} := \{ \, z \in \mathbb{C} \ ; \quad z = t_1\omega_1 + t_2\omega_2 \ , \ 0 \le t_1, t_2 \le 1 \, \}$$

   be the fundamental region of the lattice $L = \mathbb{Z}\omega_1 + \mathbb{Z}\omega_2$ with respect to a fixed basis ω_1, ω_2.

   Show: The EUCLIDian volume of the fundamental parallelogram is $|\operatorname{Im}(\overline{\omega}_1\omega_2)|$. This formula is independent of the choice of the basis.

8. The group $\mathbb{Z} + \mathbb{Z}\sqrt{2}$ is dense in $\mathbb{R}$.

9. Prove the following generalization of the First LIOUVILLE Theorem:

   Let f be an entire function, and let L be a lattice in $\mathbb{C}$. For any lattice point $\omega \in L$ we assume the existence of a polynomial P_ω with the property

   $$f(z + \omega) = f(z) + P_\omega(z) \ .$$

   Then f is a polynomial.

10. Another variant of the First LIOUVILLE Theorem:

    Let f be an entire function, and let L be a lattice in $\mathbb{C}$. For any lattice point $\omega \in L$ let there exist a number $C_\omega \in \mathbb{C}$ with the property

    $$f(z + \omega) = C_\omega f(z) \ .$$

Then

$$f(z) = Ce^{az}$$

for suitable constants C and a.

Hint. Without loss of generality, we can assume $\omega_1 = 1$ and $C_{\omega_1} = 1$. Use the FOURIER series of f. Another proof can be given by showing that f'/f is constant.

V.2 The Weierstrass ℘-function

We want to construct an as simple as possible example of a non-constant elliptic function. There is no elliptic function of order 1, so we start searching for an elliptic function of order 2. Such a function has modulo L either two poles of first order with opposite residues, or one pole of second order with residue zero. We insist on the second case and ask:

Is there any elliptic function of order two having only in 0 modulo L a pole? (This pole is thus of order two.)

Then any other pole is a lattice point. It is natural to construct such a function by means of a MITTAG-LEFFLER partial fraction series. For simplicity, we don't want to use here the theory of partial fraction series, and develop the answer from scratch.

A first trial could be

$$\sum_{\omega \in L} \frac{1}{(z - \omega)^2} \ ,$$

but this series is not absolutely convergent. For instance, in case of the lattice $L = \mathbb{Z} + \mathbb{Z}i$, we have for $\omega = m + ni \neq 0$

$$\left| \frac{1}{(z - \omega)^2} \right| = \frac{1}{|m + ni|^2} = \frac{1}{m^2 + n^2} \ .$$

But we have:

Lemma V.2.1 *The series*

$$\sum_{\substack{(m,n) \in \mathbb{Z} \times \mathbb{Z} \\ (m,n) \neq (0,0)}} \frac{1}{(m^2 + n^2)^\alpha} \ , \quad \alpha \in \mathbb{R} \ ,$$

converges iff $\alpha > 1$.

Proof. Basically, this result was already proven in connection with Proposition IV.2.6. Another proof compares the series with the integral

$$I = \int_{x^2 + y^2 \geq 1} \frac{dx \, dy}{(x^2 + y^2)^\alpha} \ ,$$

and it is easy to show that they either both converge or both diverge. Using polar coordinates

$$x = r \cos \varphi, \quad y = r \sin \varphi ,$$

we can compute the integral. The JACOBI determinant is r, hence we can rewrite the integral as

$$I = \int_0^{2\pi} \int_1^\infty \frac{r \, dr \, d\varphi}{r^{2\alpha}} = 2\pi \int_1^\infty \frac{dr}{r^{2\alpha-1}} .$$

The last integral converges iff $2\alpha - 1 > 1$. $\square$

From Lemma V.2.1 we derive:

Lemma V.2.2 *Let $L \subset \mathbb{C}$ be a lattice. The series*

$$\sum_{\omega \in L \setminus \{0\}} |\omega|^{-s} , \quad s > 2 ,$$

converges.

Proof. Let

$$L = \mathbb{Z}\omega_1 + \mathbb{Z}\omega_2 .$$

Using Lemma V.2.1 it is enough to prove an inequality of the form

$$|m\omega_1 + n\omega_2|^2 \geq \delta(m^2 + n^2)$$

with a suitable constant $\delta > 0$ which depends only on ω_1, ω_2. Equivalently, we show that the quotient

$$f(x,y) = \frac{|x\omega_1 + y\omega_2|^2}{x^2 + y^2} , \quad (x,y) \in \mathbb{R}^2 \setminus \{(0,0)\} ,$$

has a (strictly) positive minimum. Since f is homogenous it is enough to show that its has a (positive) minimum on

$$S^1 := \{ (x,y) \in \mathbb{R}^2 ; \quad x^2 + y^2 = 1 \}$$

But this clear since this set is compact. $\square$

We owe to K. WEIERSTRASS the modification of the initial "trial", by introducing summands which enforce convergence.

Lemma V.2.3 *Let $M \subset L \setminus \{0\}$ be a set of lattice points. Then the series*

$$\sum_{\omega \in M} \left[\frac{1}{(z-\omega)^2} - \frac{1}{\omega^2} \right]$$

converges normally in $\mathbb{C} \setminus M$, and defines an analytic function.

Proof. The absolute value of the general summand is

$$\left| \frac{1}{(z-\omega)^2} - \frac{1}{\omega^2} \right| = \frac{|z|\,|z-2\omega|}{|\omega|^2\,|z-\omega|^2} \ .$$

The number ω has power one in the numerator, and power four in the denominator. Usual estimate techniques give:

Let $K = \overline{U}_r(0)$ be the compact disk of radius $r > 0$ centered at 0. Then

$$\left| \frac{1}{(z-\omega)^2} - \frac{1}{\omega^2} \right| \leq 12r\,|\omega|^{-3} \quad \text{for all } z \in K$$

and for all but finitely many $\omega \in L$ (e.g. for all ω with $|\omega| \geq 2r$).

The proof of Lemma V.2.3 immediately follows from this an from V.2.2. $\square$

Definition V.2.4 (K. Weierstrass, 1862/63) *The function defined by*

$$
\boxed{
\begin{aligned}
\wp(z;L) = \wp(z) &= \frac{1}{z^2} + \sum_{\omega \in L \setminus \{0\}} \left[\frac{1}{(z-\omega)^2} - \frac{1}{\omega^2} \right] && \text{for } z \notin L \,, \\
\wp(z;L) = \wp(z) &= \infty && \text{for } z \in L \,,
\end{aligned}
}
$$

*is called the **Weierstrass $\wp$-function**[2] for the lattice L.*

The results up to now show:

Proposition V.2.5 *The Weierstrass $\wp$-function for the lattice L is a meromorphic function $\mathbb{C} \to \overline{\mathbb{C}}$. Its poles are of order two, which are located at the lattice points. So $\wp$ is analytic in $\mathbb{C} \setminus L$. The $\wp$-function is even, i.e.*

$$\wp(z) = \wp(-z) \ .$$

The Laurent expansion of $\wp$ in the origin $z_0 = 0$ is

$$\wp(z) = \frac{1}{z^2} + a_2 z^2 + a_4 z^4 + \cdots \qquad (\text{ especially } a_0 = 0! \,)$$

We will see in V.2.7 that the WEIERSTRASS $\wp$-function is an even elliptic function of order two.

Besides the $\wp$-function, its derivative also plays an important role. From V.2.3 and V.2.5 we can state:

Lemma V.2.6 *The derivative of the Weierstrass $\wp$-function*

$$\wp'(z) = -2 \sum_{\omega \in L} \frac{1}{(z-\omega)^3}$$

has poles of order 3 in the lattice points, and is analytic on $\mathbb{C} \setminus L$. As a derivative of an even function, $\wp'$ is odd,

$$\wp'(-z) = -\wp'(z) \ .$$

---

[2] This series can be traced back to EISENSTEIN, 1847, [Eis]. See also [We].

The next Proposition will show that the derivative $\wp'$ is an odd elliptic function of order three.

Proposition V.2.7 *The Weierstrass $\wp$-function for the lattice L is an even elliptic function of order two, its derivative $\wp'$ is an odd elliptic function of order three.*

Proof. A direct argument gives the double periodicity for the derivative $\wp'$: For all $\omega_0 \in L$

$$\wp'(z + \omega_0) = -2 \sum_{\omega \in L} \frac{1}{(z + \omega_0 - \omega)^3} = \wp'(z) \,.$$

This is because when ω runs through L, then $\omega - \omega_0$ also runs through L. (The "same" argument does not work for the $\wp$-function because of the correcting terms needed to ensure convergence.)

Taking primitives, the difference function

$$\wp(z + \omega_0) - \wp(z) \quad \text{for} \quad \omega_0 \in L$$

turns out to be a a constant. We show that it is zero.

We can assume that ω_0 is one of the basis elements, so $\frac{1}{2}\omega_0$ does not lie in L. We take $z = -\frac{1}{2}\omega_0$ and obtain for the constant the value

$$\wp\left(-\frac{1}{2}\omega_0 + \omega_0\right) - \wp\left(-\frac{1}{2}\omega_0\right) = \wp\left(\frac{1}{2}\omega_0\right) - \wp\left(-\frac{1}{2}\omega_0\right) = 0 \,.$$

($\wp$ is even for the last step.) This gives the double periodicity of $\wp$. $\square$

Let us now determine the zeros of $\wp'$.

Lemma V.2.8 (Invariant characterization for the zeros of $\wp'$) *A point $a \in \mathbb{C}$ is a zero of $\wp'$, iff*

$$a \notin L \quad and \quad 2a \in L \,.$$

There are exactly three zeros on the torus $\mathbb{C}/L$, all of them simple.

Proof. Let a satisfy the given property ($a \notin L$, $2a \in L$). Then

$$\begin{aligned}
\wp'(a) &= \wp'(a - 2a) &&(2a \in L) \\
&= \wp'(-a) = -\wp'(a) &&(\wp' \text{ is odd})
\end{aligned}$$

so $\wp'(a) = 0$.

We now have three zeros of $\wp'$, namely the different points modulo L:

$$\frac{\omega_1}{2}, \quad \frac{\omega_2}{2} \quad \text{and} \quad \frac{\omega_1 + \omega_2}{2} \,.$$

Applying the Third LIOUVILLE Theorem V.1.7, we see that there are no further zeros, and all above zeros are simple. $\square$

Notation.

$$e_1 := \wp\left(\frac{\omega_1}{2}\right) , \quad e_2 := \wp\left(\frac{\omega_2}{2}\right) , \quad e_3 := \wp\left(\frac{\omega_1 + \omega_2}{2}\right) .$$

Remark V.2.9 *The above "half lattice values" e_1, e_2, e_3 of the Weierstrass $\wp$-function are pairwise distinct and up to their ordering they do not depend on the choice of the basis ω_1, ω_2 of L.*

Proof. Let a be one of the three half points, $b = \wp(a)$. Since $\wp'(a) = 0$, it contributes to the b-order with multiplicity at least two. Since the b-order of $\wp$ is two, the value b cannot be taken at another point mod L. This shows that the e_i are pairwise different.

The set $\{e_1, e_2, e_3\}$ depends only on L, and not on the chosen basis. This follows from the invariant characterization in Lemma V.2.8.   □

Note that a lattice basis can be chosen in many ways. For example, ff ω_1, ω_2 is a basis, then $\omega_1, \omega_2 + \omega_1$, is a basis too.

Proposition V.2.10 *Let z and w be two arbitrary points in $\mathbb{C}$. Then*

$$\wp(z) = \wp(w)$$

iff

$$z \equiv w \quad \mod L \quad or \quad z \equiv -w \quad \mod L .$$

Proof. We fix w, and consider the elliptic function $z \mapsto \wp(z) - \wp(w)$ of order two. It has two zeros modulo L, counting multiplicities. These are obviously $z - w$ and $z = -w$ modulo L. (When w and $-w$ coincide modulo L, then w is a half lattice point and one has a double zero.)   □

This clarifies the *mapping properties* of the WEIERSTRASS $\wp$-function

$$\wp : \mathbb{C}/L \longrightarrow \overline{\mathbb{C}} .$$

There are four ramification points in $\overline{\mathbb{C}}$, namely e_1, e_2, e_3 and ∞. Each of them has exactly one pre-image in $\mathbb{C}/L$. All other points of $\overline{\mathbb{C}}$ have exactly two pre-images in $\mathbb{C}/L$.

At the end of this section, we compute the LAURENT expansion of the WEIER-STRASS $\wp$-function at $z_0 = 0$:

$$\wp(z) = \frac{1}{z^2} + \sum_{n=0}^{\infty} a_{2n} \, z^{2n} .$$

The radius of convergence for the series $\displaystyle\sum_{n=0}^{\infty} a_{2n} \, z^{2n}$, is

$$\min\{ \, |\omega| \; ; \; \omega \in L , \; \omega \neq 0 \, \} .$$

The coefficients a_{2n} are determined from TAYLOR's formula for the function

$$f(z) := \wp(z) - \frac{1}{z^2} , \qquad a_{2n} = \frac{f^{(2n)}(0)}{(2n)!} .$$

We already know $a_0 = 0$. For $n > 1$ we can inductively compute the derivatives

$$f^{(n)}(z) = (-1)^n (n+1)! \sum_{\omega \in L \setminus \{0\}} \frac{1}{(z - \omega)^{n+2}} .$$

Substituting $z = 0$ we get

$$a_{2n} = \frac{(2n+1)!}{(2n)!} \sum_{\omega \in L \setminus \{0\}} \frac{1}{\omega^{2(n+1)}} .$$

We conclude:

Proposition V.2.11 *The series*

$$G_n = \sum_{\omega \in L \setminus \{0\}} \omega^{-n} , \qquad n \in \mathbb{N} , \ n \geq 3 ,$$

converges absolutely, and one has

$$\wp(z) = \frac{1}{z^2} + \sum_{n=1}^{\infty} (2n+1) \, G_{2(n+1)} \, z^{2n}$$

in a suitable punctured neighborhood of $z = 0$ (namely in the maximal punctured open disk centered at 0, which does not contain any non-zero lattice point).

Observation. The symmetry $L \to L$, $\omega \to -\omega$, of the lattice shows that G_n vanishes for odd m.

The series G_n are the so-called *Eisenstein series*. We will study them in detail later.

Exercises for V.2

1. If $L \subset \mathbb{C}$ is a lattice, then the formula

$$\sum_{\omega \in L} \frac{1}{(z - \omega)^n}$$

    defines for any $n \geq 3$ an elliptic function of order n. What is the connection with the WEIERSTRASS $\wp$-function?

2. The WEIERSTRASS $\wp$-function has only lattice points as periods.

3. For an odd elliptic function associated to the lattice L the half-lattice points $\omega/2$, $\omega \in L$, are either zeros or poles.

4. Let f be an elliptic function of order m. Then its derivative f' is also an elliptic function of some order n, and the following inequality holds:

$$m + 1 \leq n \leq 2m .$$

Construct examples for the extreme cases $n = m + 1$ and $n = 2m$.

5. Let $L \subset \mathbb{C}$ be a lattice. We denote by $\widehat{L}$ the set of all conformal maps $\mathbb{C} \to \mathbb{C}$ of the form

$$z \longmapsto \pm z + \omega, \quad \omega \in L .$$

We identify (similar to the construction of the torus $\mathbb{C}/L$) two points in $\mathbb{C}$, iff they can be mapped into each other by suitable substitutions of $\widehat{L}$. After identification, we obtain $\mathbb{C}/\widehat{L}$, first as a set. Show that the $\wp$-function gives a bijection

$$\mathbb{C}/\widehat{L} \longrightarrow \overline{\mathbb{C}} .$$

The field of all $\widehat{L}$-invariant meromorphic functions is generated by $\wp$.

If the reader knows the fundamentals of topology, then one can say more:

When $\mathbb{C}/\widehat{L}$ is equipped with the quotient topology, then one obtains a sphere. (This can be proven using the $\wp$-function, or in a purely topological way.)

6. For two lattices L and L' the following conditions are equivalent:

(a) The intersection $L \cap L'$ is a lattice.

(b) The "lattice sum" $L + L' := \{ \omega + \omega' ; \ \omega \in L , \ \omega' \in L' \}$ is a lattice.

Then the two lattices are called *commensurable*.

Show: The fields of elliptic functions for two lattices L and L' have non-constant functions in common, iff the lattices are commensurable.

7. Any elliptic function of order ≤ 2 with period lattice L, whose pole set is contained in L, is of the form $z \to a + b\wp(z)$.

V.3 The Field of Elliptic Functions

The sum, difference, product and quotient (with non-zero denominator) of two elliptic functions are also elliptic functions. The set of all elliptic functions for a fixed lattice L is thus a field.

Notation. $K(L)$ is the field of elliptic functions for the lattice L.

Constant functions are elliptic functions. The map $\mathbb{C} \to K(L)$, associating to a complex number $C \in \mathbb{C}$ the constant function with the value C, gives an isomorphism of $\mathbb{C}$ onto the field of constant functions in $K(L)$:

$$\mathbb{C} \longrightarrow K(L) ,$$
$$C \longmapsto \text{constant function with value } C .$$

If there is no danger of confusion, we will always identify the number $C \in \mathbb{C}$ with the constant function with value C. After this identification, $\mathbb{C}$ becomes a subfield of $K(L)$.

In this section we will determine the structure of $K(L)$.

Let $f \in K(L)$ be an elliptic function, and let

$$P(w) = a_0 + a_1 w + \cdots + a_m w^m$$

be a polynomial. Then the composition $z \mapsto P(f(z))$ is also an elliptic function, denoted by $P(f)$. It is not identically zero if f is non-constant and P is not identically zero. (This is because a non-constant f takes any value.) More generally, let $R(z)$ be a rational function, i.e. a meromorphic function

$$R : \mathbb{C} \longrightarrow \overline{\mathbb{C}} \,,$$

which can be written as a quotient of two polynomials,

$$R = \frac{P}{Q} \,, \quad Q \neq 0 \,.$$

The elliptic function $\frac{P(f)}{Q(f)}$ depends only on R (and not of the chosen fraction representation), and we denote it by $R(f)$. Let $\tilde{R}$ be another rational function. Then (because f takes any value)

$$R(f) = \tilde{R}(f) \implies R = \tilde{R} \,.$$

In other words, if f is a non-constant elliptic function, then the map $R \mapsto R(f)$ gives an isomorphism from the field of rational functions onto a subfield of $K(L)$. This subfield is denoted by

$$\mathbb{C}(f) = \left\{ g \,; \quad g = R(f) \,, \ R \text{ is a rational function} \right\} \,.$$

We will now determine all *even* elliptic functions f, $(f(z) = f(-z))$, and at first only those whose pole set is contained in L. An example is the WEIERSTRASS $\wp$-function. More generally, any polynomial in $\wp$ has this property.

Proposition V.3.1 *Let $f \in K(L)$ be an even elliptic function, whose pole set is contained in L. Then f can be represented as a polynomial in $\wp$,*

$$f(z) = a_0 + a_1 \wp(z) + \cdots + a_n \wp(z)^n \quad (a_\nu \in \mathbb{C}) \,.$$

The degree of this polynomial is half of the order of f.

Proof. The case of a constant f is obvious. So let us start with a non-constant $f \in K(L)$. Then f has at least one pole in L and hence a pole at 0. The LAURENT series of the even function f in 0 has only even coefficients, and is hence of the form

$$f(z) = a_{-2n}z^{-2n} + a_{-2(n-1)}z^{-2(n-1)} + \cdots, \qquad n \geq 1, \, a_{-2n} \neq 0.$$

The LAURENT series of $\wp(z)$ is of the form (V.2.11)

$$\wp(z) = z^{-2} + \cdots$$

For the nth power we obtain

$$\wp(z)^n = z^{-2n} + \cdots$$

As f and $\wp$ the function

$$g = f - a_{-2n}\wp^n \, .$$

is an even elliptic function, whose pole set is contained in L. But the order of g is strictly smaller then the order of f. The proof now is obtained by induction. □

Proposition V.3.2 *Any **even** elliptic function is representable as a rational function in the Weierstrass $\wp$-function.*

*In other words: The field of all **even** elliptic functions for the lattice L is equal to $\mathbb{C}(\wp)$ (as a subfield of $K(L)$), and is thus isomorphic to the field of rational functions.*

Proof. Let f be a non-constant even elliptic function. If a is a pole of f which is not contained in L. The function

$$z \mapsto \left(\wp(z) - \wp(a) \right)^N f(z)$$

has in $z = a$ a removable singularity, if N is sufficiently large. Since f has only finitely many poles mod L, we find finitely many points $a_1, \ldots, a_m$ and natural numbers $N_1, \ldots, N_m$ such that

$$g(z) = f(z) \prod_{j=1}^{m} (\wp(z) - \wp(a_j))^{N_j}$$

has no poles outside L. By V.3.1 $g(z)$ is a polynomial in $\wp(z)$.   □

Any elliptic function can be written as a sum of an even and an odd function, both elliptic:

$$f(z) = \frac{1}{2}(f(z) + f(-z)) + \frac{1}{2}(f(z) - f(-z)) \, ,$$

because together with $z \mapsto f(z)$ also $z \mapsto f(-z)$ is an elliptic function. Let's look more closely to *odd* elliptic functions. The quotient of two odd elliptic functions is even. We obtain

Any odd elliptic function is the product of an even elliptic function and the odd elliptic function $\wp'$.

From V.3.2 we get the following **structure theorem for $K(L)$** :

Theorem V.3.3 *Let f be an elliptic function with period lattice L. Then there exist rational functions R and S, such that*

$$f = R(\wp) + \wp' \, S(\wp) \, ,$$

i.e.

$$K(L) = \mathbb{C}(\wp) + \mathbb{C}(\wp)\wp' \, .$$

(Hence $K(L)$ is a two-dimensional vector space over the field $\mathbb{C}(\wp)$.)
The proofs of V.3.1, V.3.2, V.3.3 are constructive.

Example. By Proposition V.3.1, the elliptic function $(\wp')^2$ is representable as a polynomial in $\wp$. Following the construction in the proof of V.3.1, we first need some LAURENT coefficients of the functions $\wp, \wp^2, \wp^3, \wp', (\wp')^2$.

(1) We already know from V.2.10

$$\wp(z) = z^{-2} + 3G_4 z^2 + 5G_6 z^4 + \cdots \quad .$$

(2) Taking termwise derivatives we obtain

$$\wp'(z) = -2z^{-3} + 6G_4 z + 20G_6 z^3 + \cdots \quad .$$

(3) Squaring $\wp$ we get

$$\wp(z)^2 = z^{-4} + 6G_4 + 10G_6 z^2 + \cdots \quad .$$

(4) Multiplying $\wp$ and $\wp^2$ we obtain

$$\wp(z)^3 = z^{-6} + 9G_4 z^{-2} + 15G_6 + \cdots \quad .$$

(5) Finally squaring $\wp'$ we get:

$$\wp'(z)^2 = 4z^{-6} - 24G_4 z^{-2} - 80G_6 + \cdots \quad .$$

We can now use the algorithm from the proof of V.3.1 to write $(\wp')^2$ as a polynomial in $\wp$. We first take the difference:

$$\wp'(z)^2 - 4\wp(z)^3 = -60G_4 z^{-2} - 140G_6 + \cdots \quad .$$

Adding $60G_4\wp(z)$, we get

$$\wp'(z)^2 - 4\wp(z)^3 + 60G_4\wp(z) = -140G_6 + \cdots \quad .$$

This is an elliptic function without poles, hence a constant. The constant is $-140G_6$.

Theorem V.3.4 (Algebraic differential equation for $\wp$) *We have:*

$$\wp'(z)^2 = 4\wp(z)^3 - g_2\wp(z) - g_3 \qquad with$$

$$g_2 = 60G_4 = 60 \sum_{\omega \in L\setminus\{0\}} \omega^{-4} \, ,$$

$$g_3 = 140G_6 = 140 \sum_{\omega \in L\setminus\{0\}} \omega^{-6} \, .$$

One can use this algebraic differential equation for $\wp$ to express all higher derivatives of $\wp$ as *polynomials* in $\wp$ and $\wp'$. (This must be possible, by V.3.3.) Differentiating the algebraic differential equation, and dividing by $\wp'$ on both sides, we get

$$2\wp''(z) = 12\wp(z)^2 - g_2 \, .$$

Differentiating again, we get

$$\wp'''(z) = 12\wp(z)\wp'(z)$$

and we can continue to obtain formulas like

$$\begin{aligned}
\wp^{IV}(z) &= 12\wp'(z)^2 + 12\wp(z)\wp''(z) \\
&= 12\wp'(z)^2 + 6\wp(z)[12\wp(z)^2 - g_2] \\
&= 12\wp'(z)^2 + 72\wp(z)^3 - 6g_2\wp(z) \\
&= 120\wp(z)^3 - 18g_2\wp(z) - 12g_3 \, ,
\end{aligned}$$

and so on. These relations induce, after taking LAURENT coefficients in all degrees, many relations between the EISENSTEIN series G_n, more precisely, the higher EISENSTEIN series G_n, $n \geq 8$, have explicit representations as polynomials in G_4 and G_6. (See also Exercise 6 to this section.)

A Appendix to Sect. V.3 : The Torus as an Algebraic Curve

By a *"polynomial in n variables"* we will understand a map

$$P : \mathbb{C}^n \longrightarrow \mathbb{C} \, ,$$

which can be written in the form

$$P(z_1, \ldots, z_n) = \sum_{\text{finite}} a_{\nu_1,\ldots,\nu_n} z_1^{\nu_1} \cdots z_n^{\nu_n} \, .$$

Here, the sum is taken over finitely many multi-indexes $(\nu_1, \ldots, \nu_n)$ consisting of non-negative integers. The corresponding coefficients $a_{\nu_1,\ldots,\nu_n} \in \mathbb{C}$ are uniquely determined by P.

Definition A.1 *A subset $X \subset \mathbb{C}^2$ is called a **plane affine curve**, iff there exists a non-constant polynomial P in two variables, such that X is exactly the zero set of this polynomial, i.e.*

$$X = \{ \, z \in \mathbb{C}^2 \, ; \quad P(z) = 0 \, \} \, .$$

Comments about the terminology.

(1) The word *plane* refers to the plane $\mathbb{C}^2$, a vector space of dimension two over $\mathbb{C}$.
(2) The word *affine* refers to the fact that the plane $\mathbb{C}^2$ sometimes occurs as *affine plane* in contrast to the projective plane (s. below)
(3) X should be considered as complex one-dimensional (and real two-dimensional).

Example of an affine curve. Let g_2 and g_3 be complex numbers. Let P be the following polynomial

$$P(z_1, z_2) = z_2^2 - 4z_1^3 + g_2 z_1 + g_3 \, , \qquad \text{and set}$$
$$X = X(g_2, g_3) = \{ \, (z_1, z_2) \in \mathbb{C}^2 \, ; \quad z_2^2 = 4z_1^3 - g_2 z_1 - g_3 \, \} \, .$$

Let's try to draw a figure. We take $g_2, g_3 \in \mathbb{R}$. Because we have only two real dimensions on this sheet of paper, we must restrict to a real figure:

$$X_{\mathbb{R}} := X \cap \mathbb{R}^2 = \{ \, (x, y) \in \mathbb{R}^2 \, ; \quad y^2 = 4x^3 - g_2 x - g_3 \, \} \, .$$

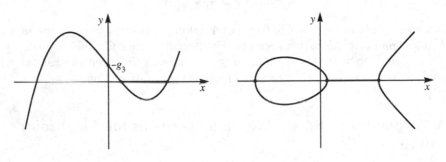

$$y = 4x^3 - g_2 x - g_3 \qquad\qquad\qquad y^2 = 4x^3 - g_2 x - g_3$$

We have to take into account that the real figure only gives partial information about the curve. It can be empty, as in the case of

$$P(X, Y) = X^2 + Y^2 + 1 \, .$$

Let us go back to our example, and *assume*, that there exists a lattice $L \subset \mathbb{C}$ with

$$g_2 = g_2(L) \quad \text{and} \quad g_3 = g_3(L) \, .$$

In Section V.8, we will see that this is the case iff $g_2^3 - 27g_3^2 \neq 0$.

From the algebraic differential equation for the $\wp$-function, it follows that the point $(\wp(z), \wp'(z))$ lies on the curve $X(g_2, g_3)$ for any $z \in \mathbb{C}$, $z \notin L$. This gives a map

$$\mathbb{C}/L \setminus \{ [0] \} \longrightarrow X(g_2, g_3) ,$$
$$[z] \longmapsto (\wp(z), \wp'(z)) .$$

Proposition A.2 *The assignment*

$$[z] \longmapsto (\wp(z), \wp'(z))$$

defines a bijective map of the punctured torus onto the plane affine curve $X(g_2, g_3)$,

$$\mathbb{C}/L \setminus \{[0]\} \overset{\sim}{\longleftrightarrow} X(g_2, g_3) .$$

Proof.
(1) *The surjectivity of the map.* Let $(u, v) \in X(g_2, g_3)$ be an arbitrary point on the curve. The $\wp$-function takes any value in $\mathbb{C}$, so there exists a $z \in \mathbb{C} \setminus L$ with $\wp(z) = u$. From the algebraic differential equation of the $\wp$-function follows

$$\wp'(z) = \pm v .$$

We thus have

$$\text{either } (\wp(z), \wp'(z)) = (u, v) , \qquad \text{or } (\wp(-z), \wp'(-z)) = (u, v) .$$

(2) *The injectivity of the map.* Let $z, w \in \mathbb{C} \setminus L$ satisfy

$$\wp(z) = \wp(w) \text{ and } \wp'(z) = \wp'(w) .$$

Then by V.2.10,

$$\text{either } z \equiv w \quad \mod L , \qquad \text{or } z \equiv -w \quad \mod L .$$

We must study only the second case. From $z \equiv -w \mod L$ we obtain

$$\wp'(z) = -\wp'(z) ,$$

hence

$$\wp'(z) = 0 .$$

But then $2z \in L$, and the second case reduces to the first case $z \equiv -z \equiv w$ mod L.   □

The affine curve $X(g_2, g_3)$ misses the partner of the point $[0]$ on the torus. We will find it in the "*projective closure*" of the curve.

The projective space

We define the *n-dimensional projective space* $\mathbb{P}^n(\mathbb{C})$ over the complex number field. It is obtained by considering equivalence classes of points in $\mathbb{C}^{n+1} \setminus \{0\}$ with respect to the following equivalence relation:

$$z \sim w \iff z = tw \text{ for a suitable } t \in \mathbb{C}^\bullet .$$

The orbit of a point z with respect to this relation will be denoted by

$$[z] = \{ tz ; \quad t \in \mathbb{C} , \ t \neq 0 \} .$$

(This should not be confused with the similar notation for the image of $z \in \mathbb{C}$ in the torus $\mathbb{C}/L$.) The *projective space* is the set of all orbits,

$$\mathbb{P}^n(\mathbb{C}) = \{ [z] ; \quad z \in \mathbb{C}^{n+1} \setminus \{0\} \} .$$

(Two points z and w lie in the same orbit, iff they lie on the same complex line through the origin in $\mathbb{C}^{n+1}$. Hence one can consider $\mathbb{P}^n(\mathbb{C})$ as the set of all lines through zero in $\mathbb{C}^{n+1}$, or as the set of all 1-dimensional subspaces of the vector space $\mathbb{C}^{n+1}$.)

We denote by

$$\mathbb{A}^n(\mathbb{C}) = \{ [z] \in \mathbb{P}^n(\mathbb{C}) ; \quad z = (z_0, \dots, z_n) \in \mathbb{C}^{n+1} , \ z_0 \neq 0 \}$$

the part of the projective space defined by "$z_0 \neq 0$". (We do not want to introduce a topology on $\mathbb{P}^n(\mathbb{C})$ here, but one should think about $\mathbb{A}^n(\mathbb{C})$ as an open, dense subset of $\mathbb{P}^n(\mathbb{C})$.)

Remark A.3 *The map*

$$\mathbb{C}^n \longrightarrow \mathbb{A}^n(\mathbb{C}) ,$$
$$(z_1, \dots, z_n) \longmapsto [1, z_1, \dots, z_n] ,$$

is bijective. Its inverse is given by the formula

$$[z_0, z_1, \dots, z_n] \longrightarrow \left(\frac{z_1}{z_0}, \dots, \frac{z_n}{z_0} \right) .$$

Proof. It is a simple check, that both maps are well defined and inverse to each other. □

We consider the complement $\mathbb{P}^n(\mathbb{C}) \setminus \mathbb{A}^n(\mathbb{C})$.

Remark A.4 *The map* $[z_1, \dots, z_n] \longmapsto [0, z_1, \dots, z_n]$ *defines a bijection*

$$\mathbb{P}^{n-1}(\mathbb{C}) \longrightarrow \mathbb{P}^n(\mathbb{C}) \setminus \mathbb{A}^n(\mathbb{C}) .$$

This is obvious, and left to the reader. □

Let us recall the essential properties of the projective space.

The n-dimensional projective space $\mathbb{P}^n(\mathbb{C})$ *is the disjoint union of an n-dimensional affine space* $\mathbb{A}^n(\mathbb{C})$ *and an* $(n-1)$*-dimensional projective space* $\mathbb{P}^{n-1}(\mathbb{C})$. *We call* $\mathbb{A}^n(\mathbb{C}) \subset \mathbb{P}^n(\mathbb{C})$ *the* **finite part**, *and the complement* $\mathbb{P}^{n-1}(\mathbb{C}) \subset \mathbb{P}^n(\mathbb{C})$ *the* **infinite part**.

Examples.

(1) $n = 0$: The 0-dimensional projective space consists of one point

$$\mathbb{P}^0(\mathbb{C}) = \{ \, [1] \, \} = \{ \, [z] \, ; \quad z \neq 0 \, \} \, .$$

(2) $n = 1$: The affine line $\mathbb{A}^1(\mathbb{C}) \subset \mathbb{P}^1(\mathbb{C})$, given by "$z_0 \neq 0$" can be bijectively mapped onto $\mathbb{C}$, A.3. The complement $\mathbb{P}^1(\mathbb{C}) \setminus \mathbb{A}^1(\mathbb{C})$ consists of a single point ($[0, 1]$).

*We can thus identify $\mathbb{P}^1(\mathbb{C})$ with the **Riemann sphere**,*

$$\mathbb{P}^1(\mathbb{C}) \longrightarrow \overline{\mathbb{C}} \, ,$$

$$[z_0, z_1] \longmapsto \begin{cases} \dfrac{z_1}{z_0} \, , & \text{if } z_0 \neq 0 \, , \\ \infty \, , & \text{if } z_0 = 0 \, . \end{cases}$$

We now introduce the notion of a *projective* plane curve.

A polynomial P is called *homogenous*, iff there exists a number $d \in \mathbb{N}_0$, such that

$$P(tz_1, \ldots, tz_n) = t^d \, P(z_1, \ldots, z_n) \, .$$

Then d is called the *degree* of P. Obviously, the homogenity means

$$a_{\nu_1, \ldots, \nu_n} \neq 0 \implies \nu_1 + \cdots + \nu_n = d \, .$$

Let $P(z_0, z_1, z_2)$ be a *homogenous* polynomial in three variables. If (z_0, z_1, z_2) is a zero of P, then this is also true for all scalar multiples (tz_0, tz_1, tz_2). The set of points

$$\widetilde{X} = \{ \, [z] \in \mathbb{P}^2(\mathbb{C}) \, ; \quad P(z) = 0 \, \}$$

is then well defined, i.e. the condition "$P(z) = 0$" does not depend on the choice of the representative z for $[z]$.

Definition A.5 *A subset $\widetilde{X} \subset \mathbb{P}^2(\mathbb{C})$ is a **plane projective curve**, iff there exists a non-constant homogenous polynomial P in three variables, such that*

$$\widetilde{X} = \{ \, [z] \in \mathbb{P}^2(\mathbb{C}) \, ; \quad P(z) = 0 \, \} \, .$$

The projective closure of a plane affine curve

Let

$$P(z_1, z_2) = \sum a_{\nu_1 \nu_2} z_1^{\nu_1} z_2^{\nu_2}$$

be a non-constant polynomial. We consider

$$d := \max\{ \, \nu_1 + \nu_2 \, ; \quad a_{\nu_1 \nu_2} \neq 0 \, . \}$$

We define

$$\widetilde{P}(z_0, z_1, z_2) := \sum a_{\nu_1 \nu_2} z_0^{d - \nu_1 - \nu_2} z_1^{\nu_1} z_2^{\nu_2} \, .$$

Then $\tilde{P}$ is a homogenous polynomial in three variables. This process is called *homogenization.*

Associated to P is a plane affine curve $X = X_P \subset \mathbb{A}^2(\mathbb{C})$.

Associated to $\tilde{P}$ is a plane projective curve $\tilde{X} = \tilde{X}_{\tilde{P}} \subset \mathbb{P}^2(\mathbb{C})$.

In this context, we have:

Remark A.6 *Let P be a non-constant polynomial in two variables. Let $\tilde{P}$ be the associated homogenous polynomial in three variables. Then the bijection*

$$\mathbb{C}^2 \overset{\sim}{\longleftrightarrow} \mathbb{A}^2(\mathbb{C}) , \quad (z_1, z_2) \longleftrightarrow [1, z_1, z_2] ,$$

maps the affine curve $X = X_P$ bijectively onto the intersection $\tilde{X} \cap \mathbb{A}^2(\mathbb{C})$ of the projective curve $\tilde{X} = \tilde{X}_{\tilde{P}}$ with the "finite part" of the projective space.

It is easy to show that there are only finitely many points of the curve $\tilde{X}$ in the infinite part (the complement of $\mathbb{A}^2(\mathbb{C})$ given by "$z_0 = 0$").

This motivates the following

Terminology: The projective curve $\tilde{X}$ is a projective closure of the affine curve X.

The polynomial P is not uniquely determined by the curve X, for instance P and P^2 have the same zero set. One can show that $\tilde{X}$ depends only on X, but not on the polynomial P. This allows us to use the more strict terminology "$\tilde{X}$ is *the* projective closure of X". If we topologize the projective space $\mathbb{P}^n(\mathbb{C})$ using the quotient topology, then $\tilde{X}$ is exactly the topological closure of X. By the way, the projective space $\mathbb{P}^n(\mathbb{C})$ is a compact topological space. The projective curve $\tilde{X} \subset \mathbb{P}^2(\mathbb{C})$ is closed, hence compact, too. It can be considered as a natural compactification of X.

Let's go back to our example

$$P(z_1, z_2) = z_2^2 - 4z_1^3 + g_2 z_1 + g_3 .$$

By homogenization, we obtain the polynomial

$$\tilde{P}(z_0, z_1, z_2) = z_0 z_2^2 - 4z_1^3 + g_2 z_0^2 z_1 + g_3 z_0^3 .$$

Let us determine all points at infinity of the associated projective curve. These points satisfy "$z_0 = 0$", and together with $\tilde{P}(z_0, z_1, z_2) = 0$ we get

$$z_0 = 0 \text{ and } z_1 = 0 .$$

All points $(0, 0, z_2)$ lie in the same orbit $[0, 0, 1]$. We thus have:

The projective curve $\tilde{X} = \tilde{X}_{\tilde{P}}$ contains exactly one point at infinity, namely $[0, 0, 1]$.

This is the missing point we were looking for.

Theorem A.7 *The map*

$$\mathbb{C}/L \longrightarrow \mathbb{P}^2(\mathbb{C}) \; ,$$

$$[z] \longmapsto \begin{cases} [1, \wp(z), \wp'(z)] \; , & \text{if } z \notin L \; , \\ [0, 0, 1] \; , & \text{if } z \in L \; , \end{cases}$$

defines a bijection of the whole torus $\mathbb{C}/L$ onto the projective curve $\tilde{X}(g_2, g_3)$. The equation of this curve is

$$z_0 z_2^2 = 4z_1^3 - g_2 z_0^2 z_1 - g_3 z_0^3.$$

Setting $z_0 = 1$ in this equation, we obtain the affine part of this curve.

The projective curve $\tilde{X}(g_2, g_3)$ with $g_2 = g_2(L), g_3 = g_3(L)$, is called the **elliptic curve** *associated to the lattice L.*

Exercises for V.3

1. For each of the elliptic functions $(\wp')^{-n}$, $1 \leq n \leq 3$, find the corresponding normal form $R(\wp) + S(\wp)\wp'$ with rational functions R and S.

2. For any $n \in \mathbb{Z}$ the function $z \to \wp(nz)$ is a rational function in $z \to \wp(z)$.

3. Using the notations of V.2.9, show

$$\wp'' \left(\frac{\omega_1}{2} \right) = 2(e_1 - e_2)(c_1 - e_3)$$

    and find the corresponding formulas for the other half lattice points.

4. Let us set $g_2 = g_2(L)$, $g_3 = g_3(L)$ for the g-invariants of a fixed lattice L. Let f be a meromorphic, non-constant function in some domain, which satisfies the same algebraic differential equation as $\wp$, i.e.

$$f'^2 = 4f^3 - g_2 f - g_3 \; .$$

    Show that f is the composition of $\wp$ with a translation, i.e. there exists an $a \in \mathbb{C}$ with $f(z) = \wp(z + a)$.

    Hint. Consider a local inverse function f^{-1} of f and $h := f^{-1} \circ \wp$.

5. The algebraic differential equation of the $\wp$-function can be rewritten as

$$\boxed{\wp'^2 = 4(\wp - e_1)(\wp - e_2)(\wp - e_3) \; .}$$

    Here, e_j, $1 \leq j \leq 3$, are the three half lattice values of the $\wp$-function, V.2.9.

6. Show the following recursion formulas for the EISENSTEIN series G_{2m} for $m \geq 4$:

$$(2m + 1)(m - 3)(2m - 1)G_{2m} = 3 \sum_{j=2}^{m-2} (2j - 1)(2m - 2j - 1) \, G_{2j} \, G_{2m-2j} \; ,$$

    for instance $G_{10} = \frac{5}{11}G_4 G_6$. Any EISENSTEIN series G_{2m}, $m \geq 4$, is thus representable as a polynomial in G_4 and G_6 with non-negative coefficients.

7.   We call a meromorphic function $f : \mathbb{C} \to \overline{\mathbb{C}}$ "real", iff $f(\overline{z}) = \overline{f(z)}$ for all $z \in \mathbb{C}$. A lattice $L \subset \mathbb{C}$ is called "real", iff $\omega \in L$ implies $\overline{\omega} \in L$ (i.e. iff L is invariant under complex conjugation).

Show tha the following properties are equivalent.

(a) $g_2(L), g_3(L) \in \mathbb{R}$.

(b) $G_n \in \mathbb{R}$ for all (even) $n \geq 4$.

(c)   The $\wp$-function is real.

(d)   The lattice L is real.

8.   A lattice is called *rectangular*, iff it admits a basis ω_1, ω_2 , such that ω_1 is real and ω_2 is purely imaginary. A lattice L is called *rhombic*, iff it admits a basis ω_1, ω_2 , such that $\omega_2 = \overline{\omega}_1$.

Show that a lattice is real, iff it is either rectangular or rhombic.

9.   The WEIERSTRASS $\wp$-function with respect to a rectangular lattice $L = \mathbb{Z}\omega_1 + \mathbb{Z}\omega_2$, where $\omega_1 \in \mathbb{R}_+^*$ and $\omega_2 \in i\mathbb{R}_+^*$, takes real values on the boundary and on the middle lines of the fundamental rectangle.

10.   Let $L = \mathbb{Z}\omega_1 + \mathbb{Z}\omega_2$ be a rectangular lattice as in Exercise 9. Show that the interior of the fundamental rectangle

$$D := \left\{ z \in \mathbb{C}; \quad z = t_1 \frac{\omega_1}{2} + t_2 \frac{\omega_2}{2}, \; 0 < t_1, t_2 < 1 \right\}$$

is mapped by the WEIERSTRASS $\wp$-function of L conformally onto the lower half-plane

$$\mathbb{H}_- := \left\{ z \in \mathbb{C} ; \quad \mathrm{Im}\, z < 0 \right\} .$$

11.   In this exercise we use the notions "extension of fields" $k \subset K$ and "algebraic dependence". The elements $a_1, \ldots, a_n$ in K are called algebraically dependent over k, iff there exists a non-zero *polynomial* P in n variables with coefficients in k, $P \in k[X_1, \ldots, X_n]$, such that $P(a_1, \ldots, a_n) = 0$. We use the following elementary facts from algebra:

Let us assume, that there exist n elements $a_1, \ldots, a_n \in K$, such that K is algebraic over the field $k(a_1, \ldots, a_n)$ generated by these elements. Then any $n + 1$ elements of K are algebraically dependent over k.

Show that any two elliptic functions (for the same lattice L) are algebraically dependent over $\mathbb{C}$.

12.   Let L be a lattice. Show that there does *not* exist any elliptic function $f \in K(L)$, with $\mathbb{C}(f) = K(L)$, i.e. such that any other elliptic function can be written as a rational function in f.

Hint. Analyze the equation $f(z) = f(w)$, and show that if there would exist an f with the above property then f would be an elliptic function of order 1.

V.4 The Addition Theorem

Let $a \in \mathbb{C}$ be fixed. Together with an elliptic function $f(z)$, the function $g(z) := f(z + a)$ is also elliptic. Then it must be possible (V.3.3) to find a formula for $\wp(z + a)$ of the form

$$\wp(z + a) = R_a\big(\wp(z)\big) + S_a\big(\wp(z)\big) \cdot \wp'(z)$$

with suitable rational functions R_a and S_a depending on a. Their explicit computation along the prescribed lines leads to the *Addition Theorem for the Weierstrass $\wp$-function.*

The first step in this construction is seperating the even and odd parts of $z \to \wp(z + a)$. The odd part is $\wp'$ times an even function. So we consider

$$\frac{\wp(z + a) + \wp(z - a)}{2} \quad \text{and} \quad \frac{\wp(z + a) - \wp(z - a)}{2\wp'(z)}$$

and look for their representations as rational functions of $\wp(z)$. This second step leads to the study of pole sets. The poles of the second function come from the poles of the numerator and the zeros of the denominator $\wp'(z)$ (eventual cancellations must be considered). The zeros of $\wp'$ are exactly the half lattice points, which are not lattice points, $\alpha \in \mathbb{C}$, $\alpha \notin L$, $2\alpha \in L$. All of them are simple zeros. But in these points the numerator also vanishes,

$$\wp(\alpha + a) = \wp(-\alpha - a) = \wp(-\alpha - a + 2\alpha) = \wp(\alpha - a) \ ,$$

so they are not poles. The functions

$$z \mapsto \frac{\wp(z + a) + \wp(z - a)}{2} \cdot \big[\wp(z) - \wp(a)\big]^2$$

and

$$z \mapsto \frac{\wp(z + a) - \wp(z - a)}{2\wp'(z)} \cdot \big[\wp(z) - \wp(a)\big]^2$$

have obviously no poles outside L, hence they are representable as *polynomials* in $\wp(z)$. In the first case the involved polynomial has degree ≤ 2, in the second one 0 (constant polynomial). In a third step one has to compute the coefficients of these polynomials by means of the LAURENT series. We skip these computations because it is very easy to verify the resulting formula directly.

Analytic form of the Addition Theorem

Theorem V.4.1 (Addition Theorem for the $\wp$-function) *Let z and w be two complex numbers, such that $z + w$, $z - w$, z and w do not lie in the lattice L. Then*

$$\boxed{\wp(z + w) = \frac{1}{4}\left[\frac{\wp'(z) - \wp'(w)}{\wp(z) - \wp(w)}\right]^2 - \wp(z) - \wp(w) \ .}$$

For a direct proof of this theorem, we first consider w as fixed, and look at the difference of both sides as function of z. It is an elliptic function with poles among the places

$$z \in L \quad \text{with} \quad z \equiv \pm w \mod L .$$

Then, a simple computation shows the vanishing of the principal and constant parts of the both sides cancel, when one takes the difference. The details are left to the reader. (A little later we will give another elegant proof.)

The formula in V.4.1 degenerates if we substitute for z the value w, but we can instead consider the limit $z \to w$, which exists and leads to a Doubling Formula .

From the expansions

$$\left.\begin{array}{l} \wp'(z) - \wp'(w) = \wp''(w)(z - w) + \cdots \\ \wp(z) - \wp(w) = \wp'(w)(z - w) + \cdots \end{array}\right\} + \text{ higher powers of } (z - w) ,$$

we get

$$\lim_{z \to w} \frac{\wp'(z) - \wp'(w)}{\wp(z) - \wp(w)} = \frac{\wp''(w)}{\wp'(w)} .$$

This gives for $2z \notin L$:

Theorem V.4.2 (Doubling Formula)

$$\wp(2z) = \frac{1}{4}\left[\frac{\wp''(z)}{\wp'(z)}\right]^2 - 2\wp(z) .$$

The expression $[\wp''(z)/\wp'(z)]^2$ can be converted by means of the algebraic differential equation of $\wp$, $\wp'^2 = 4\wp^3 - g_2\wp - g_3$, $2\wp'' = 12\wp^2 - g_2$. This leads to an explicit formula for $\wp(2z)$ as a rational function of $\wp(z)$:

$$\wp(2z) = \frac{(\wp^2(z) + \frac{1}{4}g_2)^2 + 2g_3\wp(z)}{4\wp^3(z) - g_2\wp(z) - g_3} .$$

Geometric form of the Addition Theorem

The Addition Theorem V.4.1 has a geometrical interpretation, which leads to a deeper understanding and especially to a simple, transparent proof of it.

Preliminaries. A subset $G \subset \mathbb{P}^2(\mathbb{C})$ is called a (complex projective) *line*, iff there exist two different points $[z_0, z_1, z_2]$, $[w_0, w_1, w_2]$ in $\mathbb{P}^2(\mathbb{C})$, such that G is the set of all points of the form

$$P = P(\lambda, \mu) := [\lambda z_0 + \mu w_0 , \lambda z_1 + \mu w_1 , \lambda z_2 + \mu w_2], \quad (\lambda, \mu) \in \mathbb{C}^2 \setminus \{(0,0)\} .$$

The map

$$\mathbb{P}^1(\mathbb{C}) \longrightarrow G ,$$
$$[\lambda, \mu] \longmapsto P(\lambda, \mu) ,$$

is then a bijection between the standard projective line and $G \subset \mathbb{P}^2(\mathbb{C})$. Two different points in $\mathbb{P}^2(\mathbb{C})$ are contained in exactly one projective line in $\mathbb{P}^2(\mathbb{C})$.

The torus $\mathbb{C}/L$ is, as a quotient of the additive group $\mathbb{C}$, itself a group. We can, using the bijection $\mathbb{C}/L \to \tilde{X}(g_2, g_3)$ (Appendix to V.3), transport the structure to obtain an addition on the elliptic curve such that this map gets an isomorphism. We denote this addition on $\tilde{X}(g_2, g_3)$ again by the usual plus symbol, and the neutral element $[0, 0, 1]$ of X is simply denoted by 0. The geometrical form of the Addition Theorem is:

Theorem V.4.3 *The sum of three distinct points a, b, c on the elliptic curve $\tilde{X}(g_2, g_3)$ is zero,*

$$a + b + c = 0 \; ,$$

iff a, b, c lie on a projective line.

Remark. A simple consideration shows that a projective line and a projective elliptic curve have exactly three common points in $\mathbb{P}^2(\mathbb{C})$, counting multiplicities. If these intersection points are $a, b, c \in \tilde{X}(g_2, g_3)$, then their sum is zero. Conversely, if the sum is zero, then the three points lie on a projective line. By Theorem V.4.3 the addition law in $\tilde{X}(g_2, g_3)$ is completely determined (up to some degenerate cases where two points coincide, see the comments below).

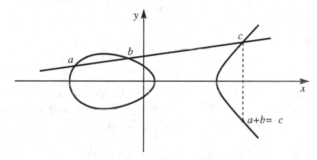

Proof of V.4.3. To avoid trivial cases we assume that none of the points a, b, c is located at 0. Three points $[u_0, u_1, u_2]$, $[v_0, v_1, v_2]$ and $[w_0, w_1, w_2]$ lie on a projective line in $\mathbb{P}^2(\mathbb{C})$, iff the vectors (u_0, u_1, u_2), (v_0, v_1, v_2), (w_0, w_1, w_2) are linearly dependent, i.e. iff the determinant of the matrix consisting of these vectors vanishes.

Let $a, b, c \in \tilde{X}(g_2, g_3)$ be three different points, which sum up to 0. Let $[u], [v], [w] \in \mathbb{C}/L$ be the corresponding points in the torus. We have the equivalence

$$a + b + c = 0 \qquad \Longleftrightarrow \qquad [u] + [v] + [w] = [0]$$

by definition of the additive structure on $\tilde{X}(g_2, g_3)$. So u, v, w sum up to $0 \in \mathbb{C}/L$, and thus $w \equiv -(u + v)$ modulo L. Theorem V.4.3 is thus equivalent to

Proposition V.4.4 *The following formula holds:*

$$\det \begin{pmatrix} 1 & \wp(u+v) & -\wp'(u+v) \\ 1 & \wp(v) & \wp'(v) \\ 1 & \wp(u) & \wp'(u) \end{pmatrix} = 0 \ .$$

(The excursion about projective curves can be ignored, if one is satisfied with this version of V.4.3.)

We postpone the proof of V.4.4.

We first show that the Addition Theorem V.4.1 can be proven starting from V.4.4. Subsequently we will give an elegant proof of V.4.4, and thus of the Addition Theorem. Let us consider three points

$$(x_1, y_1) = (\wp(u), \wp'(u)) \ ,$$
$$(x_2, y_2) = (\wp(v), \wp'(v)) \ ,$$
$$(x_3, y_3) = (\wp(u+v), -\wp'(u+v)) \ .$$

We can assume that x_1, x_2 and x_3 are pairwise different. The determinant in V.4.4 vanishes, so the three points lie on an affine line

$$Y = mX + b \ .$$

The coefficient m is

$$m = \frac{\wp'(v) - \wp'(u)}{\wp(v) - \wp(u)} \ .$$

From the algebraic differential equation of the $\wp$-function, the three points also lie on the elliptic curve $Y^2 = 4X^3 - g_2 X - g_3$, and eliminating Y we obtain that x_1, x_2 and x_3 are the zeros of the cubic polynomial

$$4X^3 - g_2 X - g_3 - (mX + b)^2 \ .$$

They are different, so there is no other root. If one compares the square coefficients on obtains

$$x_1 + x_2 + x_3 = \frac{m^2}{4} \ .$$

This is exactly the Addition Theorem in analytic form. □

One can relax the condition in V.4.3 that the three points a, b, c are pairwise distinct, if one is familiar with the concept of the tangent line at a point $a \in X$. This tangent line may intersect X in a third point c and then $a + a + c = 0$. It may also happen that the tangent has "multiplicity three". Then $c = a$ and hence $a + a + a = 0$. There are 9 points with this property.

An elegant proof of the Addition Theorem

We finally give the promised proof of the Addition Theorem in the analytic version V.4.4. We anticipate ABEL's Theorem (V.6.1), but use only the simple implication in it. It claims:

Let $a_1, \ldots, a_n$ be a system of representatives modulo L of the zeros of a non-constant elliptic function f (for the lattice L), and let $b_1, \ldots, b_m$ be a system of representatives modulo L of the poles of f. We repeat in these lists each value with its multiplicity.
Then $m = n$, and

$$(a_1 + \cdots + a_n) - (b_1 + \cdots + b_n) \in L \ .$$

For the proof of V.4.4, we fix two different points u and v not in L, such that $\wp(u)$ and $\wp(v)$ are not equal. Let us consider the elliptic function f,

$$f(z) = \det \begin{pmatrix} 1 & \wp(z) & \wp'(z) \\ 1 & \wp(v) & \wp'(v) \\ 1 & \wp(u) & \wp'(u) \end{pmatrix} \ .$$

By the LEIBNIZ formula for the determinant this function is of the form $A + B\wp(z) + C\wp'(z)$ with $C = \wp(u) - \wp(v) \neq 0$. It has its poles exactly in the lattice points (of third order). It is thus an elliptic function of order 3, its zeros are located at $z = u$ and $z = v$. By ABEL's Theorem the third zero modulo L must be $z = -(u + v)$.   □

One can, conversely, deduce the geometric form of the Addition Theorem from the analytic form of it (Exercise 2).

Exercises for V.4

1. Fill in the details of the direct proof of the Addition Theorem V.4.1, as suggested after its statement.

2. Deduce the geometric form of the Addition Theorem from its analytic form.

3. Let $L \subset \mathbb{C}$ be a lattice with the property $g_2(L) = 8$ and $g_3(L) = 0$. The point $(2, 4)$ lies on the affine elliptic curve $y^2 = 4x^3 - 8x$. Let $+$ be the addition (for points on the corresponding projective curve). Show that $2 \cdot (2, 4) := (2, 4) + (2, 4)$ is the point $(\frac{9}{4}, -\frac{21}{4})$.

   Hint. Find the (third) intersection point of the elliptic curve with its tangent in $(2, 4)$.

4. Using the notations in the proof of Proposition V.4.4 we have

$$y_3 = \frac{(x_3 - x_2)y_1 - (x_3 - x_1)y_2}{x_1 - x_2} \ .$$

   This can be considered as an analytic Addition Theorem for $\wp'$.

284   V Elliptic Functions

5. Addition Theorem for arbitrary elliptic functions

For an arbitrary elliptic function f, there exists a polynomial with complex coefficients $P \not\equiv 0$ in three variables, such that

$$P\big(f(z), f(w), f(z+w)\big) \equiv 0 \ .$$

Hint. We use some standard, elementary facts from algebra, especially facts from Exercise 11 of Sect. V.3, about extensions of fields.

A function

$$F : \mathbb{C} \times \mathbb{C} \to \mathbb{C}$$

is called analytic, iff it is continuous, and analytic with respect to each variable. The set of all these functions is denoted by $\mathcal{O}(\mathbb{C} \times \mathbb{C})$, which is a commutative, integral ring with unit. It has no zero divisors. Hence we can consider its quotient field Ω. It is a substitute for the set of all meromorphic functions of two variables, which will be introduced in the second volume.

Consider the subfield of Ω, generated over $\mathbb{C}$ by the five meromorphic functions

$$(z, w) \ \to \ \wp(z), \ \wp(w), \ f(z), \ f(w) \text{ and } f(z+w) \ .$$

This field is algebraic over the field $\mathbb{C}\big(\wp(z), \wp(w)\big)$. Use for this the fact that being algebraic is a transitive relation for field extensions.

V.5 Elliptic Integrals

By an *elliptic integral of the first kind* we understand an integral of the form

$$\int_a^z \frac{dt}{\sqrt{P(t)}} \ ,$$

where P is a polynomial of degree three or four without multiple roots.

The value of such an integral depends on both the choice of the square root and the choice of a curve connecting a and z.

Theorem V.5.1 *For any polynomial $P(t)$ of degree three or four having only simple roots, there exists a non-constant elliptic function f with the following property:*

Let $D \subset \mathbb{C}$ be an open subset where f is invertible,[3] and let be $g : f(D) \to D \subset \mathbb{C}$ be the inverse function of (the restriction) $f : D \to f(D)$, then

$$\boxed{g'(z) = 1/\sqrt{P(z)} \ .}$$

---

[3] For any point $a \in \mathbb{C}$ with $f(a) \neq \infty$, $f'(a) \neq 0$, we can find such a D to be a suitable neighborhood of a.

In a non-rigorous but pregnant way we can say:

The inverse function of an elliptic integral (of the first kind) is an elliptic function.

In the following we will prove this Theorem up to a gap that will be filled in the Sections V.7 and V.8.

In a first step, we reduce V.5.1 to the case of polynomials $P(t)$ of degree three with vanishing t^2-term.

Let us assume, that for a fixed polynomial P there exists an elliptic function f with the property quoted in V.5.1. Then, for any complex 2×2 matrix with determinant 1

$$M = \begin{pmatrix} a & b \\ c & d \end{pmatrix}$$

we can consider the new elliptic function

$$\widetilde{f} = \frac{df - b}{-cf + a} \; .$$

Obviously, the function $\widetilde{g}$ with

$$\widetilde{g}(z) := g\left(\frac{az + b}{cz + d} \right)$$

is then a local inverse of $\widetilde{f}$. Then

$$\widetilde{g}'(z) = 1/\sqrt{Q(z)} \; ,$$

with

$$Q(z) = (cz + d)^4 \; P\left(\frac{az + b}{cz + d} \right) \; ,$$

which is also a polynomial.

Remark V.5.2 *Let P be a polynomial of degree three or four without multiple roots. Then there exists a matrix M of determinant 1, such that the associated polynomial*

$$\boxed{Q(z) = (cz + d)^4 P\left(\frac{az + b}{cz + d} \right)}$$

is a polynomial of degree three with vanishing quadratic term.

Proof. We first assume $\deg P = 4$, and write it in factorized form, $P(X) = C(X - e_1)(X - e_2)(X - e_3)(X - e_4)$, $e_4 \neq 0$. Applying the matrix

$$M = \begin{pmatrix} e_4 & 0 \\ 1 & e_4^{-1} \end{pmatrix} ,$$

we get a polynomial Q of degree three without multiple roots. Then, applying a matrix of the form

$$N = \begin{pmatrix} 1 & b \\ 0 & 1 \end{pmatrix}$$

with b suitably chosen, we can delete the quadratic term.   □

So for the proof of Theorem V.5.1 we can restrict to the case of a polynomial P of the form $P(t) = at^3 + bt + c$. We can further normalize the leading coefficient $a = 4$. We use the classical notations $b = -g_2$; $c = -g_3$ to obtain the so-called WEIERSTRASS normal form:

$$\boxed{P(t) = 4t^3 - g_2 t - g_3 \ .}$$

This polynomial has three different roots if and only its discriminant

$$\Delta := g_2^3 - 27 g_3^2 \ .$$

does not vanish. This is because if P has the factorization

$$P(t) = 4(t - e_1)(t - e_2)(t - e_3) \ ,$$

then a computation shows

$$g_2^3 - 27 g_3^2 = 16(e_1 - e_2)^2 (e_1 - e_3)^2 (e_2 - e_3)^2 \ .$$

Assumption V.5.3 *There exists a lattice $L \subset \mathbb{C}$ with the property*

$$g_2 = g_2(L) \ , \quad g_3 = g_3(L) \ .$$

In Section V.8 we will prove that this Assumption is always true.

Theorem V.5.4 *We consider a lattice $L \subset \mathbb{C}$ and let P be the corresponding polynomial*

$$P(t) = 4t^3 - g_2 t - g_3 \ , \quad g_2 = g_2(L) \ , \quad g_3 = g_3(L) \ .$$

Then the function f defined by $f(z) := \wp(z)$ satisfies the property formulated in V.5.1.

Proof: This follows directly from the *algebraic differential equation* of the $\wp$-function (V.3.4)

$$\wp'^2 = 4 \wp^3 - g_2 \wp - g_3 \ .$$

Then for a local inverse function g of $f = \wp$ we have:

$$g'(t)^2 = \frac{1}{\wp'(g(t))^2} = \frac{1}{4 \wp^3(g(t)) - g_2 \wp(g(t)) - g_3} = \frac{1}{P(t)} \ . \qquad □$$

The theory of the elliptic integrals was at the beginning a purely real theory. To be fair to history, we want to say a few words about real elliptic integrals.

Let $P(t)$ be a polynomial of degree 3 or 4 with real coefficients. We assume that P has no multiple (complex) root. We also assume that the leading coefficient is positive. Then

$$P(x) > 0 \text{ for sufficiently large } x, \text{ say } x > x_0 \,.$$

For $x > x_0$ we then consider the *positive square root*

$$\sqrt{P(x)} > 0 \,,$$

and use it in the definition of the improper integral

$$E(x) = -\int_x^\infty \frac{dt}{\sqrt{P(t)}} \quad \text{for } x > x_0 \,.$$

This integral is absolutely convergent, since the comparison integral

$$\int_1^\infty t^{-s} dt$$

converges for all $s > 1$ (in our case $s = 3/2$). The function $E(x)$ is strictly increasing, since the integrand is positive. We can then consider the inverse function of $E(x)$, which is defined on a suitable real interval.

By V.4.4 we have

Theorem V.5.5 *The inverse function of the elliptic integral*

$$E(x) = -\int_x^\infty \frac{dt}{\sqrt{P(t)}} \,, \quad x > x_0 \,,$$

$$P(t) = 4t^3 - g_2 t - g_3 \,,$$

$$g_2 = g_2(L) \,, \quad g_3 = g_3(L) \quad (L \subset \mathbb{C} \text{ some lattice }) \,,$$

can be meromorphically extended to the whole $\mathbb{C}$, where it represents an elliptic function, namely the Weierstrass $\wp$-function of the lattice L.

Concretely, this means

$$\boxed{-\int_{\wp(u)}^\infty \frac{dt}{\sqrt{P(t)}} = u}$$

(where $\wp(u)$ varies in a real interval (t_0, ∞), and u varies in a corresponding suitable real interval).

Application of the theory of elliptic functions to elliptic integrals

We have already developed a formula expressing $\wp(u_1 + u_2)$ in terms of $\wp(u_1)$ and $\wp(u_2)$. This formula involves only rational operations, and taking squares.

There exists a "formula" $x = x(x_1, x_2)$, which involves the constants g_2, g_3, the variables x_1, x_2, and uses only rational operations and taking square roots, such that

$$\int_{x_1}^{\infty} \frac{dt}{\sqrt{P(t)}} + \int_{x_2}^{\infty} \frac{dt}{\sqrt{P(t)}} = \int_{x}^{\infty} \frac{dt}{\sqrt{P(t)}} \, .$$

This formula was first proven in general by EULER (1753), and is called the *Euler Addition Theorem*. Earlier, in 1718, the mathematician FAGNANO proved the special case $P(t) = t^4 - 1$, $x_1 = x_2$, of this theorem. After bringing $P(t)$ to the WEIERSTRASS normal form, and applying the Doubling Formula, we obtain FAGNANO's Doubling Formula

$$2 \int_0^x \frac{dt}{\sqrt{1-t^4}} = \int_0^y \frac{dt}{\sqrt{1-t^4}}$$

where

$$y = y(x) = \frac{2x\sqrt{1-x^4}}{1+x^4} \, .$$

By the way, this special elliptic integral is related to the arc length of the classical *lemniscate* (see also Exercise 6(c) in I.1). The given formulas can be understood as Doubling Formulas for lemniscate arcs.

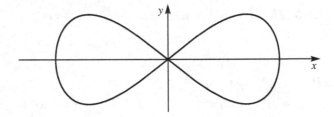

The Doubling Formula implies that one can double a lemniscate arc length using ruler and compass.

Recall that we still have to show that each pair of numbers (g_2, g_3) with non-vanishing corresponding discriminant $\Delta = g_2^3 - 27g_3^2$ "comes" from a lattice L. This will be proven at the end of Section V.8 by using methods of complex analysis. At this point, we want to make plausible, why and how starting from a polynomial $P(X) = 4X^3 - g_2 X - g_3$ we can associate a lattice, respectively a *torus*. Recall (Appendix to V.3), that there is a projective elliptic curve $\widetilde{X}(P)$ associated to P. We want to sketch why topologically $\widetilde{X}(P)$ looks like a torus.

For this, use the first coordinate projection of the affine part of the curve onto the affine complex line. This extends to give a continuous map from the projective curve onto the projective line (i.e. onto the RIEMANN sphere), $p : \widetilde{X}(P) \longrightarrow \overline{\mathbb{C}}$. Because the polynomial has degree three, one can show, that there exist exactly four (ramification) points, ∞, a, b, c say, on the sphere having exactly one pre-image point, $\widetilde{\infty}, \tilde{a}, \tilde{b}, \tilde{c}$. All other points have two pre-image points. On says that the projective curve is a 2-fold cover with four ramification points of the projective line. We divide the 4 ramification points into two pairs of two points, (∞, a) and (b, c) and join the paired points by (non-intersecting) paths. The pre-image of each of the two paths is a pair of paths connecting $\widetilde{\infty}$ with $\tilde{a}$ respectively $\tilde{b}$ with $\tilde{c}$:

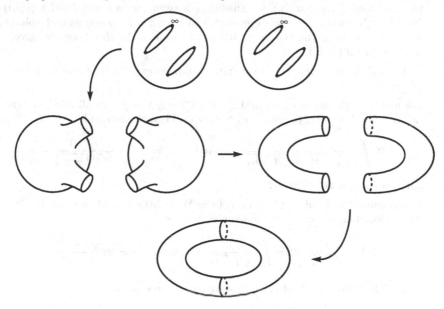

$$\tilde{X}(P)$$

$$\downarrow p$$

$$\mathbb{P}^1(\mathbb{C})$$

Let us consider the complement of the two paths in the projective line $\mathbb{P}^1(\mathbb{C}) = \mathbb{C} \cup \{\infty\}$. This complement looks like a sphere with two slits.

Now one has to show that the inverse image under p of this complement consists of *two (disjoint) connected components*. Each of them maps topologically onto the sphere with two slits. We see that the projective curve an be obtained from two copies of the double slit sphere by gluing them along the slits in a suitable way. The result is a torus as the figure tries to visualize.

In the second volume, using the theory of RIEMANN surfaces we will give a rigorous foundation for this construction. From this we will obtain a new approach to the theory of elliptic functions.

Exercises for V.5

1. The zeros e_1, e_2 and e_3 of the polynomial $4X^3 - g_2X - g_3$ are all real, iff g_2, g_3 are real, and the discriminant $\Delta = g_2^3 - 27g_3^2$ is non-negative.

2. The following exercise is in reach using the available methods:

    Let $L \subset \mathbb{C}$ be a lattice, and let $P(t) = 4t^3 - g_2t - g_3$ be the associated cubic polynomial. Let $\alpha : [0,1] \to \mathbb{C}$ be a closed path, which avoids the zeros of the polynomial. Finally, let $h : [0,1] \to \mathbb{C}$ be a continuous function with the properties

    $$h(t)^2 = \frac{1}{P(\alpha(t))} \quad \text{and} \quad h(0) = h(1) .$$

The number

$$\int_0^1 h(t)\,\alpha'(t)\,dt = \int_0^1 \frac{\alpha'(t)}{\sqrt{P(\alpha(t))}}\,dt$$

is called a *period* of the elliptic integral $\int 1/\sqrt{P(z)}\,dz$. Show that the periods of the elliptic integral lie in L. (In addition one can show that L is precisely the set of all periods of the elliptic integral.)

This fact opens a new approach to the problem, how to realize each pair of complex numbers (g_2, g_3) with non-vanishing discriminant as a pair $(g_2(L), g_3(L))$ for a suitable lattice L. This approach will play a role in the second volume in connection with the theory of RIEMANN surfaces. In this book, we give a different proof (V.8.9).

A detailed analysis delivers in concrete situations explicit formulas for a basis of L:

Assume that the zeros e_1, e_2 and e_3 of $4X^3 - g_2 X - g_3$ are all real, pairwise different, and ordered to satisfy $e_2 > e_3 > e_1$ Then the integrals

$$\omega_1 = 2i \int_{-\infty}^{e_1} \frac{1}{\sqrt{-4t^3 + g_2 t + g_3}}\,dt \quad \text{and} \quad \omega_2 = 2 \int_{e_2}^{\infty} \frac{1}{\sqrt{4t^3 - g_2 t - g_3}}\,dt$$

define a basis of the lattice L.

3.   Prove, using the Doubling Formula of the WEIERSTRASS $\wp$-function, the FAGNANO Doubling Formula for the lemniscate arcs,

$$2 \int_0^x \frac{1}{\sqrt{1-t^4}}\,dt = \int_0^y \frac{1}{\sqrt{1-t^4}}\,dt \quad \text{with} \quad y = 2x\frac{\sqrt{1-x^4}}{1+x^4}\ .$$

4.   *Show:* The rectification of the ellipse given by the equation

$$\frac{x^2}{a^2} + \frac{y^2}{b^2} = 1 \quad (0 < b \leq a)$$

leads to an integral of the form

$$\int \frac{1-k^2 x^2}{\sqrt{(1-x^2)(1-k^2 x^2)}}\,dx\ .$$

What is the meaning of k? The total length of the ellipse is

$$L = 4a \int_0^1 \frac{1-k^2 x^2}{\sqrt{(1-x^2)(1-k^2 x^2)}}\,dx = 4a \int_0^{\pi/2} \sqrt{1 - k^2 \sin^2 t}\,dt\ .$$

(This is a so-called elliptic integral of the second kind. More general, the terminology "elliptic integral" is applied to any integral of a product of a rational function with a square root of a polynomial of degree 3 or 4 with simple roots. The terminology "elliptic function" historically originates from the calculus of ellipse arcs.)

V.6 Abel's Theorem

We are now concerned with the problem whether there exists an elliptic function with prescribed zeros and poles. It is useful to ask the same question for *rational functions* on $\mathbb{C}$, i.e. for functions f of the form

$$f(z) = \frac{P(z)}{Q(z)}, \quad Q \not\equiv 0 \quad (P, Q \text{ polynomials}).$$

(Rational functions are meromorphic functions

$$f : \overline{\mathbb{C}} \longrightarrow \overline{\mathbb{C}}$$

on the RIEMANN sphere, and any meromorphic function on $\overline{\mathbb{C}}$ is rational.) The rational function

$$f(z) = z - a \quad (a \in \mathbb{C})$$

has in $z = a$ a simple zero, and in $z = \infty$ a simple pole. Rational functions can be factorized,

$$f(z) = C \frac{(z - a_1)^{\nu_1} \cdots (z - a_n)^{\nu_n}}{(z - b_1)^{\mu_1} \cdots (z - b_m)^{\mu_m}}.$$

From this we see:

A rational function on $\overline{\mathbb{C}}$ has the same number of zeros and of poles, counted with multiplicities.

This condition is also sufficient:

Let $M \subset \overline{\mathbb{C}}$ be a finite set of points. For each $a \in M$ let us fix a $\nu_a \in \mathbb{Z}$, such that the sum vanishes:

$$\sum_{a \in M} \nu_a = 0 .$$

The rational function

$$f(z) = \prod_{\substack{a \in M \\ a \neq \infty}} (z - a)^{\nu_a}$$

has in each $a \in M$ order $\nu_a \in \mathbb{Z}$, more explicitly,

    it is a zero, if $\nu_a > 0$, and then the zero order is ν_a,

    it is a pole, if $\nu_a < 0$, and then the pole order is $|\nu_a|$,

    it is neither a zero, nor a pole, if $\nu_a = 0$.

The function $f(z)$ has the wanted behavior in the finite part ($z \neq \infty$) of $\overline{\mathbb{C}}$ for the zeros and poles. The degree condition

$$\sum_{a \in M} \nu_a = 0$$

is exactly the condition needed to ensure the correct behavior also at ∞.

Let us study the problem in the case of elliptic functions. We already know that an elliptic function has as many zeros as poles, counting multiplicities. But this condition is not sufficient. For example, there is no elliptic curve of order one (having one zero, one pole).

We first introduce some notations. We are searching for elliptic functions f with prescribed zeros given by the list $a_1, \ldots, a_n$ modulo L, and poles given by the list $b_1, \ldots, b_n$ modulo L. We allow that the same value occurs several times. We make the convention that a prescribed zero, respectively pole, should have exactly the zero order, respectively pole order, given by the number of repetitions of this value in the corresponding list.

We tacitly assume that no a_j modulo L equals to any b_k modulo L.

We thus demand,

$$f(z) = 0 \quad \Longleftrightarrow \quad z \equiv a_j \mod L \quad \text{for a suitable } j \,,$$
$$f(z) = \infty \quad \Longleftrightarrow \quad z \equiv b_j \mod L \quad \text{for a suitable } j \,.$$

The zero order of f in a_j equals the number of all k with

$$a_k \equiv a_j \mod L \,,$$

correspondingly for the pole order.

Theorem V.6.1 (Abel's Theorem, N.H. Abel, 1826) *There exists an elliptic function with prescribed zeros in $a_1, \ldots, a_n$, and poles in $b_1, \ldots, b_n$, iff*

$$\boxed{a_1 + \cdots + a_n \equiv b_1 + \cdots + b_n \quad \mod L \,.}$$

Proof. We first show that the congruence condition is necessary. So let us assume that there exists an elliptic function with the prescribed behavior. We choose a point $a \in \mathbb{C}$, such that the translated fundamental parallelogram

$$\mathcal{F}_a = \{z = a + t_1 \omega_1 + t_2 \omega_2; \quad 0 \le t_1, t_2 \le 1\}$$

does not contain the zero point, and such that the boundary $\partial \mathcal{F}_a$ does not contain any zero or pole of f. Because the congruence condition does not change if we change the representatives modulo L in the prescribed lists, we can assume

$$a_j, b_j \in \overset{\circ}{\mathcal{F}}_a \ (= \text{inner part of } \mathcal{F}_a) \,.$$

We now consider the integral

$$I = \frac{1}{2\pi i} \int_{\partial \mathcal{F}_a} \zeta \frac{f'(\zeta)}{f(\zeta)} \, d\zeta \,.$$

The integrand has simple poles at $a_1, \ldots, a_n,\ b_1, \ldots, b_n$. The sum of the residues is (see also III.6.4)

$$I = a_1 + \cdots + a_n - b_1 - \cdots - b_n \,.$$

We must show $I \in L$. For this we compare the integrals along two opposite sides of the parallelogram $\mathcal{F}_a$. We are done, if we can show

$$\frac{1}{2\pi i} \left[\int_a^{a+\omega_1} \zeta \frac{f'(\zeta)}{f(\zeta)} \, d\zeta + \int_{a+\omega_1+\omega_2}^{a+\omega_2} \zeta \frac{f'(\zeta)}{f(\zeta)} \, d\zeta \right] \in L \; ,$$

and the same, when we interchange ω_1 and ω_2. For a general g we can write

$$\int_{a+\omega_1+\omega_2}^{a+\omega_2} g(\zeta) \, d\zeta = - \int_{a+\omega_2}^{a+\omega_1+\omega_2} g(\zeta) \, d\zeta = - \int_a^{a+\omega_1} g(\zeta + \omega_2) \, d\zeta \; .$$

Specializing

$$g(z) = \frac{z f'(z)}{f(z)}$$

we obtain

$$g(z) - g(z + \omega_2) = -\omega_2 \frac{f'(z)}{f(z)} \; .$$

Applying this, we can rewrite I as

$$I = -\frac{\omega_2}{2\pi i} \int_a^{a+\omega_1} \frac{f'(\zeta)}{f(\zeta)} \, d\zeta + \frac{\omega_1}{2\pi i} \int_a^{a+\omega_2} \frac{f'(\zeta)}{f(\zeta)} \, d\zeta \; .$$

Since ω_1 and ω_2 are in L, we ensure $I \in \mathbb{Z}\omega_1 + \mathbb{Z}\omega_2$ if we can show

$$\frac{1}{2\pi i} \int_a^{a+\omega} \frac{f'(\zeta)}{f(\zeta)} \, d\zeta \in \mathbb{Z} \quad \text{for} \;\; \omega = \omega_1 \;\text{or}\; \omega = \omega_2 \; .$$

The function $f(z)$ has on the line segment connecting a and $a + \omega$ neither poles, nor zeros. Therefore, we can find a rectangular open set (especially, an elementary domain) containing this line segment, where f still has neither poles, nor zeros. In this domain f has an analytic logarithm h, i.e. we can write

$$f(z) = e^{h(z)}$$

with an analytic function h (II.2.9). Then h is a primitive of f'/f, and we can use it to compute the integral

$$\int_a^{a+\omega} \frac{f'(\zeta)}{f(\zeta)} \, d\zeta = h(a + \omega) - h(a) \; .$$

But because of

$$e^{h(a+\omega)} = f(a + \omega) = f(a) = e^{h(a)}$$

the values $h(a + \omega)$ and $h(a)$ differ by an integral multiple of $2\pi i$, concluding the proof. $\square$

We now look at the *converse statement*, which is the hard part of the proof. We start by assuming the boxed relation in V.6.1, and will construct a suitable elliptic function.

(*Remark.* This function is unique up to a constant factor $\neq 0$, because the quotient of two elliptic functions with the same zeros and poles (counted with multiplicities) is constant by the First LIOUVILLE Theorem.)

The construction is based on the following

Lemma V.6.2 *Let us fix $z_0 \in \mathbb{C}$. For a given lattice $L \subset \mathbb{C}$, there exists an analytic function*

$$\sigma : \mathbb{C} \longrightarrow \mathbb{C}$$

with the following properties:

(1) $\sigma(z + \omega) = e^{az+b}\sigma(z)$, $\omega \in L$. *Here, a,b are suitable complex numbers, that may depend on ω, i.e. $a = a_\omega$ and $b = b_\omega$, but are independent of z.*

(2) σ *has a simple zero at z_0, and any other zero of σ is congruent to z_0 modulo L.*

Because the factor e^{az+b} is not 0, it is clear that together with z_0 all points of the form $z_0 + \omega$, $\omega \in L$ are also simple zeros of $\sigma(z)$.

Let us first show, how using the Lemma V.6.2 we can deduce the difficult part in ABEL's Theorem V.6.1. After this, we will give two proofs for the existence of σ.

Construction of $f(z)$ using $\sigma(z)$.

We can assume

$$a_1 + \cdots + a_n = b_1 + \cdots + b_n .$$

(For this, we replace one of the points —say a_1— by a suitable congruent one.) Then we consider the meromorphic function

$$f(z) = \frac{\prod_{j=1}^n \sigma(z_0 + z - a_j)}{\prod_{j=1}^n \sigma(z_0 + z - b_j)} .$$

Because of condition (2) in V.6.2, f has the prescribed zero and pole behavior. On the other hand, f is also *elliptic!* For this, use condition (1) in V.6.2 to infer

$$f(z + \omega) = \frac{\prod_{j=1}^n e^{a(z_0+z-a_j)+b}}{\prod_{j=1}^n e^{a(z_0+z-b_j)+b}} f(z) .$$

The fraction in the R.H.S. simplifies to 1 because of $a_1 + \cdots + a_n = b_1 + \cdots + b_n$.
□

First proof for the existence of σ (due to Weierstrass).

As an example of a WEIERSTRASS Product, we have already constructed an entire function σ having the zeros exactly in the lattice points of the lattice L. We shortly recall this construction.

It is a natural idea to search for a σ in the form of a WEIERSTRASS product

$$\sigma(z) = z \prod_{\substack{w \in L \\ w \neq 0}} \left(1 - \frac{z}{\omega}\right) e^{P_\omega(z)} .$$

One also can try to take the polynomials P_ω in the form

$$P_\omega(z) = P\left(\frac{z}{\omega}\right) .$$

with one fixed polynomial. This has to be chosen in such a way that the TAYLOR series of

$$1 - (1 - z)e^{P(z)}$$

is of the form $0 + 0z + 0z^2 + Cz^3 + \ldots$. Then the normal convergence of the WEIERSTRASS product follows from an estimate

$$\left|1 - \left(1 - \frac{z}{\omega}\right)e^{P_\omega(z)}\right| \leq \text{Const.} |\omega|^{-3} ,$$

where the constant can be chosen independently of z varying in some compact set K.

It is easy to check that the polynomial

$$P(z) = z + \frac{1}{2}z^2$$

has the desired property, and we obtain:

The infinite product

$$\sigma(z) := z \prod_{\substack{w \in L \\ w \neq 0}} \left(1 - \frac{z}{\omega}\right) \cdot \exp\left(\frac{z}{\omega} + \frac{z^2}{2\omega^2}\right)$$

converges normally in $\mathbb{C}$, *where it represents an entire function* σ. *This function* σ *has simple zeros exactly in the lattice points of the lattice* L.

We have to prove the transformation formula

$$\sigma(z + \omega_0) = e^{az+b}\sigma(z) , \qquad \omega_0 \in L .$$

Because $\sigma(z)$ and $\sigma(z + \omega_0)$ have the same (simple) zeros, the quotient $\sigma(z + \omega_0)/\sigma(z)$ is an entire function without zeros and hence of the form $e^{h(z)}$ with a suitable entire function $h(z)$ (II.2.9),

$$\sigma(z + \omega_0) = \sigma(z)e^{h(z)} .$$

We now claim:

$$h'' = 0 .$$

A simple calculation gives:

$$h'(z) = \frac{\sigma'(z + \omega_0)}{\sigma(z + \omega_0)} - \frac{\sigma'(z)}{\sigma(z)} .$$

Our claim is thus equivalent to

$$\left(\frac{\sigma'}{\sigma}\right)'(z+\omega_0) = \left(\frac{\sigma'}{\sigma}\right)'(z) ,$$

in other words

$$\left(\frac{\sigma'}{\sigma}\right)' \text{ is an elliptic function .}$$

The so-called "logarithmic derivative"

$$\frac{\sigma'(z)}{\sigma(z)} \quad (="(\log\circ\sigma)'(z)")$$

is computed as the sum of the logarithmic derivative of the factors of σ,

$$\frac{\sigma'(z)}{\sigma(z)} = \frac{1}{z} + \sum_{\omega\in L\backslash\{0\}} \left[\frac{-1/\omega}{1-z/\omega} + \frac{1}{\omega} + \frac{z}{\omega^2}\right] .$$

Taking one more derivative, we obtain the negative of the WEIERSTRASS $\wp$-function! It is elliptic, and we are done.

$$\left(\frac{\sigma'}{\sigma}\right)'(z) = -\wp(z).$$

The σ-function is thus closely related to the WEIERSTRASS $\wp$-function.

Second proof for the existence of σ.

First, we prepare the lattice L. Let ω_1, ω_2 be a basis of L. We set

$$\tau = \pm\frac{\omega_2}{\omega_1} ,$$

where the sign $\pm$ has to be chosen, such that τ lies in the upper half-plane. If $f(z)$ is an elliptic function for the lattice L, then $g(z) = f(\omega_1 z)$ is an elliptic function for the lattice $\mathbb{Z} + \mathbb{Z}\tau$, and conversely.

So there is no loss of generality, if we assume from the beginning

$$\omega_1 = 1 \text{ and } \omega_2 = \tau , \text{ Im } \tau > 0 .$$

The key for the second construction is the *theta series*

$$\vartheta(\tau, z) := \sum_{n=-\infty}^{\infty} e^{\pi i(n^2\tau + 2nz)} ,$$

considered as a function in the variable z for constant τ. We postpone the proof of its normal convergence, and show the following transformation properties of ϑ (assuming convergence!):

(1) $$\vartheta(\tau, z+1) = \vartheta(\tau, z) \qquad (\text{because of } e^{2\pi i n} = 1) ,$$

(2) $$\vartheta(\tau, z+\tau) = \sum_{n=-\infty}^{\infty} e^{\pi i (n^2 \tau + 2n\tau + 2nz)}$$

$$= e^{-\pi i \tau} \sum_{n=-\infty}^{\infty} e^{\pi i [(n+1)^2 \tau + 2nz]} .$$

As n runs through $\mathbb{Z}$, $n+1$ also runs through $\mathbb{Z}$, so we obtain

$$\boxed{\vartheta(\tau, z+\tau) = e^{-\pi i (\tau + 2z)} \vartheta(\tau, z) .}$$

By this, we have shown for the basis elements $\omega = \omega_1 = 1$ and $\omega = \omega_2 = \tau$ a transformation formula of the kind

$$\vartheta(\tau, z+\omega) = e^{a_\omega z + b_\omega} \vartheta(\tau, z) .$$

Iterated application of this proves the formula for arbitrary $\omega \in L$.
So it is useful to study the convergence.

Proof of the convergence. Let be

$$\tau = u + iv \quad (v > 0) \quad \text{and} \quad z = x + iy .$$

Then

$$\left| e^{\pi i (n^2 \tau + 2nz)} \right| = e^{-\pi (n^2 v + 2ny)} .$$

If z varies in a given compact set (especially, y remains bounded), then

$$n^2 v + 2ny \geq \frac{1}{2} n^2 v \text{ for all but finitely many } n .$$

The series

$$\sum_{n=-\infty}^{\infty} q^{n^2}, \quad q = e^{-\frac{\pi}{2} v} < 1 ,$$

is convergent, since the subseries defined by $n \geq 0$ and respectively $n < 0$ are subseries of the *geometric series*.

The normal convergence is thus granted, and $\vartheta(\tau, z)$ is an entire function in the variable z with desired transformation properties.

It remains to show that $\vartheta(\tau, z)$ has exactly one simple zero modulo L. For this, we consider the translated fundamental parallelogram $\mathcal{F}_a$, such that there is no zero of $z \to \vartheta(\tau, z)$ on its boundary. We then show

$$\frac{1}{2\pi i} \int_{\partial \mathcal{F}_a} \frac{\vartheta'(\tau, \zeta)}{\vartheta(\tau, \zeta)} d\zeta = 1 .$$

The integrand has period 1, so the integrals along the two "left" and "right" edges of the parallelogram cancel each other. Let us compare the integrals along the "upper" and "lower" edges. The integrand is the "logarithmic derivative"

$$g(z) = \frac{\vartheta'(\tau, z)}{\vartheta(\tau, z)} \quad (= \text{``}(\log \circ \vartheta)'(\tau, z)\text{''})$$

and its values on corresponding points of the edges differ by

$$g(z + \tau) - g(z) = -2\pi i .$$

This implies

$$\int_a^{a+1} g(\zeta)\, d\zeta + \int_{a+1+\tau}^{a+\tau} g(\zeta)\, d\zeta = \int_a^{a+1} [g(\zeta) - g(\zeta + \tau)]\, d\zeta = 2\pi i .$$

We obtain

$$\frac{1}{2\pi i} \int_{\partial \mathcal{F}_a} g(\zeta)\, d\zeta = 1 ,$$

as claimed.   □

By the way, the zero can be explicitly given. We obviously have (see also Exercise 3)

$$\vartheta\left(\tau, \frac{1+\tau}{2}\right) = 0 .$$

The zeros of ϑ are exactly the points equivalent to $(1 + \tau)/2$ modulo $\mathbb{Z} + \mathbb{Z}\tau$.

Remark. So far, we considered ϑ only for fixed values of $\tau \in \mathbb{H}$. But there are good reasons to let τ vary, and look at ϑ as a function on $\mathbb{H} \times \mathbb{C}$. The analytic properties of $\vartheta : \mathbb{H} \times \mathbb{C} \to \mathbb{C}$ (especially as a function of the first variable τ) will be studied in detail in VI.4.

Historical note. It is possible to build the whole theory of elliptic functions by using theta series instead of $\wp(z)$ as we did. Historically, the first approach was using theta series by ABEL (1827/28) and JACOBI (starting with 1828).

Exercises for V.6

1. Let $\sigma(z) = \sigma(z; L)$ be the WEIERSTRASS σ-function for the lattice $L = \mathbb{Z}\omega_1 + \mathbb{Z}\omega_2$. The function

    $$\zeta(z) := \zeta(z; L) := \frac{\sigma'(z)}{\sigma(z)}$$

    is called the *Weierstrass ζ-function* for the lattice L. (This function should not be confused with the RIEMANN ζ-function!) Then $-\zeta'(z) = \wp(z)$ is the WEIERSTRASS $\wp$-function for the lattice L.

    Assume $\text{Im}(\omega_2/\omega_1) > 0$.
    Show: Using the notation $\eta_\nu := \zeta(z + \omega_\nu) - \zeta(z)$ for $\nu = 1, 2$, the following relation of LEGENDRE is true:

$$\boxed{\begin{array}{c}\text{L\textsc{egendre}'s Relation}\\[4pt]\eta_1\omega_2 - \eta_2\omega_1 = 2\pi i\ .\end{array}}$$

Hint. Consider a suitable integral, which counts zeros.

2. The existence of ζ can be obtained also by different means: Setting

$$\xi(z) := -\frac{1}{z} - \sum_{\substack{\omega \in L \\ \omega \neq 0}} \left(\frac{1}{z - \omega} + \frac{1}{\omega} + \frac{z}{\omega^2} \right) ,$$

this defines an odd primitive of $\wp$. (One has $\xi(z) = -\zeta(z)$.)

3. Prove that the zeros of the J\textsc{acobi} theta series $\vartheta(\tau, z)$ are exactly the points equivalent to $\frac{1+\tau}{2}$ modulo $L_\tau = \mathbb{Z} + \mathbb{Z}\tau$.

4. For $z, a \in \mathbb{C} \setminus L$ we have the relations

$$\wp(z) - \wp(a) = -\frac{\sigma(z+a)\sigma(z-a)}{\sigma(z)^2\sigma(a)^2}$$

and

$$\wp'(a) = -\frac{\sigma(2a)}{\sigma(a)^4} \ .$$

5. **Construction of elliptic functions with prescribed principal parts**

Let f be an elliptic function for the lattice L. We choose $b_1, \ldots, b_n$ to be a system of representatives modulo L for the poles of f, and we consider for each j the principal part of f in the pole b_j:

$$\sum_{\nu=1}^{l_j} \frac{a_{\nu,j}}{(z - b_j)^\nu} \ .$$

The Second L\textsc{iouville} Theorem ensures the relation

$$\sum_{j=1}^{n} a_{1,j} = 0 \ .$$

Show:

(a) Let $c_1, \ldots, c_n \in \mathbb{C}$ be given numbers, and let $b_1, \ldots, b_n$ modulo L be a set of different points in $\mathbb{C}/L$. The function

$$h(z) := \sum_{j=1}^{n} c_j \zeta(z - b_j) \ ,$$

constructed by means of the W\textsc{eierstrass} ζ-function, is then elliptic, iff $\sum_{j=1}^{n} c_j = 0.$

(b) Let $b_1, \ldots, b_n$ be pairwise different modulo L, and let $l_1, \ldots, l_n$ be pre-scribed natural numbers. Let $a_{\nu,j}$ $(1 \le j \le n,\ 1 \le \nu \le l_j)$ be complex numbers such that $\sum a_{1,j} = 0$ and $a_{l_j,j} \ne 0$ for all j.

Then there exists an elliptic function for the lattice L, having poles modulo L exactly in the points $b_1, \ldots, b_n$, and having the corresponding principal parts respectively equal to

$$\sum_{\nu=1}^{l_j} \frac{a_{\nu,j}}{(z - b_j)^\nu} \ .$$

6.   Let $L \subset \mathbb{C}$ be a lattice, and let $b_1, b_2 \in \mathbb{C}$ with $b_1 - b_2 \notin L$. Find an elliptic function for the lattice L, having its poles exactly in b_1 and b_2, and having the corresponding principal parts

$$\frac{1}{z - b_1} + \frac{2}{(z - b_1)^2} \quad \text{and} \quad \frac{-1}{z - b_2} \ .$$

7.   We are interested in alternating $\mathbb{R}$-bilinear maps (forms)

$$A : \mathbb{C} \times \mathbb{C} \longrightarrow \mathbb{R} \ .$$

Show:

(a)   Any such map A is of the form

$$A(z, w) = h \operatorname{Im}(z\overline{w})$$

with a uniquely determined real number h. We have explicitly $h = A(1, i)$.

(b)   Let $L \subset \mathbb{C}$ be a lattice. Then A is called a *Riemannian form* with respect to L, iff h is positive, and A only takes integral values on $L \times L$. If

$$L = \mathbb{Z}\omega_1 + \mathbb{Z}\omega_2 \ , \quad \operatorname{Im} \frac{\omega_2}{\omega_1} > 0 \ ,$$

then the formula

$$A(t_1\omega_1 + t_2\omega_2, s_1\omega_1 + s_2\omega_2) := \det \begin{pmatrix} t_1 & s_1 \\ t_2 & s_2 \end{pmatrix}$$

defines a RIEMANNian form A on L.

(c)   A non-constant analytic function $\Theta : \mathbb{C} \to \mathbb{C}$ is called a *theta function* for the lattice $L \subset \mathbb{C}$, iff it satisfies an equation of the type

$$\Theta(z + \omega) = e^{a_\omega z + b_\omega} \cdot \Theta(z)$$

for all $z \in \mathbb{C}$, and all $\omega \in L$. Here, a_ω and b_ω are constants that may depend on ω, but not on z. The WEIERSTRASS σ-function for the lattice L is in this sense a theta function.

Show the existence of a RIEMANNian form A with respect to L, such that

$$A(\omega, \lambda) = \frac{1}{2\pi i}(a_\omega \lambda - \omega a_\lambda) \quad \text{for all } \omega, \lambda \in L \ .$$

Hint. For the integrality of A on $L \times L$, show that in case $\operatorname{Im}(\lambda/\omega) > 0$ the number $A(\omega, \lambda)$ is in fact equal to the number of zeros of Θ in the parallelogram

$$P = P(\omega, \lambda) = \{\ s\omega + t\lambda\ ;\quad 0 \le s, t < 1\ \}\ .$$

(Compare with Exercise 1.)

More generally, we can consider RIEMANNian forms for lattices $L \subset \mathbb{C}^n$, which are alternating $\mathbb{R}$-bilinear forms obtained as the real part of positive definite Hermitian forms on $\mathbb{C}^n$, and which take integer values on $L \times L$. In contrast to the case $n = 1$, the case $n > 1$ requires strong restrictions for the lattice L to admit a RIEMANNian form. We will come back in detail to this question in the second volume.

V.7 The Elliptic Modular Group

In this section, we do not fix a lattice, but rather focus on the *manifold of all equivalence classes of lattices.* Two lattices

$$L \subset \mathbb{C}\ ,\quad L' \subset \mathbb{C}\ ,$$

are called equivalent, in notation $L \sim L'$, iff one can obtain the lattices from each other by *rotation and scaling,* i.e. iff there exists a complex number $a \ne 0$ such that

$$L' = aL\ .$$

The fields of elliptic functions $K(L)$ and $K(L')$ with respect to L and L', are in 1-1-correspondence by means of

$$f(z) \longmapsto f(a^{-1}z),\quad g(z) \longmapsto g(az).$$

Equivalent lattices thus are "essentially equal".

Any lattice $L' \subset \mathbb{C}$ is equivalent to a lattice of the form

$$L = \mathbb{Z} + \mathbb{Z}\tau\ ,\quad \tau \in \mathbb{H}\ ,\ \text{i.e. } \operatorname{Im}\tau > 0\ .$$

When are two lattices

$$L = \mathbb{Z} + \mathbb{Z}\tau \text{ and } L' = \mathbb{Z} + \mathbb{Z}\tau'\ ,\quad \tau, \tau' \in \mathbb{H}\ ,$$

equivalent? By definition, this happens iff there exists a complex number $a \ne 0$ with the property

$$\mathbb{Z} + \mathbb{Z}\tau' = a(\mathbb{Z} + \mathbb{Z}\tau)\ .$$

Then we have in particular

$$\tau' = a(\alpha\tau + \beta) \quad \text{and}$$
$$1 = a(\gamma\tau + \delta)$$

for suitable integers α, β, γ and δ. After division, we obtain

$$\tau' = \frac{\alpha\tau + \beta}{\gamma\tau + \delta} \; .$$

The point τ' is thus obtained from τ by a special MÖBIUS transformation. Before we finish our characterization of equivalent lattices, we have to study the maps

$$\tau \longmapsto \frac{\alpha\tau + \beta}{\gamma\tau + \delta} \; , \quad \text{Im } \tau > 0 \; ,$$

for real α, β, γ and δ. We always assume that either γ or δ is non-zero. Then

$$\gamma\tau + \delta \neq 0 \; .$$

Let us compute the imaginary part of τ':

$$\begin{aligned}
\text{Im } \left(\frac{\alpha\tau + \beta}{\gamma\tau + \delta} \right) &= \frac{1}{2i} \left[\frac{\alpha\tau + \beta}{\gamma\tau + \delta} - \frac{\alpha\overline{\tau} + \beta}{\gamma\overline{\tau} + \delta} \right] \\
&= \frac{1}{2i} \frac{(\gamma\overline{\tau} + \delta)(\alpha\tau + \beta) - (\alpha\overline{\tau} + \beta)(\gamma\tau + \delta)}{|\gamma\tau + \delta|^2} \; .
\end{aligned}$$

We denote by

$$D = \alpha\delta - \beta\gamma$$

the determinant of the matrix $\begin{pmatrix} \alpha & \beta \\ \gamma & \delta \end{pmatrix}$ and obtain

Lemma V.7.1 *Let α, β, γ, δ be four real numbers, such that $\gamma \neq 0$ or $\delta \neq 0$. If τ is a point in the upper half-plane, then*

$$\text{Im } \left(\frac{\alpha\tau + \beta}{\gamma\tau + \delta} \right) = \frac{D \cdot \text{Im } \tau}{|\gamma\tau + \delta|^2} \; .$$

We only are interested in the case, where τ' also lies in the upper half-plane $\mathbb{H}$, i.e.

$$D = \alpha\delta - \beta\gamma > 0 \; .$$

Notation.

$$\mathrm{GL}_+(2, \mathbb{R}) := \left\{ M = \begin{pmatrix} \alpha & \beta \\ \gamma & \delta \end{pmatrix} \; ; \; \alpha, \beta, \gamma, \delta \in \mathbb{R} \; , \; \alpha\delta - \beta\gamma > 0 \right\} \; .$$

The set $\mathrm{GL}_+(2, \mathbb{R})$ is a group with respect to matrix multiplication, i.e.

(a) $E = \begin{pmatrix} 1 & 0 \\ 0 & 1 \end{pmatrix} \in \mathrm{GL}_+(2, \mathbb{R})$.

(b) If the matrices

$$M = \begin{pmatrix} \alpha & \beta \\ \gamma & \delta \end{pmatrix} \text{ and } N = \begin{pmatrix} \alpha' & \beta' \\ \gamma' & \delta' \end{pmatrix}$$

are in $GL_+(2, \mathbb{R})$, then the product

$$M \cdot N = \begin{pmatrix} \alpha\alpha' + \beta\gamma' & \alpha\beta' + \beta\delta' \\ \gamma\alpha' + \delta\gamma' & \gamma\beta' + \delta\delta' \end{pmatrix}$$

is also in $GL_+(2, \mathbb{R})$.

(c) If M is in $GL_+(2, \mathbb{R})$, then the inverse matrix

$$M^{-1} = \frac{1}{\det M} \begin{pmatrix} \delta & -\beta \\ -\gamma & \alpha \end{pmatrix}$$

is also in $GL_+(2, \mathbb{R})$.

To any element $M \in GL_+(2, \mathbb{R})$ we have associated an analytic map $\mathbb{H} \to \mathbb{H}$. This is compatible with the multiplication of matrices on the one side, and composition of analytic maps $\mathbb{H} \to \mathbb{H}$ on the other side, as can be checked by simple computation. (We already observed this in Chapter III, Appendix A.) These analytic maps $\mathbb{H} \to \mathbb{H}$ are in particular invertible, hence conformal. Inverse matrices lead to inverse maps. We conclude:

Proposition V.7.2 *Let M be a real matrix with positive determinant,*

$$M = \begin{pmatrix} \alpha & \beta \\ \gamma & \delta \end{pmatrix} , \quad \text{real}, \qquad \alpha\delta - \beta\gamma > 0 .$$

Then the substitution

$$\tau \longmapsto M\tau := \frac{\alpha\tau + \beta}{\gamma\tau + \delta}$$

defines a conformal map of the upper half-plane $\mathbb{H}$ onto itself. We have:

(a) $E\tau = \tau , \quad E = \begin{pmatrix} 1 & 0 \\ 0 & 1 \end{pmatrix} ,$

(b) $M(N\tau) = (M \cdot N)\tau \qquad M, N \in GL_+(2, \mathbb{R}) .$

The inverse map is given by the inverse matrix

$$M^{-1} = \frac{1}{\alpha\delta - \beta\gamma} \begin{pmatrix} \delta & -\beta \\ -\gamma & \alpha \end{pmatrix} .$$

Two matrices induce the same analytic map $\mathbb{H} \to \mathbb{H}$, iff they differ by a scalar factor.

Because the upper half-plane $\mathbb{H}$ can be mapped conformally onto the unit disk $\mathbb{E}$, fore example by means of the map

$$\tau \longmapsto \frac{\tau - \mathrm{i}}{\tau + \mathrm{i}} \, ,$$

and because we have know all conformal automorphisms of $\mathbb{E}$, III.3.10, we can easily see that every conformal automorphism of $\mathbb{H}$ is of the form described in V.7.2. (See also Exercise 6 in V.7.)

After this short digression about Möbius transformations, we come back to the characterization of equivalence classes of lattices

$$\mathbb{Z} + \mathbb{Z}\tau' = a(\mathbb{Z} + \mathbb{Z}\tau) \, .$$

The inclusion "$\subset$" in the above equality is equivalent to the existence of an *integral* matrix M with the property

$$\begin{pmatrix} \tau' \\ 1 \end{pmatrix} = aM \cdot \begin{pmatrix} \tau \\ 1 \end{pmatrix} \, .$$

The reverse inclusion is equivalent to the existence of an *integral* matrix N with the property

$$a \begin{pmatrix} \tau \\ 1 \end{pmatrix} = N \cdot \begin{pmatrix} \tau' \\ 1 \end{pmatrix} \, .$$

Putting together, we get

$$\begin{pmatrix} \tau \\ 1 \end{pmatrix} = N \cdot M \cdot \begin{pmatrix} \tau \\ 1 \end{pmatrix} \, .$$

Because τ and 1 are $\mathbb{R}$-linearly independent, we get

$$NM = E \, ,$$

in particular

$$\det N \cdot \det M = 1 \, .$$

But both determinants are *integral numbers*, so

$$\det M = \pm 1 \, .$$

By Lemma V.7.1, this determinant is also positive. This shows

$$\det M = +1 \, .$$

Definition V.7.3 *The **elliptic modular group***

$$\Gamma = \mathrm{SL}(2, \mathbb{Z}) := \{ \ M = \begin{pmatrix} \alpha & \beta \\ \gamma & \delta \end{pmatrix} \ ; \quad \alpha, \beta, \gamma, \delta \ integers \, , \ \alpha\delta - \beta\gamma = 1 \ \}$$

is the group of all integral 2×2 matrices with determinant 1.

Γ is a group, as it follows from the formula

$$M^{-1} = \begin{pmatrix} \delta & -\beta \\ -\gamma & \alpha \end{pmatrix} .$$

We showed that if the lattices $\mathbb{Z} + \mathbb{Z}\tau$ and $\mathbb{Z} + \mathbb{Z}\tau'$ (Im τ, Im $\tau' > 0$) are equivalent, then there exists a matrix

$$M \in \Gamma \quad \text{with} \quad \tau' = M\tau .$$

Conversely, from this the equivalence of the lattices follows:

One can write

$$\tau' = \frac{\alpha\tau + \beta}{\gamma\tau + \delta}$$

in the form

$$\begin{pmatrix} \tau' \\ 1 \end{pmatrix} = \begin{pmatrix} \alpha & \beta \\ \gamma & \delta \end{pmatrix} \begin{pmatrix} a\tau \\ a \end{pmatrix} \qquad (a = (\gamma\tau + \delta)^{-1}) .$$

We collect these facts:

Proposition V.7.4 *Two lattices of the form*

$$\mathbb{Z} + \mathbb{Z}\tau \quad and \quad \mathbb{Z} + \mathbb{Z}\tau' \quad with \; Im\,\tau > 0 \; and \; Im\,\tau' > 0$$

are equivalent, iff there exists a matrix $M \in \Gamma$ with the property $\tau' = M\tau$.

Two points τ and τ' in the upper half-plane are called *equivalent*, iff there exists a substitution $M \in \Gamma$, which maps τ into τ', i.e. $\tau' = M\tau$. Clearly, this is an equivalence relation.

Notations.

$$\mathbb{H} = \{\,\tau \in \mathbb{C}\,;\quad Im\,\tau > 0\,\} \quad \text{(upper half-plane)} ,$$
$$[\tau] = \{\,M\tau\,;\quad M \in \Gamma\,\} \quad \text{(orbit of } \tau \in \mathbb{H} \text{ for this equivalence relation)},$$
$$\mathbb{H}/\Gamma = \{\,[\tau]\,;\quad \tau \in \mathbb{H}\,\} \quad \text{(set of all orbits)} .$$

We showed that the equivalence classes of lattices $L \subset \mathbb{C}$ correspond one-to-one to the points of $\mathbb{H}/\Gamma$.

The meaning of the manifold $\mathbb{H}/\Gamma$.

Our next goal is to show that for any pair of complex numbers

$$(g_2, g_3) , \quad g_2^3 - 27 g_3^2 \neq 0 ,$$

there exists a lattice $L \subset \mathbb{C}$ with the property

$$g_2 = g_2(L) , \quad g_3 = g_3(L) .$$

The numbers $g_2(L), g_3(L)$ change if we replace L by an equivalent lattice. For $a \in \mathbb{C}^\bullet$ we have

$$G_k(aL) = a^{-k}G_k(L) \, ,$$

in particular

$$g_2(aL) = a^{-4}g_2(L) \qquad \text{and}$$
$$g_3(aL) = a^{-6}g_3(L) \, .$$

We would like to have an expression, which only depends on the equivalence class. For this we introduce the following *notations*:

(1) $\Delta := g_2^3 - 27g_3^2$ is called the *discriminant*,

(2) $j := \dfrac{g_2^3}{g_2^3 - 27g_3^2}$ is called the *absolute invariant*

(F. KLEIN, 1879).

Then we have

$$\Delta(aL) = a^{-12}\Delta(L)$$

and thus

$$\boxed{\; j(aL) = j(L) \quad (a \in \mathbb{C}^\bullet) \, . \;}$$

Let us assume that for any complex number $j_0 \in \mathbb{C}$ we can find a lattice $L \subset \mathbb{C}$ with

$$j(L) = j_0 \, .$$

Then we will show that any given pair (g_2, g_3) with $\Delta \neq 0$ can be realized as $(g_2(L), g_3(L))$ for a suitable lattice L.

Let us fix a pair (g_2, g_3) with $\Delta := g_2^3 - 27g_3^2 \neq 0$. First, by our assumption, there exists a lattice L with

$$\frac{g_2^3(L)}{\Delta(L)} = j(L) = \frac{g_2^3}{\Delta} \, .$$

For $a \in \mathbb{C}$ we then also have

$$\frac{g_2^3(aL)}{\Delta(aL)} = j(aL) = j(L) = \frac{g_2^3}{\Delta} \, .$$

Any non-zero complex number admits a 12th root. Hence there is an $a \neq 0$ which solves the equation

$$\Delta(aL) = a^{-12}\Delta(L) = \Delta = g_2^3 - 27g_3^2 \, .$$

We thus have

$$g_2(aL)^3 = g_2^3 \qquad \text{and} \qquad g_3^2(aL) = g_3^2 \, .$$

If we replace a by ia, then $g_2(aL)$ doesn't change ($i^{-4} = 1$), but $g_3(aL)$ changes its sign ($i^{-6} = -1$). Hence we can assume

$$g_2(aL)^3 = g_2^3 \qquad \text{and} \qquad g_3(aL) = g_3 \, .$$

If we replace a by ζa, $\zeta^6 = 1$, i.e. by multiplying it with a suitable 6th root of unity ζ, then $g_3(aL)$ doesn't change ($\zeta^{-6} = 1$), but $g_2(aL)$ changes by multiplication with $\zeta^{-4} = \zeta^2$. When ζ runs through all 6th roots of 1, then ζ^2 runs through all 3^{d} roots of 1. Hence we can assume

$$g_2(aL) = g_2 \quad \text{and} \quad g_3(aL) = g_3 \ .$$

Our problem is, as claimed, reduced to the question whether any complex number appears as the absolute invariant of a lattice. We want to attack this question by methods of complex analysis. So we consider

the EISENSTEIN series,

the discriminant,

and the absolute invariant, as functions on the upper half-plane. We thus define for any $\tau \in \mathbb{H}$:

$$G_k(\tau) := G_k(\mathbb{Z} + \mathbb{Z}\tau)$$

and analogously we introduce

$$g_2(\tau) \ , \quad g_3(\tau) \ , \quad \Delta(\tau) \ , \quad j(\tau) \ .$$

These are thus functions on the upper half-plane.

The invariance property

$$j(L) = j(aL)$$

is equivalent to the invariance of $j(\tau)$ under the action of the modular group

$$\boxed{j\left(\frac{\alpha\tau + \beta}{\gamma\tau + \delta}\right) = j(\tau) \text{ for } \begin{pmatrix} \alpha & \beta \\ \gamma & \delta \end{pmatrix} \in \Gamma \ .}$$

In the next section, we will show by *purely function theoretical methods* essentially using the above invariance property that the j-function is a *surjective* function

$$j : \mathbb{H} \longrightarrow \mathbb{C} \ .$$

We finish this section with explicit formulas for G_k as functions of τ:

$$\boxed{G_k(\tau) = \sum_{\substack{(c,d)\in\mathbb{Z}\times\mathbb{Z} \\ (c,d)\neq(0,0)}} (c\tau + d)^{-k} \qquad (k \in \mathbb{N} \ , \ k \geq 4)}$$

and from this

$$\boxed{\begin{aligned} g_2(\tau) &= 60 \, G_4(\tau) \ , \\ g_3(\tau) &= 140 \, G_6(\tau) \ , \\ \Delta(\tau) &= g_2^3(\tau) - 27g_3^2(\tau) \ , \\ j(\tau) &= \frac{g_2^3(\tau)}{\Delta(\tau)} \ . \end{aligned}}$$

Exercises for V.7

1. The elliptic modular group $\Gamma = \mathrm{SL}(2, \mathbb{Z})$ is generated by the matrices (see also VI.1.8):

    $$S := \begin{pmatrix} 0 & -1 \\ 1 & 0 \end{pmatrix} \text{ and } T := \begin{pmatrix} 1 & 1 \\ 0 & 1 \end{pmatrix}$$

    Hint. Consider the subgroup Γ_0 of Γ, which is generated by the matrices S and T. Show that a matrix $M \in \mathrm{SL}(2, \mathbb{Z})$ belongs to Γ_0, if one of its four entries vanishes. Now, if there would exist a matrix $M = \begin{pmatrix} a & b \\ c & d \end{pmatrix} \in \Gamma$, but $M \notin \Gamma_0$, then we can choose it to have a minimal value for $\mu(M) := \min\{|a|, |b|, |c|, |d|\} > 0$ under all $M \notin \Gamma_0$. But then, multiplication with a suitable matrix from Γ_0 (from left or right) reduces the value $\mu(M)$.

2. Represent the matrix $M = \begin{pmatrix} 4 & 9 \\ 11 & 25 \end{pmatrix} \in \Gamma$ in the form

    $$M = ST^{q_1} ST^{q_2} \dots ST^{q_n} \ , \quad q_\nu \in \mathbb{Z} \ , \ 1 \le \nu \le n \ ,$$

    with $S = \begin{pmatrix} 0 & -1 \\ 1 & 0 \end{pmatrix}$ and $T = \begin{pmatrix} 1 & 1 \\ 0 & 1 \end{pmatrix}$. Is such a representation unique?

3. Determine all matrices $M \in \Gamma$ which
    (a) commute with S, i.e. $MS = SM$.
    (b) commute with $ST = \begin{pmatrix} 0 & -1 \\ 1 & 1 \end{pmatrix}$.

4. Determine the smallest natural number n with

    $$(ST)^n = E = \begin{pmatrix} 1 & 0 \\ 0 & 1 \end{pmatrix} \ .$$

5. *Show:*
    (a) For the lattice $L_i = \mathbb{Z} + \mathbb{Z}i$ we have $g_3(i) = 0$ and $g_2(i) \in \mathbb{R}^\bullet$, in particular $\Delta(i) = g_2^3(i) > 0$.
    (b) For the lattice $L_\omega = \mathbb{Z} + \mathbb{Z}\omega$, $\omega := e^{2\pi i/3}$, we have $g_2(\omega) = 0$ and $g_3(\omega) \in \mathbb{R}^\bullet$, in particular $\Delta(\omega) = -27g_3^2(\omega)$.

6. Any conformal self-map of the upper half-plane is of the form

    $$\tau \longmapsto \frac{a\tau + b}{c\tau + d} \ , \quad \begin{pmatrix} a & b \\ c & d \end{pmatrix} \in \mathrm{GL}_+(2, \mathbb{R}) \ .$$

    We can moreover obtain that the determinant $ad - bc = 1$ equals 1. Then the matrix is uniquely determined up to a sign, i.e. $\mathrm{Aut}(\mathbb{H}) = \mathrm{SL}(2, \mathbb{R}) / \{\pm E\}$.

    Hint. Use the knowledge of $\mathrm{Aut}(\mathbb{E})$, III.3.10, and the conformal equivalence of $\mathbb{H}$ and $\mathbb{E}$ given by the explicit map at page 303. Because the group of all affine transformations $\tau \mapsto a\tau + b$, $a > 0$, b real, acts *transitively* on the upper half-plane, it is enough to determine the stabilizer of a special point, say i. It is then enough to show that any automorphism of $\mathbb{H}$ which fixes i is given by a special orthogonal matrix

    $$\begin{pmatrix} a & b \\ c & d \end{pmatrix} = \begin{pmatrix} \cos\varphi & -\sin\varphi \\ \sin\varphi & \cos\varphi \end{pmatrix} \ .$$

V.8 The Modular Function j

We have already established the absolute convergence of the so-called *Eisenstein series*,

$$G_k(\tau) = {\sum}'(c\tau + d)^{-k} , \quad \operatorname{Im} \tau > 0 ,$$

for all $k \geq 3$. As indicated by the dash, the sum is restricted in the sense that it is taken over all pairs $(c, d) \neq (0, 0)$ of integers.

From the theory of the $\wp$-function, we know that the discriminant

$$\Delta(\tau) = g_2^3(\tau) - 27g_3^2(\tau)$$
$$g_2 = 60 \, G_4 , \quad g_3 = 140 \, G_6 ,$$

has no zero in the upper half-plane. Besides this fact, we will not use any result from the theory of elliptic functions!

Let us show that all G_k are *analytic functions* in $\mathbb{H}$.

Lemma V.8.1 *Let $C, \delta > 0$ be real numbers. Then there exists a real number $\varepsilon > 0$ with the property*

$$|c\tau + d| \geq \varepsilon \, |ci + d| = \varepsilon \, \sqrt{c^2 + d^2}$$

for all $\tau \in \mathbb{H}$ with

$$|\operatorname{Re} \tau| \leq C , \quad \operatorname{Im} \tau \geq \delta ,$$

and all

$$(c, d) \in \mathbb{R} \times \mathbb{R} .$$

Proof. For the uninteresting case $(c, d) = (0, 0)$ there is nothing to show. So let us take $(c, d) \neq (0, 0)$. The claim of the Lemma does not change if one replaces (c, d) by (tc, td). So we can further assume

$$c^2 + d^2 = 1 .$$

The inequality reduces to

$$|c\tau + d| \geq \varepsilon \qquad (c^2 + d^2 = 1) .$$

We have

$$|c\tau + d|^2 = (c(\operatorname{Re} \tau) + d)^2 + (c \operatorname{Im} \tau)^2 ,$$

and because of this

$$|c\tau + d| \geq |c\tilde{\tau} + d| , \quad \tilde{\tau} := \operatorname{Re} \tau + i\delta .$$

The function

$$f(c, d, u) = |c(u + i\delta) + d|$$

is positive, and takes on the compact set of $\mathbb{R}^3$ defined by

$$c^2 + d^2 = 1 , \quad |u| \leq C ,$$

its positive minimum ε. $\square$

From Lemma V.2.1, the Eisenstein series converges uniformly in the claimed domain. Hence it represents an analytic function.

Proposition V.8.2 *The Eisenstein series of "weight" $k \geq 3$*

$$G_k(\tau) = {\sum}'(c\tau + d)^{-k}$$

defines an analytic function on the upper half-plane. In particular, the functions

$$g_2(\tau) = 60\, G_4(\tau) , \qquad\qquad g_3(\tau) = 140\, G_6(\tau) ,$$
$$\Delta(\tau) = g_2(\tau)^3 - 27 g_3(\tau)^2 , \qquad j(\tau) = g_2^3(\tau)/\Delta(\tau)$$

are analytic in $\mathbb{H}$.

Next, we determine the transformation behaviour of G_k under the action of the modular group. Basically, this follows from the relation "$G_k(aL) = a^{-k}G_k(L)$", but we promised to not use elliptic functions.

Remark V.8.3 *The following transformation formula holds:*

$$G_k\left(\frac{\alpha\tau + \beta}{\gamma\tau + \delta}\right) = (\gamma\tau + \delta)^k G_k(\tau) \quad for \quad \begin{pmatrix} \alpha & \beta \\ \gamma & \delta \end{pmatrix} \in \Gamma .$$

Proof. A simple calculation shows

$$c\frac{\alpha\tau + \beta}{\gamma\tau + \delta} + d = \frac{c'\tau + d'}{\gamma\tau + \delta}$$

with

$$c' = \alpha c + \gamma d , \quad d' = \beta c + \delta d .$$

As (c, d), the pair (c', d') runs through the set $\mathbb{Z} \times \mathbb{Z} \setminus \{(0, 0)\}$. This is best seen from the matrix equation

$$\begin{pmatrix} c' \\ d' \end{pmatrix} = \begin{pmatrix} \alpha & \gamma \\ \beta & \delta \end{pmatrix}\begin{pmatrix} c \\ d \end{pmatrix} , \qquad \begin{pmatrix} c \\ d \end{pmatrix} = \begin{pmatrix} \delta & -\gamma \\ -\beta & \alpha \end{pmatrix}\begin{pmatrix} c' \\ d' \end{pmatrix} .$$

$\square$

The EISENSTEIN series are especially periodic with period 1,

$$G_k(\tau + 1) = G_k(\tau) \qquad \left(\text{because of } \begin{pmatrix} 1 & 1 \\ 0 & 1 \end{pmatrix}\tau = \tau + 1 \right) .$$

The EISENSTEIN series identically vanish for odd k, as we observed before, but also the substitution $(c, d) \rightarrow (-c, -d)$ shows $G_k(\tau) = (-1)^k G_k(\tau)$.

Remark V.8.4 *For even $k \geq 4$ we have*

$$\lim_{Im\,\tau \to \infty} G_k(\tau) = 2\zeta(k) = 2\sum_{n=1}^{\infty} n^{-k} .$$

Proof. Because of the 1-periodicity of $G_k(z)$, it is enough to take the limit in the domain

$$|\mathrm{Re}\,\tau| \leq \frac{1}{2} , \quad \mathrm{Im}\,\tau \geq 1 .$$

In this region the EISENSTEIN series converges uniformly, V.8.1, so we can take the limit termwise. Obviously,

$$\lim_{Im\,\tau \to \infty} (c\tau + d)^{-1} = 0 \text{ for } c \neq 0 .$$

This implies

$$\lim_{Im\,\tau \to \infty} G_k(\tau) = \sum_{d \neq 0} d^{-k} = 2\sum_{d=1}^{\infty} d^{-k} . \qquad \square$$

For the discriminant $\Delta(\tau)$, exploiting V.8.4, we get

$$\lim_{Im\,\tau \to \infty} \Delta(\tau) = [60 \cdot 2\zeta(4)]^3 - 27 \cdot [140 \cdot 2\zeta(6)]^2 .$$

We have already computed the values of the RIEMANN ζ-function at the even natural numbers $2, 4, 6, \ldots$, III.7.14, and we got the special values

$$\zeta(4) = \sum_{n=1}^{\infty} n^{-4} = \frac{\pi^4}{90} ,$$

$$\zeta(6) = \sum_{n=1}^{\infty} n^{-6} = \frac{\pi^6}{945} .$$

This gives:

Lemma V.8.5 *One has*

$$\lim_{Im\,\tau \to \infty} \Delta(\tau) = 0 .$$

From the above results about EISENSTEIN series, we obtain

Proposition V.8.6 *The j-function is an analytic function in the upper half-plane. It is invariant under the action of the elliptic modular group:*

$$j\left(\frac{\alpha\tau + \beta}{\gamma\tau + \delta}\right) = j(\tau) \quad for \quad \begin{pmatrix} \alpha & \beta \\ \gamma & \delta \end{pmatrix} \in \Gamma .$$

We have

$$\lim_{Im\,\tau \to \infty} |j(\tau)| = \infty .$$

Using only the properties listed in V.8.6, we will derive the *surjectivity* of $j : \mathbb{H} \to \mathbb{C}$.

One should keep in mind that non-constant elliptic functions $f : \mathbb{C} \to \overline{\mathbb{C}}$, i.e. meromorphic functions with invariance under a lattice $L \subset \mathbb{C}$, are also surjective. But the theory of *modular functions* (i.e. Γ-invariant functions on the upper half-plane $\mathbb{H}$) is more complicated for two reasons:

(1) The group $\Gamma = \mathrm{SL}(2, \mathbb{Z})$ is *not commutative*.
(2) There is no compact fundamental region $K \subset \mathbb{H}$ for the action of Γ on $\mathbb{H}$, i.e. a region K such that each point in $\mathbb{H}$ can transformed into K by a modular substitution. (Else, $j(\tau)$ would be constant, as the proof of the First LIOUVILLE Theorem shows.)

We now construct a fundamental region for the action of Γ on $\mathbb{H}$. It is analogous to a fundamental parallelogram of a lattice.

Proposition V.8.7 *For any point τ in the upper half-plane there exists a modular substitution $M \in \Gamma$, such that $M\tau$ is contained in the so-called "modular figure", the fundamental region for the modular group,*

$$\mathcal{F} = \left\{\, \tau \in \mathbb{H} \;;\; |\tau| \geq 1 \,,\; |Re\,\tau| \leq 1/2 \,\right\}$$

Supplement. *One can moreover obtain that M is contained in the group generated by the two matrices*

$$T := \begin{pmatrix} 1 & 1 \\ 0 & 1 \end{pmatrix} \,, \quad S := \begin{pmatrix} 0 & -1 \\ 1 & 0 \end{pmatrix} \,.$$

(Later we will see that the full modular group Γ is generated by these two matrices, compare with VI.1.8 and Exercise 1 in V.7).

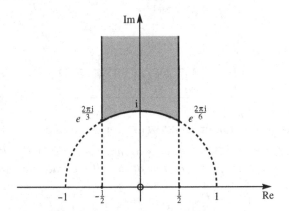

Proof. Recall the formula

$$\mathrm{Im}\, M\tau = \frac{\mathrm{Im}\, \tau}{|c\tau + d|^2} \,.$$

If (c, d) varies in a sequence of pairs of non-repeating integers, then

$$|c\tau + d| \longrightarrow \infty .$$

So there exists a matrix $M_0 \in \Gamma = \mathrm{SL}(2, \mathbb{Z})$, such that

$$\mathrm{Im}\ M_0\tau \geq \mathrm{Im}\ M\tau \text{ for all } M \in \Gamma :$$

We set

$$\tau_0 = M_0\tau .$$

The imaginary part of τ_0 does not change if we replace τ_0 by

$$\tau_0 + n = \left[\begin{pmatrix} 1 & n \\ 0 & 1 \end{pmatrix} M_0\right] \tau \qquad (n \in \mathbb{Z}) ,$$

so we can moreover require that τ_0 satisfies

$$|\mathrm{Re}\ \tau_0| \leq \frac{1}{2} .$$

Now we use the inequality

$$\mathrm{Im}\ M_0\tau \geq \mathrm{Im}\ M\tau$$

especially for the matrix

$$M = \begin{pmatrix} 0 & -1 \\ 1 & 0 \end{pmatrix} \cdot M_0 .$$

This gives

$$\mathrm{Im}\ \tau_0 \geq \mathrm{Im} \begin{pmatrix} 0 & -1 \\ 1 & 0 \end{pmatrix} \tau_0 = \frac{\mathrm{Im}\ \tau_0}{|\tau_0|^2} .$$

From this we derive

$$|\tau_0| \geq 1 .$$

If we analyze the proof, we see that we can replace the group $\mathrm{SL}(2, \mathbb{Z})$ by the subgroup generated by T and S. $\square$

We finally prove the surjectivity of the j-function.

Theorem V.8.8 (Surjectivity of the j-function) *The j-function takes on any value in $\mathbb{C}$.*

Corollary V.8.9 *For any two complex numbers g_2 and g_3 with $g_2^3 - 27g_3^2 \neq 0$ there exists a lattice $L \subset \mathbb{C}$ with the property*

$$g_2 = g_2(L) , \qquad g_3 = g_3(L) .$$

Proof of V.8.8. By the Open Mapping Theorem, $j(\mathbb{H})$ is an open set in $\mathbb{C}$. We will show that $j(\mathbb{H})$ is also closed in $\mathbb{C}$. Because $\mathbb{C}$ is connected, we then obtain $j(\mathbb{H}) = \mathbb{C}$.

So let us consider a sequence of points in $j(\mathbb{H})$, converging to a point $b \in \mathbb{C}$,

$$j(\tau_n) \to b \quad \text{for} \quad n \to \infty .$$

(We show $b \in j(\mathbb{H})$.) We can assume that the set of all τ_n is contained in the fundamental region $\mathcal{F}$. We distinguish two cases

First case: There exists a constant $C > 0$, such that

$$\operatorname{Im} \tau_n \leq C \text{ for all } n .$$

The set

$$\{ \tau \in \mathcal{F} ; \quad \operatorname{Im} \tau \leq C \}$$

is then compact. Taking a suitable subsequence, we get that (τ_n) is convergent,

$$\tau_n \to \tau \in \mathcal{F} \subset \mathbb{H} .$$

The continuity of j then gives

$$b = j(\tau) \in j(\mathbb{H}) .$$

Second case: There exists a subsequence of (τ_n), with imaginary parts converging to ∞. The j-values of this subsequence are then unbounded, V.8.6. This contradicts the convergence of $(j(\tau_n))$.

Hence the second case is not possible and we get $b \in j(\mathbb{H})$. $\square$

We will see in the next chapter that the j-function gives a *bijective* map

$$\mathbb{H}/\Gamma \longrightarrow \mathbb{C} .$$

Exercises for V.8

1. Determine a point $\tau \in \mathcal{F}$, which is equivalent modulo Γ to $\frac{5i+6}{4i+5} \in \mathbb{H}$ and respectively to $\frac{2}{17} + \frac{8}{17}i \in \mathbb{H}$.

2. The surjectivity of $j : \mathbb{H} \longrightarrow \mathbb{C}$ was motivated as follows:
    (a) $j(\mathbb{H})$ is by the Open Mapping Theorem open in $\mathbb{C}$ and non-empty.
    (b) $j(\mathbb{H})$ is closed in $\mathbb{C}$.
    This implies $j(\mathbb{H}) = \mathbb{C}$, since $\mathbb{C}$ is connected. Fill in the details.

3. The EISENSTEIN series are "real" functions, i.e. $\overline{G_k(\tau)} = G_k(-\overline{\tau})$. This implies

    $$G_k\left(\frac{\alpha(-\overline{\tau}) + \beta}{\gamma(-\overline{\tau}) + \delta}\right) = (\gamma(-\overline{\tau}) + \delta)^k \overline{G_k(\tau)} \qquad \text{and}$$

    $$j\left(\frac{\alpha(-\overline{\tau}) + \beta}{\gamma(-\overline{\tau}) + \delta}\right) = \overline{j(\tau)} \qquad \text{for} \quad \begin{pmatrix} \alpha & \beta \\ \gamma & \delta \end{pmatrix} \in \Gamma .$$

    On the vertical half-lines $\operatorname{Re} \tau = \pm\frac{1}{2}$ in $\mathbb{H}$ the EISENSTEIN series and the j-function are real. If $\tau \in \mathbb{H}$ lies on the circle line $|\tau| = 1$, then $j(\tau) = \overline{j(\tau)}$. In particular, the j-function is real on the boundary of the modular figure, and on the imaginary axis.

4. In the following exercise one may use the fact that the FOURIER expansion of the discriminant is of the form (VI.2.8)

$$\Delta(\tau) = a_1 q + a_2 q^2 + \cdots \quad , \quad a_1 \neq 0 \quad (q = e^{2\pi i \tau}) .$$

Show that for any real number j_0 there exists a τ_0 either on the boundary of the fundamental region $\mathcal{F}$ or on the imaginary axis, which satisfies $j(\tau_0) = j_0$.

Hint. Compute the limits of $j(\tau)$ for $\tau = \pm 1/2 + iy$ and respectively $\tau = 0 + iy$, when $y \in \mathbb{R}$ tends to infinity.

5. *Show:*

$$j\left(e^{\frac{2\pi i}{3}}\right) = 0 , \quad j(i) = 1 .$$

6. Prove the Supplement in V.8.7 in detail:

For any $\tau \in \mathbb{H}$ there exists a matrix M in the subgroup of $\Gamma = \mathrm{SL}(2,\mathbb{Z})$ generated by

$$T := \begin{pmatrix} 1 & 1 \\ 0 & 1 \end{pmatrix} \text{ and } S := \begin{pmatrix} 0 & -1 \\ 1 & 0 \end{pmatrix}$$

with

$$M\tau \in \mathcal{F} .$$

VI

Elliptic Modular Forms

In connection with the question which complex numbers can be written as the absolute invariant of a lattice, we were led to analytic functions with a new type of symmetries. These functions are analytic functions on the upper half-plane with a specific transformation law with respect to the action of the full elliptic modular group (or of certain subgroups) on $\mathbb{H}$, namely

$$f\left(\frac{az+b}{cz+d}\right) = (cz+d)^k f(z) .$$

Functions with such a transformation behavior are called *modular forms*.

We will see that the elliptic modular group is generated by the substitutions

$$z \longmapsto z+1 \text{ and } z \longmapsto -\frac{1}{z} .$$

It is thus enough to check the transformation behavior only for these substitutions. There is an analogy to the transformation behavior of elliptic functions under translations in a lattice L, where it was also sufficient to check the invariance under the two generating *translations* ω_1, ω_2. But in contrast to the translation lattice L, the elliptic modular group is *not* commutative. Hence the theory of modular forms is more complicated than the theory of elliptic functions. This could be already observed in the construction of a fundamental domain for the action of the modular group Γ on the upper half-plane $\mathbb{H}$, V.8.7.

In Sect. VI.2 we will find an analogue for LIOUVILLE's theorems, the so-called $k/12$-formula (valence formula). It offers information about the number of zeros of a modular form. In connection with this, we prove some structure theorems culminating in the fact that the ring of all modular forms is generated by the two EISENSTEIN series G_4 and G_6. In contrast, the field of all modular functions is generated by the j-function.

In Sect. VI.4 we introduce *theta series* as a new instrument to construct modular forms. Due to the structure theorem, we discover non-trivial identities between analytic functions. These identities have interesting number theoretical consequences, which we further pursue in Chapter VII.

Theta series are in general not modular forms for the *full* modular group, but only for subgroups of finite index. So we naturally need a more general notion of a modular

E. Freitag and R. Busam, *Complex Analysis*,   317
DOI: 10.1007/978-3-540-93983-2_VI, © Springer-Verlag Berlin Heidelberg 2009

form for such subgroups. In Sect. VI.5 we generalize this notion and also allow *half-integral weights*.

In VI.6 we will study a concrete example, namely, we determine the full ring of modular forms for the IGUSA congruence subgroup $\Gamma[4,8]$. This ring is generated by the three JACOBI theta series.

VI.1 The Modular Group and Its Fundamental Region

Recall the action of the *elliptic modular group* $\Gamma = \mathrm{SL}(2,\mathbb{Z})$ on the upper half-plane $\mathbb{H}$:

$$\Gamma \times \mathbb{H} \longrightarrow \mathbb{H} ,$$

$$(M, z) \longmapsto Mz := \frac{az+b}{cz+d} .$$

Two matrices M and N define the same substitution, i.e.

$$Mz = Nz \text{ for all } z \in \mathbb{H} ,$$

iff they differ by a sign, $M = \pm N$.

In V.8 we introduced the "modular figure"

$$\mathcal{F} := \left\{ z \in \mathbb{H}; \quad |\mathrm{Re}\, z| \leq \frac{1}{2}, \ |z| \geq 1 \right\}$$

and we proved

$$\mathbb{H} = \bigcup_{M \in \Gamma} M\mathcal{F} .$$

In this section we want to refine this, and show that this "plastering" covers the upper half-plane without overlapping, i.e. for $M, N \in \Gamma$, $M \neq \pm N$, the transformed domains $M\mathcal{F}$ and $N\mathcal{F}$ have no common *inner* points. But they may have common boundary points.

For this, we must find all $M \in \Gamma$ with the property $M\mathcal{F} \cap \mathcal{F} \neq \emptyset$. There are only finitely many of them. This follows from

Lemma VI.1.1 *Let $\delta > 0$, and let us consider*

$$\mathcal{F}(\delta) := \left\{ z \in \mathbb{H} ; \quad |x| \leq \delta^{-1} , \ y \geq \delta \right\} .$$

Then there exist only finitely many $M \in \Gamma$ with the property

$$M\mathcal{F}(\delta) \cap \mathcal{F}(\delta) \neq \emptyset .$$

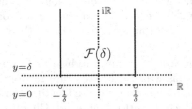

Corollary VI.1.1$_1$ *For any two compact sets $K, \tilde{K} \subset \mathbb{H}$ there exist only finitely many $M \in \Gamma$ with*

$$M(K) \cap \tilde{K} \neq \emptyset \ ,$$

since $K \cup \tilde{K} \subset \mathcal{F}(\delta)$ for a suitable $\delta > 0$.

Corollary VI.1.1₂ *Let $p \in \mathbb{H}$, and let K be a compact set in $\mathbb{H}$. Then there exist only finitely many elements*

$$M \in \Gamma \ \text{with} \ Mp \in K \ .$$

In particular, the orbit of p with respect to Γ, $\{ Mp \ ; \ M \in \Gamma \}$, is discrete in $\mathbb{H}$.

Corollary VI.1.1₃ *The **stabilizer** of p with respect to the action of Γ,*

$$\Gamma_p = \{ M \in \Gamma \ ; \quad Mp = p \}$$

is a finite group for any point $p \in \mathbb{H}$.

Proof of Lemma VI.1.1. In case of $c = 0$, the map $z \mapsto Mz$ is a translation. Since the real parts of z and Mz are bounded, there exist only finitely many such translations. So we can focus on the remaining case $c \neq 0$. Then the existence of z in the intersection gives

$$y = \operatorname{Im} z \geq \delta \quad \text{and} \quad \frac{y}{|cz + d|^2} = \operatorname{Im}(Mz) \geq \delta \ ,$$

which implies

$$y \geq \delta(cx + d)^2 + \delta c^2 y^2 \geq \delta c^2 y^2.$$

This gives

$$\frac{1}{\delta c^2} \geq y \geq \delta \ ,$$

hence $c^2 \leq 1/\delta^2$, and we obtain that there are only finitely many possible values for c. From this we then obtain the same for d.

But together with M, the inverse M^{-1} also fulfills the non-empty intersection condition. This shows that there are only finitely many values for a.

The condition $ad - bc = 1$ finally shows that b is also bounded. This finishes the proof.   □

Next, we determine the matrices $M \in \Gamma$, which fix the "right most lower vertex" ϱ of $\mathcal{F}$,

$$\varrho := e^{\pi i/3} = \frac{1}{2} + \frac{i}{2}\sqrt{3} \ .$$

We have $\varrho^2 = -\bar{\varrho} = \varrho - 1$ and $\varrho^3 = -1$.

Lemma VI.1.2 *There are exactly six matrices*

$$M \in \Gamma \text{ with } M\varrho = \varrho \, ,$$

namely

$$\pm \begin{pmatrix} 1 & 0 \\ 0 & 1 \end{pmatrix} \, , \quad \pm \begin{pmatrix} 1 & -1 \\ 1 & 0 \end{pmatrix} \, , \quad \pm \begin{pmatrix} 0 & -1 \\ 1 & -1 \end{pmatrix} \, .$$

Corollary. *Each of the equations*

$$M\varrho = \varrho^2 \, , \quad M\varrho^2 = \varrho \, , \quad M\varrho^2 = \varrho^2 \, ,$$

has exactly six solutions in Γ, namely

(1) $(M\varrho = \varrho^2)$: $\pm \begin{pmatrix} 0 & 1 \\ -1 & 0 \end{pmatrix}$, $\pm \begin{pmatrix} 1 & 0 \\ -1 & 1 \end{pmatrix}$, $\pm \begin{pmatrix} 1 & -1 \\ 0 & 1 \end{pmatrix}$,

(2) $(M\varrho^2 = \varrho)$: $\pm \begin{pmatrix} 0 & 1 \\ -1 & 0 \end{pmatrix}$, $\pm \begin{pmatrix} 1 & 1 \\ 0 & 1 \end{pmatrix}$, $\pm \begin{pmatrix} 1 & 0 \\ 1 & 1 \end{pmatrix}$,

(3) $(M\varrho^2 = \varrho^2)$: $\pm \begin{pmatrix} 1 & 0 \\ 0 & 1 \end{pmatrix}$, $\pm \begin{pmatrix} 0 & -1 \\ 1 & 1 \end{pmatrix}$, $\pm \begin{pmatrix} -1 & -1 \\ 1 & 0 \end{pmatrix}$.

The *Corollary* follows if we replace each occurrence of ϱ^2 by ϱ using

$$\varrho^2 = \begin{pmatrix} 0 & -1 \\ 1 & 0 \end{pmatrix} \varrho \, ,$$

and reduce each equation to one of the form $N\varrho = \varrho$, for instance

$$M\varrho = \varrho^2 \Longleftrightarrow \begin{pmatrix} 0 & -1 \\ 1 & 0 \end{pmatrix} M\varrho = \varrho \, . \qquad \square$$

Proof of VI.1.2. Let $M = \begin{pmatrix} a & b \\ c & d \end{pmatrix} \in \Gamma$. From the equation

$$\frac{a\varrho + b}{c\varrho + d} = \varrho \text{ or } a\varrho + b = c\varrho^2 + d\varrho$$

we deduce by means of $\varrho^2 = -\bar{\varrho} = \varrho - 1$ a linear relation in ϱ,

$$a\varrho + b = -c\bar{\varrho} + d\varrho = c\varrho - c + d\varrho.$$

This gives

$$a = c + d \, , \quad b = -c$$

and hence

$$M = \begin{pmatrix} d - b & b \\ -b & d \end{pmatrix} \, .$$

The determinant condition gives

$$b^2 - bd + d^2 = 1 \, .$$

The only solutions in integral numbers of this equation are

$$(b, d) = \pm(0, 1), \ \pm(1, 0), \ \pm(1, 1) \ .$$

☐

After this preparation, we can determine the transformed domains $M\mathcal{F}$, which are neighbors of $\mathcal{F}$.

Proposition VI.1.3 *Let $M \in \Gamma$ be a modular matrix with the property*

$$R(M) := M\mathcal{F} \cap \mathcal{F} \neq \emptyset \ .$$

Then one of the following cases occurs:

I. $\qquad M = \pm E \qquad (R(M) = \mathcal{F})$.

II. (1) $M = \begin{pmatrix} 1 & 1 \\ 0 & 1 \end{pmatrix}$ $\quad(R(M)$ *is the right vertical boundary half-line*) .

(2) $M = \begin{pmatrix} 1 & -1 \\ 0 & 1 \end{pmatrix}$ $\quad(R(M)$ *is the left vertical boundary half-line*) .

III. $\qquad M = \begin{pmatrix} 1 & -1 \\ 1 & 0 \end{pmatrix}$ $\quad(R(M)$ *is the lower boundary arc*) .

IV. *In all other cases $R(M)$ consists of a single point, namely*

$$\varrho = \frac{1}{2} + \frac{i}{2}\sqrt{3} \quad or \quad \varrho^2 = -\bar{\varrho} = \varrho - 1 = -\frac{1}{2} + \frac{i}{2}\sqrt{3} \ .$$

There are four cases, namely:

(1) $\ M\varrho \ = \varrho$
(2) $\ M\varrho^2 = \varrho$ $\qquad\left.\right\} \ (R(M) = \{\varrho\})$,

(3) $\ M\varrho^2 = \varrho^2$
(4) $\ M\varrho \ = \varrho^2$ $\qquad\left.\right\} \ (R(M) = \{\varrho^2\})$.

The list of the involved matrices can be found in Lemma VI.1.2 and its Corollary.

Proof. Again we can assume $c \neq 0$. If z is a point in the fundamental domain $\mathcal{F}$, then of course $|cz + d| \geq 1$ (in fact, this holds for any $(c, d) \in \mathbb{Z} \times \mathbb{Z}$, $(c, d) \neq (0, 0)$).

We apply this also for $Mz \in \mathcal{F}$ instead of z, and $(-c, a)$ instead of (c, d), to obtain also $1/|cz + d| = |-cMz + a| \geq 1$. This means that also the converse inequality $|cz + d| \leq 1$ is true.

Now let $z = x + iy$ be in the intersection $R(M)$. Then $|cz + d| = 1$, i.e. $(cx + d)^2 + c^2 y^2 = 1$. Using the inequality $y \geq \frac{\sqrt{3}}{2}$, which is valid in $\mathcal{F}$, we can exclude for c, d all values but 0 and ± 1. Then a can also take only the values 0 and ± 1 (replace M by M^{-1}). The determinant condition then constrains b to 0 and ± 1. Writing down all matrices of determinant 1 with entries in $\{-1, 0, 1\}$, we see that they are covered by the lists in VI.1.3.

☐

Let us mention some evident corollaries of Proposition VI.1.3.

Corollary VI.1.3$_1$ *Two different points a and b in $\mathcal{F}$ are equivalent modulo Γ, iff they lie on the boundary of $\mathcal{F}$, and additionally*

$$b = -\bar{a} \ ,$$

i.e. there are two cases:

(1)

$$\text{(a)} \qquad Re\,a \ = -\frac{1}{2} \ \ and \ \ b = a + 1 \ ,$$

$$\text{(b)} \qquad Re\,a \ = +\frac{1}{2} \ \ and \ \ b = a - 1 \ ,$$

(a and b are opposite on the vertical edges of $\mathcal{F}$.)

(2)

$$|a| = |b| = 1 \ and \ b = -\bar{a} \ .$$

(a and b are opposite at the circular arc of the boundary of $\mathcal{F}$.)

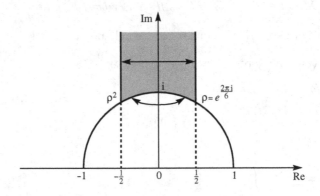

Corollary VI.1.3$_2$ *Let M and N with $M \neq \pm N$ be two elements of Γ. The regions $M\mathcal{F}$ and $N\mathcal{F}$ have only boundary points in common. In particular, inner points in $\mathcal{F}$ are inequivalent.*

A region $N\mathcal{F}$ is called a *neighbor* of $M\mathcal{F}$, (M and N both in Γ), if these regions are different, (i.e. $M \neq \pm N$) and if $M\mathcal{F} \cap N\mathcal{F} \neq \emptyset$.

It is useful to have a figure of the neighbors of $\mathcal{F}$:

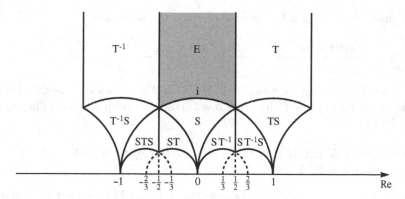

Definition VI.1.3₃ *A point $p \in \mathbb{H}$ is called an **elliptic** fixed point of $\Gamma = \mathrm{SL}(2, \mathbb{Z})$, iff the stabilizer*

$$\Gamma_p = \{\, M \in \Gamma \; ; \quad Mp = p \,\}$$

contains an element $\neq \pm E$. The order of the fixed point is

$$e = e(p) = \frac{1}{2} \# \Gamma_p \;.$$

The factor $1/2$ takes into account that M and $-M$ give the same corresponding substitution $\mathbb{H} \to \mathbb{H}$. So e is a natural number. Let $p \in \mathbb{H}$ and $M \in \Gamma$. The stabilizer of the point Mp can be simply obtained by conjugation of the stabilizer of p,

$$\boxed{\Gamma_{Mp} = M \Gamma_p M^{-1} \;.}$$

From the data in VI.1.3 we immediately extract:

Corollary VI.1.3₄ *There are exactly two equivalence classes of elliptic fixed points with respect to the action of Γ. They are represented by the two fixed points i (of order $e(\mathrm{i}) = 2$) and ϱ (of order $e(\varrho) = 3$). Especially, there exist only fixed points of order 2 and 3.*

One can generally ask, when does a matrix $M \in \mathrm{SL}(2, \mathbb{R})$ admit a fixed point in $\mathbb{H}$?

Remark VI.1.4 *A matrix $M \in \mathrm{SL}(2, \mathbb{R})$, $M \neq \pm E$, has a fixed point in $\mathbb{H}$, iff*

$$|\sigma(M)| < 2 \qquad (\sigma := \mathrm{Trace}) \;,$$

*and in case of existence, this fixed point is uniquely determined. A matrix M having the above property is called **elliptic**.*

Proof. The fixed point equation $Mz = z$ means

$$cz^2 + (d - a)z - b = 0 \;.$$

This quadratic equation has in case of $c \neq 0$ the solutions

$$z = \frac{a - d \pm \sqrt{(a-d)^2 + 4bc}}{2c} = \frac{a - d \pm \sqrt{(a+d)^2 - 4(ad - bc)}}{2c} \ .$$

If $(a + d)^2 \geq 4$, the solutions are real. If $(a + d)^2 < 4$, exactly one solution lies in the upper half-plane. (The other one is complex conjugated and lies in the lower half-plane.)   □

Remark VI.1.5 *Let $M \in \mathrm{SL}(2, \mathbb{R})$ be a matrix of finite order, i.e. $M^h = E$ for a suitable $h \in \mathbb{N}$. Then M has a fixed point in $\mathbb{H}$.*

Proof. For any 2×2 matrix M there exists an invertible complex 2×2 matrix Q with the property

$$QMQ^{-1} = \begin{pmatrix} a & b \\ 0 & d \end{pmatrix} \quad \text{(JORDAN normal form)} \ ,$$

where $a = d$ if $b \neq 0$. If M has finite order, then the (eigenvalues) a and d are roots of unity (and $b = 0$). From the determinant condition $1 = \det M = ad$, we get $d = a^{-1} = \bar{a}$. For the root of unity $a \neq \pm 1$ we surely have

$$\left| a + a^{-1} \right| = |2\mathrm{Re}\, a| < 2 \ .$$ □

From VI.1.4 and VI.1.5, in connection with VI.1.1$_3$ we can now state:

Proposition VI.1.6 *For $M \in \Gamma$ the following properties are equivalent:*

(a) *M has a fixed point in $\mathbb{H}$.*
(b) *M has finite order, i.e. $M^h = E$ for a suitable $h \geq 1$.*
(c) *M is elliptic, or $M = \pm E$.*

> *The elliptic fixed points are exactly*
> *the fixed points of elliptic substitution in Γ.*

The classification of elliptic fixed points leads to the following purely group theoretical result:

Proposition VI.1.7 *Let $M \in \Gamma$, $M \neq \pm E$, be an element of finite order. Then M is conjugated in Γ to one of the following matrices:*

$$\pm \begin{pmatrix} 1 & -1 \\ 1 & 0 \end{pmatrix}, \quad \pm \begin{pmatrix} 0 & -1 \\ 1 & -1 \end{pmatrix}, \quad \pm \begin{pmatrix} 0 & -1 \\ 1 & 0 \end{pmatrix} \ .$$

Another group theoretical result can be proved using the fundamental domain of the modular group:

Proposition VI.1.8 *The elliptic modular group is generated by the two matrices*

$$T = \begin{pmatrix} 1 & 1 \\ 0 & 1 \end{pmatrix} \quad and \quad S = \begin{pmatrix} 0 & -1 \\ 1 & 0 \end{pmatrix} .$$

For the proof, let us choose an (arbitrary) inner point $a \in \mathcal{F}$. Let $M \in$ SL$(2, \mathbb{Z})$. From V.8.7 follows that there exists a matrix N in the subgroup Γ_0 generated by S, T such that $NM(a)$ lies in $\mathcal{F}$. By VI.1.3$_2$, we must have $NM = \pm E$. But the negative of the unit matrix, lies in Γ_0:

$$S^2 = \begin{pmatrix} 0 & -1 \\ 1 & 0 \end{pmatrix}^2 = - \begin{pmatrix} 1 & 0 \\ 0 & 1 \end{pmatrix} = -E ,$$

and we are done.   □

Exercises for VI.1

1. Determine all matrices $M = \begin{pmatrix} a & b \\ c & d \end{pmatrix} \in$ SL$(2, \mathbb{R})$ with the fixed point i.

    Result.

    $$M\,i = i \quad \Longleftrightarrow \quad M \in \mathrm{SO}(2, \mathbb{R}) := \left\{ M \in \mathrm{SL}(2, \mathbb{R}) \ ; \quad M'M = E \right\} .$$

2. *Show:*
(a) The group SL$(2, \mathbb{R})$ acts transitively on the upper half-plane $\mathbb{H}$, i.e. for any two points $z, w \in \mathbb{H}$ there exists $M \in$ SL$(2, \mathbb{R})$ with $w = Mz$.

    Hint. It is enough to consider the case $w = i$. Then we can even take $c = 0$.

(b) The map
$$\mathrm{SL}(2, \mathbb{R}) \,/\, \mathrm{SO}(2, \mathbb{R}) \longrightarrow \mathbb{H} ,$$
$$M \cdot \mathrm{SO}(2, \mathbb{R}) \longrightarrow M\,i ,$$

    is bijective. (It is even a homeomorphism, if we equip its domain of definition with the quotient topology.)

3. Let $M \in$ SL$(2, \mathbb{R})$, and let l be an integer with the property $M^l \neq \pm E$. The matrix M is elliptic, iff M^l is elliptic.

4. Let $G \subset$ SL$(2, \mathbb{R})$ be a *finite* subgroup, such that its elements admit a common fixed point in $\mathbb{H}$. (One can show that any finite, or more general any compact, subgroup $G \subset$ SL$(2, \mathbb{R})$ has this property!) Show that G is cyclic.

VI.2 The $k/12$-formula and the Injectivity of the j-function

Let
$$f : U_C \longrightarrow \mathbb{C}$$
be an analytic function on an upper half-plane
$$U_C = \{ \, z \in \mathbb{H} \, ; \quad \mathrm{Im}\, z > C \, \} \, , \quad C > 0 \, .$$

We assume that f is periodic, i.e. there exists a suitable N with
$$f(z+N) = f(z) \, , \quad N \neq 0 \, , \; N \in \mathbb{R} \, .$$

The periodicity condition allows a FOURIER expansion III.5.4
$$f(z) = \sum_{n=-\infty}^{\infty} a_n e^{2\pi i n z / N} \, ,$$

which in fact corresponds to a LAURENT expansion
$$\tilde{f}(q) = \sum_{n=-\infty}^{\infty} a_n q^n \quad \left(q = e^{\frac{2\pi i z}{N}} \right)$$

in the punctured disk around the origin having radius $e^{-2\pi C / N}$.

Terminology. The function f is

(a) *non-essentially singular* at $i\infty$, iff $\tilde{f}$ is non-essentially singular at the origin.

(b) *regular* at $i\infty$, iff $\tilde{f}$ has a removable singularity at the origin.

   In case of regularity, one defines
$$f(i\infty) := \tilde{f}(0) \quad (= a_0) \, .$$

These notions do not depend on the choice of the period N. (If f is non-constant, the set of all periods is a cyclic group.)

Definition VI.2.1 *A **meromorphic modular form** of weight $k \in \mathbb{Z}$ is a meromorphic function*
$$f : \mathbb{H} \longrightarrow \overline{\mathbb{C}}$$
with the following properties:

(a) $f(Mz) = (cz+d)^k f(z)$ *for all* $M = \begin{pmatrix} a & b \\ c & d \end{pmatrix} \in \Gamma$.

   In particular, $f(z+1) = f(z)$.

(b) *There exists a number $C > 0$, such that $f(z)$ has no singularities in the domain $\mathrm{Im}\, z > C$.*

(c) *f has a non-essential singularity at* $i\infty$.

Because the negative of the unit matrix $-E$ lies in Γ, we get from (a)

$$f(z) = (-1)^k f(z) \ ,$$

i.e.

> *Any modular form of odd weight k vanishes*
> *identically.*

A meromorphic modular form f possesses thus a FOURIER expansion

$$f(z) = \sum_{n=-\infty}^{\infty} a_n q^n \ , \quad q := e^{2\pi i z} \ ,$$

where all but finitely many coefficients corresponding to indices $n < 0$ vanish. If such an f is not identically zero, then we can define

$$\mathrm{ord}(f; i\infty) : = \min\{\, n \ ; \ a_n \neq 0 \,\}$$
$$= \mathrm{ord}(\tilde{f}; 0) \ .$$

Remark VI.2.2 *A meromorphic modular form* $f \neq 0$ *has only finitely many poles and zeros in* $\mathbb{H}$ *modulo the action of* $\mathrm{SL}(2,\mathbb{Z})$ *on* $\mathbb{H}$. *The order* $\mathrm{ord}(f; a)$, $a \in \mathbb{H}$, *only depends on the* Γ*-orbit of* a.

Proof. By assumption, there exists a constant C, such that the function f has no poles in the region "Im $z > C$". After enlarging C if necessary, the same region contains no zeros, because the zeros of an analytic function cannot accumulate at a non-essential singularity when the function is $\neq 0$ in a neighborhood of this singularity.

The truncated fundamental domain $\{\, z \in \mathcal{F} \ ; \ \mathrm{Im}\, z \leq C \,\}$ is compact, it thus contains only finitely many zeros and poles. These contain a finite system of representatives modulo Γ.   □

Theorem VI.2.3 (The $k/12$-formula or the Valence Formula) *Let* $f \not\equiv 0$ *be a meromorphic modular form of weight k. Then*

$$\boxed{\ \sum_a \frac{1}{e(a)} \mathrm{ord}(f; a) + \mathrm{ord}(f; i\infty) \ = \ \frac{k}{12} \ .\ }$$

The summation index a runs through a system of representatives modulo Γ *of all poles and zeros of f, and we define*

$$e(a) = \frac{1}{2}\#\Gamma_a = \begin{cases} 3\,, & \text{if } a \sim \varrho \quad \mathrm{mod}\ \Gamma\,, \\ 2\,, & \text{if } a \sim i \quad \mathrm{mod}\ \Gamma\,, \\ 1\,, & \text{else}\,. \end{cases}$$

One can consider the $k/12$-formula as an analogue of LIOUVILLE's Theorem, which claims that any non-constant elliptic function has as many poles as zeros. Indeed, we can reformulate Theorem VI.2.3 in the important *special case $k = 0$* as follows:

The function f has in $\mathbb{H}/\Gamma \cup \{i\infty\}$ as many zeros as poles, counting multiplicities for them, and weighting them with the factors $1/e(a)$.

Proof of Theorem VI.2.3. First of all, we assume for simplicity that there are no zeros or poles on the boundary of the fundamental domain $\mathcal{F}$, excepting possible ones located at i, ϱ or ϱ^2. We then choose the constant $C > 0$ sufficiently large, such that $f(z)$ has no zeros or poles for Im $z \geq C$. Then we consider the integral

$$\frac{1}{2\pi i} \int_{\alpha} g(\zeta)\, d\zeta \,, \qquad g(z) := \frac{f'(z)}{f(z)}\,,$$

along the contour α

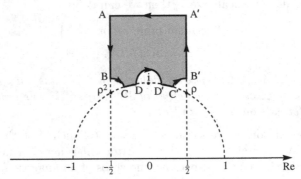

The radius of the small disks around ϱ^2, i and ϱ is $\varepsilon > 0$. Later we will take the limit $\varepsilon \to 0$. If ε is taken sufficiently small, then this integral is equal to

$$\sum_{\substack{a \bmod \Gamma \\ a \not\sim i,\, \varrho \bmod \Gamma}} \operatorname{ord}(f; a)\,.$$

Evaluation of the integral

(1) *Evaluation on the vertical boundary segments*

The function g is, together with f, also a periodic function. The integrals along the vertical boundary segments cancel each other.

(2) *Evaluation on arcs from C to D and from D′ to C′*

These arcs are mapped into each other by the substitution $z \mapsto Sz = -z^{-1}$. It is thus natural to study the transformation behavior of $g(z) = f'(z)/f(z)$ with respect to this substitution. From

$$f(-1/z) = z^k f(z)$$

we obtain

$$f'(-1/z) \cdot z^{-2} = z^k f'(z) + k z^{k-1} f(z) ,$$

hence

$$g(-1/z) = z^2 g(z) + kz .$$

Let us fix some parametrization

$$\beta : [0,1] \longrightarrow \mathbb{C}$$

for the arc from C to D. Then we obtain the parametrization $\tilde{\beta} : [0,1] \longrightarrow \mathbb{C}$

$$\tilde{\beta}(t) := -\beta(t)^{-1}$$

of the arc from C' to D'. Using these parametrizations,

$$\int_C^D g(\zeta)\, d\zeta = \int_0^1 g(\beta(t))\beta'(t)\, dt ,$$

$$\int_{D'}^{C'} g(\zeta)\, d\zeta = -\int_0^1 g(\tilde{\beta}(t))\tilde{\beta}'(t)\, dt$$

$$= -\int_0^1 g(\beta(t))\beta'(t)\, dt - k \int_0^1 \frac{\beta'(t)}{\beta(t)}\, dt .$$

Then

$$\frac{1}{2\pi i}\left[\int_C^D g(\zeta)\, d\zeta + \int_{D'}^{C'} g(\zeta)\, d\zeta \right] = -\frac{k}{2\pi i}\,(\operatorname{Log} D - \operatorname{Log} C) .$$

We are now interested in the limit of this expression for $\varepsilon \to 0$. It has the value

$$-\frac{k}{2\pi i}\,(\operatorname{Log} i - \operatorname{Log}(\varrho^2)) = \frac{k}{12} .$$

(3) *Evaluation on the upper horizontal line segment from* A *to* A'
The FOURIER series of g,

$$g(z) = \sum a_n e^{2\pi i n z} ,$$

is obtained from the FOURIER series of f using $f \cdot g = f'$. Obviously, the constant FOURIER coefficient of g is

$$a_0 = 2\pi i \operatorname{ord}(f; i\infty) .$$

This implies

$$\int_A^{A'} g(\zeta)\, d\zeta = 2\pi i \cdot \operatorname{ord}(f; i\infty) + \sum_{n \neq 0} a_n \underbrace{\int_A^{A'} e^{2\pi i n \zeta}\, d\zeta}_{= 0} .$$

There are missing the integrals along the "small" arcs around ϱ, i, ϱ^2.

(4) *Evaluation on the arc from* B *to* C

The function $g(z)$ has in $z = \varrho^2$ the power series expansion

$$g(z) = b_{-1}(z + \overline{\varrho})^{-1} + b_0 + b_1(z + \overline{\varrho}) + \cdots \quad ,$$
$$b_{-1} = \operatorname{ord}(f; \varrho^2) \ .$$

The limit for $\varepsilon \to 0$ of the integral of the function $z \mapsto g(z) - b_{-1}(z + \overline{\varrho})^{-1}$
is zero. Using the formula

$$\int \frac{d\zeta}{\zeta - a} = \mathrm{i}\alpha \ ,$$

where the integral is taken along an arc of length α with center a, we can
conclude

$$\frac{1}{2\pi\mathrm{i}} \lim_{\varepsilon \to 0} \int_{\mathrm{B}}^{\mathrm{C}} g(\zeta)\, d\zeta = -\frac{1}{6} \operatorname{ord}(f; \varrho^2) \ .$$

(5) *Evaluation on the arcs from* C' *to* B' *and from* D *to* D'

In analogy to (4) we have

$$\frac{1}{2\pi\mathrm{i}} \lim_{\varepsilon \to 0} \int_{\mathrm{C}'}^{\mathrm{B}'} g(\zeta)\, d\zeta = -\frac{1}{6} \operatorname{ord}(f; \varrho) \ ,$$

and

$$\frac{1}{2\pi\mathrm{i}} \lim_{\varepsilon \to 0} \int_{\mathrm{D}}^{\mathrm{D}'} g(\zeta)\, d\zeta = -\frac{1}{2} \operatorname{ord}(f; \mathrm{i}) \ .$$

Because of $\operatorname{ord}(f; \varrho) = \operatorname{ord}(f; \varrho^2)$, we obtain the claimed $k/12$-formula.

So far, we assumed (for simplicity) that the points ϱ^2, i and ϱ are the only
possible zeros or poles of f on the boundary of $\mathcal{F}$. The general case works in
the same way with a suitable modification of the fundamental domain $\mathcal{F}$ as
indicated in the following figure.

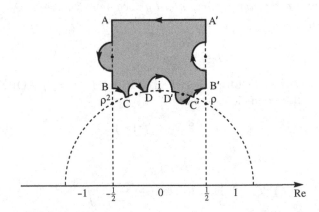

This completely proves VI.2.3.   □

Consequences of the $k/12$-formula

We first consider applications of the $k/12$-formula for *entire modular forms*. A meromorphic modular form is called *entire*, if it is regular at all points of $\mathbb{H} \cup \{i\infty\}$:

Definition VI.2.4 *An (entire) modular form of weight $k \in \mathbb{Z}$ is an analytic function $f : \mathbb{H} \to \mathbb{C}$ with the following properties:*

(a) $f(Mz) = (cz + d)^k f(z)$ *for all* $M = \begin{pmatrix} a & b \\ c & d \end{pmatrix} \in \Gamma$.

(b) f *is bounded in regions of the form "Im $z \geq C > 0$".*

By the RIEMANN Removability Theorem, condition (b) is equivalent to the regularity of f at $i\infty$.
A meromorphic modular form is entire, iff

$$\operatorname{ord}(f; a) \geq 0 \text{ for all } a \in \mathbb{H} \cup \{i\infty\} .$$

From the $k/12$-formula immediately follows:

Proposition VI.2.5 *Any entire modular form of negative weight vanishes identically. Any entire modular form of weight 0 is constant.*

The second part of this Proposition is an application of the $k/12$-formula to the function $f(z) - f(i)$.

Corollary VI.2.5$_1$ *An entire modular form of weight k, $k \in \mathbb{N}$ ($k \neq 0$), has at least one zero at $\mathbb{H} \cup \{i\infty\}$.*

If f would not have a zero, then $1/f$ would be also an entire modular form, but $1/f$ has negative weight.

If $f \neq 0$ is an entire modular form of weight k, and if $a \in \mathbb{H} \cup \{i\infty\}$ is a zero of f, then from the $k/12$-formula we obtain

$$\frac{k}{12} \geq \frac{\operatorname{ord}(f; a)}{e(a)} \geq \frac{1}{3} ,$$

where we have extended the definition of e by $e(i\infty) = 1$. From this we deduce:

Proposition VI.2.6 *There exists no entire modular form $f \neq 0$ of weight 2.*

Examples of entire modular forms are the EISENSTEIN series

$$G_k(z) = \sum_{\substack{(c,d) \in \mathbb{Z} \times \mathbb{Z} \\ (c,d) \neq (0,0)}} (cz + d)^{-k} , \quad k \geq 3 .$$

In case of $k \in \mathbb{N}$, $k \geq 4$, $k \equiv 0 \mod 2$, we have already established the behavior (V.8.4)

$$G_k(i\infty) = 2\zeta(k) .$$

Proposition VI.2.7

(1) *The Eisenstein series G_4 has a simple zero at ϱ. Besides ϱ (and the Γ-equivalent points) there is no further zero in $\mathbb{H} \cup \{i\infty\}$.*

(2) *The Eisenstein series G_6 has a simple zero at i. Excepting i (and the Γ-equivalent points) there is no further zero in $\mathbb{H} \cup \{i\infty\}$.*

The *Proof* is a direct application of the $k/12$-formula.

Corollary VI.2.7$_1$ *The functions G_4^3 and G_6^2 are $\mathbb{C}$-linearly independent.*

None of the functions G_4^3 or G_6^2 is a scalar multiple of the other one. Of course, there exists a linear combination of G_4^3 and G_6^2 which vanishes at $i\infty$. We already know this function, it is the discriminant

$$\Delta = g_2^3 - 27 g_3^2 \quad \text{with} \quad g_2 = 60 G_4 \text{ and } g_3 = 140 G_6 .$$

From the theory of elliptic functions we already know that Δ has no zero in $\mathbb{H}$. We can now prove this in another way, without using the theory of elliptic functions. From VI.2.7$_1$, we first infer that Δ is not identically zero. By the $k/12$-formula, the known zero of Δ at $i\infty$ is its only zero. We moreover obtain that this zero is simple (i.e. has first order).

Proposition VI.2.8 *Let $f \neq 0$ be an entire modular form of weight 12 (e.g. $f = \Delta$), which vanishes at $i\infty$. Then f has in $i\infty$ a simple zero, and there is no other zero in $\mathbb{H}$.*

We know that the j-function induces a surjective map

$$\hat{j} : \mathbb{H}/\Gamma \longrightarrow \mathbb{C} .$$

We are now in position to prove the injectivity of $\hat{j}$.

Theorem VI.2.9 *The j-function defines a bijection*

$$\hat{j} : \mathbb{H}/\Gamma \longrightarrow \mathbb{C} .$$

Proof. Let $b \in \mathbb{C}$. We must show that the function $f(z) = j(z) - b$ has exactly one zero modulo Γ in $\mathbb{H}$. We already know (VI.2.8) that $\operatorname{ord}(f; i\infty) = -1$. The claim easily follows from this and the $k/12$-formula. $\square$

Let us now state Theorem VI.2.9 in the language of elliptic functions:

For any complex number j_0, there exists exactly one equivalence class of lattices having j_0 as absolute invariant.

Geometrically, one can depict an image of $\mathbb{H}/\Gamma$ by identifying equivalent points of the fundamental domain $\mathcal{F}$. We can imagine the fundamental domain as a rhombus with vertices ϱ^2, i, ϱ and a fourth "missing" vertex i∞. Then gluing the adjacent upper and lower edges one obtains a figure, which is topologically equivalent to a a sphere with missing north pole. This is topologically a plane. We will later assign to $\mathbb{H}/\Gamma$ the structure of a RIEMANN surface. The map $\hat{j}$ will turn out to be a *conformal* map.

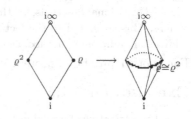

Definition VI.2.10 *A modular function is a meromorphic modular form of weight* 0.

For instance, j is a modular function. The set of all modular functions is a field denoted by $K(\Gamma)$. Any constant function is a modular function. The field $\mathbb{C}$ is thus naturally embedded as subfield of $K(\Gamma)$. Any polynomial, or more generally any rational function of a modular function is again a modular function. We have observed the same phenomenon in more detail for the field of elliptic functions (Section V.3).

Theorem VI.2.11 (Structure of the field of Modular Functions) *The field of modular functions is generated by the absolute invariant j. In other words, any modular function is a rational function in j,*

$$\boxed{K(\Gamma) = \mathbb{C}(j) \ .}$$

Proof. Let f be a modular function. Any $w \in \mathbb{C}$ can be written in the form $j(z)$ where z mod Γ is uniquely determined. Hence the function

$$R : \mathbb{C} \to \overline{\mathbb{C}}, \quad R(w) := f(z)$$

is well-defined. Let $a \in \mathbb{H}$ be a point where the derivative of j does not vanish and where the value $f(a)$ is finite. Then the Inverse Function Theorem shows that R is analytic in an open neighborhood of $j(a)$. This shows that R is analytic outside a finite set of points. From the CASORATI-WEIERSTRASS Theorem will follow that these finitely many points and also ∞ are non-essential singularities. Let a be one of the exceptional points. We take a representative b in $\mathbb{H} \cup \{i\infty\}$. When a is different from ∞, we take a small open neighborhood $V(b)$ of b in $\mathbb{H}$. Otherwise one takes $V(b) = U_C \cup \{i\infty\}$ for big enough C. The set $f(U(a))$ is not dense in $\overline{\mathbb{C}}$. The image of $V(b)$ is an open subset $U(a)$ of a in $\overline{\mathbb{C}}$. By definition $R(V(b)) = f(U(a))$. Hence the theorem of CASORATI-WEIERSTRASS shows that the exceptional points are non-essential singularities. This shows that R can be considered as a meromorphic function on the whole RIEMANN sphere $\overline{\mathbb{C}}$. But then R must be a rational function (III.A.6).   □

The field of modular functions (for the full elliptic modular group) is isomorphic to the field of rational functions, i.e. the field of meromorphic functions on the RIEMANN sphere. This is related to the fact that the quotient space $\mathbb{H}/\Gamma$, completed by the additional "infinite point" i∞, can be identified with the RIEMANN sphere.

For another proof of Theorem VI.2.11 see also Exercise 6 in Sect. VI.3.

Exercises for VI.2

1. The derivative of a modular function is a meromorphic modular form of weight 2.

2. Let f and g be entire modular forms of weight k. Then $f'g - g'f$ is an entire modular form of weight $2k + 2$.

3. The zeros of j' are exactly the points equivalent to i or ϱ modulo Γ.

   In the following three Exercises we use some fundamental topological notions, including also the notion of "quotient topology".

4. We consider $\mathbb{H}/\Gamma$ (see also V.7) as topological space by means of the quotient topology. A subset in $\mathbb{H}/\Gamma$ is open, iff its inverse image in $\mathbb{H}$ under the canonical projection is open. The j-function then induces a topological map

   $$\mathbb{H}/\Gamma \longrightarrow \mathbb{C} \ .$$

5. Show that $\mathbb{H}/\Gamma$ is topologically equivalent to $\mathbb{C}$, without using the j-function.
   Hint. Consider the fundamental domain with corresponding identifications of boundary points.

6. Let $\widehat{\Gamma}$ be the group of all self-maps of the upper half-plane of the form

   $$z \longmapsto Mz \qquad \text{and}$$
   $$z \longmapsto M(-\overline{z}) \quad \text{with } M \in \Gamma = \mathrm{SL}(2,\mathbb{Z}) \ .$$

   Show that the quotient space $\mathbb{H}/\widehat{\Gamma}$ is topologically equivalent to a closed half-plane.

VI.3 The Algebra of Modular Forms

For any $k \in \mathbb{Z}$ we denote by $[\Gamma, k]$ the vector space of all *entire* modular forms of weight k, and by $[\Gamma, k]_0$ the subspace of all *cusp forms*, which is the subspace of all $f \in [\Gamma, k]$, which vanish at the cusp i∞:

$$f(\mathrm{i}\infty) := \lim_{\mathrm{Im}\, z \to \infty} f(z) = 0 \ .$$

We obviously have:

(a) If $f_1 \in [\Gamma, k_1]$, $f_2 \in [\Gamma, k_2]$, then $f_1 f_2 \in [\Gamma, k_1 + k_2]$.

(b) The product of a cusp form with an arbitrary entire modular form is again a cusp form.

The subspace $[\Gamma, k]_0$ of all cusp forms has codimension at most one, more exactly:

Remark VI.3.1 *If $g \in [\Gamma, k]$ is not a cusp form, then*

$$[\Gamma, k] = [\Gamma, k]_0 \oplus \mathbb{C}g .$$

Proof. For any $f \in [\Gamma, k]$, the function

$$h := f - \frac{f(i\infty)}{g(i\infty)}g$$

is a cusp form, and we have

$$f = h + Cg \quad \text{with} \quad C = \frac{f(i\infty)}{g(i\infty)} \in \mathbb{C} . \qquad \square$$

As we will see, $[\Gamma, k]$ always has finite dimension. For determining a basis of $[\Gamma, k]$, it is of basic importance to prove the *existence of a cusp form $f \neq 0$ of weight* 12. The $k/12$-formula shows that such a modular form has necessarily at i∞ a zero of order one, and no other zeros in the upper half-plane, VI.2.8. There are various methods to construct such a cusp form. We already know such a cusp form, namely the discriminant Δ. This shows:

Proposition VI.3.2 *There exists a modular form $\Delta \neq 0$ of weight 12, which has in the upper half-plane no zeros, but a (necessarily simple) zero at i∞. So Δ is a cusp form. Such a function Δ is uniquely determined up to a constant factor. A possible such function is the discriminant*

$$\boxed{\Delta = (60G_4)^3 - 27 (140G_6)^2 .}$$

The importance of this cusp form of weight 12 can be extracted from the following

Proposition VI.3.3 *Multiplication with Δ gives an isomorphism*

$$[\Gamma, k - 12] \longrightarrow [\Gamma, k]_0 ,$$
$$f \longmapsto f \cdot \Delta .$$

Proof. This map is injective, since Δ does not vanish. On the other side, for $g \in [\Gamma, k]_0$, the quotient

$$f := \frac{g}{\Delta} \in [\Gamma, k - 12]$$

is a modular form with the correct transformation behavior, and is analytic in the whole upper half-plane, since Δ has no zeros there, and is also regular at i∞, since Δ has there a zero of first order. $\qquad \square$

An preliminary stage for the structure theorem is the following direct consequence of the $k/12$-formula, VI.2.6:

Any entire modular form of weight 2 vanishes identically.

Theorem VI.3.4 (Structure Theorem) *The monomials*

$$\left\{ G_4^\alpha G_6^\beta \; ; \quad \alpha, \beta \in \mathbb{N}_0 \; , \; 4\alpha + 6\beta = k \right\}$$

form a basis of $[\Gamma, k]$. *Any modular form* $f \in [\Gamma, k]$ *is thus uniquely representable as a* $\mathbb{C}$-*linear combination*

$$f = \sum_{\substack{\alpha, \beta \geq 0 \\ 4\alpha + 6\beta = k}} C_{\alpha\beta} \, G_4^\alpha \, G_6^\beta \; .$$

Supplement: *The dimension of the vector space of modular forms is finite and we have*

$$\dim_{\mathbb{C}}[\Gamma, k] = \begin{cases} \left[\frac{k}{12} \right] , & \text{if } k \equiv 2 \mod 12 , \\ \left[\frac{k}{12} \right] + 1 , & \text{if } k \not\equiv 2 \mod 12 . \end{cases}$$

Proof. By induction on k, we first show that the specified monomials generate $[\Gamma, k]$. We begin the induction with $k = 0$. Then the claim reduces to the fact (VI.2.5) that any modular form of weight 0 is a constant. Now let $f \not\equiv 0$ be a modular form of weight $k > 0$. Then $k \geq 4$. Any even $k \geq 4$ can be written in the form $k = 4\alpha + 6\beta$ with non-negative integers α, β. Then there exists a constant C, such that $f - C G_4^\alpha G_6^\beta$ is a cusp form. By VI.3.3, we can write

$$f - C G_4^\alpha G_6^\beta = \Delta \cdot g$$

with a modular form g of strictly smaller weight. By induction, we can assume that g is a $\mathbb{C}$-linear combination of monomials in G_4 and G_6 of corresponding weight. This gives a $\mathbb{C}$-linear representation of f using monomials in G_4 and G_6.

A simple combinatorial check shows that the number of monomials of weighted degree k in G_4 and G_6 coincides with the number on the R.H.S. of the dimension formula in the statement of the Theorem. The linear independence of the monomials and the validity of the dimension formula are thus equivalent. But the dimension formula can also be proven by induction using $\dim_{\mathbb{C}}[\Gamma, 0] = 1$, $\dim_{\mathbb{C}}[\Gamma, 2] = 0$ and

$$\dim_{\mathbb{C}}[\Gamma, k] = 1 + \dim_{\mathbb{C}}[\Gamma, k - 12] \quad \text{for } k \geq 4 \; .$$

The R.H.S. of the dimension formula also fulfills the same recursion.   □

The dimension of $[\Gamma, k]$ is one in the cases $k = 0, 4, 6, 8, 10, 14$.

We give a second proof for $[\Gamma, 2] = \{0\}$, which does not use the $k/12$-formula. If there would exist a non-vanishing modular form $f \in [\Gamma, 2]$, then

$$f^2 \in [\Gamma, 4] \; , \quad \text{so } f^2 = a \, G_4 \text{ with } a \in \mathbb{C}^\bullet \; ,$$
$$f^3 \in [\Gamma, 6] \; , \quad \text{so } f^3 = b \, G_6 \text{ with } b \in \mathbb{C}^\bullet \; .$$

This implies that G_4^3 and G_6^2 are linearly dependent. This contradicts the non-vanishing of Δ (VI.2.7$_1$).

The Structure Theorem VI.3.4 can be reformulated in a ring theoretical manner. For this we introduce the direct sum of all vector spaces of modular forms,

$$\mathcal{A}(\Gamma) := \bigoplus_{k \geq 0} [\Gamma, k] \; .$$

It possesses an obvious ring structure (even a structure as graded algebra over the field $\mathbb{C}$).

Theorem VI.3.5 *The map*

$$X \longmapsto G_4, \quad Y \longmapsto G_6 \; ,$$

induces an algebra isomorphism of the polynomial ring in two variables X, Y *onto the algebra of modular forms,*

$$\mathbb{C}[X, Y] \xrightarrow{\quad \sim \quad} \mathcal{A}(\Gamma) \; .$$

Exercises for VI.3

1. Let $f : \mathbb{H} \to \mathbb{C}$ be an entire modular form without zeros (in $\mathbb{H}$). Then f is a constant multiple of a power of the discriminant Δ.

2. Let $d_k = \dim_{\mathbb{C}}[\Gamma, k]$ be the dimension of the vector space of all entire modular forms of weight k. For any d_k-tuple of complex numbers $a_0, a_1, \ldots, a_{d_k-1}$ there exists exactly one modular form of weight k, having these d_k numbers as first FOURIER coefficients.

    Hint. If the first d_k FOURIER coefficients of a modular form vanish, then it is divisible by Δ^{d_k}, i.e. the quotient is again an entire modular form.

3. There exists no non-vanishing polynomial $P \in \mathbb{C}[X]$ such that $P(j) = 0$.

    Based on this result give a new proof for the fact that the EISENSTEIN series G_4 and G_6 are algebraically independent, i.e. the monomials $G_4^\alpha G_6^\beta$, $4\alpha + 6\beta = k$, are linearly independent for all k.

4. For any point $a \in \mathbb{H}$ there exists an entire modular form (even of weight 12), which vanishes at a, but does not vanish identically.

    Hint. Use the knowledge of the zeros of Δ.

5. Any meromorphic modular form is representable as a quotient of two entire modular forms.

6. Using the previous exercise and the structure theorems VI.3.4, VI.3.5, for the algebra of all modular forms, give a new proof for the fact that any modular function is a rational function of j.

VI.4 Modular Forms and Theta Series

In principle, we have determined in the previous section all (entire) modular forms. There are other possibilities to construct modular forms. The Structure Theorem then gives non-trivial identities between analytic functions. In this section we want to develop some of these identities. They often have number theoretical consequences. In VII.1 we will study in more detail number theoretical applications.

The Jacobi transformation formula for the theta function

Lemma VI.4.1 *Both series*

$$\sum_{n=-\infty}^{\infty} e^{\pi i(n+w)^2 z} \quad and \quad \sum_{n=-\infty}^{\infty} e^{\pi i n^2 z + 2\pi i n w}$$

converge normally for $(z,w) \in \mathbb{H} \times \mathbb{C}$ *. In particular, for any fixed value of z they represent analytic functions in w, and conversely.*

Of course normal convergence means that on every compact subset of $\mathbb{H} \times \mathbb{C}$ there exists a uniform majorant. The second of the above series already appeared in V.6 in connection with ABEL's Theorem. We used at that point the notation

$$\vartheta(z,w) := \sum_{n=-\infty}^{\infty} e^{\pi i n^2 z + 2\pi i n w} .$$

There, its argument was (τ, z) instead of (z, w), and as a main difference to the topic of this section, the point τ was a fixed parameter. We are now mainly interested in $\vartheta(z,w)$ as a function of z for a fixed w. In V.6 we proved the convergence of the theta series for a fixed value of its first argument. Analogous considerations lead to the normal convergence in both variables. □

Theorem VI.4.2 (Jacobi's Theta Transformation Formula, C.G.J. Jacobi, 1828) *For $(z,w) \in \mathbb{H} \times \mathbb{C}$ the following formula holds:*

$$\sqrt{\frac{z}{i}} \sum_{n=-\infty}^{\infty} e^{\pi i(n+w)^2 z} = \sum_{n=-\infty}^{\infty} e^{\pi i n^2 (-1/z) + 2\pi i n w} .$$

Here, the square root of z/i is defined by using the principal branch of the logarithm.

Proof. The function

$$f(w) := \sum_{n=-\infty}^{\infty} e^{\pi i z (n+w)^2} \quad (z \text{ being fixed})$$

has obviously the period 1, and hence it possesses a FOURIER expansion

$$f(w) = \sum_{m=-\infty}^{\infty} a_m e^{2\pi i m w}$$

with

$$a_m = \int_0^1 \sum_{n=-\infty}^{\infty} e^{\pi i z(n+w)^2 - 2\pi i m w} \, du \qquad (u = \text{Re } w) .$$

Here, we have used the complex coordinate $w = u + iv$, $u, v \in \mathbb{R}$. The imaginary part v of w can be chosen arbitrarily, we will fix it later. Because of the locally uniform convergence, we can exchange the sum and the integral. After that we substitute $u \mapsto u - n$ and obtain

$$a_m = \int_{-\infty}^{\infty} e^{\pi i (z w^2 - 2mw)} \, du .$$

We complete the exponent to a square,

$$z w^2 - 2mw = z \left(w - \frac{m}{z} \right)^2 - z^{-1} m^2 ,$$

and obtain

$$a_m = e^{-\pi i m^2 z^{-1}} \int_{-\infty}^{\infty} e^{\pi i z (w - m/z)^2} \, du .$$

Now we choose the imaginary part v of w such that $w - m/z$ becomes real. After a translation of u, we obtain

$$a_m = e^{\pi i m^2 (-1/z)} \int_{-\infty}^{\infty} e^{\pi i z u^2} \, du .$$

To compute this integral, we prove the formula

$$\int_{-\infty}^{\infty} e^{\pi i z u^2} \, du = \sqrt{\frac{z}{i}}^{-1} .$$

Both sides of the formula represent analytic functions in z, so it is enough to prove it for all purely imaginary values $z = iy$. The substitution

$$t = u \cdot \sqrt{y}$$

leads to the computation of the well-known integral (see also Exercise 17 in Sect. III.7)

$$\int_{-\infty}^{\infty} e^{-\pi t^2} \, dt = 1 . \qquad \qquad \square$$

By specialization in the JACOBI transformation formula, we get:

Proposition VI.4.3 *The function*

$$\vartheta(z) = \sum_{n=-\infty}^{\infty} e^{\pi i n^2 z}$$

represents an analytic function. It satisfies the theta transformation formulas

(a) $\vartheta(z+2) = \vartheta(z)$ *and*

(b) $\vartheta\left(-\dfrac{1}{z}\right) = \sqrt{\dfrac{z}{i}}\,\vartheta(z)$.

The theta series $\vartheta(z)$ has period 2. To obtain a form for the full modular group, besides ϑ we also consider $\tilde{\vartheta}(z) = \vartheta(z+1)$,

$$\tilde{\vartheta}(z) = \sum_{n=-\infty}^{\infty} (-1)^n e^{\pi i n^2 z} \ .$$

The function $\tilde{\vartheta}$ is a special value of the JACOBI theta function $\vartheta(z, w)$, namely

$$\tilde{\vartheta}(z) = \vartheta(z, 1/2) \ .$$

From VI.4.2 we obtain a transformation formula for $\tilde{\vartheta}$, namely

$$\tilde{\vartheta}\left(-\frac{1}{z}\right) = \sqrt{\frac{z}{i}}\,\tilde{\tilde{\vartheta}}(z) \ ,$$

where

$$\tilde{\tilde{\vartheta}}(z) := \sum_{n=-\infty}^{\infty} e^{\pi i (n+1/2)^2 z} \ .$$

We get:

Remark VI.4.4 (C.G.J. Jacobi 1833/36, 1838) *The three theta series*

$$\vartheta(z) = \sum_{n=-\infty}^{\infty} e^{\pi i n^2 z} \ ,$$

$$\tilde{\vartheta}(z) = \sum_{n=-\infty}^{\infty} (-1)^n e^{\pi i n^2 z} \ ,$$

$$\tilde{\tilde{\vartheta}}(z) = \sum_{n=-\infty}^{\infty} e^{\pi i (n+1/2)^2 z} \ ,$$

satisfy the transformation formulas:

$$\vartheta(z+1) = \tilde{\vartheta}(z) , \qquad \vartheta\left(-\frac{1}{z}\right) = \sqrt{\frac{z}{i}}\,\vartheta(z) ,$$

$$\tilde{\vartheta}(z+1) = \vartheta(z) , \qquad \tilde{\vartheta}\left(-\frac{1}{z}\right) = \sqrt{\frac{z}{i}}\,\tilde{\tilde{\vartheta}}(z) ,$$

$$\tilde{\tilde{\vartheta}}(z+1) = e^{\pi i/4}\,\tilde{\tilde{\vartheta}}(z) , \qquad \tilde{\tilde{\vartheta}}\left(-\frac{1}{z}\right) = \sqrt{\frac{z}{i}}\,\tilde{\vartheta}(z) .$$

These transformation formulas show that under the substitutions

$$z \longmapsto z+1 \text{ and } z \longmapsto -\frac{1}{z}$$

the function

$$f(z) = \left(\vartheta(z)\,\tilde{\vartheta}(z)\,\tilde{\tilde{\vartheta}}(z)\right)^{8}$$

has the same transformation law as the discriminant Δ. Hence the quotient function $f(z)/\Delta(z)$ is invariant under these generating substitutions (VI.1.8). Then it is invariant under the full modular group, since it is generated by the two special substitutions. In other words, the function f transforms like a modular form of weight 12, and it is indeed a modular form, since all three theta series are bounded in the region Im $z \geq 1$. Moreover, f is a cusp form, because the series $\tilde{\tilde{\vartheta}}(z)$ converges to 0 for Im $z \to \infty$. This leads to

Proposition VI.4.5 *For a suitable complex constant C, the following holds:*

$$\Delta(z) = C\left(\vartheta(z)\,\tilde{\vartheta}(z)\,\tilde{\tilde{\vartheta}}(z)\right)^{8} .$$

Supplement. *We will later determine for C the special value:*

$$C = \frac{(2\pi)^{12}}{2^{8}} .$$

A connection between the discriminant and pentagonal numbers

An integer of the form

$$\frac{3n^2 + n}{2} , \quad n \in \mathbb{Z} ,$$

is called *pentagonal*. The first pentagonal numbers are $0, 1, 2, 5, 7, 12, 15, 22$.

Proposition VI.4.6 *We have*

$$\Delta(z) = Ce^{2\pi i z}\left(\sum_{n=-\infty}^{\infty} (-1)^n e^{\pi i z(3n^2+n)}\right)^{24} .$$

We will later find for C the value $(2\pi)^{12}$.

Proof. The R.H.S. has the period 1, and a zero of order one at i∞. It is thus enough to show that the R.H.S. transforms as a modular form of weight 12. For this, we consider the auxiliary function

$$f(z) := \sum_{n=-\infty}^{\infty} (-1)^n e^{\pi i z (3n^2 + n)} \ .$$

This series is a special instance of the JACOBI theta series, more precisely

$$f(z) = \vartheta\left(3z, \frac{1}{2} + \frac{z}{2}\right) \quad \text{and thus} \quad f\left(-\frac{1}{z}\right) = \vartheta\left(-\frac{3}{z}, \frac{1}{2} - \frac{1}{2z}\right) \ .$$

A short computation using the theta transformation formula gives

$$f\left(-\frac{1}{z}\right) = \sqrt{\frac{z}{3i}}\, e^{\frac{\pi i}{12z}} \sum_{u=-\infty}^{\infty} e^{\pi i z \frac{u^2}{12} - \pi i \frac{u}{6}}, \quad u = 2n+1, \ n \in \mathbb{Z} \ .$$

Because the R.H.S. is invariant under $u \mapsto -u$, we have

$$f\left(-\frac{1}{z}\right) = \sqrt{\frac{z}{3i}}\, e^{\frac{\pi i}{12z}} \sum_{u=-\infty}^{\infty} e^{\pi i z \frac{u^2}{12}} \left\{ \frac{e^{-\frac{\pi i u}{6}} + e^{\frac{\pi i u}{6}}}{2} \right\},$$

where u runs through all *odd* integers. The expression

$$\frac{1}{2}\left(e^{-\frac{\pi i u}{6}} + e^{\frac{\pi i u}{6}}\right) = \cos\left(\frac{\pi u}{6}\right)$$

can be simply computed case by case modulo 6. Because u is odd,

$$u \equiv \pm 1 \text{ or } \equiv 3 \quad \text{mod } 6 \ .$$

One easily can see that in case of $u \equiv 3 \mod 6$ the expression vanishes. Because the summands do not change if we replace u by $-u$, we can sum over all $u \equiv 1 \mod 6$, and double the resulting value. Let us set $u = 6\nu + 1$, then

$$\cos\left(\frac{\pi u}{6}\right) = \cos\left(\frac{\pi}{6} + \pi\nu\right) = \frac{\sqrt{3}}{2}(-1)^\nu \ .$$

A simple computation now shows

$$f\left(-\frac{1}{z}\right) = \sqrt{\frac{z}{i}}\, e^{\left(\frac{\pi i z}{12} + \frac{\pi i}{12z}\right)} f(z) \ ,$$

which gives the claimed result.   □

The considered theta series are all special cases of a more general class of theta series, namely theta series associated to *quadratic forms*, respectively *lattices*.

Quadratic forms

In the following we denote by

$$A = A^{(n,m)} = \begin{pmatrix} a_{11} & \cdots & a_{1m} \\ \vdots & & \vdots \\ a_{n1} & \cdots & a_{nm} \end{pmatrix}$$

a matrix with n rows and m columns. In the case $m = n$ we also write for simplicity $A = A^{(n)}$. The transposed matrix of $A = A^{(n,m)}$ is

$$A' = \begin{pmatrix} a_{11} & \cdots & a_{n1} \\ \vdots & & \vdots \\ a_{1m} & \cdots & a_{nm} \end{pmatrix}.$$

Let $S = S^{(n)}$ and $A = A^{(n,m)}$. Then

$$S[A] := A'SA$$

is a $m \times m$-matrix. If S is symmetric, i.e. $S = S'$, then $S[A]$ is also symmetric. We have the rule

$$S[AB] = S[A][B] \quad (S = S^{(n)} ,\ A = A^{(n,m)} ,\ B = B^{(m,p)}) .$$

In the special case of a column vector $z = z^{(n,1)}$, the expression

$$S[z] = \sum_{1 \le \mu,\nu \le n} s_{\mu\nu} z_\mu z_\nu$$

is a 1×1 matrix, which we identify with a number. The function $z \mapsto S[z]$ is the quadratic form associated to the matrix S. A symmetric matrix is uniquely determined by its quadratic form.

A real symmetric matrix $S = S^{(n)}$ is called *positive definit*, or just *positive*, iff $S[x] > 0$ for all real column vectors $x \ne 0$. We use without proof two simple facts from Linear Algebra:

If S is a (real symmetric) positive matrix, then there exists a positive number $\delta > 0$ with the property

$$S[x] \ge \delta(x_1^2 + \cdots + x_n^2) .$$

Any positive matrix S can be written as $S = A'A$ with a suitable invertible real (quadratic) matrix A. One can arrange that the determinant of A is positive.

Surely, any matrix of the above form is positive. More generally, we have:

If $S = S^{(n)}$ is a positive matrix, and $A = A^{(n,m)}$ is a real matrix of rank m, then the matrix $S[A]$ is positive.

To any positive matrix $S = S^{(n)}$ we can associate a theta series,

$$\vartheta(S; z) = \sum_{g \in \mathbb{Z}^n} e^{\pi i S[g] z}, \qquad \text{Im } z > 0 .$$

This series is of number theoretical relevance if S is an integral matrix, because then $\vartheta(S; z)$ is a periodic function in z with period 2, having a FOURIER expansion of the form

$$\vartheta(S; z) = \sum_{m=0}^{\infty} A(S, m) e^{\pi i m z} ,$$

where

$$A(S, m) := \# \{ g \in \mathbb{Z}^n ; \quad S[g] = m \}$$

is the number of representations of a natural number m as a value of the quadratic form S. In the exercises to this section, and in Chapter VII, we will obtain number theoretical applications of the theory of modular forms for these representation numbers.

In case of the unit matrix $S = E = E^{(n)}$ this theta series splits as a CAUCHY product of n theta series $\vartheta(z)$,

$$\vartheta(E; z) = \vartheta(z)^n .$$

The convergence of $\vartheta(E; z)$ follows from this by means of the CAUCHY multiplication theorem. For any positive S there exists a suitable $\delta > 0$ such that for all column vectors x we have the estimate $S[x] \geq \delta E[x]$. Hence the convergence of $\vartheta(E; z)$ also implies the convergence of $\vartheta(S; z)$. As in the case of the series $\vartheta(z)$, we will consider also the generalized series

$$f(z, w) := \sum_{g \in \mathbb{Z}^n} e^{\pi i S[g + w] z} , \quad z \in \mathbb{H} , \ w \in \mathbb{C}^n .$$

In the case of the matrix $S = (1)$ this is exactly the JACOBI theta series.

The JACOBI transformation formula for theta functions admits the following generalization:

Theorem VI.4.7 (Jacobi's Generalized Theta Transformation Formula)
Let $S = S^{(n)}$ be a positive matrix. Then

$$\sqrt{\frac{z}{i}}^n \sqrt{\det S} \sum_{g \in \mathbb{Z}^n} e^{\pi i S[g + w] z} = \sum_{g \in \mathbb{Z}^n} e^{\pi i \left\{ S^{-1}[g](-1/z) + 2g' w \right\}} .$$

Both series converge normally in $\mathbb{H} \times \mathbb{C}^n$.

Proof. Using the estimate $S[x] \geq \delta E[x]$, we can reduce the question of normal convergence to the case of the JACOBI theta function, where we already know the positive answer. For the proof of the transformation formula, we once more consider the auxiliary function $f(w) = f(z, w)$, given by the sum in the L.H.S.,

$$f(w) = \sum_{g \in \mathbb{Z}^n} e^{\pi i S[g+w]z} \qquad (z \text{ being fixed}) .$$

It is continuous as a function of w, and analytic with respect to each complex variable w_j, $1 \leq j \leq n$. Moreover, it is periodic with period 1 in each of these variables.

Any function having these three properties can be expanded into an absolutely convergent FOURIER series

$$f(w) = \sum_{h \in \mathbb{Z}^n} a_h \, e^{2\pi i h' w} \qquad (h'w = h_1 w_1 + \cdots + h_n w_n) . \qquad (*)$$

The involved FOURIER coefficients can be computed by the formula

$$a_h = \int_0^1 \cdots \int_0^1 f(w) e^{-2\pi i h' w} \, du_1 \ldots du_n .$$

Here, $w = u + iv$ with fixed but arbitrary v. The FOURIER integral does not depend on the choice of v.

We have developed this formula only in the case $n = 1$, but this is enough for our purposes, since we can argue as follows. We first expand f into a FOURIER series with respect to the variable w_1. These "partial" FOURIER coefficients $a_{h_1}(w_2, \ldots, w_n)$ then still depend on $w_2, \ldots w_n$. Their representation by the FOURIER integral shows that they are continuous in $\mathbb{C}^{n-1}$, analytic in each of the remaining variables w_j, $2 \leq j \leq n$, and 1-periodic. We can once more expand then $a_{h_1}(w_2, \ldots, w_n)$ with respect to w_2, and repeat this process to obtain the above FOURIER series with the claimed coefficients, but with with the restriction that we have to respect the following ordering of summation:

$$f(w) = \sum_{h_1=-\infty}^{\infty} \left\{ \cdots \left\{ \sum_{h_n=-\infty}^{\infty} a_h e^{2\pi i h' w} \right\} \cdots \right\} .$$

The brackets can be omitted when we ensure the absolute convergence of the whole series $(*)$. In our case, this follows from the direct computation of the FOURIER coefficients.

The FOURIER integral will be computed as in the case of the JACOBI theta function. We briefly indicate the necessary computations:

First we observe

$$a_h = \int_{-\infty}^{\infty} \cdots \int_{-\infty}^{\infty} e^{\pi i \{S[w]z - 2h'w\}} \, du_1 \ldots du_n .$$

Quadratic completion (sometimes called "Babylonian identity") has the following natural extension

$$S[w]z - 2h'w = S[w - z^{-1}S^{-1}h]z - S^{-1}[h]z^{-1} .$$

The Principle of Analytic Continuation allows us to assume that z is purely imaginary, $z = iy$. We then set

$$v = y^{-1}S^{-1}h$$

and obtain

$$a_h = e^{-\pi S^{-1}[h]y^{-1}} \int_{-\infty}^{\infty} \cdots \int_{-\infty}^{\infty} e^{-\pi S[u]y} du_1 \ldots du_n .$$

To compute this integral, we use the linear substitution

$$u \longmapsto y^{-1/2}A^{-1}u \quad (S = A'A \text{ for a suitable } A) .$$

Its JACOBIan determinant $y^{-n/2} \det A^{-1}$ appears in the following transformation formula,

$$\int_{-\infty}^{\infty} \cdots \int_{-\infty}^{\infty} e^{-\pi S[u]y} \, du_1 \ldots du_n$$

$$= y^{-n/2} \left| \det A^{-1} \right| \int_{-\infty}^{\infty} \cdots \int_{-\infty}^{\infty} e^{-\pi(u_1^2 + \cdots + u_n^2)} \, du_1 \ldots du_n$$

$$= y^{-n/2} \sqrt{\det S^{-1}} \left[\int_{-\infty}^{\infty} e^{-\pi u^2} \, du \right]^n .$$

This completes the proof of VI.4.7.   □

As an important special case of the generalized JACOBI theta transformation formula we mention the following

Proposition VI.4.8 *The following theta transformation formula is valid:*

$$\boxed{\vartheta(S^{-1}; -z^{-1}) = \sqrt{\frac{z}{i}}^n \sqrt{\det S} \, \vartheta(S; z) .}$$

This formula has the disadvantage that besides the transformation $z \mapsto z^{-1}$ one also has to replace S by S^{-1}. Under special conditions, we can replace S^{-1} by S.

An invertible matrix $U = U^{(n)}$ is called *unimodular*, iff both U and U^{-1} have integral entries. Then $\det U = \pm 1$, and by the CRAMER formula any integral matrix with determinant ± 1 is unimodular. The set of all unimodular matrices builds the unimodular group $\mathrm{GL}(n, \mathbb{Z})$.

Two positive $n \times n$-matrices S and T are called *(unimodular) equivalent*, iff there exists a unimodular matrix U with the property $T = S[U]$. This relation is obviously an equivalence relation. The equivalence classes with respect to this relation are called *unimodular classes*.

If U is a unimodular matrix and if g runs through all integral columns then Ug runs through all integral columns as well. From this observation we immediately derive: If S and T are equivalent positive matrices, then

$$\vartheta(T;z) = \vartheta(S;z) .$$

If S is unimodular, then S and S^{-1} are equivalent, since $S = S^{-1}[S]$. The theta transformation formula gives in this case

$$\vartheta(S;-z^{-1}) = \sqrt{\frac{z}{i}}^{\,n} \vartheta(S;z) .$$

We would like to consider theta series having period 1. For this, we must restrict to positive matrices with the property

$$g \text{ integer} \implies S[g] \text{ even} .$$

Symmetric matrices which have this property are also called *even*. A symmetric matrix is even, iff it is integral and if its diagonal entries are even. This simply follows from the formula

$$S[g] = \sum_{\nu=1}^{n} s_{\nu\nu} g_{\nu}^2 + 2 \sum_{1 \leq \mu < \nu \leq n} s_{\mu\nu} g_{\mu} g_{\nu} .$$

Proposition VI.4.9 *Let $S = S^{(n)}$ be a positive unimodular even $n \times n$-matrix and n divisible by 8. Then $\vartheta(S;z)$ is an (entire) elliptic modular form of weight $n/2$.*

In any case, $\vartheta(S;z)$ has the right transformation behavior under both standard generators of the elliptic modular group. Especially, $\vartheta(S;z)/G_{n/2}(z)$ is a meromorphic function, which is invariant under these generators, and thus under any substitution of the full modular group. Thus $\vartheta(S;z)$ transforms like a modular form under the action of the full modular group. It is also clearly bounded in the region $y \geq 1$, so the regularity at $i\infty$ is ensured.   □

By the way, it can be shown that any positive, even, unimodular $n \times n$-matrix satisfies the divisibility condition $n \equiv 0 \mod 8$, (Exercise 8 to this section). An example of such a matrix for $n = 8m$ is

$$S_n = \begin{pmatrix} 2m & 1 & 1 & \dots & 1 & 1 & 1 \\ 1 & 2 & 1 & \dots & 1 & 1 & 2 \\ 1 & 1 & 2 & \dots & 1 & 1 & 2 \\ \vdots & \vdots & \vdots & & \vdots & \vdots & \vdots \\ 1 & 1 & 1 & \dots & 2 & 1 & 2 \\ 1 & 1 & 1 & \dots & 1 & 2 & 2 \\ 1 & 2 & 2 & \dots & 2 & 2 & 4 \end{pmatrix} .$$

In case $n = 16$ we can thus exhibit two unimodular matrices, namely S_{16} and

$$S_8 \oplus S_8 = \begin{pmatrix} S_8 & 0 \\ 0 & S_8 \end{pmatrix} .$$

It can be shown that these matrices are not unimodularly equivalent.

Because both vector spaces of modular forms of weight 4, and respectively 8 have dimension 1, we obtain non-trivial identities.

Proposition VI.4.10 *There hold the following identities:*

$$G_4(z) = 2\zeta(4)\, \vartheta(S_8; z) ,$$
$$G_8(z) = 2\zeta(8)\, \vartheta(S_{16}; z) = 2\zeta(8)\, \vartheta(S_8 \oplus S_8; z) .$$

The explicit constant factors arise by comparison of the constant FOURIER coefficients. For the theta series the constant coefficients are 1.   □

We will see that these (and similar) identities have number theoretical importance.

Positive matrices and lattices

A subset $L \subset \mathbb{R}^n$ is by definition a *lattice*, iff there exists an invertible real matrix $A = A^{(n)}$, such that

$$L = A\mathbb{Z}^n = \{\, Ag\,;\quad g \in \mathbb{Z}^n \,\} .$$

The matrix A is of course not uniquely determined.

Two lattices $A\mathbb{Z}^n \subset \mathbb{R}^n$ and $B\mathbb{Z}^n \subset \mathbb{R}^n$ are equal, iff there exists a unimodular matrix U with the property $B = AU$.

Lattices are discrete subgroups of $\mathbb{R}^n$, which contain a basis of $\mathbb{R}^n$. The converse is also true, as we will see in the second volume. Two lattices L and L' are said to be *congruent*, iff there exists a real orthogonal matrix

$$Q = Q^{(n)} ,\quad Q'Q = E \quad \text{(the unit matrix)} ,$$

with the property $L = QL'$. Two lattices $A\mathbb{Z}^n \subset \mathbb{R}^n$ and $B\mathbb{Z}^n \subset \mathbb{R}^n$ are *congruent*, iff there exists a unimodular matrix U and an orthogonal matrix Q with the property $B = QAU$. Then we consider the positive matrices $S = A'A$ and $T = B'B$. We have $T = S[U]$. From this we obtain:

The assignment $A \longmapsto S = A'A$ is a well defined bijection between the congruence classes of lattices $L = A\mathbb{Z}^n$ and unimodular classes of positive matrices S.

We can rewrite $S[g] = \langle Ag, Ag \rangle$, using the standard bilinear form $\langle z, w \rangle = \sum z_j w_j$ on $\mathbb{C}^n$ ($z, w \in \mathbb{C}^n$). We substitute the summation index g by $h := Ag$ and obtain the following representation of the theta series $\vartheta(S; z)$ in lattice language,

$$\vartheta(S; z) = \vartheta(L; z) := \sum_{h \in L} e^{\pi i \langle h, h \rangle z} ,\quad S = A'A ,\ L = A\mathbb{Z}^n .$$

Let us now assume that S is a matrix with integral entries. The nth FOURIER coefficient of this theta series is obviously equal to the number $A_L(n)$ of all lattice vectors

$h \in L$ with the property $n = \langle h, h \rangle$. This expression is the square of the EUCLIDian length of h. So we obtain:

The representation number $A(S,n)$, counting all representations of a natural number n by the quadratic form S is equal to the number $A_L(n)$ of all vectors of Euclidian length $\sqrt{n}$ of a lattice L associated to S.

A lattice is said to be of *type II*, iff the determinant of a generating matrix is ± 1, and the scalar product of any lattice vector with itself is always even. (A lattice is thus of type II, iff the associated quadratic forms are even and unimodular.)

Using lattices (instead of positive matrices) occasionally offers some advantages because of a higher flexibility. By the given characterization, any group L with

$$q\mathbb{Z}^n \subset L \subset (1/q)\, \mathbb{Z}^n \ , \quad \text{for some } q \in \mathbb{N} \ ,$$

is a lattice. For instance,

$$L_n = \left\{ x \in \mathbb{R}^n \ ; \quad 2x_\nu \in \mathbb{Z} \ , \ x_\mu - x_\nu \in \mathbb{Z} \ , \ \sum_{\nu=1}^{n} x_\nu \in 2\mathbb{Z} \right\} \qquad (*)$$

is a lattice in an n-dimensional space. It is of type II, if n is divisible by 8. The congruence class of this lattice corresponds in case of $n = 8m$ exactly to the unimodular class of S_n (see above). The appearance of this matrix becomes more transparent in terms of the lattice L_n.

Exercises for VI.4

1. Let f and g be two elliptic modular forms of weight k. The function $h(z) = f(z)\overline{g(z)}y^k$ is Γ-invariant.

2. Let f be a cusp form of weight k. The function

    $$h(z) = |f(z)|\, y^{k/2}$$

    has a maximum in the upper half-plane.

    Hint. Because of Exercise 1, it is enough to show that $h(z)$ has a maximal value in the fundamental domain. This follows from $\lim_{y \to \infty} h(z) = 0$.

3. Let

    $$f(z) = \sum_{n=1}^{\infty} a_n e^{2\pi i n z}$$

    be a cusp form of weight k. Prove an estimate of the kind

    $$|a_n| \le C n^{k/2} \qquad \text{(E. HECKE, 1927)}$$

    with a suitable constant C.

    Hint. Use the integral representation for the FOURIER coefficients, and apply the estimate

    $$|f(z)| \le C' y^{-k/2}$$

    for the special value $y = 1/n$.

    P. DELIGNE proved in 1974 the RAMANUJAN-PETERSSON conjecture, which ensures the much stronger estimate

    $$|a_n| \le C(\varepsilon)\, n^{(k-1)/2+\varepsilon} \text{ for any } \varepsilon > 0 \ .$$

4.   In this exercise we use the formula for the FOURIER coefficients of the EISEN-
     STEIN series, which will be obtain in VII.1,

$$G_k(z) = 2\zeta(k) + \frac{2 \cdot (2\pi i)^k}{(k-1)!} \sum_{n=1}^{\infty} \sigma_{k-1}(n)\, e^{2\pi i n z} \ .$$

Let $L \subset \mathbb{R}^m$, $m \equiv 0 \mod 8$, be a lattice of type II, and for $n \in \mathbb{N}_0$ let

$$A_L(n) \ = \ \#\{\, x \in L \, ; \quad \langle x, x \rangle = n \,\} \ .$$

Then

$$A_L(2n) \ \sim \ -\frac{m}{B_{m/2}} \sum_{d|n} d^{m/2 - 1} \ ,$$

i.e. the quotient of the L.H.S. and R.H.S. converges to 1 for $n \to \infty$.

In the cases $m = 8$ and $m = 16$ we in fact have equality in the above asymptotic
formula, but not in general. But there is a very remarkable result of SIEGEL:
[Si2] For natural numbers n,

$$\sum_L \frac{A_L(n)}{e(L)} = \sum_L \left(\frac{1}{e(L)} \right) \frac{-m}{B_{m/2}} \sum_{d|n} d^{m/2 - 1} \ .$$

Here L runs through a system of representatives of the congruence classes of
all type II lattices in $\mathbb{R}^m$. This a finite system. The involved denominator $e(L)$
is the order of the automorphism group of L. (An automorphism of L is an
orthogonal map $\mathbb{R}^n \to \mathbb{R}^n$ which stabilizes L.)

The number of these classes is 1 for $m = 8$; 2 for $m = 16$; 24 for $m = 24$ and
at least 80 millions for $m = 32$ (compare with [CS]).

5.   Let f be an arbitrary modular form of weight k. Then the FOURIER coefficients
     a_n of f satisfy an estimate of the kind

$$|a_n| \leq C n^{k-1} \qquad (\text{E. HECKE, 1927}) \ .$$

6.   Determine the number of all integral, orthogonal matrices $U = U^{(n)}$ (i.e. $U \in$
     $\mathrm{GL}(n, \mathbb{Z})$, $U'U = E$).

7.   On page 349, we defined the lattice L_n in $(*)$.
     (a) Show that the lattice L_n is of type II, iff n is divisible by 8.
     (b) Determine in the case $n \equiv 0 \mod 8$ all minimal vectors of L_n, i.e. all
          vectors $a \in L_n$ with $\langle a, a \rangle = 2$.
     (c) Show that the lattices L_{16} and $L_8 \times L_8$ are not congruent by examining
          the angles between minimal vectors. (But remarkably the lattice numbers
          $A_L(n)$ coincide.)

8.   Let a and b be real numbers. The theta series —a so-called *theta nullwert*—

$$\vartheta_{a,b}(z) := \sum_{n=-\infty}^{\infty} e^{\pi i \left((n+a)^2 z + 2bn \right)}$$

vanishes identically, iff both $a - 1/2$ and $b - 1/2$ are integers. In all other cases
it has no zero in the upper half-plane.

Hint. Express this series in terms of the JACOBI theta series $\vartheta(z, w)$, and use
the knowledge of the zeros of it (Exercise 3 in V.6).

9. The theta series $\vartheta_{a,b}$ (see Exercise 8) changes only by a constant factor, when a and b are changed by adding integers. Using the JACOBI theta transformation formula, show the following transformation formula:

$$\vartheta_{a,b}\left(-\frac{1}{z}\right) = e^{2\pi i a b}\sqrt{\frac{z}{i}}\,\vartheta_{b,-a}(z) \ .$$

10. Let n be a natural number. We consider all pairs of integers

$$(a,b) \ , \quad 0 \le a,b < 2n \ ,$$

excepting the pair $(a,b) = (n,n)$, and build the function

$$\Delta_n(z) = \prod_{\substack{(a,b)\neq(n,n)\\ 0\le a,b<2n}} \vartheta_{\left(\frac{a}{2n},\frac{b}{2n}\right)}(z) \ .$$

Show that a suitable power of Δ_n is a modular form for the full modular group.

Hint. Applying the generators of the modular group to the finite system of theta series, we obtain a permutation of this system up to elementary factors. Using Exercise 1 in Sect. VI.3, show

$$\Delta_n(z)^{24} = C\Delta(z)^{4n^2-1} \ ,$$

and also determine the involved constant C.

11. Let $S = S^{(n)}$ be a positive, even, unimodular matrix. Then $n \equiv 0 \mod 8$.
    Hint. Use the relation

$$w := 1 - \frac{1}{z} - \left(\frac{1}{1-z}-1\right)^{-1} \ ,$$

and transform $\vartheta(S;z)$ corresponding to these relations, by applying the formulas

$$\vartheta(S;z+1) = \vartheta(S;z) \ , \quad \vartheta(S;-1/z) = \sqrt{\frac{z}{i}}^{\,n}\,\vartheta(S;z) \ .$$

This gives the formula

$$\sqrt{z/i}^{\,n} = \sqrt{z/(i(1-z))}^{\,n}\sqrt{(z-1)/i}^{\,n} \ .$$

Now specialize $z = i$ in it to infer

$$1 = e^{2\pi i n/8} \ , \text{ i.e. } n \equiv 0 \mod 8 \ .$$

VI.5 Modular Forms for Congruence Groups

We would like to generalize the notion of a modular form in two directions. Firstly, we replace the modular group $\mathrm{SL}(2,\mathbb{Z})$ by a subgroup of finite index. Secondly, we also consider forms of *half-integral weight*. Examples of such modular forms are the theta series $\vartheta(S,z)$ with respect to arbitrary rational, positive definite matrices S (possibly also of odd size).

A subgroup H of a group G has *finite index*, if there are finitely many elements $g_1, \ldots, g_h \in G$ with

$$G = Hg_1 \cup \cdots \cup Hg_h$$
$$(\Longleftrightarrow G = g_1^{-1}H \cup \cdots \cup g_h^{-1}H) \ .$$

One can moreover require that the above finite union is a disjoint union. The uniquely determined number h is called the *index* of H in G.

A fundamental example for a subgroup of finite index in the elliptic modular group is the *principal congruence subgroup of level q* ($\in \mathbb{N}$),

$$\Gamma[q] := \left\{ M = \begin{pmatrix} a & b \\ c & d \end{pmatrix} \in \mathrm{SL}(2,\mathbb{Z}) \ ; \ a \equiv d \equiv 1 \mod q \ , \ b \equiv c \equiv 0 \mod q \right\} \ .$$

It is the kernel of the natural group homomorphism

$$\mathrm{SL}(2,\mathbb{Z}) \longrightarrow \mathrm{SL}(2,\mathbb{Z}/q\mathbb{Z}) \ .$$

Because the target group $\mathrm{SL}(2,\mathbb{Z}/q\mathbb{Z})$ is *finite*, $\Gamma[q]$ is a *normal subgroup of finite index* of $\mathrm{SL}(2,\mathbb{Z})$. Because of this, we have

$$N \, \Gamma[q] \, N^{-1} = \Gamma[q] \quad \text{for all } N \in \Gamma[1] = \mathrm{SL}(2,\mathbb{Z}) \ .$$

Definition VI.5.1 *A subgroup $\Gamma \subset \mathrm{SL}(2,\mathbb{Z})$ is called a* **congruence subgroup,** *iff it contains a suitable principal congruence subgroup $\Gamma[q]$, i.e.*

$$\Gamma[q] \subset \Gamma \subset \Gamma[1] \ .$$

Congruence groups are subgroups of finite index in $\mathrm{SL}(2,\mathbb{Z})$. But there are also subgroups of finite index in $\mathrm{SL}(2,\mathbb{Z})$, which are not congruence subgroups! Only *congruence subgroups* showed their importance for the theory of modular forms. Because $\Gamma[q]$ is a normal subgroup in $\Gamma[1] = \mathrm{SL}(2,\mathbb{Z})$, we have

Remark VI.5.2 *Let Γ be a congruence subgroup. Then for any $L \in \Gamma[1]$ the conjugate group $L\Gamma L^{-1}$ is also a congruence subgroup.*

Cusps of congruence subgroups

A *cusp* κ of a congruence subgroup Γ is by definition an element of $\mathbb{Q} \cup \{i\infty\}$. The group $\mathrm{SL}(2, \mathbb{Z})$ acts not only on $\mathbb{H}$ by MÖBIUS substitutions, but also on the set of cusps via the formula

$$\kappa \longmapsto \frac{a\kappa + b}{c\kappa + d}$$

with the usual conventions for computations with $i\infty$:

$$\frac{ai\infty + b}{ci\infty + d} := \frac{a}{c} \qquad (:= i\infty, \ \text{if } c = 0) ,$$

$$\frac{a\kappa + b}{c\kappa + d} := i\infty , \qquad \text{if } \kappa \neq i\infty \text{ and } c\kappa + d = 0$$

$$\text{(and in this case we have } a\kappa + b \neq 0) .$$

Two cusps are called *equivalent with respect to* Γ, iff they are in the same orbit with respect to the action of Γ. The equivalence classes with respect to this equivalence relation are called *cusp classes*.

Lemma VI.5.3 *The group* $\mathrm{SL}(2, \mathbb{Z})$ *acts transitively on the set of all cusps, i.e. for any cusp* κ *there exists an*

$$A \in \mathrm{SL}(2, \mathbb{Z}) \ \text{with } A\kappa = i\infty .$$

Corollary. *Let* Γ *be a congruence subgroup. The set of all cusp classes*

$$(\mathbb{Q} \cup \{i\infty\})/\Gamma$$

is finite.

Proof of Lemma VI.5.3. Let

$$\kappa = \frac{a}{b} , \qquad a, b \in \mathbb{Z} , \ b \neq 0 , \ \gcd(a, b) = 1 ,$$

be a cusp. We can find a matrix

$$A = \begin{pmatrix} x & y \\ -b & a \end{pmatrix} \in \mathrm{SL}(2, \mathbb{Z}) ,$$

since the equation $ax + by = 1$ has (integral) solutions x, y (because a and b are relatively prime). We obviously have $A\kappa = i\infty$.

For the proof of the corollary, let us write

$$\mathrm{SL}(2, \mathbb{Z}) = \Gamma A_1 \cup \cdots \cup \Gamma A_h .$$

Then the set

$$\{\, A_1 i\infty \,,\, \ldots \,,\, A_h i\infty \,\}$$

obviously contains a system of representatives for all cusp classes. Of course, this set may contain equivalent cusps. The number of cusp classes is thus bounded by the index of Γ in $\Gamma[1]$. $\square$

We will see in the second volume that the quotient space $\mathbb{H}/\Gamma$ carries a natural structure as a RIEMANN surface. This RIEMANN surface can be completed to a *compact* RIEMANN surface by adding finitely many points, namely the cusp classes.

Multiplier systems

We also want to introduce modular forms of half-integral weight

$$k = \frac{r}{2} \,, \quad r \in \mathbb{Z} \,.$$

For this, we need a holomorphic square root of $cz+d$. In this context, we define the square root $\sqrt{a}$ of a non-zero complex number $a \neq 0$ by the *principal value of the logarithm*,

$$\sqrt{a} := e^{\frac{1}{2}\,\mathrm{Log}\,a} \,.$$

Equivalently, $\sqrt{a}$ is uniquely specified by the following conditions:

(a) $\mathrm{Re}\,\sqrt{a} \geq 0$,
(b) $\sqrt{a} = \mathrm{i}\sqrt{|a|}$, if a is real and negative.

The function

$$z \mapsto \sqrt{cz+d} \qquad ((c,d) \in \mathbb{R} \times \mathbb{R} \setminus \{(0,0)\})$$

is analytic in the upper half-plane, since $cz + d$ cannot be real and negative in the case of $c \neq 0$.

Notation.

$$I_r(M,z) := (cz+d)^{r/2} := \sqrt{cz+d}^{\,r} \,, \quad r \in \mathbb{Z} \,.$$

Remark VI.5.4 *We have*

$$I_r(MN,z) = w_r(M,N)\, I_r(M,Nz) I_r(N,z) \,.$$

Here, the collection of all $w_r(M,N)$ is a system of numbers taking only the values ± 1. The dependence on the index r is only a dependence on r modulo 2, (i.e. there are only two cases, r even or odd), and only for even r the number system is identically 1.

Proof. A trivial computation shows that the given formula is correct for even r by setting $w_r = 1$. It then also follows for odd r by taking the square root. The sign w_r appears because of the ambiguity of the square root. For instance, we have

$$w_1(-S, -S) = \frac{I_1(-E, \mathrm{i})}{I_1(-S, \mathrm{i})^2} = \frac{\sqrt{-1}}{\sqrt{-\mathrm{i}}^2} = \frac{\mathrm{i}}{-\mathrm{i}} = -1 \quad \text{for } S := \begin{pmatrix} 0 & -1 \\ 1 & 0 \end{pmatrix}.$$

(Using our convention, our choice of the square root gives $\sqrt{-1} = \mathrm{i}$, and not $= -\mathrm{i}$.)

Definition VI.5.5 *A multiplier system of weight* [1] *$r/2$, $r \in \mathbb{Z}$, with respect to the congruence subgroup Γ is a map, which associates to each $M \in \Gamma$ a root of unity*

$$v(M) \in \mathbb{C}, \quad v(M)^l = 1,$$

of fixed order $l \in \mathbb{N}$ (independent of M), such that

$$I(M, z) = v(M) I_r(M, z)$$

becomes an automorphy factor, i.e.

$$I(MN, z) = I(M, Nz) I(N, z) \quad (M, N \in \Gamma).$$

Moreover, we require the relation $I(-E, z) = 1$ if $-E$, the negative of the unit matrix, lies in Γ.

The automorphy property can be equivalently written as

$$v(MN) = w_r(M, N)\, v(M) v(N).$$

If r is even, this means that v is a character, i.e. a homomorphism of Γ into the multiplicative group of complex numbers.

The importance of the multiplier systems can be seen also from the following observation:

Let $f : \mathbb{H} \to \mathbb{C}$ be a function with the following transformation behavior:

$$f(Mz) = I(M, z) f(z)$$

for all M in some set $\mathcal{M} \subset \Gamma$. Then this transformation behavior extends to all M in the subgroup of Γ generated by $\mathcal{M}$.

Examples.

(1) r is even.

As mentioned above, the automorphy property expresses the fact that v is a character,

$$v(MN) = v(M) v(N).$$

(The most important case is the case of the *principal character*, $v \equiv 1$.)
By definition, any multiplier system can take only finitely many values.
The kernel of the character v,

$$\Gamma_0 = \{ M \in \Gamma ; \quad v(M) = 1 \}$$

is thus a subgroup of finite index in Γ.

---

[1] It depends only on r modulo 2.

(2) r is odd.

Let Γ_ϑ be the subgroup of $\mathrm{SL}(2,\mathbb{Z})$ generated by the two matrices

$$\begin{pmatrix} 1 & 2 \\ 0 & 1 \end{pmatrix} \quad \text{and} \quad \begin{pmatrix} 0 & -1 \\ 1 & 0 \end{pmatrix} .$$

We will show in the Appendix of this section that Γ_ϑ consists of all matrices

$$\begin{pmatrix} a & b \\ c & d \end{pmatrix} \in \mathrm{SL}(2,\mathbb{Z}) , \quad \text{with} \quad a+b+c+d \quad \text{even} .$$

In particular, Γ_ϑ contains the congruence subgroup $\Gamma[2]$, so it is itself a congruence subgroup.

We have a formula

$$\vartheta(Mz) = v_\vartheta(M)\sqrt{cz+d}\,\vartheta(z) \qquad \left(v_\vartheta(M)^8 = 1\right)$$

for both generators of the theta group. Such a formula then automatically follows for all $M \in \Gamma_\vartheta$. The map

$$\Gamma_\vartheta \longrightarrow \mathbb{C}^\bullet , \quad M \longmapsto v_\vartheta(M) ,$$

is necessarily a multiplier system of weight $1/2$. This system is determined by the special values

$$v_\vartheta \begin{pmatrix} 1 & 2 \\ 0 & 1 \end{pmatrix} = 1 , \quad v_\vartheta \begin{pmatrix} 0 & -1 \\ 1 & 0 \end{pmatrix} = e^{-\pi i/4} .$$

This multiplier system is called the *theta multiplier system*. It is not easy to find an explicit formula for v_ϑ, ([Ma3]).

Now let v be an arbitrary *multiplier system of non-integral weight* with respect to the congruence subgroup Γ. The character v/v_ϑ on $\Gamma \cap \Gamma_\vartheta$ takes only finitely many values. Therefore a subgroup of finite index $\Gamma_0 \subset \Gamma \cap \Gamma_\vartheta$ exists, such that the restriction of v and v_ϑ on Γ_0 coincide, i.e.

$$v(M) = v_\vartheta(M) \text{ for all } M \in \Gamma_0 .$$

We already mentioned that only the congruence subgroups are interesting and relevant for the theory of modular forms. For the same reason, only multiplier systems with the following property are of interest:

(1) *The case of r even:* There exists a congruence subgroup $\Gamma_0 \subset \Gamma$, such that v is trivial on Γ_0 (i.e. it is the principal character).
(2) *The case of r odd:* There exists a congruence subgroup $\Gamma_0 \subset \Gamma \cap \Gamma_\vartheta$, such that the restriction of v on Γ_0 coincides with the theta multiplier system.

The conjugate multiplier system

We use (for $r \in \mathbb{Z}$) the *modified Petersson notation*

$$(f|M)(z) = (f \mid_r M)(z) := \sqrt{cz+d}^{\,-r} f(Mz) .$$

Here, f is an arbitrary function on the upper half-plane, and $M \in \mathrm{SL}(2,\mathbb{Z})$ is a modular matrix. We then have (VI.5.5)

$$f|MN = w_r(M,N)\,(f \mid M)|N .$$

The usefulness of the PETERSSON notation can be seen form the following simple

Remark VI.5.6 *A system of unit roots $\{v(M)\}_{M\in\Gamma}$ of order l is a multiplier system of weight $r/2$, iff there exists a function f defined on the upper half-plane, which is not identically zero, and satisfies the transformation formula*

$$f \mid_r M = v(M)f .$$

Proof. (1) Let us assume the existence of a function f with the specified property. We choose a point a with $f(a) \neq 0$, and make use of $(f|M)(a) = v(M)f(a)$.

(2) Conversely, let us assume that v is a multiplier system. From the equation $Ma = Na$ $(M,N \in \Gamma)$ then follows $M = \pm N$. Consider the function f on the upper half-plane which is different from 0 only at the points Ma, $M \in \Gamma$ and takes there the value $f(Ma) = I(M,a)$. Of course this function is not continuous.

Now let v be a multiplier system of weight $r/2$ with respect to the congruence subgroup Γ, and let $f : \mathbb{H} \to \mathbb{C}$ be a function with the transformation behavior

$$f \mid M = v(M)f \text{ for all } M \in \Gamma .$$

We want to show that the function $\tilde{f} := f|L^{-1}$ for an arbitrary $L \in \mathrm{SL}(2,\mathbb{Z})$ has an analogous transformation behavior with respect to the conjugated group $\tilde{\Gamma} := L\Gamma L^{-1}$. For this, let $\tilde{M} \in \tilde{\Gamma}$, i.e. $\tilde{M} = LML^{-1}$, $M \in \Gamma$. Then we have:

$$
\begin{aligned}
\tilde{f}|\tilde{M} &= (f|L^{-1})|\tilde{M} \\
&= w_r(L^{-1},\tilde{M})\, f|L^{-1}\tilde{M} \\
&= w_r(L^{-1},\tilde{M})\, f|ML^{-1} \\
&= w_r(L^{-1},\tilde{M})\, w_r(M,L^{-1})\, (f|M)|L^{-1} \\
&= v(M)\, w_r(L^{-1},\tilde{M})\, w_r(M,L^{-1})\tilde{f} \\
&= \tilde{v}(\tilde{M})\, \tilde{f} ,
\end{aligned}
$$

where we have set

$$\tilde{v}(\tilde{M}) = v(L^{-1}\tilde{M}L)\, w_r(L^{-1}, \tilde{M})\, w_r(L^{-1}\tilde{M}L, L^{-1})\,.$$

By VI.5.6, we then have:

Remark VI.5.7 *Let v be a multiplier system of weight $r/2$ with respect to the congruence subgroup Γ. Let $L \in \mathrm{SL}(2, \mathbb{Z})$ be an arbitrary matrix. Then the system $\tilde{v}$,*

$$\tilde{v}(M) := v(L^{-1}ML)\, w_r(L^{-1}, M)\, w_r(L^{-1}ML, L^{-1})$$

*is a multiplier system of weight $r/2$ with respect to the conjugated group $\tilde{\Gamma} = L\Gamma L^{-1}$ of Γ. The system $\tilde{v}$ is called the **conjugate** multiplier system.*

Supplement. *If $f : \mathbb{H} \to \mathbb{C}$ is a function with the transformation property*

$$f|M = v(M)f \text{ for } M \in \Gamma\,,$$

then the function $\tilde{f} = f|L^{-1}$ has the transformation property

$$\tilde{f}|M = \tilde{v}(M)\tilde{f} \text{ for } M \in \tilde{\Gamma}\,.$$

The notion of regularity (respectively meromorphy) in a cusp

Let v be a multiplier system of weight $r/2$ with respect to some congruence subgroup Γ, and let $f : \mathbb{H} \to \overline{\mathbb{C}}$ be a meromorphic function with the property $f|_r M = v(M)f$ for all $M \in \Gamma$. Then there exists an integer $q \neq 0$ with

$$\begin{pmatrix} 1 & q \\ 0 & 1 \end{pmatrix} \in \Gamma \quad \text{and} \quad f(z + q) = v\begin{pmatrix} 1 & q \\ 0 & 1 \end{pmatrix} f(z)\,.$$

Since v has only roots of unity as values, there exists a suitable natural number l with

$$v\begin{pmatrix} 1 & q \\ 0 & 1 \end{pmatrix}^l = 1\,.$$

This implies the existence of an integer $N \neq 0$ (e.g. $N = lq$) with the property $f(z+N) = f(z)$. Because of VI.5.7, the transformed functions $f|L^{-1}$ are also periodic. In In Section VI.2, it has been defined what it means that $f|L^{-1}$ has an inessential singularity at $i\infty$ or is regular there. We will see in VI.5.9 that this behavior only depends on the cusp class of $L^{-1}(i\infty)$.

The notion of a modular form

Definition VI.5.8 *Let Γ be a congruence subgroup, and let v be a multiplier system of weight $r/2$, $r \in \mathbb{Z}$. A **meromorphic modular form** of weight $r/2$ for the multiplier system v is a meromorphic function*

$$f : \mathbb{H} \longrightarrow \overline{\mathbb{C}}$$

with the following properties:

(1) $f \mid_r M = v(M)f$ *for all* $M \in \Gamma$.
(2) *For any* $L \in \mathrm{SL}(2,\mathbb{Z})$ *there exists a number* $C > 0$, *such that*

$$\tilde{f} := f \mid_r L^{-1}$$

is analytic in the half-plane $\mathrm{Im}\, z > C$, and has a non-essential singularity at $i\infty$.

Supplement. *If moreover f is an analytic function $f : \mathbb{H} \to \mathbb{C}$, and if $f|L^{-1}$ is regular at $i\infty$ (for all $L \in \mathrm{SL}(2,\mathbb{Z})$), then f is called an entire **modular form**. An entire modular form f is called a **cusp form**, iff we furthermore have*

$$(f|L^{-1})(i\infty) = 0 \ \text{for all} \ L \in \mathrm{SL}(2,\mathbb{Z}) \ .$$

In fact, the conditions in VI.5.8 have to be checked for only finitely many matrices L:

Remark VI.5.9 *Let $\mathcal{L}$ be a set of matrices $L \in \mathrm{SL}(2,\mathbb{Z})$, such that $L^{-1}(i\infty)$ is a system of representatives for the Γ-equivalence classes (Γ-orbits) of cusps. Then it is enough in Definition VI.5.8 to consider only matrices $L \in \mathcal{L}$. In particular, in case of the full modular group it is enough to consider only $L = E$, in concordance with Definition VI.2.4.*

Proof. Let $M^{-1}(i\infty)$ and $N^{-1}(i\infty)$ be two Γ-equivalent cusps. Then there exists a translation matrix $P = \pm \begin{pmatrix} 1 & b \\ 0 & 1 \end{pmatrix}$, and a matrix $L \in \Gamma$ with $M = PNL$. Then the functions $f|M^{-1}$ and $f|N^{-1}$ differ up to a constant factor only by a translation of the argument.   □

Notations.

$\{\Gamma, r/2, v\}$ Set of all meromorphic modular forms ,
$\cup$
$[\Gamma, r/2, v]$ Set of all entire modular forms ,
$\cup$
$[\Gamma, r/2, v]_0$ Set of all cusp forms .

If r is even, and v is the trivial multiplier system, then we neglect the argument v in the notation, and write instead for instance

$$[\Gamma, r/2] := [\Gamma, r/2, v] \ .$$

Finally, we use the notation

$$K(\Gamma) := \{\Gamma, 0\} \ .$$

The elements of $K(\Gamma)$ are Γ-invariant, and are called *modular functions*. The set of all modular functions is a *field*. It contains the (subfield of) constant functions. Directly from VI.5.9 we see:

Remark VI.5.10 *Let $L \in \mathrm{SL}(2, \mathbb{Z})$ be a modular matrix. Then the assignment*

$$f \mapsto f \underset{r}{\mid} L^{-1}$$

defines isomorphisms

$$\{\Gamma, r/2, v\} \xrightarrow{\ \sim\ } \{\tilde{\Gamma}, r/2, \tilde{v}\} \ ,$$
$$[\Gamma, r/2, v] \xrightarrow{\ \sim\ } [\tilde{\Gamma}, r/2, \tilde{v}] \ ,$$
$$[\Gamma, r/2, v]_0 \xrightarrow{\ \sim\ } [\tilde{\Gamma}, r/2, \tilde{v}]_0 \ .$$

Here, we have used the conjugate group of Γ,

$$\tilde{\Gamma} = L\Gamma L^{-1} \ ,$$

and the conjugate multiplier system $\tilde{v}$ induced by v from Γ to $\tilde{\Gamma}$ in sense of VI.5.7.

Moreover we have: Let $f : \mathbb{H} \to \overline{\mathbb{C}}$ be a function with the transformation property

$$f|M = v(M)f \text{ for all } M \in \Gamma \ .$$

Let $\Gamma_0 \subset \Gamma$ be congruence subgroups. Then f is a meromorphic modular form (respectively an entire modular form, or a cusp form) with respect to the group Γ, iff f is such a form with respect to the smaller group Γ_0.

As a simple application of this observation, we prove

Proposition VI.5.11 *Any (entire) modular form of negative weight with respect to a congruence subgroup Γ is vanishes identically. Any modular form of weight 0 is constant.*

Proof. We decompose $\mathrm{SL}(2, \mathbb{Z})$ into right cosets

$$\mathrm{SL}(2, \mathbb{Z}) = \bigcup_{\nu=1}^{k} \Gamma M_\nu \ ,$$

and we associate to any function

$$f \in [\Gamma, r/2, v]$$

the symmetrization

$$F = \prod_{\nu=1}^{k} f \mid M_\nu \; .$$

Then F is obviously a modular form of weight $kr/2$ with respect to the full modular group, and a suitable power of F has trivial multiplier system . If k is negative, then F identically vanishes by the already proven version VI.2.5 in case of the full modular group. Then at least one $f|M_\nu$, and thus also f vanishes identically. The case $k = 0$ needs a slight modification of this argument. We replace $f(z)$ by $f(z) - f(i\infty)$, and can thus assume without loss of generality that f vanishes at $i\infty$. Then the associated function F is a cusp form of weight zero, which also vanishes by the corresponding result for the full modular group.   □

Exact description of the Fourier expansion

Let Γ be a congruence subgroup, and let

$$f \in \{\,\Gamma, r/2, v\,\}$$

be a meromorphic modular form $f \not\equiv 0$. Then there exists for Γ a *smallest natural number* $R > 0$, such that the substitution $z \mapsto z + R$ is in Γ, i.e. we have either $\begin{pmatrix} 1 & R \\ 0 & 1 \end{pmatrix} \in \Gamma$ or $-\begin{pmatrix} 1 & R \\ 0 & 1 \end{pmatrix} \in \Gamma$. From the transformation behavior of f we derive $f(z + R) = \varepsilon f(z)$ with a suitable root of unity ε,

$$\varepsilon = e^{2\pi i\nu/l} \; , \quad 0 \le \nu < l \; , \quad \gcd(\nu, l) = 1 \; .$$

Setting $N = lR$ we then have in particular

$$f(z + N) = f(z) \; ,$$

and we can consider the FOURIER series

$$f(z) = \sum_{n=-\infty}^{\infty} a_n e^{2\pi inz/N} \; .$$

From the equation

$$f(z + R) = \varepsilon f(z)$$

we obtain for the FOURIER coefficients the equation

$$a_n e^{2\pi iRn/N} = e^{2\pi i\nu/l} a_n \; .$$

Equivalently,

$$a_n \neq 0 \implies n \equiv \nu \mod l \; .$$

We then introduce the new coefficients $b_n := a_{\nu+ln}$ to obtain

$$f(z)\, e^{-2\pi i \nu z/l} = \sum_{n=-\infty}^{\infty} b_n e^{2\pi i n z/R} \ .$$

We also remark that in case of $\nu \neq 0$ (i.e. $\varepsilon \neq 1$) the function f necessarily has a zero at $i\infty$ in the sense of our definition,

$$f(i\infty) := \lim_{y \to \infty} f(z) = 0 \ .$$

Definition VI.5.12 *Let $f \in \{\,\Gamma, r/2, v\,\}$, $f \not\equiv 0$. The order of f in $i\infty$ is*

$$\mathrm{ord}_\Gamma(f; i\infty) = \min\{n;\ b_n \neq 0\} \ .$$

This notion is problematically. We of course have

$$\mathrm{ord}_\Gamma(f; i\infty) \geq 0 \Longleftrightarrow f \text{ is regular at } i\infty \ ,$$

but notice that only the implication

$$\mathrm{ord}_\Gamma(f; i\infty) > 0 \implies f(i\infty) = 0$$

is true, but not also the converse!

This notion of order at $i\infty$ has the following advantage:

Let $N \in \mathrm{SL}(2, \mathbb{Z})$ be a matrix with $N(i\infty) = i\infty$. Then

$$\mathrm{ord}_\Gamma(\ f\ ;\ i\infty\ ) = \mathrm{ord}_{N\Gamma N^{-1}}(\ f \,|\, N^{-1}\ ;\ i\infty\ ) \ .$$

Corollary. *Let κ be a cusp of Γ, and let*

$$N \in \mathrm{SL}(2, \mathbb{Z}) \quad \textit{satisfy} \quad N\kappa = i\infty \ .$$

Then the definition

$$\mathrm{ord}_\Gamma(f; [\kappa]) := \mathrm{ord}_{N\Gamma N^{-1}}(f \,|\, N^{-1}; i\infty)$$

depends only on the Γ-equivalence class of κ.

We do not want to deepen this question, which —from the point of view of RIEMANN surfaces— is concerned with the association of a divisor to an arbitrary modular form. This will be done instead in the second volume, where the theory of RIEMANN surfaces will be used to solve the following problems:

(1) The association of a "divisor" to any modular form, and the generalization of the $k/12$-formula for arbitrary congruence subgroups.

(2) The proof that $[\Gamma, r/2, v]$ has a finite dimension over $\mathbb{C}$, and the computation of this dimension (in many cases).

(3) The (rough) determination of the field $K(\Gamma)$ of all modular functions.

In the second volume, we will also prove the following Proposition:

Let $S = S^{(r)}$ be a positive rational matrix. The theta series

$$\vartheta(S; z) = \sum_{g \in \mathbb{Z}^r} e^{\pi i S[g] z}$$

is a modular form with respect to a suitable congruence subgroup. More exactly, the following is true: There exists an even natural number q with the property

$$\vartheta(S; z) \in [\Gamma[q], r/2, v_\vartheta^r] \ .$$

In this volume, we will consider in Sect. VI.6 in detail a non-trivial example of a congruence subgroup.

A Appendix to VI.5 : The Theta Group

We would like to study more closely in this Appendix an important example of a congruence subgroup, the so-called theta group.

There exist modulo 2 exactly six different integral matrices with odd determinant, namely

$$\begin{pmatrix} 1 & 0 \\ 0 & 1 \end{pmatrix}, \quad \begin{pmatrix} 0 & 1 \\ 1 & 0 \end{pmatrix}, \quad \begin{pmatrix} 1 & 1 \\ 0 & 1 \end{pmatrix}, \quad \begin{pmatrix} 1 & 0 \\ 1 & 1 \end{pmatrix}, \quad \begin{pmatrix} 0 & 1 \\ 1 & 1 \end{pmatrix}, \quad \begin{pmatrix} 1 & 1 \\ 1 & 0 \end{pmatrix}.$$

The first two matrices form a group with respect to matrix multiplication. This implies:

Remark A.1 *The set of all matrices $M \in \mathrm{SL}(2, \mathbb{Z})$ with the property*

$$M \equiv \begin{pmatrix} 1 & 0 \\ 0 & 1 \end{pmatrix} \ or \ \begin{pmatrix} 0 & 1 \\ 1 & 0 \end{pmatrix} \qquad \mathrm{mod}\ 2$$

is a subgroup of $\mathrm{SL}(2, \mathbb{Z})$.

This subgroup is also called the *theta group* Γ_ϑ. A glance at the above six matrices shows that Γ_ϑ can be defined by the condition

$$a + b + c + d \equiv 0 \quad \mathrm{mod}\ 2$$

or also by

$$ab \equiv cd \equiv 0 \quad \mathrm{mod}\ 2$$

in terms of the entries a, b, c, d of a matrix in $\mathrm{SL}(2, \mathbb{Z})$.

We now introduce a set of points $\widetilde{\mathcal{F}}_\vartheta$, which will play for the theta group the same role as the fundamental domain $\mathcal{F}$ for the full modular group:

$$\widetilde{\mathcal{F}}_\vartheta := \{ z \in \mathbb{H} ; \ |z| \geq 1 , \ |x| \leq 1 \} \ .$$

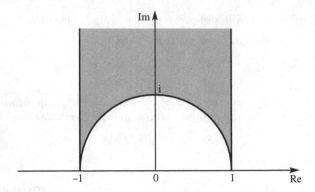

Lemma A.2 *The set $\widetilde{\mathcal{F}}_\vartheta$ is a fundamental domain for the theta group, i.e.*

$$\mathbb{H} = \bigcup_{M \in \Gamma_\vartheta} M\widetilde{\mathcal{F}}_\vartheta \ .$$

The theta group contains both matrices

$$\begin{pmatrix} 1 & 2 \\ 0 & 1 \end{pmatrix} \quad \text{and} \quad \begin{pmatrix} 0 & -1 \\ 1 & 0 \end{pmatrix} \ .$$

Let Γ_0 be the subgroup of Γ_ϑ which is generated by these two matrices.

We can show even more than claimed in the Lemma A.2:

For any point $z \in \mathbb{H}$ there exists a matrix $M \in \Gamma_0$ with the property $Mz \in \widetilde{\mathcal{F}}_\vartheta$.

The same proof as in the case of the ordinary fundamental domain $\mathcal{F}$, V.8.7 works. $\qquad\square$

We want to bring together $\mathcal{F}_\vartheta$ and the fundamental domain $\mathcal{F}$ of the full modular group. For this, let us consider the region

$$\mathcal{F}_\vartheta := \mathcal{F} \cup \begin{pmatrix} 1 & 1 \\ 0 & 1 \end{pmatrix} \mathcal{F} \cup \begin{pmatrix} 1 & 1 \\ 0 & 1 \end{pmatrix} \begin{pmatrix} 0 & -1 \\ 1 & 0 \end{pmatrix} \mathcal{F} \ .$$

The region $\begin{pmatrix} 0 & -1 \\ 1 & 0 \end{pmatrix} \mathcal{F}$ is obviously characterized by the inequalities

$$|z| \leq 1 \ , \quad |z \pm 1| \geq 1 \ .$$

We define

$$S := \begin{pmatrix} 0 & -1 \\ 1 & 0 \end{pmatrix} \ , \quad T := \begin{pmatrix} 1 & 1 \\ 0 & 1 \end{pmatrix} \ .$$

In the figure the regions $S\mathcal{F}$ and $\mathcal{F}_\vartheta$ are sketched:

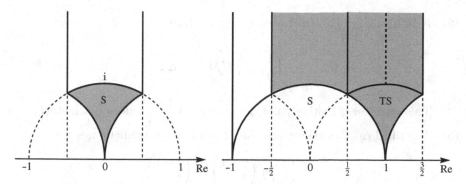

If one shifts the part of $\mathcal{F}_\vartheta$, defined by the equation "$x \geq 1$", by means of the translation $z \mapsto z - 2$ to the left, then one obtains exactly the region $\tilde{\mathcal{F}}_\vartheta$. This gives us:

Proposition A.3 *The region*

$$\mathcal{F}_\vartheta = \mathcal{F} \cup \begin{pmatrix} 1 & 1 \\ 0 & 1 \end{pmatrix} \mathcal{F} \cup \begin{pmatrix} 1 & 1 \\ 0 & 1 \end{pmatrix} \begin{pmatrix} 0 & -1 \\ 1 & 0 \end{pmatrix} \mathcal{F}$$

is a fundamental domain for the theta group,

$$\mathbb{H} = \bigcup_{M \in \Gamma_\vartheta} M \mathcal{F}_\vartheta \ .$$

Our arguments give even more, namely the fact that $\mathcal{F}_\vartheta$ is a fundamental domain for the subgroup Γ_0. We show now the equality of these groups.

Proposition A.4 *The theta group Γ_ϑ is generated by the two matrices* $\begin{pmatrix} 1 & 2 \\ 0 & 1 \end{pmatrix}$ *and* $\begin{pmatrix} 0 & -1 \\ 1 & 0 \end{pmatrix}$.

Proof. We first remark that the negative unit matrix belongs to Γ_0,

$$\begin{pmatrix} 0 & -1 \\ 1 & 0 \end{pmatrix}^2 = - \begin{pmatrix} 1 & 0 \\ 0 & 1 \end{pmatrix} \ .$$

Now let $M \in \Gamma_\vartheta$ be an arbitrary matrix. We fix some inner point a in the fundamental domain $\mathcal{F}$. By A.2, we can find a matrix $N \in \Gamma_0$ with the property

$$NM(a) \in \mathcal{F}_\vartheta \ .$$

There are now three possibilities:

(1) $NM(a) \in \mathcal{F}$. In this case we have by VI.1.3

$$M = \pm N^{-1} \ ,$$

and this gives $M \in \Gamma_0$.

(2) $NM(a) \in \begin{pmatrix} 1 & 1 \\ 0 & 1 \end{pmatrix} \mathcal{F}$. In this case we then have

$$\begin{pmatrix} 1 & -1 \\ 0 & 1 \end{pmatrix} NM = \pm E .$$

This case can definitively not occur, since $\begin{pmatrix} 1 & -1 \\ 0 & 1 \end{pmatrix}$ does *not* lie in Γ_ϑ.

(3) The third case, as the second one, leads to a contradiction, since

$$\begin{pmatrix} 1 & 1 \\ 0 & 1 \end{pmatrix} \begin{pmatrix} 0 & -1 \\ 1 & 0 \end{pmatrix} = \begin{pmatrix} 1 & -1 \\ 1 & 0 \end{pmatrix}$$

does *not* lie in Γ_ϑ.   □

Further properties of the theta group

Using the above list of all six mod 2 different integral matrices with odd determinant, considered mod 2, one can easily show the following property of the theta group:

Proposition A.5

(1) *We have the decomposition*

$$\Gamma = \Gamma_\vartheta \cup \Gamma_\vartheta \begin{pmatrix} 1 & 1 \\ 0 & 1 \end{pmatrix} \cup \Gamma_\vartheta \begin{pmatrix} 1 & -1 \\ 1 & 0 \end{pmatrix} .$$

The theta group is thus a subgroup of $\Gamma = \mathrm{SL}(2, \mathbb{Z})$ of index 3.

(2) *The principal congruence subgroup $\Gamma[2]$ of level 2, is a subgroup of index two in Γ_ϑ, i.e.*

$$\Gamma_\vartheta = \Gamma[2] \cup \Gamma[2] \cdot \begin{pmatrix} 0 & -1 \\ 1 & 0 \end{pmatrix} .$$

(3) *The conjugates of Γ_ϑ are*

(a) $\tilde{\Gamma}_\vartheta := \begin{pmatrix} 1 & 1 \\ 0 & 1 \end{pmatrix} \Gamma_\vartheta \begin{pmatrix} 1 & -1 \\ 0 & 1 \end{pmatrix}$

$$= \left\{ M \in \Gamma ; \quad M \equiv \begin{pmatrix} 1 & 0 \\ 0 & 1 \end{pmatrix} \text{ or } \begin{pmatrix} 1 & 0 \\ 1 & 1 \end{pmatrix} \quad \mathrm{mod}\ 2 \right\} .$$

(b) $\tilde{\tilde{\Gamma}}_\vartheta := \begin{pmatrix} 0 & -1 \\ 1 & 1 \end{pmatrix} \Gamma_\vartheta \begin{pmatrix} 1 & 1 \\ -1 & 0 \end{pmatrix}$

$$= \left\{ M \in \Gamma ; \quad M \equiv \begin{pmatrix} 1 & 0 \\ 0 & 1 \end{pmatrix} \text{ or } \begin{pmatrix} 1 & 1 \\ 0 & 1 \end{pmatrix} \quad \mathrm{mod}\ 2 \right\} .$$

In particular, Γ_ϑ is not a normal subgroup of Γ, since the three conjugate groups Γ_ϑ, $\tilde{\Gamma}_\vartheta$ and $\tilde{\tilde{\Gamma}}_\vartheta$ are pairwise different.

(4) *We have*

$$\Gamma_\vartheta \cap \tilde{\Gamma}_\vartheta \cap \tilde{\tilde{\Gamma}}_\vartheta = \Gamma[2] .$$

From this, we derive an interesting consequence for the congruence subgroup of level 2.

Proposition A.6 *The principal congruence subgroup $\Gamma[2]$ is generated by the three matrices*

$$\begin{pmatrix} 1 & 2 \\ 0 & 1 \end{pmatrix} , \quad \begin{pmatrix} 1 & 0 \\ 2 & 1 \end{pmatrix} , \quad and \quad -\begin{pmatrix} 1 & 0 \\ 0 & 1 \end{pmatrix} .$$

Proof. Let Γ_0 be the subgroup of $\Gamma[2]$ generated by the above three matrices. It is enough to show:

$$\Gamma_\vartheta \stackrel{!}{=} \mathcal{M} := \Gamma_0 \cup \Gamma_0 \begin{pmatrix} 0 & -1 \\ 1 & 0 \end{pmatrix} .$$

This means two different things:

(1) The generators $\begin{pmatrix} 1 & 2 \\ 0 & 1 \end{pmatrix}$ and $\begin{pmatrix} 0 & -1 \\ 1 & 0 \end{pmatrix}$ of Γ_ϑ are contained in $\mathcal{M}$.

   This is trivial.

(2) $\mathcal{M}$ is a group.

   This follows from the obvious property

$$\Gamma_0 \begin{pmatrix} 0 & -1 \\ 1 & 0 \end{pmatrix} = \begin{pmatrix} 0 & -1 \\ 1 & 0 \end{pmatrix} \Gamma_0 ,$$

   and from

$$\begin{pmatrix} 0 & -1 \\ 1 & 0 \end{pmatrix}^2 = -\begin{pmatrix} 1 & 0 \\ 0 & 1 \end{pmatrix} . \qquad \square$$

Exercises for VI.5

1.   The group $\mathrm{SL}(2, R)$ can be defined for any associative ring R with unit $1 = 1_R$. Show that for the finite commutative ring $R = \mathbb{Z}/q\mathbb{Z}$ the two matrices

$$\begin{pmatrix} 0_R & -1_R \\ 1_R & 0_R \end{pmatrix} \quad and \quad \begin{pmatrix} 1_R & 1_R \\ 0_R & 1_R \end{pmatrix}$$

   generate $\mathrm{SL}(2, R)$.

2.   The natural group homomorphism

$$\mathrm{SL}(2, \mathbb{Z}) \longrightarrow \mathrm{SL}(2, \mathbb{Z}/q\mathbb{Z})$$

   is surjective. In particular,

$$[\Gamma : \Gamma[q]] = \# \, \mathrm{SL}(2, \mathbb{Z}/q\mathbb{Z}) .$$

3. Let p be a prime number. The group $\mathrm{GL}(2, \mathbb{Z}/p\mathbb{Z})$ has $(p^2-1)(p^2-p)$ elements.

    Hint. How many first columns are possible. How often can a given column be extended to an invertible matrix?

    Deduce from this that the group $\mathrm{SL}(2, \mathbb{Z}/p\mathbb{Z})$ has $(p^2-1)p$ elements.

4. Let p be a prime, and let m be a natural number. The kernel of the natural homomorphism
$$\mathrm{GL}(2, \mathbb{Z}/p^m\mathbb{Z}) \longrightarrow \mathrm{GL}(2, \mathbb{Z}/p^{m-1}\mathbb{Z})$$
is isomorphic to the additive group of all 2×2 matrices with entries in $\mathbb{Z}/p\mathbb{Z}$. Using this, show:
$$\# \mathrm{GL}(2, \mathbb{Z}/p^m\mathbb{Z}) = p^{4m-3}(p^2-1)(p-1) ,$$
$$\# \mathrm{SL}(2, \mathbb{Z}/p^m\mathbb{Z}) = p^{3m-2}(p^2-1) .$$

5. Let q_1 and q_2 be two relatively prime natural numbers. The Chinese Remainder Theorem claims that the natural homomorphism $\mathbb{Z}/q_1q_2\mathbb{Z} \to \mathbb{Z}/q_1\mathbb{Z} \times \mathbb{Z}/q_2\mathbb{Z}$ is an isomorphism. Deduce from this that the natural homomorphism
$$\mathrm{GL}(2, \mathbb{Z}/q_1q_2\mathbb{Z}) \longrightarrow \mathrm{GL}(2, \mathbb{Z}/q_1\mathbb{Z}) \times \mathrm{GL}(2, \mathbb{Z}/q_2\mathbb{Z})$$
is an isomorphism.

6. Use Exercises 2, 4 and 5 to obtain the index formula
$$[\Gamma : \Gamma[q]] = q^3 \prod_{p|q}\left(1 - \frac{1}{p^2}\right) .$$

7. A subset $\mathcal{F}_0 \subset \mathbb{H}$ is called a fundamental domain of a congruence subgroup Γ_0, iff the following two conditions are satisfied:

    (a) There exists a subset $S = S(\mathcal{F}_0) \subset \mathcal{F}_0$ of LEBESGUE measure 0, such that $\mathcal{F}_0 \setminus S$ is open, and any two points of $\mathcal{F}_0 \setminus S$ are inequivalent with respect to Γ_0.

    (b) One has
$$\mathbb{H} = \bigcup_{M \in \Gamma_0} M\mathcal{F}_0 .$$

    Let
$$\Gamma = \bigcup_{\nu=1}^{h} \Gamma_0 M_\nu$$
be the decomposition of the full modular group into right cosets with respect to the subgroup Γ_0, and let $\mathcal{F}$ be the standard modular figure. Then
$$\mathcal{F}_0 = \bigcup_{\nu=1}^{h} M_\nu \mathcal{F}$$
is a fundamental domain of Γ_0.

8. The (invariant) volume

$$v(\mathcal{F}_0) := \int_{\mathcal{F}_0} \frac{dx\,dy}{y^2}$$

is independent of the choice of a fundamental domain $\mathcal{F}_0$ for a congruence subgroup Γ_0. We moreover have

$$v(\mathcal{F}_0) = [\Gamma : \Gamma_0] \cdot \frac{\pi}{3} \ .$$

Hint. Let T be the union of the set S with the set of all points in $\mathcal{F}_0$, which are $\mathrm{SL}(2,\mathbb{Z})$-equivalent to some point at the boundary of the standard modular figure. Then split the open set $\mathcal{F}_0 \setminus T$ into countably many disjoint fragments, using some net of squares, such that each fragment can be transformed into the inner of the fundamental domain of the modular group by a modular substitution.

9. Let Γ_0 be a normal subgroup of finite index of the full modular group. The corresponding factor group G acts on the field of modular functions $K(\Gamma_0)$ by

$$f(z) \longmapsto f^g(z) := f(Mz) , \qquad f \in K(\Gamma_0) , \ g \in G , \ M \in \Gamma \text{ represents } g \ .$$

The fixed field is

$$K(\Gamma) = K(\Gamma_0)^G \ .$$

In particular, $K(\Gamma_0)$ is algebraic over $K(\Gamma)$.

10. From Exercise 9 follows that any two functions, which are modular for some arbitrary subgroup of finite index in the modular group, are algebraically dependent.

By the way: The theorem of a primitive element shows that there exists a modular function f with the property

$$K(\Gamma_0) = \mathbb{C}(j)[f] \ .$$

One can show that the map

$$\mathbb{H}/\Gamma_0 \longrightarrow \mathbb{C} \times \overline{\mathbb{C}} \ ,$$
$$[z] \longmapsto (j(z), f(z)) \ ,$$

is injective, and that its image is an algebraic curve, more exactly, its intersection with $\mathbb{C} \times \mathbb{C}$ is an affine curve. We will prove this in the second volume, by means of the theory of RIEMANN surfaces.

11. Let q be a natural number. Show that

$$\Gamma_0[q] \quad := \left\{ M = \begin{pmatrix} a & b \\ c & d \end{pmatrix} \in \mathrm{SL}(2,\mathbb{Z}) , \quad c \equiv 0 \mod q \right\} ,$$

$$\Gamma^0[q] \quad := \left\{ M = \begin{pmatrix} a & b \\ c & d \end{pmatrix} \in \mathrm{SL}(2,\mathbb{Z}) , \quad b \equiv 0 \mod q \right\}$$

are congruence subgroups. These groups are conjugated in the full modular group.

We have:

$$\tilde{\Gamma}_\vartheta = \Gamma^0[2], \quad \tilde{\tilde{\Gamma}}_\vartheta = \Gamma_0[2] \ .$$

12. Let p be a prime number. The group $\Gamma_0[p]$ has exactly two cusp classes, which can be represented by 0, and $i\infty$.

VI.6 A Ring of Theta Functions

The **theta group**

$$\Gamma_\vartheta = \left\{ \begin{pmatrix} a & b \\ c & d \end{pmatrix} \in \mathrm{SL}(2,\mathbb{Z}) \; ; \quad a+b+c+d \text{ even} \right\}$$

is generated by the two matrices

$$\begin{pmatrix} 1 & 2 \\ 0 & 1 \end{pmatrix} \quad \text{and} \quad \begin{pmatrix} 0 & -1 \\ 1 & 0 \end{pmatrix},$$

as we already know. From the well-known formulas

$$\vartheta(z+2) = \vartheta(z) \quad \text{and} \quad \vartheta\left(-\frac{1}{z}\right) = \sqrt{\frac{z}{i}}\,\vartheta(z)$$

follows that the theta series

$$\vartheta(z) := \sum_{n=-\infty}^{\infty} \exp \pi i n^2 z$$

transforms like a modular form of weight $1/2$ with respect to some suitable multiplier system v_ϑ. We will not need any explicit formula for v_ϑ.

Using the system of representatives from A.5 for the congruence classes of Γ_ϑ in Γ one can show:

Lemma VI.6.1 *The theta group has two cusp classes, which are represented by* $i\infty$ *and* 1.

The theta series $\vartheta(z)$ has three conjugated forms. Besides ϑ, the other two are

$$\tilde{\vartheta}(z) = \sum_{n=-\infty}^{\infty} (-1)^n \exp \pi i n^2 z \;,$$

$$\tilde{\tilde{\vartheta}}(z) = \sum_{n=-\infty}^{\infty} \exp \pi i (n+1/2)^2 z \;,$$

and we have already met them in Sect. VI.4. Let us recall the theta transformation formulas, VI.4.4:

$$\vartheta(z+1) = \tilde{\vartheta}(z) \;, \quad \tilde{\vartheta}(z+1) = \vartheta(z) \;, \quad \tilde{\tilde{\vartheta}}(z+1) = e^{\pi i/4}\,\tilde{\tilde{\vartheta}}(z) \;,$$

$$\tilde{\vartheta}\left(-\frac{1}{z}\right) = \sqrt{\frac{z}{i}}\,\tilde{\tilde{\vartheta}}(z) \quad \text{and} \quad \tilde{\tilde{\vartheta}}\left(-\frac{1}{z}\right) = \sqrt{\frac{z}{i}}\,\tilde{\vartheta}(z) \;.$$

The three series are regular at $i\infty$. Because of this, ϑ is regular at both cusps of Γ_ϑ, and thus is an entire modular form of weight $1/2$. The other two conjugate forms are also (entire) modular forms of weight $1/2$ with respect to the corresponding multiplier systems, i.e

$$\vartheta \in \left[\, \varGamma_\vartheta \, , \, 1/2 \, , \, v_\vartheta \, \right] \, ,$$

$$\tilde{\vartheta} \in \left[\, \widetilde{\varGamma}_\vartheta \, , \, 1/2 \, , \, \widetilde{v}_\vartheta \, \right] \, , \quad \widetilde{\varGamma}_\vartheta = \begin{pmatrix} 1 & 1 \\ 0 & 1 \end{pmatrix} \varGamma_\vartheta \begin{pmatrix} 1 & -1 \\ 0 & 1 \end{pmatrix} \, ,$$

$$\tilde{\tilde{\vartheta}} \in \left[\, \widetilde{\tilde{\varGamma}}_\vartheta \, , \, 1/2 \, , \, \widetilde{\tilde{v}}_\vartheta \, \right] \, , \quad \widetilde{\tilde{\varGamma}}_\vartheta = \begin{pmatrix} 0 & 1 \\ -1 & 0 \end{pmatrix} \widetilde{\varGamma}_\vartheta \begin{pmatrix} 0 & -1 \\ 1 & 0 \end{pmatrix} \, .$$

The values of the three involved conjugated multiplier systems can be computed for any explicitly given matrix. This can be done most simply by representing this matrix using the generators of the full modular group, and then applying the above formulas.

The intersection of the three conjugates of the theta group is the principal congruence subgroup of level 2,

$$\varGamma_\vartheta \cap \widetilde{\varGamma}_\vartheta \cap \widetilde{\tilde{\varGamma}}_\vartheta \ = \ \varGamma[2]$$

$$:= \ \mathrm{Kernel} \left(\, \mathrm{SL}(2, \mathbb{Z}) \longrightarrow \mathrm{SL}(2, \mathbb{Z}/2\mathbb{Z}) \, \right) \, .$$

We know that the principal congruence subgroup of level 2 is generated by the three matrices

$$\begin{pmatrix} 1 & 2 \\ 0 & 1 \end{pmatrix} \, , \quad \begin{pmatrix} 1 & 0 \\ 2 & 1 \end{pmatrix} \, , \quad \begin{pmatrix} -1 & 0 \\ 0 & -1 \end{pmatrix} \, .$$

The three conjugated multiplier systems do not coincide on $\varGamma[2]$. They coincide on a smaller subgroup introduced by J.-I. IGUSA,

$$\varGamma[4, 8] := \left\{ \begin{pmatrix} a & b \\ c & d \end{pmatrix} \in \varGamma \, ; \ a \equiv d \equiv 1 \quad \mathrm{mod} \ 4 \, ; \ b \equiv c \equiv 0 \quad \mathrm{mod} \ 8 \right\} \, .$$

The group generated by $\varGamma[4, 8]$ and $-E$ (negative of the unit matrix) will be denoted by $\widetilde{\varGamma}[4, 8]$. Then

$$\widetilde{\varGamma}[4, 8] = \left\{ \begin{pmatrix} a & b \\ c & d \end{pmatrix} \in \varGamma \, ; \ a \equiv d \equiv 1 \quad \mathrm{mod} \ 2 \, ; \ b \equiv c \equiv 0 \quad \mathrm{mod} \ 8 \right\} \, .$$

Both groups define the same transformation groups.

Lemma VI.6.2 (J. Igusa) *The group $\varGamma[4, 8]$ is a nomal subgroup of the full modular group. The group*

$$\varGamma[2] \, / \, \widetilde{\varGamma}[4, 8]$$

is isomorphic to the group

$$\mathbb{Z}/4\mathbb{Z} \times \mathbb{Z}/4\mathbb{Z} \, .$$

An isomorphism is given by

$$\begin{pmatrix} 1 & 2 \\ 0 & 1 \end{pmatrix} \longleftrightarrow (1, 0) \, , \quad \begin{pmatrix} 1 & 0 \\ 2 & 1 \end{pmatrix} \longleftrightarrow (0, 1) \, .$$

Corollary. *The three multiplier systems* v_ϑ, $\tilde{v}_\vartheta$ *and* $\tilde{\tilde{v}}_\vartheta$ *coincide on the group* $\Gamma[4,8]$, *where their possible values are only* ± 1. *Even powers of them are in particular trivial.*

Proof of VI.6.2. Any element of $\Gamma[2]$ can be written in the form

$$M = \pm \begin{pmatrix} 1 & 2 \\ 0 & 1 \end{pmatrix}^x \begin{pmatrix} 1 & 0 \\ 2 & 1 \end{pmatrix}^y K \, ,$$

where K is a suitable element in the commutator group of $\Gamma[2]$. As one can quickly check by using generators, K lies in $\Gamma[4,8]$. This implies that M is in $\Gamma[4,8]$, iff the sign $\pm$ in the above representation is a plus, and both x and y are divisible by 4.

We can now consider the homomorphism

$$\mathbb{Z} \times \mathbb{Z} \longrightarrow \Gamma[2] \,/\, \widetilde{\Gamma}[4,8] \, ,$$

$$(a,b) \longmapsto \begin{pmatrix} 1 & 2 \\ 0 & 1 \end{pmatrix}^a \begin{pmatrix} 1 & 0 \\ 2 & 1 \end{pmatrix}^b \, .$$

One can easily check that its kernel is exactly $4\mathbb{Z} \times 4\mathbb{Z}$.

For the proof of the Corollary note that any two of the three multiplier systems differ by a character. But characters are trivial on commutators. It remains to check that the three multiplier systems coincide on the elements

$$\begin{pmatrix} 1 & 2 \\ 0 & 1 \end{pmatrix}^{4x} \begin{pmatrix} 1 & 0 \\ 2 & 1 \end{pmatrix}^{4y} \, . \qquad \qquad \square$$

Theorem VI.6.3 *The vector space* $\left[\Gamma[4,8], r/2, v_\vartheta^r\right]$ *is generated by the monomials*

$$\vartheta^\alpha \, \tilde{\vartheta}^\beta \, \tilde{\tilde{\vartheta}}^\gamma \, , \qquad \alpha + \beta + \gamma = r \, , \qquad \alpha, \, \beta, \, \gamma \in \mathbb{N}_0 \, .$$

One has the Jacobi theta relation

$$\boxed{\vartheta^4 = \tilde{\vartheta}^4 + \tilde{\tilde{\vartheta}}^4 \, .}$$

Because of this, one can restrict to monomials satisfying the supplementary condition $\alpha < 4$ *to generate all modular forms. This restricted family of monomials is then linearly independent, so it is a basis. In particular,*

$$\dim_{\mathbb{C}} \left[\Gamma[4,8], r/2, v_\vartheta^r\right] = \begin{cases} 3 \, , & \text{if } r = 1 \, ; \\ 6 \, , & \text{if } r = 2 \, ; \\ 10 \, , & \text{if } r = 3 \, ; \\ 4r - 2 \, , & \text{if } r \geq 4 \, . \end{cases}$$

Using ring theoretical structures, Theorem VI.6.3 can be reformulated in a more elegant way. For this we consider the graded ring of modular forms

$$\mathcal{A}(\Gamma[4,8]) := \bigoplus_{r \in \mathbb{Z}} \left[\, \Gamma[4,8]\, , \; r/2\, , \; v_\vartheta^r \,\right] \; .$$

Then Theorem VI.6.3 states:

Structure Theorem VI.6.3′ *One has:*

$$\mathcal{A}(\Gamma[4,8]) = \mathbb{C}\left[\, \vartheta, \tilde\vartheta, \tilde{\tilde\vartheta} \,\right] \; .$$

The defining relation of the ring in the R.H.S. is the Jacobi theta relation

$$\vartheta^4 = \tilde\vartheta^4 + \tilde{\tilde\vartheta}^4 \; .$$

If $\mathbb{C}[X, Y, Z]$ is the ring of polynomials in three variables X, Y, Z, and if the ring morphism

$$\mathbb{C}[X, Y, Z] \longrightarrow \mathcal{A}(\Gamma[4,8])\, , \quad X \mapsto \vartheta\, , \quad Y \mapsto \tilde\vartheta\, , \quad Z \mapsto \tilde{\tilde\vartheta}\, ,$$

is given by the above substitutions of X, Y, Z, then this morphism is surjective, and its kernel is generated by $X^4 - Y^4 - Z^4$.

Theorem VI.6.3 is a special case of much deeper results of J. IGUSA, [Ig1, Ig2]. We will give an elementary proof, which can be treated in an introductory seminar on modular forms.

For the proof, we use the fact that the finite *commutative* group

$$G \; = \; \Gamma[2] \,/\, \widetilde\Gamma[4,8]$$

acts on the vector space $\left[\, \Gamma[4,8]\, , \; r/2\, , \; v_\vartheta^r \,\right]$ via

$$f(z) \longmapsto f^M(z) := v_\vartheta^{-r}(M)\, (cz + d)^{-r/2}\, f(Mz) \; .$$

Let us briefly explain what this means.

Let G be a group, and let V be a vector space over the field $\mathbb{C}$ of complex numbers. We say that G *acts (operates) linearly* on V, iff there is given a map

$$V \times G \longrightarrow V\, ,$$
$$(f, a) \mapsto f^a\, ,$$

having the following properties:

(1) $f^e = f$, where e is the neutral element of G,

(2) $(f^a)^b = f^{ab}$ for all $f \in V$, and all $a, b \in G$,

(3) $(f + g)^a = f^a + g^a$, $(\lambda f)^a = \lambda f^a$ for all $f, g \in V$, and all $a \in G$, $\lambda \in \mathbb{C}$.

Let now

$$\chi : G \longrightarrow \mathbb{C}^{\bullet}$$

be a character, i.e. a homomorphism from G into the multiplicative group of all non-zero complex numbers. We define a subspace V^χ of V. It consists of all

$$f \in V \quad \text{with} \quad f^a = \chi(a)\, f \text{ for all } a \in G .$$

Because of (3), V^χ is a linear subspace of V.

Remark VI.6.4 *Let G be a finite* **commutative** *group, which acts linearly on the $\mathbb{C}$-vector space V. Then*

$$V = \bigoplus_{\chi \in \widehat{G}} V^\chi ,$$

where $\widehat{G}$ is the group of all characters of G.

Proof. Let $f \in V$. The element

$$f^\chi := \sum_{a \in G} \chi(a)^{-1}\, f^a$$

lies obviously in V^χ, since we can check its transformation property

$$(f^\chi)^b = \sum \chi(a)^{-1} f^{ab} = \sum \chi(b)\chi(b)^{-1}\chi(a)^{-1} f^{ab} = \sum \chi(b)\chi(ab)^{-1} f^{ab}$$
$$= \chi(b)\, f^\chi .$$

Claim. We have

$$f = \frac{1}{\#G} \sum_{\chi \in \widehat{G}} f^\chi ,$$

where χ runs through all characters of G.

The proof of the Claim follows directly from the formula

$$\sum_{\chi \in \widehat{G}} \chi(a) = \begin{cases} 0 , & \text{if } a \neq e , \\ \#G , & \text{if } a = e . \end{cases}$$

This well-known (more general) formula follows from the structure theorem for finite ABELian groups.

If

$$f = \sum_{\chi} h^\chi , \qquad h^\chi \in V^\chi ,$$

is some decomposition of f into eigenforms, then we have from the above character relations

$$h^\chi = \frac{1}{\#G} \sum_{a \in G} \chi(a)^{-1}\, f^a ,$$

so the decomposition is unique. □

We will use Remark VI.6.4 only in the case

$$G = \mathbb{Z}/4\mathbb{Z} \times \mathbb{Z}/4\mathbb{Z} \ .$$

Here, the formula is trivial, since one can explicitly write down the characters: Since any element of G has order 1, 2 or 4, the characters of $G = \mathbb{Z}/4\mathbb{Z} \times \mathbb{Z}/4\mathbb{Z}$ can take only the values 1, -1, i or $-$i. Obviously, these values can be prescribed arbitrarily for the generators $(1,0)$ and $(0,1)$ of G, and we obtain all 16 characters of G.

Then we can split $[\Gamma[4,8], r/2, v_\vartheta^r]$ with respect to these 16 characters, i.e. we have

$$[\,\Gamma[4,8]\,,\ r/2\,,\ v_\vartheta^r\,] = \bigoplus_v [\,\Gamma[2]\,,\ r/2\,,\ vv_\vartheta^r\,] \ .$$

Here v runs through 16 characters of $\Gamma[2]$ with the property

$$v(\pm M) = 1 \text{ for all } M \in \Gamma[4,8] \ .$$

These characters are determined by their values on the generating matrices

$$\begin{pmatrix} 1 & 2 \\ 0 & 1 \end{pmatrix} \quad \text{and} \quad \begin{pmatrix} 1 & 0 \\ 2 & 1 \end{pmatrix} ,$$

and these values are arbitrary fourth roots of unity. We encode them by pairs $[a,b]$ of numbers with

$$a = v \begin{pmatrix} 1 & 2 \\ 0 & 1 \end{pmatrix} \quad \text{and} \quad b = v \begin{pmatrix} 1 & 0 \\ 2 & 1 \end{pmatrix} \ .$$

Two multiplier systems of the same weight $r/2$ differ only by a character. We apply this information to the three fundamental multiplier systems, so $\widetilde{v}_\vartheta / v_\vartheta$ and $\widetilde{\widetilde{v}}_\vartheta / v_\vartheta$ are characters. Simple computations give:

$$\widetilde{v}_\vartheta / v_\vartheta = [1, -\mathrm{i}] \ , \quad \widetilde{\widetilde{v}}_\vartheta / v_\vartheta = [\mathrm{i}, 1] \ .$$

Next, we use the fact that $\Gamma[2]$ is a normal subgroup in Γ. This means that for $N \in \Gamma$ the map $f \mapsto f|N^{-1}$ induces an isomorphism

$$[\,\Gamma[2]\,,\ r/2\,,\ vv_\vartheta^r\,] \longrightarrow [\,\Gamma[2]\,,\ r/2\,,\ v^{(N,r)}v_\vartheta^r\,] \ .$$

Here, for any fixed $r \in \mathbb{Z}$, and $N \in \Gamma$, the assignment $v \mapsto v^{(N,r)}$ is in fact a permutation of all 16 characters. This permutation depends only on r mod 4. We then obtain four ($r = 0, 1, 2, 3$) representations of the modular group

$$\mathrm{SL}(2, \mathbb{Z}/2\mathbb{Z}) \quad (\cong S_3)$$

with values in the permutation group of the 16 characters. It is easy to compute explicitly these representations. Using the formula

$$v^{(N,r)} = v^N \left(\frac{v_\vartheta^N}{v_\vartheta} \right)^r , \qquad v_\vartheta^N(M) := v_\vartheta(NMN^{-1}) ,$$

one can compute $v^{(N,r)}$ for concrete values of N. A short computation, which is left to the reader, gives for instance:

Lemma VI.6.5

(1) *If* $N = \begin{pmatrix} 1 & 1 \\ 0 & 1 \end{pmatrix}$, *then*

$$[a,b]^{(N,r)} = [a, (-\mathrm{i})^r a b^{-1}] .$$

(2) *If* $N = \begin{pmatrix} -1 & 0 \\ 1 & -1 \end{pmatrix}$, *then*

$$[a,b]^{(N,r)} = [\mathrm{i}^r a^{-1} b, b] .$$

The above two matrices generate $\mathrm{SL}(2,\mathbb{Z})$, since we can write

$$\begin{pmatrix} 0 & -1 \\ 1 & 0 \end{pmatrix} = \begin{pmatrix} 1 & 1 \\ 0 & 1 \end{pmatrix} \begin{pmatrix} -1 & 0 \\ 1 & -1 \end{pmatrix} \begin{pmatrix} 1 & 1 \\ 0 & 1 \end{pmatrix} .$$

Lemma VI.6.6 *The three basis theta series* ϑ, $\tilde{\vartheta}$ *and* $\tilde{\tilde{\vartheta}}$ *have no zeros in the upper half-plane.*

Proof. The eighth power of their product is a constant times the discriminant. $\square$

Another proof can be given using the knowledge of the zeros of the JACOBI theta function $\vartheta(z,w)$ as a function of w. They are exactly the points equivalent to $\frac{z+1}{2}$ with respect to the lattice $\mathbb{Z} + z\mathbb{Z}$. A third, direct proof will we given in VII.1.

Lemma VI.6.7 *The group* $\Gamma[2]$ *has three cusp classes, which can be represented by* $\mathrm{i}\infty$, 0 *and* 1.

The proof, which follows the lines of VI.6.1, can be skipped. $\square$

The theta series $\tilde{\vartheta}$ has a zero of order 1 at $\mathrm{i}\infty$, where the order is measured by using the parameter $q := e^{\pi \mathrm{i} z/4}$. Since any modular form in $[\Gamma[4,8], r/2, v_\vartheta^r]$ has period 8, it allows a q-expansion and we get:

Lemma VI.6.8 *The map* $f \mapsto f \cdot \tilde{\vartheta}$ *defines an isomorphism between* $\left[\Gamma[2] , r/2 , kvv_\vartheta^r \right]$ *and the subspace of all forms vanishing at* $\mathrm{i}\infty$ *of the space*

$$\left[\Gamma[2] , (r+1)/2 , v^* v_\vartheta^{r+1} \right] , \qquad v^* = v \frac{\tilde{\tilde{v}}_\vartheta}{v_\vartheta} .$$

Forced zeros

We assume

$$v \begin{pmatrix} 1 & 2 \\ 0 & 1 \end{pmatrix} \neq 1 \ .$$

Then all forms in $\left[\, \Gamma[2] \, , \, r/2 \, , \, vv_\vartheta^r \, \right]$ are forced to vanish at the cusp i∞, as we see from the equation

$$f(z+2) = v \begin{pmatrix} 1 & 2 \\ 0 & 1 \end{pmatrix} f(z)$$

by passing to the limit $y \to \infty$.

Definition VI.6.9 *We will use the following terminology:*

(1) *A form f in*

$$\left[\, \Gamma[2] \, , \, r/2 \, , \, vv_\vartheta^r \, \right]$$

*has a **forced zero** at* i∞, *iff*

$$v \begin{pmatrix} 1 & 2 \\ 0 & 1 \end{pmatrix} \neq 1 \ .$$

(2) *Let $N \in \mathrm{SL}(2,\mathbb{Z})$ be a modular matrix. The form f has a forced zero at the cusp $N^{-1}(\mathrm{i}\infty)$, iff the transformed form $f|N^{-1}$ has a forced zero at* i∞.

Remark VI.6.10 *If the form $f \in \left[\, \Gamma[2] \, , \, r/2 \, , \, v_\vartheta^r \, \right]$ has a forced zero at a cusp, then one of the three modular forms*

$$f/\vartheta \, , \quad f/\tilde\vartheta \, , \quad f/\tilde{\tilde\vartheta}$$

is an entire modular form (regular also at the cusps).

Lemma VI.6.5 implies:

Proposition VI.6.11 *Only in the case $r \equiv 0 \mod 4$ and $v = 1$ the forms in $\left[\, \Gamma[2] \, , \, r/2 \, , \, vv_\vartheta^r \, \right]$ do not have forced zeros at any cusp.*

We now prove by induction on r that the space $\left[\, \Gamma[2] \, , \, r/2 \, , \, vv_\vartheta^r \, \right]$ is generated by the monomials $\vartheta^\alpha \, \tilde\vartheta^\beta \, \tilde{\tilde\vartheta}^\gamma$.

Let us pick a form f in this space. If it has a forced zero, then we can divide by one of the basis forms to reduce it to the case of a space of lower weight. If it has no forced zero, then r is divisible by 4, and v is trivial. In this case, ϑ^r also belongs to the same space as the form f. Then the difference of f and a suitable scalar multiple of ϑ^r vanishes at the cusp i∞. Then we divide by $\tilde{\tilde\vartheta}$ and apply induction.

A slight refinenement of this argumentation also gives the defining relations. The difference $\vartheta^4 - \tilde{\vartheta}^4$ obviously vanishes of order at least four at i∞, so it can be divided by $\tilde{\tilde{\vartheta}}^4$. The quotient is a modular form of weight 0, hence constant. In this way, one can show the JACOBI theta relation

$$\vartheta^4 = \tilde{\vartheta}^4 + \tilde{\tilde{\vartheta}}^4 \ .$$

Any modular form in $\left[\Gamma[4,8] \ , \ r/2 \ , \ v_\vartheta^r \ \right]$ is thus a $\mathbb{C}$-linear combination of monomials

$$\vartheta^\alpha \tilde{\vartheta}^\beta \tilde{\tilde{\vartheta}}^\gamma \ , \quad \alpha + \beta + \gamma = r \ , \quad 0 \le \alpha \le 3 \ .$$

The number of these monomials is $\begin{cases} 3 \ , & \text{if } r = 1 \ , \\ 4r - 2 \ , & \text{if } r \ge 2 \ . \end{cases}$

On the other side, the same inductive proof also gives the dimensions of all 16 constituents of $\left[\ \Gamma[4,8] \ , \ r/2 \ , \ v_\vartheta^r \ \right]$. Summing them, we obtain exactly the number of the above monomials. Hence, the set of these monomials gives basis. The details are left to the reader.

Exercises for VI.6

1. Show that the set $\Gamma[q, 2q]$ of all matrices

    $$\begin{pmatrix} a & b \\ c & d \end{pmatrix} \in \Gamma[q] \ , \quad \frac{ab}{q} \equiv \frac{cd}{q} \equiv 0 \mod 2 \ ,$$

    is a congruence subgroup for any natural number q.

2. The group $\mathrm{SL}(2, \mathbb{Z}/2\mathbb{Z})$ and the symmetric group S_3 both have six elements. Because any two non-commutative groups with six elements are isomorphic, the groups $\mathrm{SL}(2, \mathbb{Z}/2\mathbb{Z})$ and S_3 must be isomorphic. Give an explicit isomorphism between them.

    Hint. There is a canonical action of $\mathrm{SL}(2, \mathbb{Z})$ on the three basis theta series.

3. There exists a congruence subgroup of index 2 in the full elliptic modular group.

4. Determine all congruence subgroups of level 2, i.e. all subgroups Γ with $\Gamma[2] \subset \Gamma \subset \Gamma[1]$. In each case find a system of representatives for the left cosets of $\Gamma[1]$ modulo Γ and of Γ modulo $\Gamma[2]$.

5. A monomial $\vartheta^\alpha \tilde{\vartheta}^\beta \tilde{\tilde{\vartheta}}^\gamma$, $\alpha + \beta + \gamma = r$, is a modular form with respect to the theta group and the multiplier system v_ϑ^r, iff $\beta \equiv \gamma \equiv 0 \mod 8$. Show that these monomials define a basis for $\left[\ \Gamma_\vartheta \ , \ r/2 \ , \ v_\vartheta^r \ \right]$. In particular, the graded ring

    $$\mathcal{A}(\Gamma_\vartheta) = \bigoplus_{r \in \mathbb{Z}} \left[\ \Gamma_\vartheta \ , \ r/2 \ , \ v_\vartheta^r \ \right]$$

    is a polynomial ring generated by the two algebraically independent modular forms

    $$\vartheta \ , \quad (\tilde{\vartheta} \, \tilde{\tilde{\vartheta}})^8 \ .$$

    This implies:

    $$\dim_{\mathbb{C}} \left[\ \Gamma_\vartheta \ , \ r/2 \ , \ v_\vartheta^r \ \right] = 1 + \left[\frac{r}{8} \right] \ .$$

6.   Express the EISENSTEIN series G_4 and G_6 as polynomials in ϑ^4, $\tilde{\vartheta}^4$ and $\tilde{\tilde{\vartheta}}^4$.

   Hint. The searched polynomials are homogenous of degrees 2 and respectively 3. There are not too many possibilities, if one also recalls the transformation behavior of the three theta series with respect to the generators of the modular group.

7.   In the special case of the group $\Gamma[4,8]$, our instruments are good enough to solve the following (not quite simple) exercise.

   The map
   $$\mathbb{H} \ / \ \Gamma[4,8] \longrightarrow \mathbb{C} \times \mathbb{C} \ ,$$
   $$[z] \longmapsto \left(\frac{\tilde{\tilde{\vartheta}}(z)}{\vartheta(z)} \ , \ \frac{\tilde{\vartheta}(z)}{\vartheta(z)} \right) \ ,$$

   is injective. Its image is contained in the affine curve given by the equation $X^4 + Y^4 = 1$, because of the JACOBI theta relation. The complement of the image consists of exactly 8 points, which are defined by $XY = 0$.

VII

Analytic Number Theory

Analytic number theory contains one of the most beautiful applications of complex analysis. In the following sections we will treat some of the distinguished pearls of this fascinating subject.

We have already seen in Sect. VI.4 that quadratic forms or the corresponding lattices can serve to construct modular forms. The FOURIER coefficients of the theta series associated to quadratic forms or to lattices have number theoretical importance. They appear as representation numbers for quadratic forms, and respectively as numbers of lattice points of a given norm. Due to the general structure theorems for modular forms, one can bring together theta series and EISENSTEIN series. We will compute the FOURIER coefficients of the EISENSTEIN series, obtaining thus number theoretical applications. In particular, in Sect. VII.1 we will find the number of representations of a natural number as a sum of four or eight squares by purely function theoretical means.

Starting with the second section, we will be concerned with DIRICHLET series, including the RIEMANN ζ-function. There is a strong connection between modular forms and DIRICHLET series (Sect. VII.3). We prove HECKE's Theorem, claiming a one-to-one correspondence between DIRICHLET series with a functional equation of special type and FOURIER series with special transformation property under the substitution $z \mapsto -1/z$ and with certain asymptotic growth conditions. This correspondence will be obtained by means of the MELLIN transform of the Γ-function. As an application, we obtain in particular the analytic continuation of the ζ-function into the plane, and also its functional equation.

Sections VII.4, VII.5, VII.6 contain a proof of the *Prime Number Theorem* with a weak form for the error term. We were trying to prove the Prime Number Theorem using as few instruments as possible. For this reason, we have once more collected some facts about the ζ-function, which have to be used, with rather simple proofs. Indeed, less than the content of Sect. VII.3 is necessary to proceed. For instance, for the purpose of proving the Prime Number Theorem there is no need of the analytic continuation of the ζ-function into the whole plane, or of the functional equation. It is sufficient to have the continuation slightly beyond the vertical line Re $s = 1$, and there are very simple arguments for this fact. The functional equation would be

E. Freitag and R. Busam, *Complex Analysis*,   381
DOI: 10.1007/978-3-540-93983-2_VII, © Springer-Verlag Berlin Heidelberg 2009

needed for a more precise estimate of the error term, but we do not want to go so far. Instead, we refer to the specialized literature [Lan], [Pr], [Ed].

VII.1 Sums of Four and Eight Squares

Let k be a natural number. We are interested in determining how many times one can represent a natural number n as a sum of k squares of integers:

$$A_k(n) := \#\{ \; x = (x_1,\dots,x_k) \in \mathbb{Z}^k \; ; \quad x_1^2 + \cdots + x_k^2 = n \; \} \; .$$

We will determine these representation numbers in the special cases $k = 4$ and $k = 8$, namely

$$A_4(n) = 8 \sum_{\substack{4 \nmid d \,,\, d \mid n \\ 1 \le d \le n}} d \; ,$$

and

$$A_8(n) = 16 \sum_{\substack{d \mid n \\ 1 \le d \le n}} (-1)^{n-d} d^3 \; .$$

As we mentioned already, one has

$$\left(\sum_{m=-\infty}^{\infty} q^{m^2} \right)^k = \sum_{n=0}^{\infty} A_k(n) \, q^n \; .$$

So we will first characterize the function

$$\left(\sum_{m=-\infty}^{\infty} q^{m^2} \right)^k$$

(it converges for $|q| < 1$) by function theoretical properties, then we will use this characterization to express it for $k = 4$ and $k = 8$ in terms of EISENSTEIN series. The above formulas for the representation numbers are direct consequences of analytic identities. The case $k = 4$ is more intricate than the case $k = 8$, since for $k = 4$ the involved EISENSTEIN series (of weight 2) does not converge absolutely.

The number theoretical identities will appear as identities between modular forms, more exactly between theta series and EISENSTEIN series. We will deduce the needed identities by as simple methods as possible, in particular, we will avoid the short cut through the relatively complicated structure theorem VI.6.3.

The Fourier series of the Eisenstein series

Recall the partial fraction series decomposition of the cotangent and the negative its derivative

$$\pi \cot \pi z = \frac{1}{z} + \sum_{n=1}^{\infty} \left[\frac{1}{z+n} + \frac{1}{z-n} \right]$$

$$\frac{\pi^2}{(\sin \pi z)^2} = \sum_{n=-\infty}^{\infty} \frac{1}{(z+n)^2} \ .$$

Both series converge normally in $\mathbb{C} \setminus \mathbb{Z}$. They represent analytic functions in the upper half-plane, and are 1-periodic. Thus they admit FOURIER expansions.

Lemma VII.1.1 *Setting* $q = e^{2\pi i z}$, *Im* $z > 0$, *we have*

$$\sum_{n=-\infty}^{\infty} \frac{1}{(z+n)^2} = (2\pi i)^2 \sum_{n=1}^{\infty} n q^n \ .$$

Proof. One has

$$\pi \cot \pi z = \pi \frac{\cos \pi z}{\sin \pi z} = \pi i \frac{q+1}{q-1} = \pi i - \frac{2\pi i}{1-q} = \pi i - 2\pi i \sum_{n=0}^{\infty} q^n \ .$$

Differentiating with respect to z, we obtain

$$\frac{\pi^2}{(\sin \pi z)^2} = (2\pi i)^2 \sum_{n=1}^{\infty} n q^n \ ,$$

and the claim follows. $\square$

Taking repeated derivatives with respect to the variable z, we obtain:

Corollary VII.1.2 *For any natural number* $k \geq 2$ *we have*

$$(-1)^k \sum_{n=-\infty}^{\infty} \frac{1}{(z+n)^k} = \frac{1}{(k-1)!} (2\pi i)^k \sum_{n=1}^{\infty} n^{k-1} q^n \ .$$

We now rewrite the EISENSTEIN series

$$G_k(z) = \sum_{(c,d) \neq (0,0)} \frac{1}{(cz+d)^k} \qquad (k \geq 4, \ k \equiv 0 \mod 2)$$

as

$$G_k(z) = 2\zeta(k) + 2 \sum_{c=1}^{\infty} \left\{ \sum_{d=-\infty}^{\infty} \frac{1}{(cz+d)^k} \right\} \ .$$

By VII.1.2, we have (after replacing z by cz, and n by d)

$$G_k(z) = 2\zeta(k) + \frac{2(2\pi i)^k}{(k-1)!} \sum_{c=1}^{\infty} \sum_{d=1}^{\infty} d^{k-1} \, q^{cd} \ .$$

We now claim, that the series

$$\sum_{c=1}^{\infty} \sum_{d=1}^{\infty} d^{k-1} \, q^{cd} \qquad (|q| < 1)$$

converges normally in $\mathbb{H}$ for $k \geq 2$, inclusively $k = 2$. First, we rearrange the series, such that all terms with the same q-exponent cd are grouped together. We obtain the series

$$\sum_{n=1}^{\infty} \left\{ \sum_{\substack{d \mid n \\ 1 \leq d \leq n}} d^{k-1} \right\} q^n \, ,$$

which converges for $|q| < 1$ due to the trivial estimate

$$\sum_{\substack{d \mid n \\ 1 \leq d \leq n}} d^{k-1} \leq n \cdot n^{k-1} = n^k \, .$$

The above rearrangement can also be done with $|q|$ instead of q. This shows the claimed normal convergence.

Conversely, the same rearrangements show, that the series

$$G_k(z) := \sum_{c=-\infty}^{\infty} \left\{ \sum_{\substack{d=-\infty \\ d \neq 0, \text{ if } c=0}}^{\infty} (cz + d)^{-k} \right\}$$

converge for all $k \geq 2$. Notice, that the case $k = 2$ is also included. We thus can define an EISENSTEIN series G_2 of weight 2, but the brackets are necessary! This series is of course *not* a modular form, since any modular form of weight 2 vanishes (by Proposition VI.2.6). In the following, we will study G_2 in detail.

Notation. $\sigma_k(n) := \sum_{\substack{d \mid n \\ 1 \leq d \leq n}} d^k$ for all $k \in \mathbb{N}_0$ and $n \in \mathbb{N}$.

Proposition VII.1.3 (The Fourier series of the Eisenstein series) *For any even $k \in \mathbb{N}$ we have:*

$$G_k(z) := \sum_{c=-\infty}^{\infty} \left\{ \sum_{\substack{d=-\infty \\ d \neq 0, \text{ if } c=0}}^{\infty} (cz + d)^{-k} \right\}$$

$$= 2\zeta(k) + \frac{2 \cdot (2\pi \mathrm{i})^k}{(k-1)!} \sum_{n=1}^{\infty} \sigma_{k-1}(n) \, q^n \, .$$

All involved series are normally convergent in $\mathbb{H}$.

The Eisenstein series G_2

Since the series

$$\sum_{(c,d)\neq(0,0)} |cz+d|^{-2}$$

does *not* converge, we must be very cautious when using rearrangements of G_2. Such rearrangements have been used for the proof of

$$G_k\left(-\frac{1}{z}\right) = z^k G_k(z) , \qquad k > 2 .$$

The above formula is false for $k = 2$, and we will need a correction term! Doing this, we are working with an interesting convergent series, which is not absolutely convergent. All involved rearrangements must be done carefully!

We have

$$G_2\left(-\frac{1}{z}\right) = \sum_{c=-\infty}^{\infty}\left\{ \sum_{\substack{d=-\infty \\ d\neq 0 \text{ if } c=0}}^{\infty} \left(\frac{-c}{z}+d\right)^{-2} \right\}$$

$$= z^2 \sum_{c=-\infty}^{\infty}\left\{ \sum_{\substack{d=-\infty \\ d\neq 0 \text{ if } c=0}}^{\infty} (-c+dz)^{-2} \right\} .$$

Now, we can substitute in the inner sum the index d by $-d$, and obtain

$$G_2\left(-\frac{1}{z}\right) = z^2 \sum_{c=-\infty}^{\infty}\left\{ \sum_{\substack{d=-\infty \\ d\neq 0 \text{ if } c=0}}^{\infty} (dz+c)^{-2} \right\} .$$

We now exchange the symbols c and d to obtain

$$G_2\left(-\frac{1}{z}\right) = z^2\, G_2^*(z)$$

where

$$G_2^*(z) := \sum_{d=-\infty}^{\infty}\left\{ \sum_{\substack{c=-\infty \\ c\neq 0 \text{ if } d=0}}^{\infty} (cz+d)^{-2} \right\} .$$

This series is formally obtained from $G_2(z)$, by exchanging the order of summation. The series G_2 and G_2^* are different (there is no absolute convergence to allow rearrangements), and indeed we have:

Proposition VII.1.4

$$G_2^*(z) = G_2(z) - \frac{2\pi i}{z} \ .$$

Consequence.

$$\boxed{G_2\left(-\frac{1}{z}\right) = z^2 G_2(z) - 2\pi i z.}$$

The basic idea for the following delicate proof goes back to G. Eisenstein, [Eis], see also [Hu1, Hu2] or [Se], pp. 95/96. Following this idea, we introduce the series

$$H(z) := \sum_{c=-\infty}^{\infty} \left\{ \sum_{\substack{d=-\infty \\ c^2+d(d-1)\neq 0}}^{\infty} \frac{1}{(cz+d)(cz+d-1)} \right\}, \quad \text{and}$$

$$H^*(z) = \sum_{d=-\infty}^{\infty} \left\{ \sum_{\substack{c=-\infty \\ c\neq 0 \ \text{if} \ d\in\{0,1\}}}^{\infty} \frac{1}{(cz+d)(cz+d-1)} \right\}.$$

Then we have:

$$H(z) - G_2(z) = \sum_{c=-\infty}^{\infty} \left\{ \sum_{\substack{d=-\infty \\ d\neq 0 \ \text{and} \ d\neq 1 \ , \ \text{if} \ c=0}}^{\infty} \frac{1}{(cz+d)^2(cz+d-1)} \right\} - 1 \ .$$

The expressions

$$\frac{1}{(cz+d)^2(cz+d-1)} \quad \text{and} \quad \frac{1}{(cz+d)^3}$$

are not very different, since for any fixed z one can find an $\epsilon > 0$, such that

$$\frac{\epsilon}{|cz+d|^2|cz+d-1|} \leq \frac{1}{|cz+d|^3} \quad \text{or equivalently} \quad \epsilon \leq \left| 1 - \frac{1}{cz+d} \right| \ .$$

The series $\sum_{(c,d)\neq(0,0)} |cz+d|^{-3}$ converges, as we already know. This implies first the convergence of $H(z)$. In the formula for the difference $H(z) - G_2(z)$ we can exchange c and d, so we obtain

$$H(z) - G_2(z) = H^*(z) - G_2^*(z)$$

or:

Lemma VII.1.5 *We have*

$$G_2(z) - G_2^*(z) = H(z) - H^*(z) \ .$$

We will now separately sum up the series $H(z)$ and $H^*(z)$:

Lemma VII.1.6

$$\text{(a)} \qquad H\,(z) = 2\,, \qquad\qquad and$$
$$\text{(b)} \qquad H^*(z) = 2 - 2\pi i/z \,.$$

For the summation of both series we use the formula

$$\frac{1}{(cz+d)(cz+d-1)} = \frac{1}{cz+d-1} - \frac{1}{cz+d} \,,$$

together with the repeated use of the following simple principle, the so-called "telescopic trick": Let $a_1, a_2, \dots$ be a convergent sequence of complex numbers. Then

$$\sum_{n=1}^{\infty}(a_n - a_{n+1}) = a_1 - \lim_{n\to\infty} a_n \quad \text{(Telescopic series)} \,.$$

Using this principle, we immediately have

$$\sum_{\substack{d=-\infty \\ c^2+d(d-1)\neq 0}}^{\infty} \left(\frac{1}{cz+d-1} - \frac{1}{cz+d}\right) = \begin{cases} 0\,, & \text{if } c \neq 0\,, \\ 2\,, & \text{if } c = 0\,, \end{cases}$$

and from this, of course, $H(z) = 2$.

Somewhat more laboriously, we now sum up $H^*(z)$. We have

$$H^*(z) = \sum_{d=-\infty}^{\infty} \left\{ \sum_{\substack{c=-\infty \\ c\neq 0,\text{ if } d\in\{0,1\}}}^{\infty} \left[\frac{1}{cz+d-1} - \frac{1}{cz+d}\right] \right\}$$

$$= \lim_{N\to\infty} \sum_{d=-N+1}^{N} \left\{ \sum_{\substack{c=-\infty \\ c\neq 0,\text{ if } d\in\{0,1\}}}^{\infty} \left[\frac{1}{cz+d-1} - \frac{1}{cz+d}\right] \right\}$$

$$= \lim_{N\to\infty} \left\{ \begin{aligned} &\sum_{d=-N+1}^{-1}\sum_{c=-\infty}^{\infty} \left[\frac{1}{cz+d-1} - \frac{1}{cz+d}\right] \\ &+ \sum_{d=2}^{N}\sum_{c=-\infty}^{\infty} \left[\frac{1}{cz+d-1} - \frac{1}{cz+d}\right] \\ &+ \sum_{\substack{c=-\infty \\ c\neq 0}}^{\infty} \left[\frac{1}{cz-1} - \frac{1}{cz}\right] + \sum_{\substack{c=-\infty \\ c\neq 0}}^{\infty} \left[\frac{1}{cz} - \frac{1}{cz+1}\right] \end{aligned} \right\}$$

$$= \lim_{N\to\infty} \sum_{c=-\infty,\, c\neq 0}^{\infty} \left[\frac{1}{cz-N} - \frac{1}{cz+N}\right] + 2 \,.$$

The series

$$\sum_{c=-\infty,\, c\neq 0}^{\infty} \left[\frac{1}{cz-N} - \frac{1}{cz+N} \right]$$

can be related to the partial fraction series decomposition of the cotangent by a simple step,

$$\sum_{\substack{c=-\infty \\ c\neq 0}}^{\infty} \left[\frac{1}{cz-N} - \frac{1}{cz+N} \right] = \frac{2}{z} \cdot \sum_{c=1}^{\infty} \left[\frac{1}{c-N/z} - \frac{1}{c+N/z} \right]$$

$$= \frac{2}{z} \cdot \left[\pi \cot\left(-\pi \frac{N}{z} \right) + \frac{z}{N} \right].$$

We have now to take the limit of the above expression for $N \to \infty$,

$$\frac{2\pi}{z} \lim_{N\to\infty} \cot\left(-\pi \frac{N}{z} \right) = \frac{2\pi}{z} \lim_{N\to\infty} i \frac{e^{-2\pi i N/z}+1}{e^{-2\pi i N/z}-1} = -\frac{2\pi i}{z}.$$

This proves Lemma VII.1.6.   □

A function theoretical characterization of ϑ^r

The theta series

$$\vartheta(z) = \sum_{n=-\infty}^{\infty} e^{\pi i n^2 z}$$

converges in the upper half-plane, where it represents an analytic function. We collect the properties (VI.6) of this theta series, which we will need in the following.

Remark VII.1.7 *The theta series $\vartheta(z)$ has the following properties:*

(a) $\vartheta(z+2) = \vartheta(z)$, $\vartheta\left(-\frac{1}{z} \right) = \sqrt{\frac{z}{i}}\, \vartheta(z)$,

(b) $\lim\limits_{y\to\infty} \vartheta(z) = 1$,

(c) $\lim\limits_{y\to\infty} \sqrt{\frac{z}{i}}^{\,-1} \vartheta\left(1 - \frac{1}{z} \right) e^{-\frac{\pi i z}{4}} = 2$.

Proof. The transformation formula (a) is a special case of the JACOBI Theta Transformation Formula. The property (b) is trivial, and (c) follows using the transformation formula

$$\vartheta\left(1 - \frac{1}{z} \right) = \sqrt{\frac{z}{i}} \sum_{n=-\infty}^{\infty} e^{\pi i z \left(n + \frac{1}{2} \right)^2},$$

once more a direct consequence of the JACOBI Theta Transformation Formula VI.4.2. □

This leads to a *function theoretical characterization of ϑ^r*:

Proposition VII.1.8 *Let $r \in \mathbb{Z}$, and let $f : \mathbb{H} \to \mathbb{C}$ be an analytic function with the properties:*

$$\text{(a)} \quad f(z+2) = f(z) , \qquad f\left(-\frac{1}{z}\right) = \sqrt{\frac{z}{i}}^{\,r} f(z) ,$$

$$\text{(b)} \quad \lim_{y \to \infty} f(z) \ exists ,$$

$$\text{(c)} \quad \lim_{y \to \infty} \sqrt{\frac{z}{i}}^{\,-r} f\left(1 - \frac{1}{z}\right) e^{-\frac{\pi i r z}{4}} \ exists .$$

Then

$$f(z) = constant \cdot \vartheta(z)^r .$$

(*The constant is of course* $\lim_{y \to \infty} f(z)$.)

For the proof, we consider the auxiliary function

$$h(z) = \frac{f(z)}{\vartheta(z)^r} .$$

We know (and will see once more) that the theta function $\vartheta(z)$ has no zeros in the upper half-plane. The function h is thus analytic in the upper half-plane. Now Proposition VI.1.7 follows immediately from VI.1.6 and the following

Proposition VII.1.9 *Let there be given an analytic function $h : \mathbb{H} \to \mathbb{C}$ with the property*

$$h(z+2) = h(z) , \quad h(-1/z) = h(z) .$$

We also assume that both limits

$$a := \lim_{y \to \infty} h(z) \qquad and \qquad b := \lim_{y \to \infty} h(1 - 1/z)$$

exist. Then h is a constant function.

Proof of VI.1.8 and VI.1.7. The conditions (a), (b) and (c) express the fact, that h is an entire modular form of weight 0 with respect to the theta group, so it is a constant. □

Because of the beautiful number theoretical applications of this Proposition, it may be of interest to also have an alternative direct proof of it, and of the non-vanishing of the theta series in the upper half-plane.

We denote by

$$\Gamma_\vartheta = \left\langle \begin{pmatrix} 1 & 2 \\ 0 & 1 \end{pmatrix} , \begin{pmatrix} 0 & -1 \\ 1 & 0 \end{pmatrix} \right\rangle$$

the subgroup of $\mathrm{SL}(2, \mathbb{Z})$ generated by the two matrices $\begin{pmatrix} 1 & 2 \\ 0 & 1 \end{pmatrix}$ and $\begin{pmatrix} 0 & -1 \\ 1 & 0 \end{pmatrix}$.

In Appendix A we have introduced this subgroup, and also remarked, that this group is exactly the group of all matrices in $\mathrm{SL}(2, \mathbb{Z})$, whose entries satisfy $a + b + c + d \equiv 0 \mod 2$. We will not use this last characterization, so we can take as a definition of the theta group the definition using the above two generators.

In Appendix A we have also introduced the region

$$\mathcal{F}_\vartheta := \mathcal{F} \cup \begin{pmatrix} 1 & 1 \\ 0 & 1 \end{pmatrix} \mathcal{F} \cup \begin{pmatrix} 1 & 1 \\ 0 & 1 \end{pmatrix} \begin{pmatrix} 0 & -1 \\ 1 & 0 \end{pmatrix} \mathcal{F} ,$$

and in a few lines we could prove:

For any point $z \in \mathbb{H}$ there exists a matrix $M \in \Gamma_\vartheta$ with $Mz \in \mathcal{F}_\vartheta$.

These are all ingredients needed to prove Proposition VI.1.8:

We consider the function

$$H(z) = \big(h(z) - a\big)\big(h(z) - b\big) .$$

It is an analytic function, which is invariant under both substitutions $z \mapsto z+2$ and $z \mapsto -1/z$. From our assumption,

$$\lim_{y \to \infty} H(z) = 0 , \qquad \lim_{\substack{z \to 1 \\ z \in \mathcal{F}_\vartheta}} H(z) = 0 .$$

In particular, $|H(z)|$ takes its maximum in $\mathcal{F}_\vartheta$. On the other side, $H(z)$ is invariant under the action of the theta group Γ_ϑ. This implies, that $|H(z)|$ has a maximum in the entire half-plane $\mathbb{H}$. By the Maximum Principle, $H(z)$ is a constant function. This constant is of course zero, so h can take only the values a and b. But h is continuous, and its domain of definition $\mathbb{H}$ is connected, hence h is constant.

We also insert a direct proof of the fact that ϑ has no zero in the upper half-plane $\mathbb{H}$. Because of the transformation formulas it is enough to prove this in $\mathcal{F}_\vartheta$, and this means that the three functions

$$\vartheta(z) , \quad \vartheta(z + 1) , \quad \text{and} \quad \vartheta\left(1 - \frac{1}{z}\right)$$

have no zero in $\mathcal{F}$.

We extract from the theta series ϑ the constant term 1 which belongs to $n = 1$ and estimate the rest by the series of the absolute values:

$$|\vartheta(z) - 1| \le 2 \sum_{n=1}^{\infty} e^{-\pi n^2 y} \le 2 \sum_{n=1}^{\infty} e^{-\frac{\pi}{2}\sqrt{3}n^2}$$

$$\le 2 \sum_{n=1}^{\infty} e^{-\frac{\pi}{2}\sqrt{3}n} = \frac{2e^{-\frac{\pi}{2}\sqrt{3}}}{1 - e^{-\frac{\pi}{2}\sqrt{3}}} = 0,14\ldots < 0,2 .$$

(The lowest points of the modular figure $\mathcal{F}$ are $\pm\frac{1}{2} + \frac{i}{2}\sqrt{3}$.)

For the second series, the same argument works. For the third series we observe that $\sqrt{z/i}^{-1}\vartheta(1 - 1/z)$ up to a trivial factor equals

$$\sum_{n=-\infty}^{\infty} e^{\pi i z(n^2+n)} = 2\sum_{n=0}^{\infty} e^{\pi i z(n^2+n)} = 2 \cdot (1 + e^{2\pi i z} + \cdots) ,$$

and use the same type of argument.

Representations of a natural number as the sum of eight squares

By means of the EISENSTEIN series G_4, we will construct now a function $f(z)$, which is analytic in the upper half-plane, which has the characteristic transformation properties of $\vartheta^8(z)$, namely

$$f(z + 2) = f(z) , \qquad f(-1/z) = z^4 f(z) ,$$

and which admits the (finite) limits

(a) $\lim_{y\to\infty} f(z)$ and

(b) $\lim_{y\to\infty} z^{-4} f(1 - 1/z) e^{-2\pi i z} .$

One might think that G_4 already satisfies this full list of properties. Anyway, the transformation formulas are true, and the existence of the limit

$$\lim_{y\to\infty} G_4(z) = 2\zeta(4)$$

can be extracted from its FOURIER expansion. What about the limit in (b)? We have

$$G_4\left(1 - \frac{1}{z}\right) = G_4\left(-\frac{1}{z}\right) = z^4 G_4(z) ,$$

hence

$$z^{-4} G_4\left(1 - \frac{1}{z}\right) e^{-2\pi i z} = G_4(z) e^{-2\pi i z} .$$

This expression has no limit for $y \to \infty$, because on the one side G_4 has a non-zero limit, and on the other side

$$\left|e^{-2\pi i z}\right| = e^{2\pi y}$$

is unbounded for $y \to \infty$. So we have to search for something else. We next observe the following fact:

Lemma VII.1.10 *The function*

$$g_k(z) := G_k\left(\frac{z+1}{2}\right) , \quad k > 2 ,$$

satisfies the transformation formulas

$$g_k(z+2) = g_k(z) , \quad g_k(-1/z) = z^k g_k(z) .$$

(We are interested in the case $k = 4$.)

Proof. The periodicity is clear, since the EISENSTEIN series has period one. Let us look closer to the second formula. Obviously,

$$\frac{-\frac{1}{z}+1}{2} = A\left(\frac{z+1}{2}\right) \quad \text{with} \quad A := \begin{pmatrix} 1 & -1 \\ 2 & -1 \end{pmatrix} \in SL(2,\mathbb{Z}) .$$

From this we deduce

$$g_k\left(-\frac{1}{z}\right) = \left(2 \cdot \frac{z+1}{2} - 1\right)^k g_k(z)$$

$$= z^k \, g_k(z) .$$

(This trick is due to J. ELSTRODT, personal communication.)   □

Can it be, that $g_4(z)$ has the characteristic properties of $\vartheta^8(z)$? The transformation formulas are satisfied, and we moreover have

$$\lim_{y\to\infty} g_4(z) = 2\zeta(4) .$$

But once more, the condition (b) is false, since

$$g_4\left(1 - \frac{1}{z}\right) = G_4\left(-\frac{1}{2z}\right) = (2z)^4 G_4(2z) ,$$

and the same argument as for the first candidate G_4 shows that the limit in (b) does not exist. Now the idea is to take a linear combination of the two candidates,

$$f(z) := a\, G_4(z) + b\, G_4\left(\frac{z+1}{2}\right) , \quad a,b \in \mathbb{C} .$$

The transformation formulas

$$f(z+2) = f(z) , \quad f\left(-\frac{1}{z}\right) = z^4 f(z) ,$$

are satisfied, and the limit

$$\lim_{y\to\infty} f(z) = 2(a+b)\zeta(4)$$

exists. The idea is now to find suitable constants a and b such that the limit (b) exists. The above computations already give

$$z^{-4} f\left(1 - \frac{1}{z}\right) e^{-2\pi i z} = e^{-2\pi i z} \cdot \left(a\, G_4(z) + 16b\, G_4(2z) \right) .$$

All we still need is to know that G_4 is a power series in $q = e^{2\pi i z}$:

$$G_4(z) = a_0 + a_1 q + a_2 q^2 + \cdots .$$

This implies

$$z^{-4} f\left(1 - \frac{1}{z}\right) e^{-2\pi i z} = q^{-1}\left[\, a_0(a + 16b) + \text{higher } q\text{-powers}\,\right] .$$

We impose on a and b the condition

$$a + 16b = 0 ,$$

obtaining the power series expansion

$$z^{-4} f\left(1 - \frac{1}{z}\right) e^{-2\pi i z} = c_0 + c_1 q + \cdots ,$$

since the factor q^{-1} has been absorbed by the higher q-powers. Because of the relation

$$y \longrightarrow \infty \Longleftrightarrow q \longrightarrow 0,$$

we get the existence of the needed limit in (b).

This means that

$$a = -16b \implies f(z) = \text{constant } \vartheta^8(z) .$$

We would like to have the constant equal to 1, and consider for this once more the limit for $y \to \infty$. Since $\lim_{y \to \infty} \vartheta(z) = 1$, we would like to have

$$2(a + b)\zeta(4) = 1 .$$

Connecting this with $a + 16b = 0$, and solving the linear system, we get

$$b = -\frac{1}{30\zeta(4)} \quad \text{and} \quad a = \frac{16}{30\zeta(4)} .$$

Using these numbers, we have proven the identity of analytic functions

$$\vartheta^8(z) = a\, G_4(z) + b\, G_4\left(\frac{z+1}{2}\right) .$$

Recall

$$\zeta(4) = \frac{\pi^4}{90} .$$

(By the way: This value also follows from our above identity by comparing the first coefficient in the q-expansions of both sides.)

We are led to the following result:

Theorem VII.1.11

$$\vartheta^8(z) = \frac{3}{\pi^4}\left(16G_4(z) - G_4\left(\frac{z+1}{2}\right)\right).$$

Using now for the EISENSTEIN series its q-expansion VII.1.3, we obtain

$$1 + \sum_{n=1}^{\infty} A_8(n)\, e^{\pi i n z} = 1 + 16^2 \sum_{n=1}^{\infty} \sigma_3(n)\, e^{2\pi i n z} - 16 \sum_{n=1}^{\infty} \sigma_3(n)(-1)^n e^{\pi i n z}.$$

Theorem VII.1.12 (C.G.J. Jacobi, 1829) *For $n \in \mathbb{N}$ the following formula holds:*

$$A_8(n) = 16 \sum_{d\,|\,n} (-1)^{n-d} d^3.$$

For odd n we have just identified the corresponding coefficients of q^n, and for even n the identification needs a further short check, which is skipped.   □

Representations of a natural number as the sum of four squares

Using the EISENSTEIN series

$$G_2(z) = \sum_c \left\{ \sum_{d\neq 0,\ \text{if } c=0} (cz + d)^{-2} \right\}$$

we would like to construct a function f having the characteristic properties of ϑ^4. The main transformation property is

$$f\left(-\frac{1}{z}\right) = \sqrt{\frac{z}{i}}^{\,4} f(z) = -z^2 f(z).$$

A linear combination of the form $f(z) = a\, G_2\left(\frac{z+1}{2}\right) + b\, G_2(z)$ is hopeless for this purpose, it would not work since it would in the best case lead to a transformation behavior of the type

$$f\left(-\frac{1}{z}\right) = z^2 f(z).$$

But another approach can be used, namely

$$f(z) = a\, G_2(z/2) + b\, G_2(2z).$$

Such an f has period 2, because

$$G_2(z+1) = G_2(z) \implies f(z+2) = f(z).$$

We also know the transformation formula from the Consequence in VII.1.4,

$$G_2\left(-\frac{1}{z}\right) = z^2 G_2(z) - 2\pi i z \ .$$

This gives:

$$f\left(-\frac{1}{z}\right) = a\, G_2\left(-\frac{1}{2z}\right) + b\, G_2\left(-\frac{1}{z/2}\right)$$

$$= a(2z)^2 G_2(2z) - 4\pi i\, az + b\left(\frac{z}{2}\right)^2 G_2\left(\frac{z}{2}\right) - \pi i\, bz \ .$$

Now we impose for a and b the condition

$$b = -4a$$

and obtain the desired formula

$$f\left(-\frac{1}{z}\right) = z^2\left(\,4a\, G_2(2z) + \frac{b}{4} G_2\left(\frac{z}{2}\right)\,\right) = -z^2\left(\,a\, G_2\left(\frac{z}{2}\right) + b\, G_2(2z)\,\right)$$

$$= -z^2 f(z) \ .$$

We now have to prove for the function

$$f(z) = a\big(G_2(z/2) - 4G_2(2z)\big)$$

the existence of both limits

$$\lim_{y \to \infty} f(z) \quad \text{and} \quad \lim_{y \to \infty} z^{-2} f\left(1 - \frac{1}{z}\right) e^{-\pi i z} \ .$$

The existence of the first limit is trivial, since (VII.1.4, Corollary)

$$\lim_{y \to \infty} G_2(z) = 2\zeta(2) \ ,$$

which follows from the q-expansion. Normalizing the limit for f to 1, the constant a necessarily has the value

$$a = -\frac{1}{6\zeta(2)} \ .$$

It remains to study $f(1 - 1/z)$, in particular

(a) $G_2\left(2 - \frac{2}{z}\right)$ and

(b) $G_2\left(\frac{1 - 1/z}{2}\right) \ .$

At our disposal we have only the formulas

$$G_2(z+1) = G_2(z) \quad \text{and} \quad G_2\left(-\frac{1}{z}\right) = z^2 G_2(z) - 2\pi i \, z \ .$$

Part (a) is simple, since

$$G_2\left(2 - \frac{2}{z}\right) = G_2\left(-\frac{2}{z}\right) = \left(\frac{z}{2}\right)^2 G_2\left(\frac{z}{2}\right) - \pi i \, z \ .$$

The limit in (b) is more tricky. First, let us observe that the matrix $\begin{pmatrix} 1 & -1 \\ 2 & -1 \end{pmatrix}$ has determinant 1. If z lies in the upper half-plane, then the same is true also for $(z-1)(2z-1)^{-1}$. A simple computation now gives

$$G_2\left(\frac{z-1}{2z-1}\right) = \left(\frac{2z-1}{z-1}\right)^2 G_2\left(-\frac{2z-1}{z-1}\right) + 2\pi i \frac{2z-1}{z-1} \ .$$

Using

$$-\frac{2z-1}{z-1} = -2 - \frac{1}{z-1} \ ,$$

we obtain

$$
\begin{aligned}
G_2\left(\frac{z-1}{2z-1}\right) &= \left(\frac{2z-1}{z-1}\right)^2 G_2\left(-\frac{1}{z-1}\right) + 2\pi i \frac{2z-1}{z-1} \\
&= (2z-1)^2 G_2(z-1) - 2\pi i \frac{(2z-1)^2}{z-1} + 2\pi i \frac{2z-1}{z-1} \\
&= (2z-1)^2 G_2(z) \qquad - 2\pi i \, (4z-2) \ .
\end{aligned}
$$

In this equation, we replace z by $z/2 + 1/2$ and obtain

$$G_2\left(\frac{-1/z+1}{2}\right) = z^2 G_2\left(\frac{z}{2} + \frac{1}{2}\right) - 4\pi i \, z \ .$$

In conclusion,

$$
\begin{aligned}
z^{-2} f\left(1 - \frac{1}{z}\right) &= z^{-2} a\left[z^2 G_2\left(\frac{z}{2} + \frac{1}{2}\right) - 4\pi i z - z^2 G_2\left(\frac{z}{2}\right) + 4\pi i \, z \right] \\
&= a\left[G_2\left(\frac{z}{2} + \frac{1}{2}\right) - G_2\left(\frac{z}{2}\right) \right] \ .
\end{aligned}
$$

We know that $G_2(z)$ can be written as a power series in $e^{2\pi i z}$. This implies that $G_2(z/2 + 1/2)$ and $G_2(z/2)$ can both be written as power series in $h := e^{\pi i z}$, and both have then the same coefficient in h^0, namely $2\zeta(2)$. This gives the h-expansion

$$z^{-2} f(1 - 1/z) = a_1 h + a_2 h^2 + \cdots$$

with some (uninteresting) coefficients $a_1, a_2, \ldots$.

Using
$$y \longrightarrow \infty \Longleftrightarrow h \longrightarrow 0 \ ,$$
we get
$$z^{-2} f(1 - 1/z) h^{-1} \longrightarrow a_1 \quad \text{for } y \longrightarrow \infty \ .$$

This is exactly what we wanted to have. Putting all together, we have finished the proof of the formula
$$f(z) = \frac{1}{6\zeta(2)} \cdot \left(4G_2(2z) - G_2\left(\frac{z}{2}\right) \right) = \vartheta^4(z) \ .$$

Now, we either know
$$\zeta(2) = \frac{\pi^2}{6} \ ,$$
or we recover it from the proven identity by comparing the first FOURIER coefficients. We can conclude:

Theorem VII.1.13
$$\vartheta^4(z) = \frac{4G_2(2z) - G_2(z/2)}{\pi^2} \ .$$

From this, we can derive the number theoretical identities we were interested in.

Theorem VII.1.14 (C.G.J. Jacobi, 24.4.1828) *For $n \in \mathbb{N}$ we have*

$$A_4(n) = \#\left\{ x \in \mathbb{Z}^4 \ ; \quad x_1^2 + x_2^2 + x_3^2 + x_4^2 = n \right\} = 8 \sum_{\substack{4 \nmid d \mid n \\ 1 \leq d \leq n}} d \ .$$

The proof follows directly from VII.1.13 by using the q-expansion VII.1.3 of G_2 and identifying coefficients.

Corollary VII.1.15 (J.L. Lagrange, 1770) *Any natural number can be represented as a sum of four squares of integers.*

Exercises for VII.1

1. The function $f(z) = j'(z)\Delta(z)$ is an *entire* modular form (compare with VI.2, Exercise 1). Find a representation of it as a polynomial in G_4 and G_6.

2. The function $G'_{12}\Delta - G_{12}\Delta'$ is a modular form of weight 26 (VI.2, Exercise 3). Express this function as a polynomial in G_4 and G_6.

3. Represent G_{12} as a polynomial in G_4 and G_6 by using the Structure Theorem VI.3.4 and the formulas for the FOURIER coefficients of the EISENSTEIN series. Compare the result with the recursion formulas in Exercise 6 from Sect. V.3.

4. How many vectors x with the property $\langle x, x \rangle = 10$ exist in the lattice L_8 (see VI.4)? Compute this number
    (a) directly,
    (b) using the identity $G_4(z) = 2\zeta(4)\vartheta(L_8; z)$.

5. The FOURIER coefficients $\tau(n)$ of

    $$\frac{\Delta(z)}{(2\pi)^{12}} = \tau(1)q + \tau(2)q^2 + \cdots$$

    are all integral. The same also holds for the FOURIER coefficients $c(n)$ of

    $$1728\, j(z) = 1/q + c(0) + c(1)q + c(2)q^2 + \cdots \quad .$$

    Compute explicitly the first coefficients, and check

    $$(2\pi)^{-12}\Delta(z) = q - 24q^2 + 252q^3 - 1472q^4 + 4830q^5 - 6048q^6 + \cdots \quad ,$$

    $$1728 j(z) = 1/q + 744 + 196\,884q + 21\,493\,760q^2 + 864\,299\,970q^3 + \cdots \quad .$$

    RAMANUJAN's Conjecture states:

    $$|\tau(n)| \le C n^{11/2+\varepsilon} \text{ for any } \varepsilon > 0 \ (C = C(\varepsilon)) .$$

    It was generalized by H. PETERSSON to arbitrary cusp forms. We have already mentioned in the Exercises to Sect. VI.4 that this conjecture was proven by P. DELIGNE. By the way, the following estimate holds

    $$|\tau(n)| \le n^{11/2}\sigma_0(n) .$$

6. The DEDEKIND η-function is defined by

    $$\eta(z) = e^{\frac{\pi i z}{12}} \prod_{n=1}^{\infty} \left(1 - e^{2\pi i n z}\right) .$$

    Prove that this product converges normally in the upper half-plane, where it represents an analytic function. Compute its logarithmic derivative, and show

    $$\frac{\eta'(z)}{\eta(z)} = \frac{i}{4\pi} G_2(z) .$$

    From the transformation formula

    $$G_2(-1/z) = z^2 G_2(z) - 2\pi i z$$

    deduce then that both functions

    $$\eta\left(-\frac{1}{z}\right) \quad \text{and} \quad \sqrt{\frac{z}{i}}\, \eta(z)$$

    have the same logarithmic derivative. Hence, they coincide up to a constant factor. This factor is 1, as it can be seen by specializing $z = i$. This gives:

    $$\boxed{\eta\left(-\frac{1}{z}\right) = \sqrt{\frac{z}{i}}\, \eta(z) .}$$

    On the other side, we trivially have

    $$\boxed{\eta(z + 1) = e^{\frac{\pi i}{12}} \eta(z) .}$$

7. Prove the identity

$$\Delta(z) = (2\pi)^{12}\eta^{24}(z) \ .$$

8. For $|q| < 1$ prove the identity

$$\sum_{n=-\infty}^{\infty} (-1)^n q^{\frac{n(3n+1)}{2}} = \prod_{n=1}^{\infty}(1 - q^n) \ .$$

Hint. Use VI.4.6 and the previous Exercise.

9. A *partition* of the natural number n is by definition a k-tuple $(x_1, \ldots x_k)$ (k arbitrary) consisting of natural numbers (≥ 1) with the properties

$$n = x_1 + x_2 + \cdots x_k \ , \qquad x_1 < x_2 < \cdots < x_n \ .$$

Let A_n be the number of all partitions with even k, and
Let B_n be the number of all partitions with odd k.
(The sum $A_n + B_n$ is thus the number of all partitions of n.) *Show*

$$\prod_{n=1}^{\infty}(1 - q^n) = \sum_{n=0}^{\infty}(A_n - B_n)\, q^n \ .$$

From this and the previous Exercise, we deduce the EULER Pentagonal Number Theorem (L. EULER, 1754/55)

$$A_n = B_n \qquad \text{for } n \neq \frac{3m^2 + m}{2} \ ,$$

$$A_n = B_n + 1 \qquad \text{for } n = \frac{3m^2 + m}{2} \ , \ m \text{ even} \ ,$$

$$A_n = B_n - 1 \qquad \text{for } n = \frac{3m^2 + m}{2} \ , \ m \text{ odd} \ .$$

VII.2 Dirichlet Series

An ordinary DIRICHLET series is a series of the type

$$\sum_{n=1}^{\infty} a_n n^{-s} \ , \ a_n \in \mathbb{C} \ , \ s \in \mathbb{C} \ .$$

If we set all coefficients to one, $a_n = 1$, then we obtain the most famous among all DIRICHLET series, the RIEMANN ζ-function

$$\zeta(s) := \sum_{n=1}^{\infty} n^{-s} \ .$$

We know that this series converges absolutely for Re $(s) > 1$.

Definition VII.2.1 *A Dirichlet series*

$$D(s) = \sum_{n=1}^{\infty} a_n n^{-s} \ , \ a_n \in \mathbb{C} \ , \ s \in \mathbb{C},$$

is called (somewhere absolutely) **convergent** *, iff there exists a complex number s_0, such that the series*

$$\sum_{n=1}^{\infty} \left| a_n n^{-s_0} \right|$$

converges in the usual sense.

We use the traditional convention (RIEMANN, LANDAU) for the notation of the complex variable s, and set

$$s = \sigma + \mathrm{i}t, \ s_0 = \sigma_0 + \mathrm{i}t_0, \ \dots \quad .$$

We have

$$\left| n^{-s} \right| = \left| e^{-s \log n} \right| = \left| e^{-(\sigma + \mathrm{i}t) \log n} \right| = n^{-\sigma}$$

and

$$n^{-\sigma} \leq n^{-\sigma_0} \text{ for } \sigma \geq \sigma_0 \ .$$

Hence, if the DIRICHLET series converges absolutely in s_0, then it converges absolutely and uniformly in the half-plane $\sigma \geq \sigma_0$.

Definition VII.2.2 *A right half-plane*

$$\{ \ s \in \mathbb{C} \ ; \quad \sigma > \tilde{\sigma} \ \}$$

is called a **convergence half-plane** *of a Dirichlet series, iff the series converges* **absolutely** *for all s in this half-plane. Here, we also allow the value*

$$\tilde{\sigma} = -\infty \ ,$$

where the "convergence half-plane" becomes the whole plane.

The union of all convergence half-planes for a given DIRICHLET series $D(s)$ is also a convergence half-plane of $D(s)$, i.e. is of the form $\{s \in \mathbb{C}; \ \sigma > \sigma_0\}$. It is the biggest of all convergence half-planes for $D(s)$, and it is called *the* convergence half-plane of $D(s)$, or more exactly the *half-plane of absolute convergence*.

So let $\{ \ s \in \mathbb{C} \ ; \ \sigma > \sigma_0 \ \}$ be the convergence half-plane. Then $D(s)$ converges absolutely for all s with $\sigma > \sigma_0$, but not for any s with $\sigma < \sigma_0$. The behavior of $D(s)$ on the vertical line $\sigma = \sigma_0$ cannot be made more specific in this general setting. The involved value σ_0 is also called the *convergence abscissa* of $D(s)$. (More precisely, σ_0 is the *abscissa of absolute convergence*, see Exercise 1.) Of course, $D(s)$ represents an analytic function in the convergence half-plane. The convergence half-plane of the RIEMANN ζ-function is

$$\mathrm{Re} \ (s) > \sigma_0 = 1 \ .$$

Definition VII.2.3 *A sequence* $a_1, a_2, a_3, \ldots$ *of complex numbers has* **(at most) polynomial growth,** *iff there exist constants* $C > 0$ *and* N, *such that*

$$|a_n| \leq C\, n^N$$

for all n.

Remark VII.2.4 *Assume that the sequence* $a_1, a_2, a_3, \ldots$ *has polynomial growth. Then the associated Dirichlet series* $D(s)$ *converges. (The converse is also true). More exactly, with the above notations we have* $\sigma_0 \leq 1 + N$.

Example. In the case of the ζ-function we can take $N = 0$.

The *Proof of VII.2.4* follows from the estimate

$$\left| a_n n^{-s} \right| \leq C\, n^{-(\sigma - N)}$$

and from the convergence behavior of the RIEMANN ζ-function.   □

Similarly to the case of power series, the coefficients of a DIRICHLET series are uniquely determined by the corresponding function.

Proposition VII.2.5 (Uniqueness property) *Let*

$$D(s) = \sum_{n=1}^{\infty} a_n n^{-s} \ , \ a_n \in \mathbb{C} \ , \ s \in \mathbb{C} \ ,$$

be a Dirichlet series, which vanishes in some convergence half-plane. Then

$$a_n = 0 \ \text{ for all } n \ .$$

Proof. We assume the contrary, and let k be the smallest index, such that a_k does not vanish. Then

$$D(s)k^s = \sum_{n=k}^{\infty} a_n \left(\frac{n}{k} \right)^{-s} = a_k + \cdots$$

and thus

$$a_k = \lim_{\sigma \to \infty} D(\sigma)\, k^\sigma = 0$$

which is a contradiction.   □

DIRICHLET series can serve to encode *multiplicative properties* of a sequence of numbers in an analytic form. Let

$$D(s) = \sum_{n=1}^{\infty} a_n n^{-s}$$

be a DIRICHLET series. For any non-empty set $A \subset \mathbb{N}$ of natural numbers we consider the partial series

$$D_A(s) = \sum_{n \in A} a_n n^{-s} \ .$$

Lemma VII.2.6 *Let* $A, B \subset \mathbb{N}$ *be two non-empty sets of natural numbers, and let* (a_n) *be a sequence of complex numbers. We assume that the following conditions are satisfied:*

(1) *The multiplication map*

$$A \times B \longrightarrow \mathbb{N} , \quad (a, b) \mapsto ab ,$$

is injective.

(2) $a_{n \cdot m} = a_n \cdot a_m$ *for* $n \in A$, $m \in B$.

If C *is the image of* $A \times B$ *by the multiplication map, then*

$$D_C(s) = D_A(s) \cdot D_B(s)$$

in the convergence half-plane of $D(s)$.

By induction on the number N, we obtain then

$$D_C(s) = D_{A_1}(s) \cdot \cdots \cdot D_{A_N}(s) ,$$

for subsets $A_1, \ldots, A_N \subset \mathbb{N}$, such that the multiplication map

$$A_1 \times \cdots \times A_N \longrightarrow C ,$$
$$(n_1, \ldots, n_N) \longmapsto n_1 \cdot \cdots \cdot n_N ,$$

is bijective, and the coefficients satisfy the multiplicative property

$$a_{n_1 \cdots n_N} = a_{n_1} \cdot \cdots \cdot a_{n_N}$$

for all $n_1 \in A_1, \ldots, n_N \in A_N$. The *Proof* of the Lemma is a trivial consequence of the CAUCHY Product Theorem,

$$\left(\sum_{\mu \in A} a_\mu \mu^{-s} \right) \left(\sum_{\nu \in B} b_\nu \nu^{-s} \right) = \sum_{(\mu, \nu) \in A \times B} a_{\mu\nu} (\mu\nu)^{-s} = \sum_{n \in C} a_n n^{-s} .$$

We treat an important special case:

Let

$$p_1 = 2; \; p_2 = 3; \; p_3 = 5; \; \ldots$$

be the sequence of prime numbers. We denote by

$$A_n := \{ p_n^\nu ; \quad \nu = 0, 1, 2, \ldots \}$$

the set of all (natural) powers of the nth prime number. The multiplication

$$A_1 \times \cdots \times A_N \longrightarrow \mathbb{N} ,$$
$$(n_1, \ldots, n_N) \longmapsto n_1 \cdot \cdots \cdot n_N ,$$

is injective. Let us then denote by B the set of all natural numbers which do *not* contain in their prime number decomposition any primes among $p_1, \ldots, p_N$. Then the map

$$A_1 \times \cdots \times A_N \times B \longrightarrow \mathbb{N} \,,$$

$$(n_1, \ldots, n_N, m) \longmapsto n_1 \cdots \cdots n_N \cdot m \,,$$

is bijective. For this, we use the theorem of *unique factorization into primes*. Let us now assume that the coefficients of a DIRICHLET series satisfy the condition

$$a_{n \cdot m} = a_n \cdot a_m$$

for arbitrary, relatively prime natural n, m, i.e. $(n, m) = 1$. Moreover, we require $a_1 = 1$. Then, from Lemma VII.2.6 we have

$$D(s) = \prod_{n=1}^{N} \left(\sum_{\nu=0}^{\infty} a_{p_n^\nu} p_n^{-\nu s} \right) D_B(s) \,.$$

The condition $n \in B$, i.e. $(n, p_1 \cdots \cdots p_N) = 1$ then implies either $n = 1$, or $n \geq N$. We obtain

$$\lim_{N \to \infty} D_B(s) = 1$$

and hence

$$D(s) = \lim_{N \to \infty} \prod_{n=1}^{N} \left(\sum_{\nu=0}^{\infty} a_{p_n^\nu} p_n^{-\nu s} \right) \,.$$

The following question arises: Is this an absolutely convergent product in the sense of VI.1.3? Equivalently, we ask for the convergence of

$$\sum_{n=1}^{\infty} \left| \sum_{\nu=0}^{\infty} a_{p_n^\nu} p_n^{-\nu s} - 1 \right| \leq \sum_{n=1}^{\infty} \sum_{\nu=1}^{\infty} \left| a_{p_n^\nu} p_n^{-\nu s} \right| \,.$$

But the series in the R.H.S. is just a rearrangement of a partial series of

$$\sum_{n=1}^{\infty} \left| a_n n^{-s} \right| \,.$$

We thus obtain a result, which goes back to L. EULER (1737):

Proposition VII.2.7 *Let*

$$D(s) = \sum_{n=1}^{\infty} a_n n^{-s}$$

be a (somewhere absolutely) convergent Dirichlet series, whose coefficients have the multiplicative property

$$a_1 = 1 \text{ and } a_{n \cdot m} = a_n \cdot a_m \text{ for } (n, m) = 1 .$$

Then

$$D(s) = \prod_{p \text{ prime}} D_p(s) \text{ with } D_p(s) := \sum_{n=0}^{\infty} a_{p^n} p^{-ns} .$$

The infinite product is indexed by all prime numbers p, and the convergence is normal in the convergence half-plane of D(s).

We can in particular apply this Proposition for the RIEMANN ζ-function. One has:

$$\zeta_p(s) = \sum_{\nu=0}^{\infty} p^{-\nu s} = \frac{1}{1 - p^{-s}} \qquad \text{(geometric series)}$$

in the convergence half-plane. This leads to the *Euler product expansion* of the ζ-function in the half-plane $\sigma > 1$.

Proposition VII.2.8 (L. Euler, 1737) *We have*

$$\boxed{\; \zeta(s) = \prod_{p \text{ prime}} \left(1 - p^{-s}\right)^{-1} \qquad (\sigma > 1) . \;}$$

In particular, $\zeta(s)$ has no zero in the convergence half-plane

$$\{ \, s = \sigma + \mathrm{i}t \; ; \; \sigma, t \in \mathbb{R} \, , \; \sigma > 1 \, \} .$$

Exercises for VII.2

1. We have introduced in this section the half-plane of convergence $\sigma > \sigma_0$ of a DIRICHLET series

    $$D(s) = \sum_{n=1}^{\infty} a_n n^{-s}.$$

    More precisely, we dealt with the half-plane of absolute convergence. Show that there also exists a maximal right half-plane for the *ordinary* convergence,

    $$\{ \, s \in \mathbb{C} \; ; \quad \mathrm{Re}\, s > \sigma_1 \, \} \quad (\sigma_1 \geq -\infty)$$

    if the series $D(s)$ converges (possibly not absolutely) in at least one point.

    Then $D(s)$ converges normally in this half-plane, and represents there an analytic function. The series does not converge for any s with $\sigma < \sigma_1$.

    Hint. Use ABEL's partial summation.

    Supplement. If the DIRICHLET series $D(s)$ converges in at least one point, i.e. if $\sigma_1 < \infty$ exists, then it converges absolutely in at least one point, i.e. $\sigma_0 < \infty$ exists, and we have

    $$\sigma_0 \geq \sigma_1 \geq \sigma_0 - 1 .$$

    Give an example for the case $\sigma_0 = 1$, $\sigma_1 = 0$.

2. The FOURIER coefficients of the normalized EISENSTEIN series

$$\frac{(k-1)!}{2(2\pi i)^k} G_k(z) = \sum_{n=0}^{\infty} a(n)\, e^{2\pi i n z}\ , \quad k \geq 4\ ,$$

satisfy the equations

(a) $a(n)\, a(m) = a(nm)\ ,$ if $(n,m) = 1\ ,$

(b) $a(p^{\nu+1}) = a(p)\, a(p^\nu) - p^{k-1} a(p^{\nu-1})\ .$

Deduce from this

$$\sum_{n=1}^{\infty} a(n) n^{-s} = \prod_p \sum_{\nu=0}^{\infty} a(p^\nu)\, p^{-\nu s}$$

$$= \prod_p \frac{1}{(1-p^{-s})(1-p^{k-1-s})}$$

$$= \zeta(s)\, \zeta(s+1-k) \text{ for } \sigma > k\ .$$

3. Let p be a prime number. For any integer ν, $1 \leq \nu < p$, there exists a uniquely determined integer μ, $1 \leq \mu < p$, such that the matrix

$$\begin{pmatrix} 1 & \nu \\ 0 & p \end{pmatrix} \begin{pmatrix} 0 & 1 \\ -1 & 0 \end{pmatrix} \begin{pmatrix} 1 & \mu \\ 0 & p \end{pmatrix}^{-1}$$

has integral entries, and lies thus in the modular group. The assignment $\nu \mapsto \mu$ is a permutation of the numbers $1, \ldots, p-1$.

Hint. By a direct computation, one obtains as a condition on μ the congruence

$$\nu \mu \equiv -1 \mod p\ .$$

Make use of the fact that $\mathbb{Z}/p\mathbb{Z}$ is a field.

4. Let f be an elliptic modular form (with respect to the full modular group) of weight k. The function

$$(T(p)f)(z) := p^{k-1} f(pz) + \frac{1}{p} \sum_{\nu=0}^{p-1} f\left(\frac{z+\nu}{p}\right)$$

is for any prime number p again a modular form of weight k. We thus obtain for any prime p an operator (a linear map)

$$T(p) : [\Gamma, k] \longrightarrow [\Gamma, k]\ .$$

Hint. The periodicity of $T(p)f$ is trivial. For the transformation rule under the involution $z \mapsto -1/z$ make use of Exercise 3.

The operators $T(p)$ were introduced by E. HECKE (1935). (Compare with [He3].) These HECKE operators turned out to be of fundamental importance for deeper research in the theory of modular forms.

5.   Let

$$f(z) = \sum_{n=0}^{\infty} a(n)\, e^{2\pi i n z}$$

be a modular form of weight k, and let

$$T(p)f(z) = \sum_{n=0}^{\infty} b(n)\, e^{2\pi i n z}$$

be its image under $T(p)$. Additionally we define

$$a(n) := 0 \text{ for non-integral numbers } n \ .$$

Show:

$$b(n) = a(pn) + p^{k-1} a(n/p) \ .$$

6.   From the explicit knowledge of the FOURIER coefficients of all EISENSTEIN series, show that the EISENSTEIN series are eigenforms for all HECKE operators $T(p)$, i.e.

$$T(p)G_k = \lambda_k(p) G_k \ , \quad \lambda_k(p) \in \mathbb{C} \ .$$

7.   Let $f \in [\Gamma, k]$, $f(z) = \sum_{n=0}^{\infty} a(n)\, e^{2\pi i n z}$ be an eigenform of *all* operators $T(p)$,

$$T(p)f = \lambda(p)f \ .$$

Let f be normalized by the condition $a(1) = 1$. Show that $a(p) = \lambda(p)$. Using Exercise 5 show

$$a(n) = a(pm) + p^{k-1} a(n/p) \ ,$$

and then derive from this

$$\boxed{\begin{aligned} a(p)\, a(p^\nu) &= a(p^{\nu+1}) + p^{k-1} a(p^{\nu-1}) \ , \qquad &\text{and} \\ a(m)\, a(n) &= a(mn) \ , \qquad &\text{if } (m,n) = 1 \ . \end{aligned}}$$

Hint. The second relation must be checked only for powers of prime numbers, $m = p^\nu$. For this, use induction on ν and the first relation.

8.   Let $f \in [\Gamma, k]$ be a normalized eigenform of *all* $T(p)$. (Recall that "normalized" means $a(1) = 1$.) We consider the DIRICHLET series

$$D(s) = \sum_{n=1}^{\infty} \frac{a(n)}{n^s} \ ,$$

$$D_p(s) = \sum_{\nu=0}^{\infty} \frac{a(p^\nu)}{p^{\nu s}} \ .$$

Show that these series converge absolutely for $\sigma > k$ (respectively for $\sigma > k/2 + 1$, if f is a cusp form.) Using the relations from Exercise 7, show:

$$D(s) = \prod_p D_p(s) \quad \text{with} \quad D_p(s) = \frac{1}{1 - a(p)p^{-s} + p^{k-1-2s}} \ .$$

In the next section we will see that the DIRICHLET series $D(s)$ has a meromorphic continuation into the whole plane, where it satisfies a certain simple functional equation.

9. The operators $T(p)$ map cusp forms to cusp forms. Because of this, the discriminant $\Delta(z)$ is an eigenform of all $T(p)$. As a special case of Exercise 8 we obtain (for $\sigma > 7$)

$$\sum_{n=1}^{\infty} \frac{\tau(n)}{n^s} = \prod_p \frac{1}{1 - \tau(p)p^{-s} + p^{11-2s}} \cdot$$

Here, $\tau(n)$ is the RAMANUJAN τ-function, i.e. $\tau(n)$ is the nth FOURIER coefficient of $\Delta/(2\pi)^{12}$. The relations for $\tau(n)$ reflected by the above product representation were conjectured by S. RAMANUJAN (1916), and proven by L.J. MORDELL (1917).

The RAMANUJAN Conjecture, which was formulated in Exercise 1 of Sect. VII.1, is equivalent to the fact that both zeros of the polynomial

$$1 - \tau(p)X + p^{11}X^2$$

are complex conjugated.

10. (1) Let $f \in [\Gamma, k]_0$ be a cusp form, let p be a prime number, and set $\widetilde{f} = T(p)f$. The functions

$$g(z) = |f(z)|\, y^{k/2} \quad \text{and} \quad \widetilde{g}(z) = \left|\widetilde{f}(z)\right| y^{k/2}$$

have maxima m, $\widetilde{m}$ in $\mathbb{H}$ (see Exercise 2 in VI.4).

Show:

$$\widetilde{m} \le p^{\frac{k}{2}-1}(1+p)m \ .$$

We further assume that $f \not\equiv 0$ is an eigenform of the HECKE operator $T(p)$ with eigenvalue $\lambda(p)$.

Show:

$$|\lambda(p)| \le p^{\frac{k}{2}-1}(1+p) \ .$$

(2) If, contrarily, $f \in [\Gamma, k]$ is *not* a cusp form, with the property $T(p)f = \lambda(p)f$, then by Exercise 5

$$\lambda(p) = 1 + p^{k-1} \ .$$

Deduce from this (J. ELSTRODT, 1984, [El]):

The Eisenstein series G_k, $k \ge 4$, $k \equiv 0 \mod 2$, is up to a constant factor the only non-cusp form, which is the eigenform of at least one Hecke operator $T(p)$.

VII.3 Dirichlet Series with Functional Equations

We want to build a bridge between DIRICHLET series with functional equation and modular forms. Our exposition closely follows the line of the original paper [He2] of E. HECKE (1936) "On the determination of DIRICHLET series by their functional equation" (translated title).

Definition VII.3.1 *Let $R(s)$ be a meromorphic function in the complex plane. Then $R(s)$ is called **decaying** (to zero) in a given vertical strip*

$$a \leq \sigma \leq b$$

iff for any $\varepsilon > 0$ there exists a number $C > 0$ with the property

$$|R(s)| \leq \varepsilon \text{ for } a \leq \sigma \leq b \, , \, |t| \geq C \, .$$

We are particularly interested in functions that are decaying in any vertical strip. The constant C from the definition may of course depend on a, b.

We fix now three parameters, namely two positive real numbers

$$\lambda > 0 \text{ and } k > 0 \, ,$$

and a sign ε,

$$\varepsilon = \pm 1 \, .$$

We associate to these parameters two spaces of functions, namely

(a) a space $\{\, \lambda \, , \, k \, , \, \varepsilon \,\}$ of DIRICHLET series,
(b) and a space $[\, \lambda \, , \, k \, , \, \varepsilon \,]$ of FOURIER series.

It will turn out that these spaces are isomorphic.

Definition VII.3.2 *The space*

$$\{\, \lambda \, , \, k \, , \, \varepsilon \,\} \qquad (\lambda > 0 \, , \, k > 0 \, , \, \varepsilon = \pm 1)$$

is the set of all Dirichlet series

$$D(s) = \sum_{n=1}^{\infty} a_n n^{-s}$$

with the following properties:

(1) *The Dirichlet series converges (somewhere).*
(2) *The function represented by the Dirichlet series, and defined on its convergence half-plane, can be meromorphically extended into the whole plane. The extension is then analytic outside k, and at $s = k$ it has a removable singularity or a simple pole.*

(3) *The following functional equation is satisfied:*

$$R(s) = \varepsilon R(k - s) \text{ with } R(s) := \left(\frac{2\pi}{\lambda}\right)^{-s} \Gamma(s)D(s) \ .$$

(4) *The meromorphic function $R(s)$ is decaying in any vertical strip.*

Observation. The function

$$s \cdot (s - k) \cdot R(s)$$

is analytic in the right half-plane $\sigma > 0$. Because of the functional equation it is up to a sign invariant under the involution $s \mapsto k - s$. So it is an *entire* function.

Next, we introduce the corresponding space of FOURIER series. It is a space of FOURIER series with period λ.

Definition VII.3.3 *The space*

$$[\,\lambda\,,\,k\,,\,\varepsilon\,] \qquad (\lambda > 0\,,\ k > 0\,,\ \varepsilon = \pm 1)$$

is the set of all Fourier series

$$f(z) = \sum_{n=0}^{\infty} a_n e^{\frac{2\pi i n z}{\lambda}}$$

with the following properties:

(1) *The sequence (a_n) has polynomial growth. In particular, $f(z)$ converges in the upper half-plane, where it represents an analytic function.*

(2) *The following functional equation is satisfied:*

$$f\left(-\frac{1}{z}\right) = \varepsilon \left(\frac{z}{i}\right)^k f(z) \ .$$

Here, $(z/i)^k$ is defined by the principal branch of the logarithm.

Theorem VII.3.4 (E. Hecke, 1936) *The correspondence*

$$f(z) = \sum_{n=0}^{\infty} a_n e^{\frac{2\pi i n z}{\lambda}} \ \longmapsto\ D(s) = \sum_{n=1}^{\infty} a_n n^{-s}$$

defines an isomorphism

$$[\lambda, k, \varepsilon] \xrightarrow{\ \sim\ } \{\lambda, k, \varepsilon\} \ .$$

The residue of D at $s = k$ is

$$\mathrm{Res}(D; k) = a_0 \varepsilon \left(\frac{2\pi}{\lambda}\right)^k \Gamma(k)^{-1} \ .$$

In particular, D is an entire function, iff a_0 vanishes.

Remark preparing the Proof. On the R.H.S. of the correspondence there appear only coefficients a_n for positive values of n, on the L.H.S. the coefficient a_0 appears in addition. We must have this in mind, when we construct the inverse correspondence. In any case, the correspondence is injective, since it is a linear map between complex vector spaces, and in its kernel are only constant functions, which must vanish to satisfy the functional equation.

Proof the Theorem.

First part. Let $f \in [\lambda, k, \varepsilon]$. For the existence of the analytic continuation, and for the functional equation of $D(s)$ we must construct a function theoretic bridge from $f(z)$ to $D(s)$, This will be enabled by the Γ-integral

$$\Gamma(s) := \int_0^\infty t^{s-1} e^{-t}\, dt \qquad (\mathrm{Re}\ s > 0)\ .$$

Let us substitute the variable of integration,

$$t \longmapsto \frac{2\pi n}{\lambda} t\ ,$$

to obtain

$$\left(\frac{2\pi}{\lambda}\right)^{-s} \Gamma(s)\, n^{-s} = \int_0^\infty t^{s-1} e^{-\frac{2\pi n}{\lambda} t}\, dt\ .$$

Multiplying this equation with a_n, and summing over n, we get

$$R(s) = \left(\frac{2\pi}{\lambda}\right)^{-s} \Gamma(s)\, D(s) = \sum_{n=1}^\infty a_n \left[\int_0^\infty t^{s-1} e^{-\frac{2\pi n}{\lambda} t}\, dt\right]\ .$$

This representation is valid in some right half-plane (namely in the intersection of the convergence half-plane of $D(s)$ with the convergence half-plane of the Γ-integral).

We want now to exchange summation and integration. Because of the polynomial growth we can estimate $|a_n|$ by a power of n, $|a_n| \le n^K$. The modulus of t^{s-1} is $t^{\mathrm{Re}\ s-1}$. We can now apply the Monotone Convergence Theorem of B. LEVI for the LEBESGUE integral, and this insures that we can exchange the order of summation and integration. To avoid this result from measure theory, one can proceed alternatively and approximate the improper integral by a proper integral in order to be able to exchange summation and integration as a consequence of uniform convergence. The details are left to the reader. (Let us remark, that an analogous complication appeared in the proof of the analyticity of the Γ-integral.)

After performing the exchange of summation and integration, we obtain the searched function theoretical bridge from $f(z)$ to $D(s)$:

$$R(s) = \int_0^\infty t^s\, [\, f(\mathrm{i}t) - a_0\,]\, \frac{dt}{t}\ .$$

As in the case of the Γ-function, the involved integral is in general improper with respect to both integration limits. So we split it in two improper integrals,

$$R_\infty(s) := \int_1^\infty t^s \, [\, f(\mathrm{it}) - a_0 \,] \, \frac{dt}{t} \quad \text{and} \quad R_0(s) = \int_0^1 t^s \, [\, f(\mathrm{it}) - a_0 \,] \, \frac{dt}{t} \, ,$$

such that

$$R(s) = R_0(s) + R_\infty(s) \, .$$

The integral $R_\infty(s)$ converges in the whole plane and represents an entire function. This is because of the fact that $f(\mathrm{it}) - a_0$ decays exponentially for $t \to \infty$, since

$$e^{\frac{2\pi t}{\lambda}} \, [\, f(\mathrm{it}) - a_0 \,]$$

is bounded for $t \to \infty$ (since the corresponding q-power series is bounded near zero).

The study of $f(\mathrm{it})$ for $t \to 0$ turns out to be slightly more complicated. The functional equation for $f(\mathrm{it})$ is a good help,

$$f\left(\frac{\mathrm{i}}{t}\right) = \varepsilon t^k f(\mathrm{it}) \, ,$$

since the roles of ∞ and 0 are exchanged in the above equation. It is then natural to substitute $t \mapsto 1/t$ in $R_0(s)$, and to use the functional equation. The result is

$$R_0(s) = \int_1^\infty t^{-s} \, [\, \varepsilon t^k f(\mathrm{it}) - a_0 \,] \, \frac{dt}{t} \, ,$$

and we can transform it to obtain

$$R_0(s) = \varepsilon \int_1^\infty t^{k-s} [\, f(\mathrm{it}) - a_0 \,] \, \frac{dt}{t} + \varepsilon a_0 \int_1^\infty t^{k-s} \, \frac{dt}{t} - a_0 \int_1^\infty t^{-s} \, \frac{dt}{t} \, .$$

The first integral can be expressed in terms of R_∞, and the other two integrals can be computed. We obtain

$$R_0(s) = \varepsilon R_\infty(k - s) - a_0 \left[\frac{\varepsilon}{k - s} + \frac{1}{s} \right] \, ,$$

and hence

$$R(s) = R_\infty(s) + \varepsilon R_\infty(k - s) - a_0 \left[\frac{\varepsilon}{k - s} + \frac{1}{s} \right] \, .$$

We have already recognized $R_\infty(s)$ to be an entire function, hence the above equation gives the meromorphic continuation of $R(s)$ (and correspondingly of $D(s)$) into the whole plane, and the location of the poles is clear. The functional equation for $R(s)$ is also obvious from this formula. From the integral representation we deduce that $R(s)$ is bounded in vertical strips outside some neighborhood of its poles. By partial integration, $u(t) = f(\mathrm{i}t) - a_0$, $v(t) = t^{s-1}$, it is easy to show that $R_\infty(s)$, and thus also $R(s)$, is decaying in any vertical strip (compare with Lemma VII.6.10).

Second part. We must construct the inverse map

$$\{\lambda, k, \varepsilon\} \longrightarrow [\lambda, k, \varepsilon] \ .$$

It is natural to realize this by an inverse of the integral representation formula for $R(s)$. The direct integral representation formula was based on the Γ-integral. So we are searching for an inverse formula for the Γ-integral. Such a formula is known as the *Mellin integral*, and we will deduce it in the following. Before doing this, we need to remind the asymptotic behavior of $\Gamma(s)$ for Im $s \to \infty$ which can be read from the STIRLING formula. As we know, the Γ-function is bounded —away from the poles— in any vertical strip. An essentially stronger result can be worked out using the STIRLING formula. In this formula, the essential term is the function

$$s^{s-\frac{1}{2}} = e^{(s-\frac{1}{2})\,\mathrm{Log}\,s} \qquad \text{(Log } s \text{ is the principal value) .}$$

We study this function in a vertical strip $a \le \sigma \le b$ and we impose the supplementary condition $|t| \ge 1$ to be away from the poles. Because of the law $\Gamma(\bar{s}) = \overline{\Gamma(s)}$, it is enough to focus on points in the upper half-plane, i.e. for $t \ge 1$. We can write

$$\mathrm{Log}\,s = \log|s| + \mathrm{i}\,\mathrm{Arg}\,s \ ,$$

and using

$$\lim_{t\to\infty} \mathrm{Arg}\,s = \frac{\pi}{2} \qquad \text{(in the vertical strip) ,}$$

we immediately understand the asymptotic behavior of

$$\left| s^{s-\frac{1}{2}} \right| = e^{\mathrm{Re}\,\left[(s-\frac{1}{2})\,\mathrm{Log}\,s\right]} \ ,$$

since we have

$$\mathrm{Re}\,\left[\left(s - \frac{1}{2}\right)\mathrm{Log}\,s\right] = \left(\sigma - \frac{1}{2}\right)\log|s| - t\,\mathrm{Arg}\,s \ .$$

The Γ-function is thus exponentially decaying for $|t| \to \infty$ in any vertical strip of finite width. More precisely:

Lemma VII.3.5 *Let ε be an arbitrarily small positive number, $0 < \varepsilon < \pi/2$. In any vertical strip*

$$a \le \sigma \le b \ ; \quad |t| \ge 1 \ ,$$

the Γ-function satisfies an estimate of the form

$$|\Gamma(s)| \le Ce^{-(\pi/2-\varepsilon)|t|}$$

with a suitable positive number $C = C(a, b, \varepsilon)$.

Now let σ be an arbitrary real number, which is not a pole of the Γ-function. We consider the improper integral

$$\int_{-\infty}^{\infty} \frac{\Gamma(\sigma + it)}{z^{\sigma+it}} \, dt \, .$$

Here,

$$z^{\sigma+it} = e^{(\sigma+it)\,\mathrm{Log}\,z}$$

is once more defined by the principal value of the logarithm. Taking into account the asymptotic behavior of the Γ-function on a vertical strip, and the relation

$$\left| z^{\sigma+it} \right| = e^{\sigma \log|z| - t\,\mathrm{Arg}\,z} \, ,$$

we obtain the absolute convergence of the integral under the condition

$$|\mathrm{Arg}\,z| < \frac{\pi}{2} \, ,$$

i.e. in the right half-plane Re $z > 0$.

Let us now specialize σ to the values

$$-\frac{1}{2}, \quad -\frac{3}{2}, \quad -\frac{5}{2}, \quad \dots \quad .$$

Using the functional equation and the resulting asymptotic behavior of the function $\Gamma(z)$ for Re $(z) \to -\infty$ we deduce

$$\lim_{k\to\infty} \int_{-\infty}^{\infty} \frac{\Gamma(\frac{1}{2} - k + it)}{z^{\frac{1}{2}-k+it}} \, dt = 0 \, .$$

The Residue Theorem now easily implies for $\sigma > 0$

$$i \int_{-\infty}^{\infty} \frac{\Gamma(\sigma + it)}{z^{\sigma+it}} \, dt = 2\pi i \sum_{n=0}^{\infty} \mathrm{Res}\left(\frac{\Gamma(s)}{z^s} ; \, s = -n \right) = 2\pi i \sum_{n=0}^{\infty} \frac{(-z)^n}{n!} \, .$$

We finally have obtained the MELLIN Inversion Formula for the Γ-Integral.

Lemma VII.3.6 (R.H. Mellin, 1910) *Under the assumption*

$$\sigma > 0 \ and \ Re \ z > 0$$

the following formula is true:

> ### The Mellin Inversion Formula
>
> $$e^{-z} = \frac{1}{2\pi} \int_{-\infty}^{\infty} \frac{\Gamma(\sigma + it)}{z^{\sigma+it}} \, dt \, .$$

Using this inversion formula, we can realize the promised function theoretical bridge from $D(s)$ to $f(z)$. So let us start with a DIRICHLET series $D(s)$. With a constant a_0, which still has to be defined, we consider the function

$$f(z) := \sum_{n=0}^{\infty} a_n e^{\frac{2\pi i n z}{\lambda}} \ .$$

It converges by VII.2.4 in the upper half-plane, and we have

$$f(iy) - a_0 = \frac{1}{2\pi} \sum_{n=1}^{\infty} a_n \int_{-\infty}^{\infty} \frac{\Gamma(s)}{\left(\frac{2\pi}{\lambda} n y\right)^s} \, dt$$

with $s = \sigma + it$, σ sufficiently large. It is now easy to show, using the asymptotic behavior of the Γ-function on vertical lines, that the summation and integration commute, and we are directly led to the desired formula

$$f(iy) - a_0 = \frac{1}{2\pi} \int_{-\infty}^{\infty} \frac{R(s)}{y^s} \, dt, \quad \sigma \text{ sufficiently large.}$$

Our aim is now to derive the functional equation for $f(iy)$ from the functional equation for $R(s)$ (see VII.3.2). Regarding the growth conditions imposed on $R(s)$, we know that it decays to zero in any vertical strip of the plane. Because of this, we can arbitrarily move the abscissa σ in the above formula also into the negative region, if we pick up the residues of the involved poles. Actually we want to move the abscissa σ to $k - \sigma$. We then pick up the residues of both poles at $s = k$ and $s = 0$:

$$f(iy) - a_0 = \frac{1}{2\pi} \int_{-\infty}^{\infty} \frac{R(k-s)}{y^{k-s}} \, dt + \operatorname{Res}\left(\frac{R(s)}{y^s}; s = 0\right) + \operatorname{Res}\left(\frac{R(s)}{y^s}; s = k\right) \ .$$

We can now give the precise value of a_0:

$$a_0 := -\operatorname{Res}\left(\frac{R(s)}{y^s}; s = 0\right) = -\operatorname{Res}(R(s); s = 0) \ .$$

The functional equation $R(k - s) = \varepsilon R(s)$ gives now immediately

$$f\left(\frac{i}{y}\right) = \varepsilon y^k f(iy) \ ,$$

and by analytic continuation

$$f\left(-\frac{1}{z}\right) = \varepsilon \left(\frac{z}{i}\right)^k f(z) \ .$$

$\square$

An *Example.*

(1) We consider the space

$$\left[2, \frac{1}{2}, 1\right] \ .$$

Functions in this space have in particular the transformation behavior

$$f(z+2) = f(z) \quad \text{and} \quad f\left(-\frac{1}{z}\right) = \sqrt{\frac{z}{i}}\, f(z) \ .$$

For instance, $\vartheta(z)$ is such a function. We claim

Proposition VII.3.7

$$\left[2, \frac{1}{2}, 1\right] = \mathbb{C} \cdot \vartheta(z) \ .$$

Proof. We use the results about the determination of modular forms of half-integral weight with respect to the theta group (see Appendix A). The substitutions of this group are generated by

$$z \mapsto z + 2 \quad \text{and} \quad z \mapsto -\frac{1}{z} \ .$$

The vector space $[\,\Gamma_\vartheta \,,\, 1/2 \,,\, v_\vartheta\,]$ is 1-dimensional. This follows for instance from the general Structure Theorem VI.6.3. We have only to show that any $f \in [2, 1/2, 1]$ also lies in this space, i.e. it is regular at all cusps of the theta group[1]. For this, we can use the following information. In the FOURIER series

$$f(z) = \sum_{n=0}^{\infty} a_n e^{\pi i n z}$$

the coefficients have polynomial growth. We will prove in the next two lemmas that this is enough to ensure regularity.

Lemma VII.3.8 *The map*

$$(a_n)_{n \geq 0} \longmapsto f(z) = \sum_{n=0}^{\infty} a_n e^{\frac{2\pi i n}{\lambda} z}$$

induces a bijection between

(1) *the set of all sequences $(a_n)_{n \geq 0}$ with polynomial growth, and*

(2) *the set of all analytic functions $f(z)$ in the upper half-plane with the properties*

   (a) *$f(z + \lambda) = f(z)$,*

   (b) *$f(z)$ is bounded in the region $y \geq 1$,*

---

[1] The theta group has two cusp classes.

(c)   *and there exist constants A, B with the property*

$$|f(z)| \le A \left(\frac{1}{y}\right)^{B} \quad \text{for all } y \le 1 .$$

(f thus has polynomial growth when approaching the real axis, and this growth is uniform with respect to the x-variable.)

Proof. We have

$$|f(z)| \le \sum_{n=0}^{\infty} |a_n| e^{-\frac{2\pi n y}{\lambda}} .$$

Since (a_n) has by assumption polynomial growth, we can estimate it by n^K for suitable natural K. The function

$$\sum_{n=0}^{\infty} n^K q^n , \quad |q| < 1 ,$$

is a rational function in q, as one can see by repeated differentiation of the geometric series (induction on K). The order of its pole at $q = 1$ is $b := K+1$. Then we have

$$\left| \sum_{n=0}^{\infty} n^K q^n \right| \le \frac{C}{|q-1|^b} \quad \text{for } |q-1| \le 1$$

with a suitable constant $C > 0$. We now substitute $q \mapsto e^{-\frac{2\pi y}{\lambda}}$, and obtain

$$|f(z)| \le \frac{C}{\left| e^{-\frac{2\pi y}{\lambda}} - 1 \right|^b} \quad \text{for } 0 < y < 1 .$$

The expression in the R.H.S. has polynomial growth in $1/y$ for $y \to 0$.   □

Lemma VII.3.9 *Let*

$$f(z) = \sum_{n=0}^{\infty} a_n e^{\frac{2\pi i n z}{\lambda}}$$

*be a Fourier series, whose coefficients a_n have polynomial growth. Assume that the function $f(z)$ has the transformation properties of a modular form in $[\, \Gamma \,,\, r/2 \,,\, v \,]$ with respect to some suitable congruence subgroup, and some suitable multiplicator system v. Then $f(z)$ is a modular form, it is thus regular at **all** cusps.*

Proof. We must show that $f(z)$ is regular at all cusps, i.e. that

$$g(z) = (cz + d)^{-r/2} f(Mz) \qquad (M \in \mathrm{SL}(2, \mathbb{Z}))$$

is bounded in $y \ge 1$. By assumption, f is regular at the cusp $i\infty$. We can thus assume that $M(i\infty)$ is different from $i\infty$. We know the periodicity of the function $g(z)$, and can take λ as a period. Then there exists a FOURIER expansion

$$g(z) = \sum_{n=-\infty}^{\infty} b_n e^{\frac{2\pi i n z}{\lambda}} .$$

The claim is equivalent to

$$b_{-n} = \int_0^1 g(\lambda z)e^{2\pi i n z}\, dx = 0 \quad \text{for } n > 0 .$$

We want to pass to the limit $y \to \infty$ in this integral. Since the exponential term is rapidly decaying, it is enough to show that $g(z)$ grows polynomially. This is an immediate consequence of the definition of $g(z)$ in connection with Lemma VII.3.8. Observe that the imaginary part of Mz converges to zero for $z \to i\infty$.

From our main result, we obtain

$$\dim\left\{ 2,\, \frac{1}{2},\, 1 \right\} = 1 .$$

This delivers a characterization of the corresponding DIRICHLET series,

$$1 + 2\sum_{n=1}^{\infty} e^{\frac{2\pi i n^2 z}{2}} \longmapsto 2\sum_{n=1}^{\infty}(n^2)^{-s} = 2\zeta(2s) .$$

We now obtain the famous functional equation of the RIEMANN ζ-function, and its unique characterization by this functional equation.

Theorem VII.3.10 (B. Riemann, 1859) *The Riemann ζ-function*

$$\zeta(s) = \sum_{n=1}^{\infty} n^{-s} \quad (\sigma > 1)$$

can be meromorphically extended into the whole plane $\mathbb{C}$. This extension is analytic in $\mathbb{C}\setminus\{1\}$, and there is a simple pole at $s = 1$ with residue 1. Defining

$$\xi(s) := \pi^{-s/2}\Gamma\left(\frac{s}{2}\right)\zeta(s) ,$$

the following functional equation

$$\xi(s) = \xi(1 - s) .$$

is satisfied. The function $\xi(s)$ is a meromorphic function, which is decaying in any vertical strip.

The relation $\xi(s) = \xi(1 - s)$ is equivalent to the historically first version of the functional equation,

$$\zeta(1-s) \;=\; 2^{1-s}\pi^{-s}\cos\frac{\pi s}{2}\,\Gamma(s)\,\zeta(s)\,, \qquad s\in\mathbb{C}\,,$$

given by RIEMANN in 1859. One can read off this formula the trivial zeros $\zeta(-2k)=0$ $(k\in\mathbb{N})$. For the original functional equation of the ζ-function see also Exercise 3 in VII.5.

Conversely to Theorem VII.3.10, we have (E. HECKE, 1936):

Supplement. *Let $D(s)$ be an analytic function in a right half-plane $\sigma > \sigma_0$ with the following properties:*

(1) $D(2s)$ *can be represented as a Dirichlet series in the right half-plane $\sigma > \sigma_0$.*

(2) $D(s)$ *can be meromorphically extended to the whole $\mathbb{C}$, and it has at $s=1$ a simple pole with residue 1.*

(3) $D(s)$ *satisfies the functional equation*

$$R(s) = R(1-s) \qquad with \qquad R(s) := \pi^{-s/2}\Gamma\left(\frac{s}{2}\right)D(s)\,.$$

(4) $R(s)$ *is decaying in any vertical strip.*

Then the function $D(s)$ identically coincides with the Riemann ζ-function.

The first characterization of the RIEMANN ζ-function by its functional equation and growth conditions goes back to H. HAMBURGER (1921, 1922), but he used modified assumptions. Especially, he used the stronger assumption, that the function $D(s)$ itself can be represented as a DIRICHLET series, and not just $D(2s)$.

Exercises for VII.3

1. Let D be a meromorphic function in the whole plane, which can be represented as a DIRICHLET series in a suitable right half-plane. We assume that there exists a natural number k, such that the following functional equation is satisfied,

$$R(s) = (-1)^k R(2k-s) \text{ with } R(s) = (2\pi)^{-s}\Gamma(s)D(s)\,.$$

We also assume that $R(s)$ is decaying in any vertical strip and that $D(s)$ is analytic excepting the point $s=2k$, which is either regular, or a simple pole.

Show: In the case $k=1$ the function D vanishes identically. In the cases $k=2,3,4$ we have

$$D(s) = C\zeta(s)\zeta(s+1-2k)\,, \qquad C\in\mathbb{C}\,.$$

2. Let D be a meromorphic function in the whole plane, and which can be represented as a DIRICHLET series in a suitable right half-plane. We assume that there exists a natural number r, such that the following functional equation is satisfied,

$$R(s) = R(r/2-s) \text{ with } R(s) = \pi^{-s}\Gamma(s)D(s)\,.$$

We also assume that $R(s)$ is decaying in any vertical strip and that $D(s)$ is analytic excepting the point $s = r/2$, which is either a removable singularity or a simple pole.

Show in the cases $r < 8$, that this DIRICHLET series is up to a constant factor of the form

$$D_r(s) = \sum_{n=1}^{\infty} A_r(n) n^{-s} \ ,$$

where $A_r(n)$ is the number of representations of n as a sum of r squares.

In the case $r = 1$ we have $D_1(s) = 2\zeta(2s)$. The DIRICHLET series $D_2(s)$ can also be written in the form

$$\zeta_K(s) := D_2(s) = \sum_{a \in \mathbb{Z} + i\mathbb{Z}} |a|^{-2s} \ .$$

(This is the ζ-function associated to the GAUSS number field $K = \mathbb{Q}(\sqrt{-1})$.)

3. Let D be a meromorphic function in the whole plane, which can be represented as a DIRICHLET series in a suitable right half-plane, $D(s) = \sum_{n=1}^{\infty} a_n n^{-s}$. We assume

$$a_1 = 1 \quad \text{and} \quad \lim_{n \to \infty} \frac{a_n}{n^{11}} = 0$$

and that D satisfies the functional equation

$$R(s) = R(12 - s), \quad \text{with} \quad R(s) = (2\pi)^{-s} \Gamma(s) D(s).$$

We also assume that $R(s)$ is decaying in any vertical strip. Show that the function $n \to a_n$ is exactly the RAMANUJAN τ-function, i.e. $a_n = \tau(n)$ for all $n \geq 1$. (See also Exercise 5 in VII.1.)

4. Verify the identities

$$f(z) := \sum_{n=-\infty}^{\infty} (-1)^n (n + 1/2) e^{\pi i z (n+1/2)^2} = 2 \sum_{n=0}^{\infty} (-1)^n (n + 1/2) e^{\pi i z (n+1/2)^2}$$

$$= 4 \sum_{n=-\infty}^{\infty} (n + 1/4) e^{4\pi i z (n+1/4)^2} = -\frac{1}{4\pi} \left. \frac{\partial \vartheta(4z, w)}{\partial w} \right|_{w=1/4} \ ,$$

and deduce from the JACOBI Theta Transformation Formula VI.4.2 the identity

$$f\left(-\frac{1}{z}\right) = \left(\frac{z}{i}\right)^{3/2} f(z) \ .$$

So we have $f \in [\, 8 \, , 3/2 \, , 1 \,]$.

5. Let

$$\chi(n) = \begin{cases} 0 & \text{if } n \text{ is even} \ , \\ 1 & \text{if } n \equiv 1 \mod 4 \ , \\ -1 & \text{if } n \equiv 3 \mod 4 \ . \end{cases}$$

Deduce from the previous Exercise, that the DIRICHLET series

$$L(s) = \sum_{n=1}^{\infty} \chi(n) n^{-s} \qquad (\sigma > 1)$$

can be analytically extended to the whole plane, where it satisfies the functional equation

$$R(s) = R(1-s) \quad \text{with} \quad R(s) = \left(\frac{\pi}{4}\right)^{-s/2} \Gamma\left(\frac{s+1}{2}\right) L(s) .$$

6.   Using Exercises 2 and 5, prove the identity

$$\boxed{\zeta_K(s) = 4\,\zeta(s)\,L(s) .}$$

We have the following number theoretical applications of this identity:

(a)   The number of representations of a natural number n as a sum of two squares of integers is given by

$$\boxed{A_2(n) := 4\sum_{d|n}\chi(d) = 4\sum_{\substack{d|n \\ d\equiv 1 \mod 4}} 1 - 4\sum_{\substack{d|n \\ d\equiv 3 \mod 4}} 1 .}$$

One can write this also as an identity of power series as follows:

$$\boxed{\left(\sum_{n=-\infty}^{\infty} e^{\pi i n^2 z}\right)^2 = 1 + 4\sum_{n=0}^{\infty}(-1)^n \frac{e^{\pi i(2n+1)z}}{1 - e^{\pi i(2n+1)z}} .}$$

(b)   The following formula holds:

$$L(s) = \prod_{p \text{ prime}} \left(1 - \frac{\chi(p)}{p^s}\right)^{-1} .$$

Deduce from Exercise 6 that the function $L(s)$ has no zero at $s = 1$, and from this:

There exist infinitely many prime numbers p with the property $p \equiv 1$ mod 4, and respectively with the property $p \equiv 3$ mod 4 .

This is a special case of the DIRICHLET Prime Number Theorem, which affirms that any arithmetic progression $\{\,a+kb\,;\ k \in \mathbb{N}\,\}$ contains infinitely many prime numbers when a and b are relatively prime. One can prove our special case also directly by elementary methods. The use of function theory for this special case shows the direction of the general proof, which is based on showing that a DIRICHLET series of the form

$$L(s) = \sum_{n=1}^{\infty} \chi(n)\, n^{-s}$$

does not vanish at $s = 1$. Here, χ is an arbitrary DIRICHLET character. The formula in Exercise 6 also possesses a generalization. Instead of the GAUSSIAN number field, one can take an arbitrary imaginary quadratic number field.

VII.4 The Riemann ζ-function and Prime Numbers

The theory of the prime number distribution is based on the RIEMANN ζ-*function*

$$\zeta(s) := \sum_{n=1}^{\infty} n^{-s} \qquad (n^s := \exp(s \log n)) \ .$$

As we know, this series converges normally in the half-plane Re $s > 1$, where it represents an analytic function. The bridge to prime numbers is given by the *Euler product formula* for the ζ-function (L. EULER, 1737): For Re $(s) > 1$ we have (see also VII.2.8)

$$\zeta(s) = \prod_{p \in \mathbb{P}} \left(1 - p^{-s}\right)^{-1} \quad \left(:= \prod_{\nu=1}^{\infty} \left(1 - p_\nu^{-s}\right)^{-1}\right),$$

where $\mathbb{P} := \{p_1, p_2, p_3, \dots\}$ is the set of all prime numbers in their natural ordering, $p_1 = 2$, $p_2 = 3$, $p_3 = 5$, ...

For the sake of completeness, we sketch one more direct proof. Using the geometric series

$$(1 - p^{-s})^{-1} = \sum_{\nu=0}^{\infty} p^{-\nu s}$$

one can show (CAUCHY product of series):

$$\prod_{k=1}^{m} \left(1 - p_k^{-s}\right)^{-1} = \prod_{k=1}^{m} \sum_{\nu=0}^{\infty} \frac{1}{p_k^{\nu s}} = \sum_{\nu_1,\dots,\nu_m=0}^{\infty} (p_1^{\nu_1} \cdots p_m^{\nu_m})^{-s} \ .$$

From the fact, that any natural number can be uniquely decomposed into prime factors (Fundamental Theorem of Elementary Number Theory), we have

$$\prod_{k=1}^{m} \left(1 - p_k^{-s}\right)^{-1} = \sum_{n \in \mathcal{A}(m)} n^{-s} \ ,$$

where $\mathcal{A}(m)$ is the set of all natural numbers that have only the primes $p_1, \dots, p_m$ in their prime factor decomposition.

For any natural number N, there exists a natural number m, such that $\{1, \dots, N\}$ is contained in $\mathcal{A}_m$. From this,

$$\lim_{m \to \infty} \prod_{k=1}^{m} (1 - p_k^{-s})^{-1} = \sum_{n=1}^{\infty} n^{-s} \ .$$

Finally, the estimate

$$\sum_{p} \left|1 - (1 - p^{-s})^{-1}\right| \leq \sum_{p} \sum_{m} \left|p^{-ms}\right| \leq \sum_{n=1}^{\infty} \left|n^{-s}\right|$$

implies the normal convergence of the EULER product for Re $(s) > 1$.   □

The ζ-function has in the convergence half-plane Re $(s) > 1$ no zero, since none of the factors vanishes there.

We once more (see also VII.2.8) formulate the fundamental convergence properties of the ζ-function, and its representability as an EULER product:

Proposition VII.4.1 *The series*

$$\zeta(s) := \sum_{n=1}^{\infty} n^{-s}$$

converges in the half-plane $\{\ s \in \mathbb{C} \ ; \quad Re\ (s) > 1\ \}$ *normally, where it represents an analytic function, the **Riemann ζ-function**. It has in this half-plane a representation as a (normally convergent) Euler product,*

$$\boxed{\zeta(s) = \prod_{p\in\mathbb{P}} \left(1 - p^{-s}\right)^{-1} .}$$

In particular, we have

$$\zeta(s) \neq 0 \text{ for } Re\ (s) > 1 .$$

The logarithmic derivative of the Riemann ζ-function

The derivative of $s \mapsto 1 - p^{-s}$ is $(\log p)p^{-s}$, the logarithmic derivative being thus

$$\frac{(\log p)p^{-s}}{1 - p^{-s}} = (\log p) \sum_{\nu=1}^{\infty} p^{-\nu s} .$$

This implies

$$-\frac{\zeta'(s)}{\zeta(s)} = \sum_{p}(\log p) \sum_{\nu=1}^{\infty} p^{-\nu s} .$$

The double series converges absolutely, because of $|p^{-\nu s}| = p^{-\nu Re\ (s)}$. Reordering the terms with respect to pure prime powers $n = p^{\nu}$, we obtain:

Lemma VII.4.2 *In the convergence half-plane, we have the formula:*

$$-\frac{\zeta'(s)}{\zeta(s)} = \sum_{n=1}^{\infty} \Lambda(n)\ n^{-s} \quad with$$

$$\Lambda(n) = \begin{cases} \log p , & if\ n = p^{\nu}\ ,\ p\ prime\ , \\ 0 , & else . \end{cases}$$

Our aim is to understand the asymptotic behaviour of the *summatoric function*

$$\psi(x) := \sum_{n \leq x} \Lambda(n)$$

by function theoretical methods. The function $\Lambda(n)$ is also called the *Mangoldt function,* and ψ is the *Tschebyscheff function.* During the following estimates it is convenient to use the LANDAU symbols "O" and "o".

Let $f, g : [x_0, \infty[\to \mathbb{C}$ be two functions. The notation

$$\boxed{f(x) = O\big(g(x)\big)}$$

has the following meaning:

There exist a constant $K > 0$, and a value $x_1 > x_0$, such that

$$|f(x)| \leq K\,|g(x)| \quad \text{for all} \quad x \geq x_1 \;.$$

In particular,

$$f(x) = O(1) \Longleftrightarrow f \text{ is bounded for } x \geq x_1 \;, \; x_1 \text{ suitable.}$$

The notation

$$f(x) = o(g(x))$$

has the following meaning:

For any $\varepsilon > 0$ there exists a number $x(\varepsilon) \geq x_0$, such that

$$|f(x)| \leq \varepsilon\,|g(x)| \quad \text{for} \quad x \geq x(\varepsilon) \;.$$

In particular,

$$f(x) = o(1) \Longleftrightarrow \lim_{x \to \infty} f(x) = 0 \;.$$

Finally, if $h(x)$, $x > x_0$, is a third function, then we write

$$f(x) = h(x) + O\big(g(x)\big) \qquad \text{instead of} \qquad f(x) - h(x) = O\big(g(x)\big) \;,$$
$$f(x) = h(x) + o\big(g(x)\big) \qquad \text{instead of} \qquad f(x) - h(x) = o\big(g(x)\big) \;.$$

Lemma VII.4.3 *Let Θ be the function*

$$\Theta(x) := \sum_{\substack{p \in \mathbb{P} \\ p \leq x}} \log p \;.$$

Then we have

$$\psi(x) = \Theta(x) + O\big((\log x)\sqrt{x}\big) \;.$$

The function $\Theta(x)$ is called the *Tschebyscheff theta function*.

Proof. Any term $\log p$ can be estimated by $\log x$, so it is enough to show

$$\# \{ (\nu, p) ; \quad 2 \leq \nu , \ p^\nu \leq x \} = O(\sqrt{x}) .$$

Since

$$p^\nu \leq x \Rightarrow p \leq \sqrt[\nu]{x} \text{ and } \nu \leq \frac{\log x}{\log p} \leq \frac{\log x}{\log 2} ,$$

the above number can be estimated by

$$\sum_{2 \leq \nu \leq \frac{\log x}{\log 2}} \sqrt[\nu]{x} = \sqrt{x} + \sum_{3 \leq \nu \leq \frac{\log x}{\log 2}} \sqrt[\nu]{x} .$$

The sum in the R.H.S. can be further estimated by

$$\frac{\log x}{\log 2} \sqrt[3]{x} = O(\sqrt{x}) .$$

Here, we have used that

$$\log x = O(x^\varepsilon) \text{ for any } \varepsilon > 0 . \qquad \square$$

The aim of the following sections is to prove the *Prime Number Theorem*:

Theorem VII.4.4 (Prime Number Theorem) *We have:*

$$\boxed{\Theta(x) = \sum_{\substack{p \leq x \\ p \in \mathbb{P}}} \log p = x + o(x) .}$$

Remark VII.4.4$_1$ *Because of* $\log x \cdot \sqrt{x} = o(x)$, *using Lemma VII.4.3 we see that Theorem VII.4.4 is equivalent to*

$$\psi(x) = \sum_{n \leq x} \Lambda(n) = x + o(x) .$$

The Prime Number Theorem is traditionally formulated in an other way:

Theorem VII.4.5 *Let* $\pi(x) := \# \{ p \in \mathbb{P} ; \quad p \leq x \}$ *. Then the following holds:*

$$\boxed{\begin{array}{c} \textbf{\textit{Prime Number Theorem}} \\ \lim_{x \to \infty} \left(\pi(x) \Big/ \frac{x}{\log x} \right) = 1 . \end{array}}$$

Even if there is no function theoretical relevance for this reformulation of the Prime Number Theorem, we show for the sake of completeness how to deduce it from VII.4.4.

We thus show: Theorem VII.4.4 $\Rightarrow$ Theorem VII.4.5 .

Let us define $r(x)$ by

$$\sum_{p \leq x} \log p = x(1 + r(x)),$$

so we have $r(x) \to 0$ for $x \to \infty$ by VII.4.4 .

Trivially, the following holds:

$$\sum_{p \leq x} \log p \leq \pi(x) \log x ,$$

and hence

$$\pi(x) \geq \frac{x}{\log x}(1 + r(x)) .$$

Slightly more complicated is the

Estimate of $\pi(x)$ from above.

Let us choose a number q, $0 < q < 1$. From the trivial estimate $\pi(x^q) \leq x^q$, we obtain for any $x > 1$

$$\sum_{p \leq x} \log p \geq \sum_{x^q \leq p \leq x} \log p \geq \log(x^q) \cdot \# \{ p ; \quad x^q \leq p \leq x \}$$

$$= q \log(x)(\pi(x) - \pi(x^q)) \geq q \log(x)(\pi(x) - x^q) .$$

This gives

$$\pi(x) \leq \frac{x}{\log x}(1 + r(x))q^{-1} + x^q .$$

This inequality can now be specified for a suitable value of q, namely for $q = 1 - 1/\sqrt{\log x}$ ($x \geq 2$). Then we have

$$\pi(x) \leq \frac{x}{\log x}(1 + R(x))$$

with

$$R(x) = -1 + (1 + r(x))\left(1 - \frac{1}{\sqrt{\log x}}\right)^{-1} + (\log x)x^{-1/\sqrt{\log x}} .$$

One easily derives $R(x) \to 0$ for $x \to \infty$.

Appendix : Error estimates

One can naturally ask the question, whether it is possible to find besides the qualitative information $r(x) = o(1)$ explicit error estimates. Indeed, our function theoretical method will give the following

Proposition VII.4.6 (Error estimates) *There exists a natural number N, such that*

$$\Theta(x) = x(1 + r(x)) , \qquad\qquad r(x) = O(1/\sqrt[N]{\log x}) ,$$
$$\pi(x) = \frac{x}{\log x}(1 + R(x)) , \qquad\qquad R(x) = O(1/\sqrt[N]{\log x}) .$$

(We will explicitly find $N = 128$.)

Using other methods, one can even prove that $N = 1$ also works. One has even

$$\frac{C_1}{\log x} \leq R(x) \leq \frac{C_2}{\log x} \quad (\, C_1, C_2 \text{ being suitable constants}) \, .$$

Better asymptotic formulas for $\pi(x)$ can be obtained by replacing $x/\log x$ by the *integral logarithm*

$$\mathrm{Li}(x) := \int_2^x \frac{1}{\log t} \, dt \, .$$

By partial integration, it is easy to show

$$\mathrm{Li}(x) = \frac{x}{\log x} \left(1 + s(x)\right) \, , \qquad s(x) = O(1/\log x) \, .$$

In the Theorem VII.4.5 and in VII.4.6 it is thus possible to replace $x/\log x$ by $\mathrm{Li}(x)$. It turns out that one can *better* approximate $\pi(x)$ by $\mathrm{Li}(x)$, namely we have (see also [Pr] or [Sch])

$$\pi(x) = \mathrm{Li}(x) + O\bigl(x \exp(-C\sqrt{\log x})\bigr)$$

with a positive constant C.

A stronger error estimate is conjectured:

Conjecture. *For any* $\varepsilon > 0$ *the **relative** error of the asymptotic* $\pi(x) \sim \mathrm{Li}(x)$ *is of order*

$$O\left(x^{-\frac{1}{2} + \varepsilon}\right) \, .$$

Equivalent to this conjecture is the famous RIEMANN Conjecture:

The Riemann Conjecture

$$\zeta(s) \neq 0 \text{ for Re } (s) > \frac{1}{2} \, .$$

This conjecture, stated in 1859 by B. RIEMANN, could not be decided despite the huge effort of the mathematical community. It is known, that there are infinitely many zeros on the critical line $\sigma = 1/2$.

The following figure gives the analytic landscape of the absolute value of $\zeta(s)^{-1}$, the zeros of ζ appear as poles.

The figure figures the first six non-trivial zeros $\varrho_n = \frac{1}{2} + it_n$ of the ζ-function with $t_n > 0$: The imaginary parts are

$$t_1 = 14.134725\ldots \, , \qquad t_2 = 21.022040\ldots \, ,$$
$$t_3 = 25.010856\ldots \, , \qquad t_4 = 30.424878\ldots \, ,$$
$$t_5 = 32.935057\ldots \, , \qquad t_6 = 37.586176\ldots \, .$$

The pole of the ζ-function at $s = 1$ appears in the figures as the only global minimum of $|1/\zeta(s)|$. The "thick tower" on the left side of the figure illustrates the trivial zero of the ζ-function at $s = -2$.

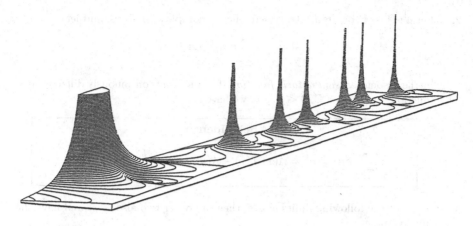

It may be that B. RIEMANN's initial intention was to prove the Prime Number Theorem through a proof of his conjecture. But the Prime Number Theorem was proven, independently, by J. HADAMARD and C. DE LA VALLÉE-POUSSIN in 1896. Both proofs are based on a weaker form of the RIEMANN Conjecture.

> *The ζ-function has no zero on the vertical line $\sigma = 1$.*

In the next section we will prove this. Using a so-called TAUBER theorem, one can then deduce from this the Prime Number Theorem. In the last section, we will prove a TAUBER Theorem, which also delivers a weak form for the error term in the Prime Number Theorem.

In his famous work [Ri2], B. RIEMANN made six conjectures concerning the ζ-function, one of them —now called the RIEMANN conjecture— is still open. In 1900 D. HILBERT included the RIEMANN Conjecture as problem number 8 in his famous list of 23 unsolved problems. For further remarks on the history of the Prime Number Theorem and the RIEMANN Conjecture see also the short exposition at the end of this chapter, page 444.

Exercises for VII.4

1. The MÖBIUS μ-function is defined by the equation

$$\frac{1}{\zeta(s)} = \prod_p \left(1 - p^{-s}\right) = \sum_{n=1}^{\infty} \frac{\mu(n)}{n^s} \ .$$

Show:

$$\mu(n) = \begin{cases} 1 & \text{if } n = 1 \ , \\ (-1)^k & \text{if } n = p_1 \cdots p_k \text{ is a product of } k \text{ } different \text{ primes } p_1, \ldots, p_k \\ 0 & \text{else} \ . \end{cases}$$

2.   Let $a : \mathbb{N} \to \mathbb{C}$ be an arbitrary sequence of complex numbers, and let

$$A(x) := \sum_{n \leq x} a(n) \quad (A(0) = 0)$$

be the associated *summatoric function*. Then for any continuously differentiable function $f : [x, y] \to \mathbb{C}, 0 < y < x$, we have

---

Abel's Identity

$$\sum_{y < n \leq x} a(n) f(n) = A(x)\, f(x) - A(y)\, f(y) - \int_y^x A(t)\, f'(t)\, dt \ .$$

---

3.   If one of the following limits exists, then the other two also exist and have the same value:

$$\lim_{x \to \infty} \frac{\psi(x)}{x} \ , \qquad \lim_{x \to \infty} \frac{\Theta(x)}{x} \ , \qquad \lim_{x \to \infty} \frac{\pi(x)}{x / \log(x)} \ .$$

4.   For Re $s > 2$ we have:

$$\frac{\zeta(s-1)}{\zeta(s)} = \sum_{n=1}^{\infty} \frac{\varphi(n)}{n^s} \ .$$

Here we use the notation

$$\varphi(n) = \#(\mathbb{Z}/n\mathbb{Z})^* \ .$$

$\varphi(n)$ is thus equal to the number of all cosets modulo n, which are relatively prime to n. The function $\varphi : \mathbb{N} \to \mathbb{N}$ defined by the above formula is the EULER (indicator) φ-function.

5.   Show that the series

$$\sum_{p \text{ prime}} \frac{1}{p}$$

diverges.

Hint. Assume the contrary, and deduce from this that the series

$$\sum \log(1 - p^{-s})$$

converges uniformly for $1 \leq \sigma \leq 2$. This would imply that $\zeta(\sigma), \sigma > 1$, is bounded for $\sigma \to 1$.

6.   *Show* $\zeta(\sigma) < 0$ for $0 < \sigma < 1$.

7.   Let p_n be the nth prime number with respect to the natural ordering. Then the Prime Number Theorem VII.4.5 is equivalent to

$$\lim_{n \to \infty} \frac{p_n}{n \log n} = 1 \ .$$

VII.5 The Analytic Continuation of the ζ-function

In the next Proposition we formulate the properties of the RIEMANN ζ-function needed for the proof of the Prime Number Theorem.

Proposition VII.5.1

I. The function $s \mapsto (s-1)\zeta(s)$ can be analytically extended to an open set, which contains the closed half-plane $\{s \in \mathbb{C} \; ; \quad \mathrm{Re}\,(s) \geq 1\}$. It has the value 1 at $s = 1$, i.e. ζ has a simple pole with residue 1 at $s = 1$.

II. Estimates in the half-plane $\{s \in \mathbb{C} \; ; \quad \mathrm{Re}\,(s) \geq 1\}$

(1) Estimate from above. For any $m \in \mathbb{N}_0$ there exists a constant C_m, such that the mth derivative of the ζ-function satisfies the estimate

$$\left|\zeta^{(m)}(s)\right| \leq C_m\,|t| \qquad \text{for } |t| \geq 1 \text{ and } \sigma > 1 \qquad (s = \sigma + it)\,.$$

(2) Estimate from below. There exists a constant $\delta > 0$ with the property

$$|\zeta(s)| \geq \delta\,|t|^{-4} \qquad \text{for } |t| \geq 1 \text{ and } \sigma > 1\,.$$

The ζ-function has in particular no zeros on the vertical line $\mathrm{Re}\,(s) = 1$. (We already know that ζ has no zeros in the open half-plane $\mathrm{Re}\,(s) > 1$.)

The proof of VII.5.1 will follow after a series of intermediate results (Lemmas VII.5.2 to VII.5.5).

I. We have proven at an other place (VII.3.10) much more: The function $s \mapsto (s-1)\zeta(s)$ has an analytic continuation to the whole $\mathbb{C}$, and satisfies a functional equation. For the Prime Number Theorem this result is not really necessary, and since we can show the analytic continuation of ζ slightly beyond the vertical line $\mathrm{Re}\,(s) = 1$ much simpler, we include the simple proof.

Lemma VII.5.2 For $t \in \mathbb{R}$ we define

$$\beta(t) := t - [t] - 1/2 \qquad \left([t] := \max\{n \in \mathbb{Z},\, n \leq t\}\right)\,.$$

Then we have $\beta(t+1) = \beta(t)$ and $|\beta(t)| \leq \frac{1}{2}$.
The integral

$$F(s) := \int_1^\infty t^{-s-1}\beta(t)\,dt$$

converges absolutely for $\mathrm{Re}\,(s) > 0$, and represents in this right half-plane an analytic function F. For $\mathrm{Re}\,(s) > 1$ it holds:

$$\zeta(s) = \frac{1}{2} + \frac{1}{s-1} - sF(s)\,. \qquad (*)$$

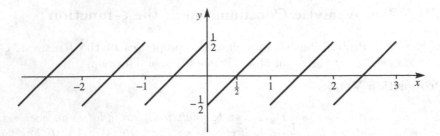

Remark. If one uses the R.H.S. in (∗) to define $\zeta(s)$ for Re $(s) > 0$, then this is the meromorphic continuation of ζ in the right half-plane Re $(s) > 0$. The only singularity is a simple pole at $s = 1$, and we obtain a new proof for

$$\lim_{s \to 1}(s-1)\zeta(s) = \mathrm{Res}(\zeta; 1) = 1 .$$

Proof of Lemma VII.5.2. From the estimate

$$\left|t^{-s-1}\beta(t)\right| \le t^{-\sigma-1} \qquad (\sigma = \mathrm{Re}\,(s))$$

we infer the convergence of the integral for Re $(s) > 0$, and the analyticity of F. (Compare this with the corresponding argumentation for the Γ-function.) Using partial integration, one proves for any natural number $n \in \mathbb{N}$ the formula

$$\int_n^{n+1} \beta(t)\frac{d}{dt}\,(t^{-s})\,dt = \frac{1}{2}\left((n+1)^{-s}+n^{-s}\right) - \int_n^{n+1} t^{-s}\,dt .$$

Summing these expressions for n from 1 to $N-1$, $N \ge 2$, we obtain after a small computation

$$\sum_{n=1}^N n^{-s} = \frac{1}{2} + \frac{1}{2}N^{-s} + \int_1^N t^{-s}dt - s\int_1^N t^{-s-1}\beta(t)\,dt$$

$$= \frac{1}{2} + \frac{1}{2}N^{-s} + \frac{N^{1-s}-1}{1-s} - s\int_1^N t^{-s-1}\beta(t)\,dt$$

$$= \frac{1}{2} + \frac{1}{2}N^{-s} + \frac{N^{1-s}}{1-s} + \frac{1}{s-1} - s\int_1^N t^{-s-1}\beta(t)\,dt .$$

Passing to the limit $N \to \infty$, and observing that

$$N^{-s},\; N^{1-s} \to 0 \text{ for } N \to \infty \quad (\text{for } \sigma > 1) ,$$

we obtain the identity claimed in Lemma VII.5.2.

II. (1) *The estimate from above.*

In the region $\sigma \ge 2$ the ζ-function is bounded:

$$|\zeta(s)| = \left|\sum n^{-s}\right| \leq \sum n^{-\sigma} \leq \zeta(2) .$$

The same argument shows that also the derivatives of ζ are bounded in this region, since one can differentiate the ζ-series termwise. So we can restrict to the vertical strip $1 < \sigma \leq 2$. Then it is enough to show that

$$\left|\zeta^{(m)}(s)\right| \leq C_m |s| \quad (1 < \sigma \leq 2, \ |t| \geq 1) .$$

For this, we use the integral representation in Lemma VII.5.2. (It is also possible to use the integral representation from Sect. VII.3.)

By the product formula, the higher derivatives of $sF(s)$ are a linear combination of $F^{(\nu)}(s)$ and $sF^{(\mu)}(s)$. So it is enough to show that each of the higher derivatives of F is bounded in the vertical strip $1 < \sigma \leq 2$. We have

$$F^{(m)}(s) = \int_1^\infty (-\log t)^m t^{-s-1} \beta(t) \, dt .$$

Using an estimate of the kind

$$|\log(t)| \leq C'_m t^{\frac{1}{2m}} \qquad (|t| \geq 1) \qquad \text{with a suitable constant } C'_m ,$$

in connection with $|\beta(t)| \leq 1$, we obtain

$$\left|F^{(m)}(s)\right| \leq C'_m \int_1^\infty t^{-\frac{3}{2}} dt < \infty .$$

Observation. This proof also shows that $F^{(m)}(s)$ is bounded in half-planes of the form $\sigma > \delta > 0$. Using the integral representation from Sect. VII.3, one can also remove the condition "$\delta > 0$". Of course, one can also replace "$|t| > 1$" by "$|t| \geq \varepsilon > 0$".

II. (2) *The estimate from below.*

We need a simple inequality.

Lemma VII.5.3 *Let a be a complex number of modulus 1. Then*

$$Re\,(a^4) + 4Re\,(a^2) + 3 \geq 0 .$$

Proof. From the binomial formula

$$(a + \bar{a})^4 = a^4 + \bar{a}^4 + 4(a^2 + \bar{a}^2) + 6$$

it follows

$$Re\,(a^4) + 4Re\,(a^2) + 3 = 8(Re\,a)^4 \qquad (\text{for } a\bar{a} = 1) . \qquad \square$$

We apply the inequality VII.5.3 to $a = n^{-it/2}$, and obtain

$$Re\,(n^{-2it}) + 4Re\,(n^{-it}) + 3 \geq 0 .$$

After multiplication on both sides with $n^{-\sigma}$, and with a non-negative real number b_n, followed by summation over n, we obtain:

Lemma VII.5.4 *Let b_1, b_2, b_3, ... be a sequence of non-negative numbers, such that the series*

$$D(s) = \sum_{n=1}^{\infty} b_n n^{-s} \qquad (\sigma > 1)$$

converges. Then we have

$$Re\, D(\sigma + 2it) + 4Re\, D(\sigma + it) + 3D(\sigma) \geq 0 \;.$$

Corollary. *Let*

$$Z(s) := e^{D(s)} \;,$$

then we have

$$|Z(\sigma + it)|^4 \, |Z(\sigma + 2it)| \, |Z(\sigma)|^3 \geq 1 \;.$$

We want to show that this Lemma can be applied to $\zeta(s) = Z(s)$. For this we consider

$$b_n := \begin{cases} 1/\nu & \text{if } n = p^\nu \,, \; p \text{ prime} \,, \\ 0 & \text{else} \,. \end{cases}$$

Then one has

$$D(s) = \sum_p \sum_\nu \frac{1}{\nu} p^{-\nu s} = \sum_p -\log(1 - p^{-s}) \,,$$

and because of this

$$e^{D(s)} = \prod_p (1 - p^{-s})^{-1} = \zeta(s) \;. \qquad \qquad \square$$

We obtain after a trivial transformation

Lemma VII.5.5 *For $\sigma > 1$ we have*

$$\left| \frac{\zeta(\sigma + it)}{\sigma - 1} \right|^4 \, |\zeta(\sigma + 2it)| \, [\zeta(\sigma)(\sigma - 1)]^3 \geq (\sigma - 1)^{-1} \;.$$

From this estimate we directly deduce that ζ has no zero on the vertical line $Re\,(s) = 1$:

If otherwise $\zeta(1 + it) = 0$ for some $t \neq 0$, then the L.H.S. of the above inequality would converge for $\sigma \to 1_+$ to a finite value

$$|\zeta'(1 + it)|^4 \, |\zeta(1 + 2it)| \;,$$

while the R.H.S. goes to ∞.

The following investigation of $|\zeta(s)|$ will give finally the estimate from below claimed in part II. (2) of Proposition VII.5.1.

We can once more restrict to the vertical strip $1 < \sigma \leq 2$, since for $\sigma > 2$ the function $|\zeta(s)|$ is already bounded from below by a positive constant,

$$|\zeta(s)| \geq 1 - |\zeta(s) - 1| \geq 1 - \sum_{n=2}^{\infty} n^{-2} > 0 \ .$$

To obtain an estimate from below of $|\zeta(s)|$, $1 < \sigma \leq 2$, we rewrite the inequality VII.5.5:

$$|\zeta(s)| \geq (\sigma - 1)^{3/4} |\zeta(\sigma + 2it)|^{-1/4} \left[\zeta(\sigma)(\sigma - 1)\right]^{-3/4} \ .$$

The function $\sigma \mapsto \zeta(\sigma)(\sigma - 1)$ is continuous on the interval $1 \leq \sigma \leq 2$, and it has no zeros on it. The modulus of this function is thus bounded from below by a positive constant. Using the already proven estimate

$$|\zeta(\sigma + it)| \leq C_0 |t| \qquad (|t| \geq 1)$$

we get

$$|\zeta(s)| \geq A(\sigma - 1)^{3/4} |t|^{-1/4} \qquad (1 < \sigma \leq 2 \ , \ |t| \geq 1) \qquad (*)$$

with a suitable constant A.

Let ε, $0 < \varepsilon < 1$, be a suitably small number, that we will chosen below. We define

$$\sigma(t) := 1 + \varepsilon |t|^{-5} \qquad (\in \ ]0,1[\text{ for } |t| \geq 1) \ .$$

We prove now the claimed inequality $|\zeta(s)| \geq \delta |t|^{-4}$ separately for $\sigma \geq \sigma(t)$ and for $\sigma \leq \sigma(t)$.

First case. $\sigma \geq \sigma(t)$. From the definition of $\sigma(t)$ and the estimate $(*)$, we obtain directly

$$|\zeta(\sigma + it)| \geq A(\varepsilon |t|^{-5})^{3/4} |t|^{-1/4} = A\varepsilon^{3/4} |t|^{-4} \ .$$

Second case. $\sigma \leq \sigma(t)$. Then one has

$$\zeta(\sigma + it) = \zeta(\sigma(t) + it) - \int_{\sigma}^{\sigma(t)} \zeta'(x + it) \ dx \ ,$$

and hence

$$|\zeta(\sigma + it)| \geq |\zeta(\sigma(t) + it)| - \left| \int_{\sigma}^{\sigma(t)} \zeta'(x + it) \ dx \right| \ .$$

Using the already proven estimate from above of $|\zeta'(s)|$, we deduce the existence of a further constant B, which does not depend on ε, such that

$$|\zeta(\sigma + it)| \geq |\zeta(\sigma(t) + it)| - B(\sigma(t) - 1) |t|$$
$$\geq A(\sigma(t) - 1)^{3/4} |t|^{-1/4} - B(\sigma(t) - 1) |t|$$
$$= (A\varepsilon^{3/4} - B\varepsilon) |t|^{-4} \ .$$

We can choose ε such that $\delta := A\varepsilon^{3/4} - B\varepsilon > 0$. Now the claimed estimate is proven. $\qquad \Box$

Exercises for VII.5

1. Show that the RIEMANN ζ-function has in the punctured plane $\mathbb{C} \setminus \{1\}$ the LAURENT expansion

$$\zeta(s) = \frac{1}{s-1} + \gamma + a_1(s-1) + a_2(s-1)^2 + \ldots \quad .$$

Here, γ is the EULER-MASCHERONI constant. (See also IV.1.9 or Exercise 3 in IV.1.)

2. A further elementary method for the meromorphic continuation of the ζ-function in the half-plane $\sigma > 0$ is to consider the functions

$$P(s) := \left(1 - 2^{1-s}\right)\zeta(s) = \sum_{n=1}^{\infty} \frac{(-1)^{n-1}}{n^s} \quad ,$$

$$Q(s) := \left(1 - 3^{1-s}\right)\zeta(s) = \sum_{n=1}^{\infty} a_n n^{-s}, \qquad a_n = \begin{cases} 1 & \text{if } n \not\equiv 0 \mod 3, \\ -2 & \text{if } n \equiv 0 \mod 3. \end{cases}$$

Show that $P(s)$ and $Q(s)$ converge in the half-plane $\sigma > 0$, and deduce from this that the ζ-function can be extended meromorphically into the half-plane $\sigma > 0$, with exactly one singularity at $s = 1$, which is a simple pole with residue 1.

3. The functional equation of the ζ-function can also be written in the following form:

$$\zeta(1-s) = 2(2\pi)^{-s}\Gamma(s)\cos\left(\frac{\pi s}{2}\right)\zeta(s) \ .$$

Deduce from this:

In the half-plane $\sigma \le 0$, the function $\zeta(s)$ has exactly the zeros $s = -2k,\ k \in \mathbb{N}$. All other zeros of the ζ-function are located in the vertical strip $0 < \operatorname{Re} s < 1$.

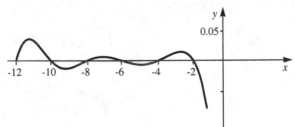

4. The function

$$\Phi(s) := s(s-1)\pi^{-s/2}\Gamma(s/2)\zeta(s)$$

has the following properties:

(a) Φ is an entire function.
(b) $\Phi(s) = \Phi(1-s)$.
(c) Φ is real on the lines $t = 0$ and $\sigma = 1/2$.
(d) $\Phi(0) = \Phi(1) = 1$.
(e) The zeros of Φ are located in the *critical strip* $0 < \sigma < 1$. Moreover, the zeros are placed symmetrically with respect to the real line and the critical line $\sigma = 1/2$ as symmetry axis.

5.   The following special case of the HECKE Theorem was already known to B. RIEMANN (1859):

$$\xi(s) := \pi^{-s/2}\Gamma\left(\frac{s}{2}\right)\zeta(s) = \sum_{n=-\infty}^{\infty}\int_0^\infty e^{-\pi n^2 t}t^{s/2}\,\frac{dt}{t}$$

$$= \frac{1}{2}\int_1^\infty\left(\vartheta(it)-1\right)\left(t^{s/2}+t^{(1-s)/2}\right)\frac{dt}{t}-\frac{1}{s}-\frac{1}{1-s}\ .$$

Deduce directly this special case, and use it to prove the meromorphic continuation and the functional equation.

6.   For $\sigma > 1$ the following integral representation holds (B. RIEMANN, 1859),

$$\Gamma(s)\cdot\zeta(s) = \int_0^\infty\frac{t^{s-1}e^{-t}}{1-e^{-t}}\,dt\ .$$

7.   We also owe to B. RIEMANN (1859) another proof for the analytic continuation of the ζ-function, and for its functional equation.

Let us consider the path γ sketched in the figure, $\gamma = \gamma_1 \oplus \gamma_2 \oplus \gamma_3$:

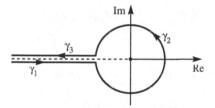

(RIEMANN considered instead this path reflected along the imaginary axis.) Both curves γ_1 and γ_3 are contained in the real axis. We say that γ_1 is at the lower bank and γ_3 is at the upper bank. For the integration along the "upper bank" γ_3 we define z^{s-1} using the principal branch of the logarithm of z. For the integration along the other two pieces γ_1 and γ_2 we define it such that $\gamma(t)^{s-1}$ remains continuous. This means in particular that on the lower bank z^{s-1} is defined not by the principal value $\operatorname{Log} z$, but by $\operatorname{Log} z - 2\pi i$ instead. (The integral over γ is thus strictly speaking a sum of three integrals.) Show that the integral

$$I(s) = \frac{1}{2\pi i}\int_\gamma\frac{z^{s-1}e^z}{1-e^z}\,dz$$

defines an entire function of s. The identity

$$\zeta(s) = \Gamma(1-s)\,I(s)$$

first holds for $\sigma > 1$. Since the integral converges in the whole plane it can be used for the meromorphic continuation. of $\zeta(s)$ into the whole plane.

VII.6 A Tauberian Theorem

Theorem VII.6.1 *Let there be given a sequence a_1, a_2, a_3, ... of non-negative real numbers, such that the Dirichlet series*

$$D(s) := \sum_{n=1}^{\infty} a_n n^{-s}$$

converges for Re $(s) > 1$. Assume the following:

I.   *The function $s \mapsto (s-1)D(s)$ can be analytically extended to an open set containing the closed half-plane $\{ s \in \mathbb{C} ;\quad \text{Re } (s) \geq 1 \}$, such that the extension D has at $s = 1$ a simple pole with the residue*

$$\varrho = \text{Res}(D; 1) .$$

II.   *The following estimates hold:*
    There exist constants C, κ with the property

$$|D(s)| \leq C \, |t|^{\kappa} \quad and \quad |D'(s)| \leq C \, |t|^{\kappa} \quad for \ \sigma > 1 \ , \ |t| \geq 1 .$$

Then the following holds:

$$\sum_{n \leq x} a_n = \varrho x (1 + r(x)) , \qquad\qquad where$$

$$r(x) = O\left(1/ \sqrt[N]{\log x} \right) , \qquad N = N(\kappa) \in \mathbb{N} \ suitably \ chosen .$$

(For instance one can take $N(\kappa) = 2^{[\kappa]+2}$.)

Remark VII.6.2 *The Dirichlet series*

$$D(s) = -\zeta'(s)/\zeta(s) = \sum_{n=1}^{\infty} \Lambda(n) \, n^{-s} , \qquad with$$

$$\Lambda(n) := \begin{cases} \log p & if \ n = p^{\nu} \ , \ p \ prime \ , \\ 0 & else \ , \end{cases}$$

satisfies the hypothesis of Theorem VII.6.1,

since the coefficients $\Lambda(n)$ are non-negative real numbers, and the series converges for Re $(s) > 1$. The function $s \mapsto (s-1)D(s)$ can be analytically extended to an open neighborhood of the half-plane $\{ s \in \mathbb{C} ;$ Re $s \geq 1 \}$, since this is true for the RIEMANN ζ-function, and since ζ has no zero in the region Re $(s) \geq 1$ (including the line Re $(s) = 1$).

The estimates for $D(s)$ and $D'(s)$ follow directly from the estimates in VII.5.1 for the RIEMANN ζ-function. (One can take $\kappa = 5$, which gives rise to $N(\kappa) = 2^7 = 128$.)

So we get:

From the Tauberian Theorem VII.6.1 in connection with the results VII.2.1 for the Riemann ζ-function, the Prime Number Theorem theorem follows.

For the *Proof* of the TAUBERian Theorem VII.6.1, it is useful to consider "higher" summatory functions, which are defined by

$$A_k(x) = \frac{1}{k!} \sum_{n \leq x} a_n \, (x - n)^k \qquad (k = 0, 1, 2, \dots) \, .$$

Then we have

$$A'_{k+1}(x) = A_k(x) \, , \qquad A_{k+1}(x) = \int_1^x A_k(t) \, dt \, ,$$

and

$$A_0(x) = A(x) = \sum_{n \leq x} a_n \, .$$

We will now determine the asymptotic behavior of $A_k(x)$ *for all* k (and not only for $k = 0$). For this, we will prove the following Lemmas:

We define $r_k(x)$ by

$$A_k(x) = \varrho \frac{x^{k+1}}{(k+1)!} (1 + r_k(x)) \, .$$

Lemma VII.6.3 *Let* $k \geq 0$. *Then*

$$r_{k+1}(x) = O\left(1/ \sqrt[N]{\log x}\right)$$

implies

$$r_k(x) = O\left(1/ \sqrt[2N]{\log x}\right) \, .$$

Lemma VII.6.4 *In case of* $k > \kappa + 1$ *we have*

$$r_k(x) = O(1/\log x) \, .$$

The above two Lemmas together imply

$$r_k(x) = O\left(1/ \sqrt[N_k]{\log x}\right) \qquad \text{with}$$

$$N_k := \begin{cases} 1 & \text{for } k > \kappa + 1 \, , \\ 2^{[\kappa]+2-k} & \text{for } k \leq \kappa + 1 \, . \end{cases}$$

In case of $k = 0$, this the TAUBERian Theorem VII.6.1.

Proof of Lemma VII.6.3. Since the function $x \mapsto A_k(x)$ is an increasing function[2], we have

$$cA_k(x) \le \int_x^{x+c} A_k(t)\, dt \quad \text{for any } c > 0 \,.$$

We will apply this inequality for $c = hx$, $x \ge 1$, with a suitable number $h = h(x)$, $0 < h < 1$, that we will choose later. The R.H.S. is equal to

$$A_{k+1}(x + hx) - A_{k+1}(x) = \frac{\varrho}{(k+2)!} \Big[(x+hx)^{k+2}(1 + r_{k+1}(x+hx)) $$
$$ - x^{k+2}(1 + r_{k+1}(x)) \Big] \,.$$

This implies

$$1 + r_k(x) \le \frac{(1+h)^{k+2}\big(1 + r_{k+1}(x+hx)\big) - \big(1 + r_{k+1}(x)\big)}{h(k+2)} \,.$$

Setting

$$\varepsilon(x) := \sup_{0 \le \xi \le 1} |r_{k+1}(x + \xi x)|$$

we obtain

$$r_k(x) \le \frac{(1+h)^{k+2}\big(1 + \varepsilon(x)\big) - \big(1 - \varepsilon(x)\big)}{h(k+2)} - 1$$
$$= \frac{\big[(1+h)^{k+2} + 1\big]\varepsilon(x)}{h(k+2)} + \frac{(1+h)^{k+2} - \big[1 + (k+2)h\big]}{h(k+2)} \,.$$

Now we choose for h the special value $h = h(x) = \sqrt{\varepsilon(x)}$. For sufficiently large values of x we have $h(x) < 1$. Together with h, the expression $(1+h)^{k+2} + 1$ is bounded from above. The first term in the estimate for r_k can be thus estimated from above by $\varepsilon(x)/h = \sqrt{\varepsilon(x)}$ up to some constant factor. The second term is a polynomial in h, whose free coefficient vanishes. It can be thus estimated by $h = \sqrt{\varepsilon(x)}$ up to some constant factor. Obviously, we have

$$\varepsilon(x) = O\left(1/\sqrt[N]{\log x}\right) \,.$$

We obtain an estimate of the form

$$r_k(x) \le K\sqrt{\varepsilon(x)} \,,$$

where K is a constant depending on k. For an O-estimate of $r_k(x)$ we need also an estimate of the absolute value, i.e. an estimate of r_k from below. Using the estimate

---

[2] When one wants to deduce from the growth behavior of a function the growth behavior of its derivative, it is necessary to know something about the oscillatory character of the derivative, e.g. that it is monotone.

$$cA_k(x) \geq \int_{x-c}^{x} A_k(t)\, dt = A_{k+1}(x) - A_{k+1}(x-c) \qquad \text{for } 0 < c < x$$

one obtains in the same way

$$r_k(x) \geq -K\sqrt{\varepsilon(x)} \qquad \text{(after we possibly enlarge } K) .$$

This implies

$$r_k(x) = O(\sqrt{\varepsilon(x)}) \qquad \text{and hence} \qquad r_k(x) = O\left(1/\sqrt[N]{\log x}\right) . \qquad \square$$

The rest of this section is devoted to the

Proof of Lemma VII.6.4.(This lemma completes the proof of the TAUBER Theorem VII.6.1, and hence also of the Prime Number Theorem.) Let us first notice:

Remark VII.6.5 *Let* $k \geq 1$ *be an integer, and let* $x > 0$, *and* $\sigma > 1$. *Then the following integral converges:*

$$\int_{\sigma-i\infty}^{\sigma+i\infty} \frac{|x^{s+k}|}{|s(s+1)\cdots(s+k)|}\, ds .$$

Here, the improper integral over the line Re $(s) = \sigma$ is in general defined by

$$\int_{\sigma-i\infty}^{\sigma+i\infty} f(s)\, ds := i \int_{-\infty}^{\infty} f(\sigma + it)\, dt .$$

The proof of VII.6.5 is trivial, since the integrand can be estimated by $1/\sigma^2$ up to some constant factor.

On the vertical line Re $s = \sigma$, the series $D(s)$ is dominated by the following series, which is independent of t,

$$\sum_{n=1}^{\infty} |a_n| n^{-\sigma} .$$

From VII.6.5, and from the LEBESGUE Limit Theorem, we obtain:

Corollary VII.6.6 *The integral*

$$\int_{\sigma-i\infty}^{\sigma+i\infty} \frac{D(s)x^{s+k}}{s(s+1)\cdots(s+k)}\, ds \qquad (k \in \mathbb{N} ,\ x > 0)$$

is absolutely convergent for $\sigma > 1$. *Integration and summation can be exchanged. The integral is thus equal to*

$$\sum_{n=1}^{\infty} a_n x^k \int_{\sigma-i\infty}^{\sigma+i\infty} \frac{(x/n)^s}{s(s+1)\cdots(s+k)}\, ds .$$

Exercise. Prove that summation and integral can be exchanged without using the LEBESGUE Limit Theorem by approximating the integral by proper integrals.

Let us now compute the integral in Corollary VII.6.6.

Lemma VII.6.7 *For any $k \in \mathbb{N}$, and any $\sigma > 0$ (we need only $\sigma > 1$) we have*

$$\frac{1}{2\pi i} \int_{\sigma - i\infty}^{\sigma + i\infty} \frac{a^s}{s(s+1)\cdots(s+k)}\, ds = \begin{cases} 0 & \text{for } 0 < a \le 1\,, \\ \frac{1}{k!}(1 - 1/a)^k & \text{for } a \ge 1\,. \end{cases}$$

Proof. Let

$$f(s) = \frac{a^s}{s(s+1)\cdots(s+k)}\,.$$

(1) $(0 < a \le 1)$. The integral of $f(s)$ along the path $\gamma := \gamma_1 \oplus \gamma_2$

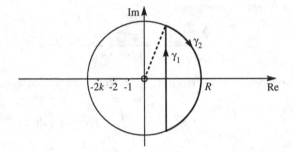

vanishes by the CAUCHY Integral Theorem. Because of the condition "$0 < a \le 1$" the function a^s is bounded uniformly in R on the integration contour. Passing to the limit $R \to \infty$, we obtain

$$\int_{\sigma - i\infty}^{\sigma + i\infty} f(s)\, ds = 0\,.$$

(2) $(a \ge 1)$. In this case we use the path $\tilde{\gamma} = \tilde{\gamma}_1 \oplus \tilde{\gamma}_2$,

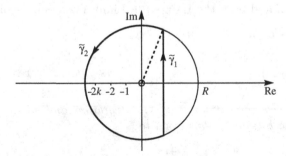

since on this integration contour a^s $(a \ge 1)$ is bounded uniformly in R. The Residue Theorem then implies

$$\frac{1}{2\pi i}\int_{\sigma-i\infty}^{\sigma+i\infty} = \sum_{\nu=0}^{k}\operatorname{Res}(f;-\nu) = \sum_{\nu=0}^{k}\frac{(-1)^{\nu}a^{-\nu}}{\nu!(k-\nu)!} = \frac{1}{k!}(1-1/a)^{k}\ . \qquad\square$$

From VII.6.6 and VII.6.7 we infer now a "function theoretical formula" for the generalized summatory function in case of $k \geq 1$.

Lemma VII.6.8 *In case of $k \geq 1$, we have for all $\sigma > 1$*

$$A_k(x) = \frac{1}{2\pi i}\int_{\sigma-i\infty}^{\sigma+i\infty}\frac{D(s)x^{s+k}}{s(s+1)\cdots(s+k)}\,ds\ .$$

We apply the Lemma for a fixed value of σ, for instance $\sigma = 2$. The estimate

$$|D(s)| \leq C\,|t|^{\kappa} \qquad (|t| \geq 1\,,\ 1 < \sigma \leq 2)$$

then holds by continuity also for $\sigma = 1$. This implies for a fixed value of x,

$$\left|\frac{D(s)x^{s+k}}{s(s+1)\cdots(s+k)}\right| \leq \text{Constant } |t|^{\kappa-k-1} \qquad |t| \geq 1\,,\ 1 \leq \sigma \leq 2\,,$$

$$\leq \text{Constant } |t|^{-2}\,, \qquad \text{if } k > \kappa+1\ .$$

Using the CAUCHY Integral Theorem, we can then move the integration contour (Re $(s) = 2$) to Re $(s) = 1$, if we avoid the singularity $s = 1$ by a small detour.

So let L be the following path.

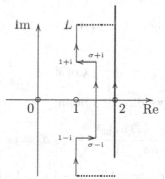

We obtain:

Lemma VII.6.9 *In case of $k > \kappa+1$, we have*

$$A_k(x) = \frac{1}{2\pi i}\int_L\frac{D(s)x^{s+k}}{s(s+1)\cdots(s+k)}\,ds\,, \qquad where$$

$$\int_L = \int_{1-i\infty}^{1-i} + \int_{1-i}^{\sigma-i} + \int_{\sigma-i}^{\sigma+i} + \int_{\sigma+i}^{1+i} + \int_{1+i}^{1+i\infty}\ .$$

Next, we estimate both improper integrals from $1-i\infty$ to $1-i$, and respectively from $1+i$ to $1+i\infty$. For this, we use

Lemma VII.6.10 (B. Riemann, H. Lebesgue) *Let*

$$I =]a, b[, \quad -\infty \le a < b \le \infty ,$$

be a (not necessarily finite) interval, and let $f : I \to \mathbb{C}$ be a function with the following properties:

(a)   *f is bounded.*
(b)   *f is continuously differentiable.*
(c)   *f and f' are absolutely integrable (from a to b).*

Then the function $t \mapsto f(t)x^{\mathrm{i}t}$ $(x > 0)$ is also absolutely integrable, and we have

$$\int_a^b f(t)x^{\mathrm{i}t} \, dt = O(1/\log x) .$$

Proof. We choose sequences

$$a_n \to a, \ b_n \to b, \quad a < a_n < b_n < b .$$

Then

$$\int_a^b f(t)x^{\mathrm{i}t} \, dt = \lim_{n\to\infty} \int_{a_n}^{b_n} f(t)x^{\mathrm{i}t} \, dt$$

$$= \frac{1}{\mathrm{i}\log x} \lim_{n\to\infty} \left(\left[f(t)x^{\mathrm{i}t} \right]_{a_n}^{b_n} - \int_{a_n}^{b_n} f'(t)x^{\mathrm{i}t} \, dt \right) .$$

By assumption, $f(t)$ is bounded, and $\left| f'(t)x^{\mathrm{i}t} \right| = |f'(t)|$ is integrable. This implies

$$\left| \int_a^b f(t)x^{\mathrm{i}t} \, dt \right| \le \text{Constant} \left| \frac{1}{\log x} \right| . \qquad \square$$

From Lemma VII.6.10, we directly obtain

$$\frac{1}{2\pi\mathrm{i}} \int_{1+\mathrm{i}}^{1+\mathrm{i}\infty} \frac{D(s)x^{s+k}}{s(s+1)\cdots(s+k)} \, ds = O(x^{k+1}/\log x) ,$$

and also the same for the integral from $1-\mathrm{i}\infty$ to $1-\mathrm{i}$. Both improper integrals contribute in Lemma VII.6.4 only to the error $r_k(x)$!

Let us now focus on the integral along the vertical line segment from $\sigma - \mathrm{i}$ to $\sigma + \mathrm{i}$. (At this point we still have $\sigma > 1$).

By Lemma VII.6.10,

$$\frac{1}{2\pi\mathrm{i}} \int_{\sigma-\mathrm{i}}^{\sigma+\mathrm{i}} \frac{D(s)x^{s+k}}{s(s+1)\cdots(s+k)} \, ds = O\left(x^{k+1} \frac{x^{\sigma-1}}{\log x} \right) .$$

Observe that $x^{\sigma-1}/\log x$ is *not* of the order $O(1/\log x)$ for $\sigma > 1$. But we have:

$$x^{\sigma-1}/\log x = O\left(\frac{1}{\log x}\right), \text{ for } \sigma \leq 1 .$$

It is thus a natural idea to move the integration contour to the left!

We know that $s \mapsto (s-1)\,D(s)$ *can be analytically extended to an open neighborhood of the closed half-plane* $\{\ s \in \mathbb{C}\ , \ \mathrm{Re}\,(s) \geq 1\}$.

There exists a number σ, $0 < \sigma < 1$, such that the closed rectangle with vertices in $\sigma - \mathrm{i}$, $2 - \mathrm{i}$, $2 + \mathrm{i}$ and $\sigma + \mathrm{i}$ is contained in this neighborhood.

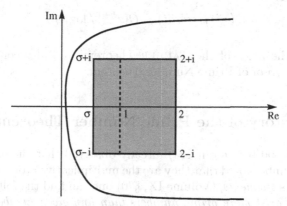

By the Residue Theorem, we have

$$\int_E * = \int_F * \ + \mathrm{Res}\left(\frac{D(s)x^{s+k}}{s(s+1)\cdots(s+k)} \; ; \; s = 1\right) .$$

Here, E and respectively Γ are the following integration paths:

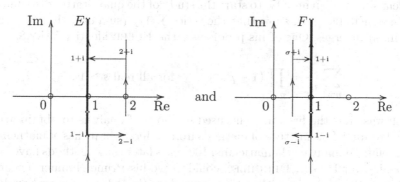

and

Since $D(s)$ has a pole of order one with residue ϱ at $s = 1$, the residue in the above formula is

$$\frac{\varrho}{(k+1)!}\,x^{k+1} .$$

This is exactly the main term in the asymptotic formula for $A_k(x)$ in the Lemma VII.6.4, that we are proving. All other terms "must be absorbed" by the error term. For the integral from $\sigma - \mathrm{i}$ to $\sigma + \mathrm{i}$ we have already seen it (using $\sigma \leq 1$). It remains to consider the integrals over the horizontal line segments from $\sigma + \mathrm{i}$ to $1 + \mathrm{i}$, and respectively from $\sigma - \mathrm{i}$ to $1 - \mathrm{i}$. Let us for instance show

$$\int_{\sigma+\mathrm{i}}^{1+\mathrm{i}} \frac{D(s)x^{s+k}}{s(s+1)\cdots(s+k)}\, ds = O(x^{k+1}/\log x) \ .$$

This integral can be estimated by

$$O\Big(x^k \int_{\sigma}^{1} x^t\, dt\Big) = O(x^{k+1}/\log x) \ .$$

This finishes the proof of the TAUBER Theorem, and also concludes at the same time the proof of Prime Number Theorem.   □

A short history of the Prime Number Theorem

EUCLID (circa 300 before our era) already observed that there are infinitely many prime numbers, and that they are the multiplicative atoms of the natural numbers. In his *Elements*, (Volume IX, §20) one can find the following Proposition: *"There exist more prime numbers than any given number of primes."* The proof of EUCLID is as simple as ingenious, so that it can be found in most monographs on the elementary number theory.

After EUCLID there is a huge gap in the mathematical literature. There cannot be found investigations about the distribution of the primes. First in 1737, as EULER gave new proofs for the infinity of the set of prime numbers, this problem obtained an impulse to start the study of the quantitative distribution of primes. EULER could show that the series $\sum 1/p$ (sum over the inverses of all primes) diverges. One of his proofs uses the EULER identity VII.2.8,

$$\sum_{n=1}^{\infty} \frac{1}{n^s} = \prod_{p \in \mathbb{P}} \left(1 - p^{-s}\right)^{-1} \qquad \text{for all real } s > 1 \ .$$

EULER was thus the first one who used methods of analysis to obtain arithmetical results! This mixture of methods from different branches of mathematics discomfited many mathematicians. 100 years later these methods have been accepted, after P.G.L. DIRICHLET could prove his Prime Number Theorem about prime numbers in arithmetic progressions (in 1837), using real-analytic methods as EULER before. In the meantime, C.F. GAUSS (1792/1793, being fifteen years old!) and A.M. LEGENDRE (1798, 1808) were searching for a "simple" function $f(x)$, which approximates the *prime number function*

$$\pi(x) := \#\{\ p \in \mathbb{P}\ ;\ p \leq x\ \} \ ,$$

in the sense that the relative error for $x \to \infty$ tends to zero, i.e.

$$\lim_{x \to \infty} \frac{\pi(x) - f(x)}{f(x)} = 0 \; .$$

Using tables of prime numbers from logarithm tables, they were led to conjectures, which are equivalent to the fact that

$$f(x) = \operatorname{Li} x := \int_2^x \frac{dt}{\log t} \; , \qquad \text{and respectively} \qquad f(x) = \frac{x}{\log x}$$

are such functions, but they could not prove this. However they were able to recover the result of EULER, that there are "infinitely fewer" prime numbers than natural numbers, or in a rigorous statement

$$\boxed{\lim_{x \to \infty} \frac{\pi(x)}{x} = 0 \; .}$$

A remarkable progress in the theory of the distribution of prime number was marked by the work of P.L. TSCHEBYSCHEFF around 1850. For sufficiently large values of x he could show:

$$\boxed{0.92129\ldots \cdot \frac{x}{\log x} < \pi(x) < 1.10555\ldots \cdot \frac{x}{\log x} \; ,}$$

i.e. $\pi(x)$ grows similar as $x/\log x$. His proof uses methods of elementary number theory. Moreover, using the ζ-function (only for real values of the argument s), he could show: If the limit

$$l := \lim_{x \to \infty} \frac{\pi(x)}{x/\log x}$$

exists, then it is $l = 1$.

The Prime Number Theorem was proven first in 1896, independently and almost simultaneously by J. HADAMARD and C. DE LA VALLÉE-POUSSIN. In the proofs, they both used essentially (besides HADAMARD's methods for transcendental functions) the fact that the function ζ has no zeros in certain regions including the closed half-plane $\operatorname{Re} s \geq 1$. The ζ-function for *complex* arguments had been introduced almost three decades earlier, in 1859, by B. RIEMANN in his famous paper "On the number of primes smaller than a given bound".[3]

RIEMANN could not prove the Prime Number Theorem, but he realized the connection between $\pi(x)$ resp. $\psi(x)$, and the non-trivial zeros of the ζ-function. He found "explicit formulae" for $\psi(x)$. One of these formulas is equivalent to

---

[3] Original German title: *"Über die Anzahl der Primzahlen unter einer gegebenen Grösse"*.

$$\psi(x) = x - \sum_{\varrho} \frac{x^{\varrho}}{\varrho} - \frac{\zeta'(0)}{\zeta(0)} - \frac{1}{2}\log\left(1 - x^{-2}\right) ,$$

where ϱ runs through the set of all non-trivial zeros of the ζ-function. From this formula it becomes plausible, that one can find the Prime Number Theorem written in the form $\psi(x) \sim x$ with an explicit error estimate, if one can find a number $\sigma_0 < 1$, such that all zeros lie in the region $\sigma \leq \sigma_0$. Unfortunately, the existence of such a bound is still open! The famous RIEMANN Conjecture states more, namely that one can choose $\sigma_0 = 1/2$. Because of the functional equation, this means that all non-trivial zeros of the ζ-function are located on the critical line $\sigma = 1/2$. A bound which is better then $\sigma_0 = 1/2$ cannot exist, since we know (G.H. HARDY, 1914) that there are infinitely many zeros on the critical line. A. SELBERG could prove in 1942, that the number $M(T)$ of all zeros ϱ on the critical line with $0 < \mathrm{Im}\,\varrho < T$, $T \geq T_0$, fulfill the estimate

$$M(T) > AT\log T$$

with a positive constant A. Already in 1905, H. VON MANGOLDT proved an asymptotic formula conjectured by RIEMANN for the number $N(T)$ of all zeros ϱ of the ζ-function in the critical strip $0 < \sigma < 1$ with $0 < \mathrm{Im}\,\varrho < T$:

$$N(T) = \frac{T}{2\pi}\log\frac{T}{2\pi} - \frac{T}{2\pi} + O(\log T) .$$

From this and from SELBERG's result it follows that a non-zero percentage of all non-trivial zeros lie on the critical strip. J.B. CONREY proved in 1989, that at least 40% of all non-trivial zeros lie on the critical line.

We also mention, that A. SELBERG and P. ERDŐS succeeded in 1948 (published in 1949) to give "elementary" proofs of the Prime Number Theorem, i.e. such proofs that do not use methods of complex analysis.

In 1903 only the first 15 zeros were located (J. P. GRAM). Using modern computers one could verify the RIEMANN conjecture for the first 10^{13} zeros. All computed zeros have been simple zeros.

These numerical investigations and many theoretical results are signs of evidence and support for the truth of RIEMANN's Hypothesis. Despite of a million dollar prize and extreme efforts of the mathematical community, a general proof of the RIEMANN Hypothesis (or a disproof) is still missing.

Exercises for VII.6

1.   Let $\mu(n)$ be the MÖBIUS μ-function. *Show:*

$$\sum_{n \leq x} \mu(n) = o(x) .$$

Hint. Apply the TAUBERian Theorem to

$$\zeta^{-1}(s) + \zeta(s) = \sum\left((\mu(n) + 1)n^{-s}\right) .$$

2.   Show
$$\frac{1}{2\pi i}\int_{2-i\infty}^{2+i\infty}\frac{y^s}{s^2}\,ds=\begin{cases}0\,, & \text{if }0<y<1\,,\\ \log y\,, & \text{if }y\ge1\,.\end{cases}$$

3.   For all $x\ge1$ and $c>1$ one has
$$\frac{1}{x}\sum_{n\le x}\Lambda(n)(x-n)=-\frac{1}{2\pi i}\int_{c-i\infty}^{c+i\infty}\frac{x^s}{s(s+1)}\frac{\zeta'(s)}{\zeta(s)}\,ds\,.$$

4.   Prove the following generalization of the HECKE Theorem:
Let $f:\mathbb{H}\to\mathbb{C}$ be an analytic function. We assume that both $f(z)$, and
$$g(z):=\left(\frac{z}{i}\right)^{-k}f\left(-\frac{1}{z}\right)$$
can be expanded into FOURIER series, where the involved coefficients have polynomial growth,
$$f(z)=\sum_{n=0}^{\infty}a_ne^{\frac{2\pi inz}{\lambda}}\,,\qquad g(z)=\sum_{n=0}^{\infty}b_ne^{\frac{2\pi inz}{\lambda}}\,.$$
Show, that both DIRICHLET series
$$D_f(s)=\sum_{n=1}^{\infty}a_nn^{-s},\quad D_g(s)=\sum_{n=1}^{\infty}b_nn^{-s}$$
can be meromorphically extended to the whole plane, where the following relation is satisfied
$$R_f(s)=R_g(k-s)\text{ with }R_f(s)=\left(\frac{2\pi}{\lambda}\right)^{-s}\Gamma(s)D_f(s)$$
and analogously for R_g. The functions $(s-k)D_f(s)$ and $(s-k)D_g(s)$ are entire, and one has:
$$\text{Res}(D_f;k)=a_0\left(\frac{\lambda}{2\pi}\right)^k\Gamma(k)^{-1}\,,\qquad\text{Res}(D_g;k)=b_0\left(\frac{\lambda}{2\pi}\right)^k\Gamma(k)^{-1}\,.$$
Examples are modular forms for arbitrary congruence subgroups.

5.   Let $S=S^{(r)}$ be a symmetric, rational, positive definite matrix. The EPSTEIN ζ-function
$$\zeta_S(s):=\sum_{g\in\mathbb{Z}^r\setminus\{0\}}S[g]^{-s}\qquad(\sigma>r/2)$$
can be analytically extended to the whole plane with the exception of a simple pole at $s=r/2$. The following functional equation holds:
$$R(S;s)=\left(\sqrt{\det S}\right)^{-1}R\left(S^{-1};\frac{r}{2}-s\right)\quad\text{with}\quad R(S;s)=\pi^{-s}\Gamma(s)\zeta_S(s)\,.$$
The residue in the pole is
$$\text{Res}(\zeta_S;r/2)=\frac{\pi^{r/2}}{\sqrt{\det S}\,\Gamma(r/2)}\,.$$

Hint. Apply the theta transformation formula from Exercise 4. The number λ has to be determined such that $2\lambda S$ and $2\lambda S^{-1}$ are integral matrices.

Remark. The EPSTEIN ζ-function can also be defined for arbitrary *real* $S > 0$, but in general the resulting series is not an ordinary DIRICHLET series. However the statements about analytic continuation, functional equation, and residue are still valid. Once more the proof can be given by HECKE's method.

6.   Show that the Prime Number Theorem, e.g. in the version

$$\psi(x) = x + o(x) \ ,$$

implies that $\zeta(1 + it) \neq 0$ for all $t \in \mathbb{R}^\bullet$. The Prime Number Theorem, and the proposition "$\zeta(1 + it) \neq 0$ for all $t \in \mathbb{R}^\bullet$" are thus equivalent.

7.   At the end, a curiosity:

A rather trivial asymptotic behavior can be obtained for the summatoric function

$$A_r(1) + A_r(2) + \cdots + A_r(n) \sim V_r n^{r/2} \ ,$$

where V_r is the volume of the r-dimensional unit ball. If we put a compact unit cube (of volume 1) centered at any lattice point g inside the r-dimensional ball of radius $\sqrt{n}$, then we obtain a covering of the r-dimensional ball by unit cubes. It looks like a plastering of the ball, which is slightly deformed at the boundary.

Deduce now from the Theorems of HECKE and TAUBER the well known formula for the volume of the unit ball,

$$V_r = \frac{\pi^{r/2}}{\Gamma\left(\frac{r}{2} + 1\right)} \ .$$

VIII

Solutions to the Exercises

VIII.1 Solutions to the Exercises of Chapter I

Exercises in Sect. I.1

1. If a complex number z is given in the normal form $z = a + ib$, $a, b \in \mathbb{R}$, then $a = \operatorname{Re} z$ is the real part, and $b = \operatorname{Im} z$ the imaginary part of z. If it does not have this "explicit" form, one frequently has to bring it into this form:

$$\frac{i-1}{i+1} = \frac{i-1}{i+1} \cdot \frac{-i+1}{-i+1} = \frac{2i}{2} = i \,,$$

i.e. explicitly

$$\operatorname{Re} \frac{i-1}{i+1} = 0 \,, \quad \operatorname{Im} \frac{i-1}{i+1} = 1 \,.$$

Similarly,

$$\frac{3+4i}{1-2i} = -1 + 2i \,.$$

Because of $i^4 = 1$, the values of i^n for integer n lie among $1, i, -1, -i$, depending whether n is of the form $4k, 4k+1, 4k+2, 4k+3$, $k \in \mathbb{Z}$. Because of

$$\varrho := \frac{1+i}{\sqrt{2}} = \cos \frac{\pi}{4} + i \sin \frac{\pi}{4} \,,$$

it is an eighth (primitive) root of 1. The values of ϱ^n thus depend only on n modulo 8. Computing the powers ϱ^n for n from 0 to 7, we obtain the real parts

$$1, \quad \sqrt{2}/2, \quad 0 - \sqrt{2}/2, \quad -1, \quad \sqrt{2}/2, \quad \sqrt{2}/2.$$

We recognize $(1 + i\sqrt{3})/2$ as a sixth unit root, and proceed in the same way. The number $(1 - i)/\sqrt{2}$ is also an eighth unit root. The sum of all unit roots of order 8 is zero.

The value of the last expression is 2.

E. Freitag and R. Busam, *Complex Analysis*,
DOI: 10.1007/978-3-540-93983-2_VIII, © Springer-Verlag Berlin Heidelberg 2009

2. The modulus is always easy to compute, using e.g. $|z| = \sqrt{z\bar{z}}$. The argument is often harder to isolate, since inverse trigonometric functions are involved. A general closed formula will be given in Exercise 21, Sect. I.2. For instance, for real positive values of a, we have

$$\mathrm{Arg}\,\frac{1+ia}{1-ia} = \arccos\frac{1-a^2}{1+a^2} = 2\arctan a \ .$$

3. A simple proof using $|\mathrm{Re}\,z| \leq |z|$ is:

$$|z+w|^2 = (z+w)(\bar{z}+\bar{w}) = |z|^2 + 2\mathrm{Re}\,(z\bar{w}) + |w|^2$$
$$\leq |z|^2 + 2\,|z|\,|w| + |w|^2 = (|z|+|w|)^2 \ .$$

The equality holds, iff $z\bar{w}$ is real and non-negative.

4. All claims follow by direct computations. For instance,

$$\langle z, w\rangle^2 + \langle iz, w\rangle^2 = \big(\mathrm{Re}\,(z\bar{w})\big)^2 + \big(-\mathrm{Im}\,(z\bar{w})\big)^2 = |z\bar{w}|^2 = |z|^2\,|w|^2 \ ,$$

where we used $\langle iz, w\rangle = -\mathrm{Im}\,(z\bar{w})$.

The formula

$$\frac{\langle z, w\rangle}{|z|\,|w|} + i\,\frac{\langle iz, w\rangle}{|z|\,|w|} = \frac{\bar{z}w}{|z|\,|w|}$$

shows, that $\omega(z, w)$ is exactly the principal value of the argument of w/z.

5. Start with the double sum

$$\sum_{\nu=1}^{n}\sum_{\mu=1}^{n}|z_\nu\bar{w}_\mu - z_\mu\bar{w}_\nu|^2 = \sum_{\nu=1}^{n}\sum_{\mu=1}^{n}(z_\nu\bar{w}_\mu - z_\mu\bar{w}_\nu)(\bar{z}_\nu w_\mu - \bar{z}_\mu w_\nu) \ ,$$

and split it into 4 double sums, which can then be written as products of simpler sums.

6. (a) G_0 is a line, G_+ and G_- are the half-planes having G_0 as boundary.

(b) K is a circle.

(c) L is a lemniscate looking like ∞.

7. We search for solutions $x+iy$, $x, y \in \mathbb{R}$. The substitution into the equation gives $c = a + ib = z^2 = (x+iy)^2$, i.e. the two equations $x^2 - y^2 = a$ and $2xy = b$, in two unknowns x and y. Together with $x^2 + y^2 = |c|$ we get $2x^2 = |c| + a$ and $2y^2 = |c| - a$. This determines x and y up to sign. There are thus in principle 4 possibilities, the two correct ones are singled out by the condition $2xy = b$, i.e. xy has the same sign as b. The solutions are

$$z = \pm\left(\sqrt{\tfrac{1}{2}(|c|+a)} + i\varepsilon\,\sqrt{\tfrac{1}{2}(|c|-a)}\right) \ , \quad \varepsilon = \begin{cases} 1 & \text{if } b \geq 0 \ , \\ -1 & \text{if } b < 0 \ . \end{cases}$$

One solves the quadratic equation $z^2 + \alpha z + \beta = 0$ by the Babylonian identity

$$z^2 + \alpha z + \beta = \left(z + \frac{\alpha}{2}\right)^2 + \frac{4\beta - \alpha^2}{4} \ .$$

8. See Proposition I.1.7.

9.   The solutions are $z_\nu = e^{i\left(\frac{\pi}{6} + \frac{2\pi}{3}\nu\right)}$, $\nu = 0, 1, 2$.

10. If the coefficients are all real, then $P(\bar{z}) = \overline{P(z)}$.

11. (a) We have $\mathrm{Im}\,\dfrac{-1}{z} = \mathrm{Im}\,\dfrac{-\bar{z}}{z\bar{z}} = \dfrac{\mathrm{Im}\,z}{|z|^2}$.

   (b) Use $|w|^2 = w\bar{w}$ to check the equations.

12. After squaring the inequalities become trivial.

13. If $z = x + iy \in \mathbb{C}$ then we must have $\varphi(z) = x + \varphi(i)y = x \pm jy$, where j is a
   imaginary unit in $\tilde{\mathbb{C}}$, and in fact this formula indeed defines an isomorphism.
   In the special case $\tilde{\mathbb{C}} = \mathbb{C}$ we obtain the two automorphisms $z \mapsto z$ and $z \mapsto \bar{z}$,
   which leave $\mathbb{R}$ elementwise invariant.

   If φ is an automorphism of the field of real numbers, then $\varphi(1)$ is the neutral
   element with respect to the multiplication, i.e. $\varphi(1) = 1$. This gives $\varphi(x) = x$
   for all rational x. An automorphism of $\mathbb{R}$ maps squares to squares, and thus
   positive numbers into positive numbers. For an arbitrary real x, and for arbitrary rational a, b with $a < x < b$ we then deduce $a = \varphi(a) < \varphi(x) < \varphi(b) = b$.
   This shows $\varphi(x) = x$.

14. The intersection point of the line through -1 and $z = x + iy$ with the imaginary
   axis is computed as

$$i\lambda = \frac{iy}{1 + x} \ .$$

   Conversely, intersecting the line through $i\lambda$ and -1 with the unit circle we
   obtain the intersection point

$$x = \frac{1 - \lambda^2}{1 + \lambda^2}\ , \quad y = \frac{2\lambda}{1 + \lambda^2}\ .$$

15. (a) Write z in polar coordinates, $z = re^{i\varphi}$, then

$$\frac{1}{\bar{z}} = \frac{1}{r}e^{i\varphi} \ .$$

   The point $1/\bar{z}$ lies on the line through 0 and z, and has the absolute
   value $1/r$. From this we derive the following geometric construction. Let
   $0 < |z| < 1$. We take the line through z which is perpendicular to the line
   through 0 and z. and intersect it with the unit circle. The tangents in the
   intersection points then intersect in $1/\bar{z}$ (see also the right figure at page
   16).

   (b) Construct $1/\bar{z}$, and reflect it with respect to the real axis.

16. (a) For $a, b \in W(n)$ we trivially have $ab \in W(n)$ and $a^{-1} \in W(n)$.

   (b) One can take $\zeta = \exp(2\pi i/n)$. The map $n \mapsto \zeta^n$ is then a surjective
   homomorphism $\mathbb{Z} \to W(n)$ with kernel $n\mathbb{Z}$.

   An element ζ^d is a primitive root of unity of order n, iff n and d are
   relatively prime. The number of all primitive unit roots of order n is thus

$$\varphi(n) := \#\{d;\ \ 1 \le d \le n\ ,\ \gcd(d, n) = 1\}\ .$$

17. Check first that $\mathcal{C}$ is stable with respect to addition and multiplication of matrices. So $\mathcal{C}$ is a ring. The map

$$\mathbb{C} \longrightarrow \mathcal{C}, \quad a + ib \longmapsto \begin{pmatrix} a & -b \\ b & a \end{pmatrix}$$

is an isomorphism.

One can independently of the knowledge of $\mathbb{C}$ compute, that $\mathcal{C}$ satisfies the axioms for "'*the*"' complex number field.

18. The factor ring $K := \mathbb{R}[X]/(X^2 + 1)$ is a field, since $\mathbb{R}$ is a field and $X^2 + 1$ is a prime prime element in $\mathbb{R}[X]$. This must be proved. Let us denote by 1_K the image of 1, and by i_K the image of X in K. Then $K = \mathbb{R}1_K + \mathbb{R}i_K$. The axioms for a "field of complex numbers" are now easily checked, e.g. $i_K^2 = -1_K$.

19. As in Exercise 17, a direct computation shows that $\mathcal{H}$ is a ring. The unit element is the unit matrix. The formula

$$\begin{pmatrix} z & -w \\ \overline{w} & \overline{z} \end{pmatrix}^{-1} = \frac{1}{|z|^2 + |w|^2} \begin{pmatrix} \overline{z} & w \\ -\overline{w} & z \end{pmatrix}$$

shows that $\mathcal{H}$ is a skew field.

20. The bilinearity is clear. We only have to show that there are no zero divisors. For this, use the conjugation $\overline{(z, w)} := (\overline{z}, -w)$ on $\mathcal{C}$. A straightforward computation gives $\overline{u}(uv) = \mu(u)v$, where $\mu(u)$ is the sum of the squares of the eight components involved in u with respect an obvious $\mathbb{R}$-basis. Form $uv = 0$ we then have either $v = 0$ or $\mu(u) = 0$. In the last case then $u = 0$.

Exercises in Sect. I.2

1. Assuming convergence, and passing to the limit on both sides of the defining recursive relation we obtain only ± 1 as possible limits. If z_0 lies in the right half-plane $x_0 > 0$, then inductively all z_n also lie in the right half-plane. Correspondingly if z_0 lies in the left half-plane. A special case occurs when z_0 lies on the imaginary axis. Then the recursive values z_n are either purely imaginary, or undefined (if $z_n = 0$ then z_{n+1} is undefined). Hence the sequence cannot converge to ± 1. Without loss of generality we can assume that the initial value z_0 lies in the right half-plane. The auxiliary sequence (w_n) then satisfies the recursion $w_{n+1} = w_n^2$, and has $|w|_0 < 1$. So it is a null sequence. From $|z_n + 1| \geq 1$ we obtain that $+1$ is the limit.

2. Reduction to the case $a = 1$ (Exercise 1).

3. If (z_n) is a CAUCHY sequence, then the sequences (x_n) and (y_n) are also CAUCHY sequences, and conversely.

4. (a) Simple estimates lead to

$$|\exp(z) - 1| \leq \sum_{\nu=1}^{\infty} \frac{|z|^\nu}{\nu!} = \exp(|z|) - 1 = |z| \left(1 + \sum_{\nu=1}^{\infty} \frac{|z|^\nu}{(\nu+1)!} \right)$$

$$\leq |z| \sum_{\nu=0}^{\infty} \frac{|z|^\nu}{\nu!} = |z| \exp |z| \ .$$

(b) Estimate the remainder term by means of the geometric series.

5.   We solve exemplarily the equation $\cos z = a$. This becomes a quadratic equation for $q := \exp(iz)$, namely $q^2 - 2aq + 1 = 0$. The solutions are

$$z \equiv -i \log \left(a \pm \sqrt{a^2 - 1} \right) \quad \mod 2\pi \ .$$

For concrete values of a the square root can also be made explicit by Exercise 7, Sect. I.1.

6.   Part (a) is trivial. The remaining claims follow from (a) and the corresponding properties of cos and sin.

7.   Part (a) is valid, since the coefficients involved in the defining power series are real. Part (b) follows from Exercise 6 (a), and the Addition Theorem.

   The inequality $|\sin z| \leq 1$ is equivalent with $|y| \leq \operatorname{Arcsinh}|\cos x|$. Then take n to be for instance $[\log 20\,000] + 1$.

8.   Express sin and cos in terms of the exponential function.

9.   The inverse map is given by

$$a_0 = S_0 \ , \quad a_n = S_n - S_{n-1} \quad (n \geq 1) \ .$$

10. One has

$$\sum_{\nu=0}^{n} a_\nu = b_0 - b_{n+1} \ .$$

11. The convergence is ensured e.g. by the quotient criterion. The required functional equation is equivalent to

$$\sum_{\nu=0}^{n} \binom{\alpha}{\nu} \binom{\beta}{n-\nu} = \binom{\alpha + \beta}{n} .$$

   Use induction on n for this.

12. Use termwise differentiation of the geometric series, applied k times.

13. Substitute the defining sum for A_ν in the R.H.S. and look for cancellations.

14. Use ABEL's partial summation (Exercise 13).

15. We use ABEL's partial summation once more, and we use the notations of Exercise 13. From the assumptions, (A_n) and (b_n) converge. Apply now Exercise 14 (b).

   We want now to show directly even more, namely the *absolute* convergence. The sequence (A_n) is bounded, so it is enough to show the convergence of $\sum |b_n - b_{n+1}|$. But since (b_n) is monotone we can remove the modulus. The claim follows from the convergence of (b_n).

16. Assume $\sum a_n$ converges absolutely. For

$$A_n := \sum_{k=0}^{n} a_k \ , \quad B_n := \sum_{k=0}^{n} b_k \ , \quad C_n := \sum_{k=0}^{n} c_k$$

we then have

$$C_n = a_0 B_n + a_1 B_{n-1} + \cdots + a_n B_0 = \sum_{j=0}^{n} a_j B_{n-j} \ .$$

With $B = \lim B_n$ we get

$$|A_n B - C_n| \leq \sum_{j=0}^{n} |a_j| \, |B - B_{n-j}|$$

Let $\varepsilon > 0$ be arbitrary. Since the series (A_n) converges absolutely, there exists an N with the property $\sum_{j>N} |a_j| < \varepsilon$. Since (B_n) converges, the sequence $(B - B_n)$ is bounded, $|B - B_n| \leq M$. For $n > N$ we then have

$$|A_n B - C_n| \leq \sum_{j=0}^{N} |a_j| \, |B - B_{n-j}| + C\varepsilon \ .$$

Setting $M' = M + \sum_{j=0}^{\infty} |a_j|$ we get for sufficiently large n

$$|A_n B - C_n| \leq M'\varepsilon.$$

17. Without loss of generality we can assume $S = 0$. Let $\varepsilon > 0$ be arbitrary. Then there exists a natural number N with the property $|S_n| \leq \varepsilon$ for $n > N$. Then the formula

$$\sigma_n = \frac{S_0 + \cdots + S_N}{n+1} + \frac{S_{N+1} + \cdots + S_n}{n+1}$$

implies

$$|\sigma_n| \leq \frac{|S_0| + \cdots + |S_N|}{n+1} + \varepsilon \frac{n-N}{n+1} \ ,$$

and the claim follows.

18. Replace in the geometric sum formula

$$\sum_{\nu=0}^{n} q^n = \frac{1 - q^{n+1}}{1 - q}$$

the value of q by $\exp(\pi i \varphi)$, and split into real and imaginary parts.

19. We have

$$\frac{z^n - 1}{z - 1} = \prod_{\nu=1}^{n-1} (z - \zeta^\nu) \ , \quad z \neq 1 \ .$$

Passing to the limit $z \to 1$, we obtain

$$\prod_{\nu=1}^{n-1} (1 - \zeta^\nu) = n \ .$$

This is the needed formula, after expressing the sine in terms of the exponential function, $\sin z = \big(\exp(\mathrm{i}z) - \exp(-\mathrm{i}z)\big)/2\mathrm{i}$.

20. We consider only (b):

$$\big(\mathrm{i}(\mathrm{i} - 1)\big)^{\mathrm{i}} = e^{3\pi/4} e^{\mathrm{i} \log \sqrt{2}}, \quad \mathrm{i}^{\mathrm{i}} (\mathrm{i} - 1)^{\mathrm{i}} = e^{-5\pi/4} e^{\mathrm{i} \log \sqrt{2}} \ .$$

The absolute values of the two numbers are different.

21. It is enough to restrict to the case $|z| = 1$. Then $|x| \leq 1$, and there exists an $\alpha \in [0, \pi]$ with $\cos \alpha = x$, i.e. $\alpha = \arccos x$. This implies $\sin \alpha = \pm y$. In case of $z = -1$ we have $\alpha = \pi$, and $\mathrm{Arg}\, z = \pi$. In case of $z \neq -1$ we distinguish the cases of z being in the upper closed or in the lower half-plane. Then we have $\mathrm{Arg}\, z = \alpha$, and respectively $\mathrm{Arg}\, z = -\alpha$.

22. The number $k(z, w)$ must be determined such that the R.H.S. is a complex number in the vertical strip $-\pi < y \leq \pi$.

23. The argumentation uses the "formula" $\left(e^z\right)^z = e^{\left(z^2\right)}$ (for $z = 1 + 2\pi i n$), but this "formula" is in general false.

Exercises in Sect. I.3

The Exercises 1 to 5, 7 and 8 have the purpose to recall facts from real analysis. These are very basic facts, so in case of open questions we refer to introductory text books.

6. For the first part of the exercise use the identity

$$\exp z - \left(1 + \frac{z}{n}\right)^n = \frac{z}{n}\left(\frac{\exp(z/n) - 1}{z/n} - 1\right)\sum_{\nu=0}^{n-1}\exp\left(\frac{z}{n}\right)^{n-1-\nu}\left(1 + \frac{z}{n}\right)^\nu,$$

which immediately gives the estimate

$$\left|\exp z - \left(1 + \frac{z}{n}\right)^n\right| \leq |z|\left|\frac{\exp(z/n) - 1}{z/n} - 1\right|\exp|z|.$$

The involved quotient converges to the derivative of exp at 0, which is 1. The whole expression then converges to 0.

The same proof also works in general.

9. We assume the existence of a function f with properties (a) and (b). Then we have

$$1 = f(1)^2 = f(1)f(1) = f(1 \cdot 1) = f(1).$$

Because of

$$-1 = f(-1)^2 = f(-1)f(-1) = f((-1)(-1)) = f(1),$$

we get $-1 = 1$, which is a contradiction.

10. It is enough to prove (a). Let $a \in \mathbb{C}$, $a \neq 0$, be fixed. The function

$$g(z) := \frac{f(a)f(z)}{f(az)} \quad (z \neq 0)$$

takes only the values ± 1. By continuity, g is constant. The value of this constant is $g(1) = f(1) = \pm 1$. After replacing f by $\pm f$ we can can assume $f(1) = 1$. This implies $f(a)f(z) = f(az)$, and we apply the previous exercise.

11. Apply Exercise 10 for the function $f(z) = \sqrt{|z|}\exp(\,i\varphi(z)/2\,)$.

12. Apply Exercise 10 for the function $f(z) = \exp(l(z)/2)$.

13. One argues as in Exercise 9.

14. Compare with Exercise 10.

Exercises in Sect. I.4

1. We will exemplarily prove the LEIBNIZ product formula. By assumption,

    $$f(z) = f(a) + \varphi(z)(z - a) , \quad g(z) = g(a) + \psi(z)(z - a) ,$$

    where the functions φ, ψ are continuous at a, taking respectively the values $\varphi(a) = f'(a)$, $\psi(a) = g'(a)$. Taking the product $f(z)g(z) = f(a)g(a) + \chi(z)(z - a)$, we are lead to the function χ with

    $$\chi(z) = \varphi(z)\, g(a) + f(a)\, \psi(z) + \varphi(z)\psi(z)\, (z - a) .$$

    The function χ is also continuous at a, where it takes the value

    $$(fg)'(a) = \chi(a) = \varphi(a)g(a) + f(a)\psi(a) = f'(a)g(a) + f(a)g'(a) .$$

2. All listed functions are continuous. The function $f(z) = z\,\mathrm{Re}\, z$ is complex differentiable only at the origin, where the derivative is 0. The function $f(z) = \bar{z}$ is nowhere complex differentiable. This can be seen by considering the difference quotient on parallels to the axes. The function $f(z) = z\bar{z}$ is complex differentiable only at the origin, where the derivative is 0. The last function in (a) is nowhere differentiable.

    The complex differentiability of the exponential function can be reduced by means of the functional equation to the differentiability at the origin. The claim is reduced to the case of the real exponential function by means of the estimate:

    $$\left| \frac{\exp z - 1}{z} - 1 \right| \le \frac{\exp |z| - 1}{|z|} - 1 .$$

    This follows immediately by from the power series.

3. We assume that f takes only real values. Then the difference quotient

    $$\frac{f(a + h) - f(a)}{h}$$

    is real for a real h, and purely imaginary for a purely imaginary h. The derivative $f'(a)$ is thus both real and purely imaginary, i.e. it is zero. Then the partial derivatives of f vanish everywhere, so f is constant.

4. This follows using the difference quotient.

5. Let $z, a \in D$ be different, $z \ne a$. We set $b = f(a)$, and $w = f(z)$. Then

    $$\frac{f(z) - f(a)}{z - a} = \frac{w - b}{g(w) - g(b)} = \frac{1}{\dfrac{g(w) - g(b)}{w - b}} .$$

    Now we pass to the limit with $z \to a$. Since f is continuous, this implies $w \to b$, and the claim is proven.

6. The logarithm is the inverse function of the exponential function. Now apply Exercises 2(b) and 5.

Exercises in Sect. I.5

1. The CAUCHY-RIEMANN differential equations are satisfied for $f(z) = z\operatorname{Re} z$ only in the origin, for $f(z) = \bar{z}$ nowhere, for $f(z) = z\bar{z}$ only in the origin, for $f(z) = z/|z|$ ($z \neq 0$) nowhere, and finally for $f(z) = \exp z$ in the complex plane.

2. The CAUCHY-RIEMANN differential equations are satisfied only on the coordinate axes. In particular, there is no open set where these equations are valid.

3. Use the formulas from the Exercises to Sect. I.2.

4. The function f is obviously analytic on the complement of $\{0\}$ in $\mathbb{C}$. It is unbounded in a neighborhood of the origin, as we can see by considering $z = \varepsilon(1 + i)$. So it cannot be analytic in the whole plane. We remark also that the two partial derivatives exist in 0, they are both zero, so the CAUCHY-RIEMANN equations are satisfied in the entire $\mathbb{C}$. This is because the restriction of f to both axes is rapidly decaying, when one approaches 0.

5. We consider among the 10th roots of unity each second one,

$$a_j = \exp(2\pi i(2j + 1)/10) , \qquad 0 \leq j < 5 .$$

Then we slit the plane along the half-lines ta_j ($t \geq 1, 0 \leq j < 5$).

6. The parts (a) and (b) follow from the CAUCHY-RIEMANN differential equations in connection with Remark I.5.5. Finally, let $f = u + iv$ be analytic satisfying (c). Then $|f|^2 = u^2 + v^2$ is constant. We can assume that this constant is non-zero. Differentiating this expression with respect to x and y, and making use the CAUCHY-RIEMANN differential equations, we obtain the system

$$uu_x - vu_y = 0 , \qquad uu_y + vu_x = 0 .$$

This gives $u_x = u_y = 0$.

7. The searched functions are $z^3 + 1$, $1/z$, $z \exp z$ and $\sqrt{z}$ (principal value).

8. From the chain rule,

$$\frac{\partial U}{\partial r} = \frac{\partial u}{\partial x} \cos \varphi + \frac{\partial u}{\partial y} \sin \varphi .$$

A repeated application of the chain rule gives

$$\frac{\partial^2 U}{\partial r^2} = \left(\frac{\partial^2 u}{\partial x^2} \cos \varphi + \frac{\partial^2 u}{\partial x \partial y} \sin \varphi \right) \cos \varphi + \left(\frac{\partial^2 u}{\partial x \partial y} \cos \varphi + \frac{\partial^2 u}{\partial y^2} \sin \varphi \right) \sin \varphi .$$

A double application of the chain rule, assisted by the product formula, gives

$$\frac{\partial^2 U}{\partial \varphi^2} = -r \sin \varphi \left[-\frac{\partial^2 u}{\partial x^2} r \sin \varphi + \frac{\partial^2 u}{\partial x \partial y} r \cos \varphi \right] - \frac{\partial u}{\partial x} r \cos \varphi$$
$$+ r \cos \varphi \left[-\frac{\partial^2 u}{\partial x \partial y} r \sin \varphi + \frac{\partial^2 u}{\partial y^2} r \cos \varphi \right] - \frac{\partial u}{\partial y} r \sin \varphi .$$

The claimed formula now follows by putting all these facts together.

9. Using the previous exercise one can show that the formula $u(x, y) = a \log r + b$ with real constants a, b is the general solution.

10. As in Exercise 8, make repeated use of the chain rule.

11. Differentiate $f(z)\exp(-Cz)$.

12. If the function χ is differentiable, then we easily deduce

$$\chi'(x) = C\chi(x) \quad \text{with } C = \chi'(0) .$$

By the previous exercise, we then have $\chi(x) = A\exp(Cx)$. This expression has modulus 1 for all x, and satisfies the functional equation. This is possible, iff $A = 1$ and C is purely imaginary.

The differentiability of χ follows from the main theorem of differential and integral calculus, because of

$$\chi(x)\int_0^a \chi(t)\,dt = \int_0^a \chi(x+t)\,dt = \int_0^{x+a} \chi(t)\,dt - \int_0^x \chi(t)\,dt ,$$

where a has to be suitably chosen, such that the integral in the L.H.S. does not vanish.

14. The image is the slit annulus

$$f(D) = \{\, w \in \mathbb{C} ; \quad 1 < |w| < \exp b , \ -\pi < \operatorname{Arg} w < \pi \,\} .$$

15. Using the notation $z = r\exp(\mathrm{i}\varphi)$, $f = u + \mathrm{i}v$, we have

$$u = \frac{1}{2}\left(r + \frac{1}{r}\right)\cos\varphi , \quad v = \frac{1}{2}\left(r - \frac{1}{r}\right)\sin\varphi .$$

Because of this, the image of the circle C_r for $r \neq 1$ is an ellipse with focal points ± 1, and half-axes $\frac{1}{2}\left(r + \frac{1}{r}\right)$ and respectively $\frac{1}{2}\left|r - \frac{1}{r}\right|$. In case of $r = 1$, the ellipse degenerates to the interval $[-1, 1]$.

Analogously, the image of the half-line is computed to be a branch of the hyperbola

$$\frac{u^2}{\cos^2\varphi} - \frac{v^2}{\sin^2\varphi} = 1 .$$

The function f maps both D_1 and D_2 bijectively, in fact even conformally, onto the slit plane $\mathbb{C}\setminus[-1,1]$.

16. (a) We know that sin is surjective. Because of the periodicity, the sine is also surjective after restricting it to the vertical strip $-\pi/2 \le \operatorname{Re} z \le \pi/2$. The boundary lines are mapped onto $]-\infty, -1]$ and $[1, \infty[$. A simple computation shows that the only real values of sine corresponding to interior points $-\pi/2 < \operatorname{Re} z < \pi/2$ of the vertical strip lie in $]-1, 1[$.

    (b) Use the representation

$$\tan z = \mathrm{i}\frac{1 - \exp(2\mathrm{i}z)}{1 + \exp(2\mathrm{i}z)} .$$

The tangent is thus a composition of the four maps

$$z \mapsto 2\mathrm{i}z , \quad \exp z , \quad \frac{1-z}{1+z} , \quad \mathrm{i}z .$$

Successively computing images, we then obtain the four domains:

$$-\pi < \operatorname{Im} z < \pi \text{ , } \mathbb{C}_- \text{ , } \mathbb{C}\setminus\{t \in \mathbb{R}, \ |t| \geq 1\} \text{ , } \mathbb{C}\setminus\{\mathrm{i}t, \ t \in \mathbb{R}, \ |t| \geq 1\} \text{ .}$$

All four maps are conformal.

The inverse map of f is obtained by taking the inverse of each of the four intermediate maps.

17. If z is a point in the upper half-plane, then z is closer to i than to $-\mathrm{i}$. In particular, we see that $f(z) \in \mathbb{E}$. Analogously, one can show that for $w \in \mathbb{E}$ we have $g(w) := \mathrm{i}\frac{1+w}{1-w} \in \mathbb{H}$. The maps f and g are inverse maps.

18. Let us assume the property (b). After composing T with a suitable similarity substitution, we can assume $T(1) = 1$. The triangle with vertices in $0, 1, \mathrm{i}$ is then mapped by T also in a triangle, which has the same angles. This implies that $T(\mathrm{i})$ is purely imaginary, since the $90°$-angle is preserved. We have $T(\mathrm{i}) = \pm\mathrm{i}$ since the $45°$-angles are also preserved. Because of orientation reasons, the plus sign is the correct one. So T is the identity.

19. We tacitly use the fact that any real polynomial $u : \mathbb{R} \times \mathbb{R} \to \mathbb{R}$, uniquely extends to a complex polynomial $\mathbb{C} \times \mathbb{C} \to \mathbb{C}$. This extension will also denoted by u. In this sense , f is well defined. It is clear that f is analytic. We only have to show that $\operatorname{Re} f(x + \mathrm{i}y) = u(x, y)$. For the proof we can make use of the fact, that any harmonic function u in $\mathbb{C}$ is the real part of an analytic function. The proof of this fact shows that this function is a polynomial z if u is a polynomial in x and y. Hence one can restrict to the case

$$u(x, y) = \operatorname{Re}\,(x + \mathrm{i}y)^n = \sum_{\nu+2\mu=n} (-1)^\nu \binom{n}{\nu} x^\nu y^{2\mu}.$$

Now the claim follows from an elementary identity for binomial coefficients.

20. This is just a reformulation of the CAUCHY-RIEMANN differential equations, I.5.3.

21. The CAUCHY-RIEMANN differential equations are satisfied only in $+1$.

VIII.2 Solutions to the Exercises of Chapter II

Exercises in Sect. II.1

1. A possible parametrization defined on the interval $[0, 4]$ is

$$\alpha(t) = \mathrm{i}^k + \left(\mathrm{i}^{k+1} - \mathrm{i}^k\right)(t - k) \text{ , } \quad k \leq t \leq k+1 \text{ , } \quad k = 0, 1, 2, 3 \text{ .}$$

The line integral is then computed as

$$\sum_{k=0}^{3} \int_k^{k+1} \frac{\mathrm{i} - 1}{1 + (\mathrm{i} - 1)(t - k)} \, dt = 4\mathrm{i} \int_0^1 \frac{2}{(2t - 1)^2 + 1} \, dt = 2\pi\mathrm{i} \text{ .}$$

2. The image of α is a half-circle from the point $z = 1$ to the point $z = -1$ inside the upper half-plane, the whole circle being centered at 0. The path β is a piecewise linear path, its image consists of the segments from 1 to $-\mathrm{i}$, and from $-\mathrm{i}$ to -1. The line integrals have the values $\pi\mathrm{i}$ and respectively $-\pi\mathrm{i}$.

3.   The substitution law gives $\varphi(a) = c$, $\varphi(b) = d$,

$$\int_{\alpha \circ \varphi} f(\eta) \, d\eta = \int_a^b f((\alpha \circ \varphi)(t)) \; (\alpha \circ \varphi)'(t) \, dt$$

$$= \int_a^b f(\alpha(\varphi(t))) \; \alpha'(\varphi(t)) \, \varphi'(t) \, dt$$

$$= \int_{\varphi(a)}^{\varphi(b)} f(\alpha(s)) \, \alpha'(s) \, ds$$

$$= \int_c^d f(\alpha(s)) \, \alpha'(s) \, ds = \int_\alpha f(\zeta) \, d\zeta \; .$$

4.   The image of α is a lying "figure eight".

5.   The integrand has the primitive $F(z) = \frac{1}{2} \exp(z^2)$. The common value of both line integrals is thus $F(1+\mathrm{i}) - F(0)$.

6.   A primitive is $-\cos z$. The value of the integral is $1 - \cos(-1+\mathrm{i})$.

7.   Such an affine map is

$$\varphi(t) = \frac{d-c}{b-a}t + \frac{cb-ad}{b-a} \; .$$

8.   We have $\left| e^{\mathrm{i}z^2} \right| = e^{-R^2 \sin 2t}$, and use then the estimate

$$\sin(2t) \geq \frac{4}{\pi}t \quad \text{for} \quad 0 \leq t \leq \frac{\pi}{4} \; .$$

9.   Splitting the real and imaginary parts, we can reduce the claim to the corresponding standard approximation of real integrals by RIEMANN sums.

10.   The given formula immediately follows by splitting the integrand into real and imaginary parts.

11.   The oriented angle between two "vectors" $z, w \in \mathbb{C}^\bullet$ is nothing else but the argument $\mathrm{Arg}(w/z)$. Let now $z = \alpha'(0)$, $w = \beta'(0)$. The chain rule gives $(f \circ \alpha)'(0) = f'(a)\alpha'(0)$, $(f \circ \beta)'(0) = f'(a)\beta'(0)$, which implies

$$\mathrm{Arg}\frac{(f \circ \beta)'(0)}{(f \circ \alpha)'(0)} = \mathrm{Arg}\frac{\beta'(0)}{\alpha'(0)} \; .$$

Exercises in Sect. II.2

1.   The subsets in (b), (c), (e) and (f) are domains.

2.   If any two points of D can be joined by a polygonal path in D, then it is arcwise connected, and thus connected by II.2.2. For the proof of the converse, we fix a point $a \in D$ and consider the set $U \subset D$ of all points, which can be connected inside D by a polygonal path with a. The function $f : D \to \mathbb{C}$, which has the value 1 on U, and 0 on the complement $D \setminus U$, is constant in any disk contained in D. So it is in particular locally constant. Since D is connected, we deduce $U = D$.

3. Any two points in the punctured disk can be joined by a polygonal path inside the disk. (For this, at most two segments are needed.)

   If $f : D' \to \mathbb{C}$ is a locally constant function, then using the first part we can extend it to a locally constant function on D.

4. After composition with a translation and a contraction, we can assume that the closed unit disk is contained in D. Then computing the integral of f along the segment from -1 to $+1$, and respectively along a half-circle of the unit disk, we obtain a real value, and respectively an imaginary value.

5. The integrals in (a) and (b) are zero. The estimate (c) follows from the standard estimate II.1.5, (2), because of $|4 + 3z| \geq 4 - 3\,|z| \geq 1$. In fact, the integral is 0.

6. The value of the integral is $2\pi i$.

7. One can split the curve α, and correspondingly β, into the pieces in the upper and lower half-plane, $\alpha = \alpha^+ \oplus \alpha^-$, $\alpha^+ = \alpha|[0, 1/2]$, $\alpha^- = \alpha|[1/2, 1]$. The CAUCHY Integral Theorem can be used to show that the integrals along the two pieces are correspondingly equal. For this, slit the plane along the positive, and respectively the negative imaginary axis.

   The formula in (b) is obtained from (a) and II.1.7.

8. The statements are evident.

9. Only the region (b) is star-shaped. All three domains are "sickle-shaped domains". We can decide as follows whether a sickle-shaped domain is star-shaped. We draw the tangents from the two vertices to the concave boundary circle. We have a star domain, if they intersect inside the sickle-shaped domain. The possible star centers lie in the convex region which is bounded by pieces of the concave boundary circle and pieces of the two tangents.

10. The sickle-shaped domain is mapped by the conformal map $z \mapsto 1/(1 - z) = w$ onto the (convex) vertical strip $1/2 < \operatorname{Re} w < 1$.

11. We express the line integral of f using the given parameter representation, and use

$$\operatorname{Im} \frac{R + re^{it}}{(R - re^{it})\,e^{it}} i e^{it} = \frac{R^2 - r^2}{R^2 - 2Rr\cos t + r^2} \;.$$

   The value of the integral is obtained by the partial fraction decomposition. For the second integral, use instead of f the function $1/(R - z)$.

12. Let $Q(z) = P(z) - a_n z^n$. The triangle inequality implies for $|z| \geq \varrho$

$$|Q(z)| \leq \left(\sum_{\nu=0}^{n-1} |a_\nu| \right) |z|^{n-1} \leq \frac{|a_n|}{2}|z|^n \;.$$

   Once more applying the triangle inequality, we get

$$|a_n|\,|z|^n - |Q(z)| \leq |P(z)| \leq |a_n|\,|z|^n + |Q(z)| \;.$$

   Together, the inequalities imply the claim.

13. The standard estimate for line integrals and the above polynomial growth lemma lead to

$$2\pi \leq 2\pi R \frac{2\,|a_0|}{R\,|a_n|\,R^n} \;.$$

   But this inequality is false for large values of R.

14. The absolute value of $f(z)$ rapidly decays to zero on α_2 and α_4 for $|R| \to \infty$. The (arc) length of the two vertical edges is constant equal to a. The standard estimate then gives the needed limit behavior.

Taking the real part of $I(a)$, we deduce the *Corollary*.

15. Use the parameter representation of the line integral, and

$$e^{i(t+\pi)} = -e^{it} \ , \quad \text{and} \quad f\left(e^{i(t+\pi)} \right) = f\left(e^{it} \right) \ .$$

16. (a) The function $\tilde{l}(z) - l(z)$ takes values in $2\pi i\mathbb{Z}$. Continuity and connectivity considerations show that it is a constant.

(b) This is a local property, so we can assume that there is an analytic branch of the logarithm in D, which has the derivative $1/z$. Use now for instance the Implicit Function Theorem, or Exercise 5 in Sect. I.4. By (a), the function l differs from this analytic function by an additive constant.

(c) In (b) we have already seen one direction. Now let l be a primitive of $1/z$. After changing l by an additive constant we can assume that l is in some suitable small open set a branch of the logarithm. The equation $\exp(l(z)) = z$ then holds in the entire D.

(d) The principal value of the logarithm is defined by taking the inverse function of the restriction of the exponential function on the strip $-\pi < y < \pi$. Restricting instead on $0 < y < 2\pi$ we obtain an analytic branch of the logarithm, which is analytic in the plane which is slit along the positive real half-axis. This branch and the principal branch coincide in the upper half-plane. In the lower half-plane they differ by $2\pi i$.

17. Using the CAUCHY Integral Theorem and the cited estimate it is easy to show that the integral of $\exp(iz^2)$ along the positive real axis is equal to the integral along the half-line $t \exp(\pi i/4)$, $t \geq 0$, and thus the latter integral is up to the constant factor $(1+i)/\sqrt{2}$ equal to the integral of the real function $\exp(-t^2)$.

Exercises in Sect. II.3

1. The values of the integrals (a) to (d) are respectively 0, $\pi i/\sqrt{2}$, $e^2\pi i$, and 0. The computation of the integrals in (b) and (d) is best done by partial fraction decomposition. The integral in (e) is 0 in case of $|b| > r$, and $2\pi i \sin b$ in case of $|b| < r$.

2. The values are:
$$\frac{-ie^i}{2} \ , \quad \frac{ie^{-i}}{2} \ , \quad \frac{-ie^i}{2} + \frac{ie^{-i}}{2} \ , \quad 2 \ .$$

3. The values of the integrals are
$$2\pi i n \ , \quad 2\pi i(-1)^m \binom{n+m-2}{n-1} \frac{1}{(b-a)^{n+m-1}} \ .$$

4. The value of the integral is obtained by splitting $\frac{1}{1+z^2} = \frac{1}{2i}\left(\frac{1}{z-i} - \frac{1}{z+i}\right)$. For the estimate, use the standard method II.1.5, (2).

5. The value of the integral is $-2\pi i$.

6. The function g is by assumption bounded, hence constant. From $0 = g' = f' \exp(f)$ we deduce that f is also constant, since $\mathbb{C}$ is connected.

7. By periodicity we deduce that the image of f coincides with the image of its restriction to the compact parallelogram $\{t\omega + t'\omega'; \ 0 \leq t, t' \leq 1\}$. This restriction is continuous. Hence f is bounded, and by LIOUVILLE's Theorem it is constant.

8. We write P in the form $P(z) = C \prod(z - \zeta_\nu)$ and use the product formula for the logarithmic derivative.

    For the proof of GAUSS-LUCAS' Theorem we can assume that the zero ζ of P' is not a zero of P. Let

    $$m_\nu = \frac{1}{|\zeta - \zeta_\nu|^2} \quad \text{and} \quad m = \sum_{\nu=1}^{n} \frac{1}{m_\nu} \, .$$

    From the formula for P'/P follows

    $$\zeta = \sum_{\nu=1}^{n} \lambda_\nu \zeta_\nu \quad \text{with} \quad \lambda_\nu = \frac{m_\nu}{m} \, .$$

9. After canceling all common linear factors, we can assume that P and Q have no common root. Let s be a zero of order n of Q. We subtract from R the "partial fraction" $C(z - s)^{-n}$ with a suitable constant $C \in \mathbb{C}$ determined such that the numerator of

    $$\frac{P(z)}{Q(z)} - \frac{C}{(z-s)^n} = \frac{P(z) - CQ_1(z)}{Q(z)} \quad \text{with} \quad Q_1(z) = \frac{Q(z)}{(z-s)^n}$$

    vanishes at s. This is possible, since the polynomial Q_1 does not vanish at s. After subtraction, we can simplify numerator and denominator by the factor $(z - s)$. A simple induction leads to the desired result.

    In case of real polynomials P, Q, using $2R(z) = R(z) + \overline{R(\bar{z})}$ we can write R as a linear combination of polynomials with real coefficients and "partial fractions" of the form $(z - a)^{-n} + (z - \bar{a})^{-n}$, which are real for real values of z.

10. A simple algebraic transformation gives for $m = 1$

    $$\frac{F_1(z) - F_1(a)}{z - a} - \frac{1}{2\pi i} \int_\alpha \frac{\varphi(\zeta)}{(\zeta - a)^2} \, d\zeta = \frac{z - a}{2\pi i} \int_\alpha \frac{\varphi(\zeta)}{(\zeta - a)^2(\zeta - z)} \, d\zeta \, .$$

    This expression goes to zero for $z \to a$. The general case is obtained by induction on m via the identity

    $$\frac{1}{(\zeta - z)^m} = \frac{1}{(\zeta - z)^{m-1}(\zeta - a)} + \frac{z - a}{(\zeta - z)^m(\zeta - a)} \, .$$

11. By MORERA's Theorem it is enough to show that the integral of f along any triangle path is zero, when the corresponding triangle is contained in D. By taking subdivisions we can assume that the considered triangles are contained either in the closed upper half-plane, or in the closed lower half-plane. By a simple approximation argument, and using the continuity of f we can assume that the considered triangles are contained in the open upper or lower half-plane. Now we can apply the CAUCHY Integral Theorem for triangles.

12. The function $\widetilde{f}$ is continuous, and its restrictions to D_+ and to D_- are analytic. Now we can apply Exercise 11.

13. The exercise is a simple consequence of the LEIBNIZ rule.

14. The continuity of φ is problematic only in diagonal points (a, a). We can choose an $r > 0$, such that the closed disk centered at a is contained in D. If z and ζ $(z \neq \zeta)$ lie in the inner part of this disk, then we have the CAUCHY Integral Formula

$$\frac{f(\zeta) - f(z)}{\zeta - z} = \frac{1}{2\pi i} \oint_{|\eta - a| = r} \frac{f(\eta)\, d\eta}{(\eta - \zeta)(\eta - z)} \; .$$

We take the limits $\zeta \to a$, $z \to a$, and notice that they commute with the integral. By the Generalized CAUCHY Integral Formula we obtain in the R.H.S. the value $f'(a)$.

For the proof of the second part we can assume that D is a disk. Using the first part and II.2.7$_1$, the function $f(\zeta) := \varphi(\zeta, z)$ admits a primitive. Applying II.3.4 the derivative of this primitive is analytic.

15. From the factorization $f^2 + g^2 = (f + ig)(f - ig)$ we see that $f + ig$ has no zero in $\mathbb{C}$. Then there exists an entire function h such that $f + ig = e^{ih}$. As a consequence, $f - ig = e^{-ih}$. We obtain two equations for the functions f and g, which can be solved.

16. If the image of f is not dense, then we can find a disk $U_r(a)$ in the complement. Then consider $1/(f(z) - a)$.

VIII.3 Solutions to the Exercises of Chapter III

Exercises in Sect. III.1

1. Continuity is a local property, so we can assume that the series converges uniformly. Now use the standard method from real analysis.

2. The claim follows from III.1.3 by induction on k.

3. By the HEINE-BOREL Theorem it is enough to construct for any point $a \in D$ an ε-neighborhood, where the derivatives are (simultaneously) bounded. Fix an $a \in D$. We choose $\varepsilon > 0$ small enough, such that the closed disk of radius 2ε centered at a is contained in D. For $z \in U_\varepsilon(a)$ we obtain by means of the Generalized CAUCHY Integral Formula the estimate

$$|f_n'(z)| = \frac{1}{2\pi} \left| \oint_{|\zeta - a| = 2\varepsilon} \frac{f_n(\zeta)\, d\zeta}{(\zeta - z)^2} \right| \le \frac{1}{2\pi} 4\pi\varepsilon \frac{M(\overline{U_\varepsilon(a)})}{\varepsilon^2} \; .$$

4. In the region $|z| \le r < 1$ the general term of the series is dominated by $(1 - r)^{-1} r^{2\nu}$.

5. In $z = 0$ the series does not converge absolutely (harmonic series).

   The general term converges to 0, so it is allowed for convergence purposes to group pairs of successive terms. Grouping the first plus second term, then the third plus fourth term, and so on, we obtain a new series which can be majorized on any compact set $K \subset \mathbb{C} \setminus \mathbb{N}$ by a series with general term of the form $C(K)n^{-2}$.

6.   The series is a telescopic series (see also Exercise 10, Sect. I.2). Its limit is $1/(1-z)$.

7.   If the given series is convergent, then the sequence $\sin(nz)/2^n$ must be bounded. If z is for instance in the upper half-plane this means that $\exp(ny)/2^n$ is bounded, i.e. $y \leq \log 2$.

   The additional question must be negatively answered for the same reason. The series converges only for real values of z.

8.   The integral of f_r vanishes by the CAUCHY Integral Theorem. Specializing $r = 1 - 1/n$ we obtain a sequence which converges uniformly to f.

Exercises in Sect. III.2

1.   The radii of convergence are respectively 0, e, e and $1/b$.

2.   We have to show that the sequence $(nc_n\varrho'^n)$ is bounded for any ϱ' with the property $0 < \varrho' < \varrho$, if the sequence $(c_n\varrho^n)$ is bounded. But the sequence $n(\varrho'/\varrho)^n$ is bounded, since $\sqrt[n]{n} \to 1$ for $n \to \infty$, hence $\sqrt[n]{n}\varrho'/\varrho < 1$ for all but finitely many n.

   In the second part we have to show the continuity of the series $\sum c_n\varphi_n(z)$. Using the first part we obtain its normal convergence, because in the region $|z| \leq \varrho < r$ we can dominate it by $\sum n|c_n|\varrho^n$.

3.   We can consider the following series, all of them having the radius of convergence equal to 1.

(a) $\sum n^{-2}z^n$.

(b) $\sum_{} z^n$.

(c)   The series $\sum n^{-1}z^{2n}$ converges for $z = \pm i$, and diverges for $z = \pm 1$.

4.   The examples (a) and (b) in Exercise 3 also work here.

5.   We restrict to (c). One obtains the TAYLOR series by means of the partial fraction decomposition

$$\frac{1}{z^2 - 5z + 6} = \frac{1}{z-3} - \frac{1}{z-2}$$

   using the geometric series (and not by taking derivatives).

6.   We only treat one direction in (a). If the mentioned limit exists, then for any real $\alpha > 1/R$ we have, excepting finitely many n, the inequality $|a_n| \leq |a_0|\,\alpha^n$. Then for any ϱ with $0 < \varrho < 1/\alpha$ the sequence $(a_n\rho^n)$ is bounded, more than this, it is a null sequence. The radius of convergence of the given series is thus at least $1/\alpha$, and since $1/\alpha < R$ was arbitrary, it is at least R.

7.   By general results about power series, the convergence radius is at least r, and of course, it cannot be strictly larger.

   As an example for (b) one can take the principal value of the logarithm. Let the point a be located in the second quadrant ($\operatorname{Im} a > 0, \operatorname{Re} a < 0$). The radius of convergence is easily seen to be $r = |a|$. The convergence disk contains a piece of the lower half-plane.

8.   This is a trivial consequence of the Identity Theorem III.3.2.

9.   Starting with a power series, one computes the solution

$$e^{z^2/2} \quad \text{and respectively} \quad \frac{5e^{2z} - 2z - 1}{4} \, .$$

10.   The radius of convergence is equal to the minimal modulus of a zero of cos, i.e. $\pi/2$. The coefficients can be recursively computed. We have $E_0 = 1$ and

$$E_{2n} = -\sum_{\nu=0}^{n-1} (-1)^{n-\nu} \binom{2n}{2\nu} E_{2\nu} \quad (n \geq 1) \, .$$

The integrality inductively follows from this recursion.

11.   The convergence radius is $\pi/2$. The first six coefficients are 0, 1, 0, $1/3$, 0, $2/15$.

12.   (a) Assume that each boundary point is regular. Using the HEINE-BOREL covering theorem, we can find an $\varepsilon > 0$ and for any boundary point ϱ an analytic function g_ε on $U_\varepsilon(\varrho)$, which coincides on $U_\varepsilon(\varrho) \cap D$ with P. Two functions g_ϱ, g'_ϱ coincide in the intersection $U_\varepsilon(\varrho) \cap U_\varepsilon(\varrho')$ Hence there exists an analytic extension of P into the domain $D \cup \bigcup_\varrho U_\varepsilon(\varrho)$. This domain contains an open disk of radius $R > r$, so the convergence radius of P would be at least R.

      (b) A comparison with the geometric series shows that the radius of convergence is at least 1. Of course it cannot be bigger. Concerning the singular points on the unit circle, we consider for any natural k, a unit root ζ of order 2^k (i.e. $\zeta^{2^k} = 1$). Then the power series is unbounded on the segment $t\zeta$, $0 \leq t < 1$, hence ζ is a singular point. The set of all these roots of unity is a dense subset of the unit circle, hence all boundary points are singular.

13.   The mentioned series has the convergence radius ∞, as a comparison with the exponential series shows. The differential equation is easily checked for this series. Observe that the differential equation gives a recursion formula for the TAYLOR coefficients.

14.   Comparison with the exponential series.

15.   One integrates termwise the double series

$$\frac{1}{z} f(z) \overline{f(z)} = \frac{1}{z} \sum_{n,m} a_n \overline{a_m} z^n \overline{z}^m$$

along the circle of radius ϱ. The terms corresponding to $m \neq n$ vanish. A trivial estimate of the integral gives the required result. The weaker CAUCHY inequality is obtained by forgetting all terms but one in the GUTZMER inequality.

16.   In case of $m = 0$, the claim reduces to LIOUVILLE's Theorem, II.3.7. The proof of the general case follows analogously. One can prove by means of the Generalized CAUCHY Integral Formula that the $(n+1)^{\text{th}}$ derivative of f vanishes. Or one uses Exercise 15.

17.   Use the bijectivity of f to show $\lim_{|z| \to \infty} |f(z)| = \infty$. Then the auxiliary analytic function $g : \mathbb{C}^\bullet \to \mathbb{C}^\bullet$, $g(w) := 1/f(1/w)$ can be continuously, and hence analytically, extended into $w = 0$. From this we get that f is of polynomial growth for $|z| \to \infty$. Exercise 16 shows that f is a polynomial of degree m.

Then $f(f(z))$ is a polynomial of degree $m^2 = 1$. We obtain as only solutions $f(z) = z$ and $f(z) = -z$. (The proof can be drastically simplified, if we assume the knowledge of $\text{Aut}(\mathbb{C})$. See also the Exercises 5, 6, 7 in the Appendix A to the Sections III.4 and III.5.)

18. Use the quotient criterion.

Exercises in Sect. III.3

1. The power series P of $\sin \frac{1}{1-z}$ has in the convergence disk $|z| < 1$ infinitely many zeros. It thus coincides at infinitely many points with $Q \equiv 0$.

2. (a) The zero set of $f_1(z) - 2z$ has the accumulation point 0, hence we deduce $f_1(z) = 2z$, but this function has not the required properties. Hence there is no f_1.

   (b) We necessarily have $f_2(z) = z^2$.

   (c) The nth TAYLOR coefficient would be $n!$. But the power series $\sum n! \, z^n$ converges only in 0.

   (d) The series $f_4(z) = \sum z^n / n^2$ has the required property.

3. All higher derivatives in zero are real.

4. This exercise should once more make clear that being *"discrete in"* is in our terminology a *relative notion*. The set $\{\, 1/n \,;\, n \in \mathbb{N} \,\}$ is discrete in $\mathbb{C}^*$, but not in $\mathbb{C}$. The property (b) can also be reformulated as follows: A subset $M \subset D$ is discrete in D, iff it is closed in D and the topology inherited from D is the discrete topology on M.

5. It is enough to show that there exists a sequence of compact sets K_n, which exhausts D, i.e. $D = \bigcup_n K_n$. It is easy to show that D is the union of all compact disks $K \subset D$, which have rational radius and which are centered at points with rational components. The system of all these discs is countable.

6. The zero set is discrete. Use now Exercise 5.

7. The function f vanishes on the image of g, which is open for a non-constant g.

8. The Maximum Principle implies $|f(z)/g(z)| \leq 1$ and $|g(z)/f(z)| \leq 1$ and thus $|f(z)/g(z)| = 1$. The Open Mapping Theorem implies then that f/g is constant.

9. Apply Lemma III.3.8 for the function $f \circ g^{-1}$.

10. The maximal values are respectively e, 2, $\sqrt{5}$, 3. The function in (d) is not analytic.

11. It can be assumed that D is a disk. Then u is the real part of an analytic function. By the open mapping theorem its image is an open set. The projection of this open set onto the real axis is an open interval.

12. Since the closure is compact, there exists a maximum. This maximal value cannot be taken at an interior point, unless f is constant.

13. Because of Proposition III.3.9 it can be assumed that one of the two fixed points is 0, i.e. $f(0) = 0$. The SCHWARZ Lemma implies $|f(z)/z| \leq 1$. The function $g(z) = f(z)/z$ has a maximal modulus if f has a fixed point different from 0. By the Maximum Principle g is constant.

14. If $b \in \mathbb{C}$ is a boundary point of the image of the polynomial P, then there exists a sequence (a_n), such that $P(a_n) \to b$. By the Lemma on polynomial growth (Exercise 12 in Sect. II.2, page 90) the sequence (a_n) is bounded. Passing to a subsequence, we can assume that it converges, $a_n \to a$. Then $P(a) = b$, so b lies in the image of P. As a conclusion, the image of P is closed. By the Open Mapping Theorem, this image is also open. Since it is connected it is the whole $\mathbb{C}$.

15. We give an indirect proof and consider $g = 1/f$. By assumption, we have $|g(a)| > |g(z)|$ for all z at the boundary of the disk. Using the maximum principle we get a contradiction.

    If f is a non-constant function on a domain $D \supset \overline{U}_r(a)$, then there exists an $\varepsilon > 0$ with $|f(z) - f(a)| \geq 2\varepsilon$ for $|z - a| = r$. Then, using the first part of the exercise it is easy to see that $U_\varepsilon(f(a))$ is contained in the image of f.

16. One verifies (a) by direct computation. Applying the SCHWARZ lemma to $\varphi_{f(a)} \circ f \circ \varphi_a$ one obtains (b).

17. Let $f : \mathbb{C} \to \mathbb{C}$ be entire, and $|f(z)| \leq C$. For arbitrary $a \in \mathbb{C}$ and $r > 0$ we apply the SCHWARZ lemma to the function $\frac{1}{2C}(f(rz + a) - f(z))$. Dividing by r and taking the limit $r \to \infty$ we obtain $f'(a) = 0$.

Exercises in Sect. III.4

1. (a) From (α) we deduce (β) by the Removability Condition (III.4.2), and then trivially also (γ). If (γ) is satisfied, then by RIEMANN Removability once more we see that $g(z) := (z - a)f(z)$ has a removable singularity. Then in the power series of g we can factorize $(z - a)$.

   (b) If the limit exists, then the function g from part (a) has a removable singularity, and we can use the Supplement of III.4.4.

2. Use the Supplement of III.4.4.

3. As in both previous exercises, use the characterization of the order from the supplement in III.4.4.

4. The functions (b), (c) and (d) have removable singularities at the origin.

5. The pole orders are respectively 2, 7 and 3.

6. Let U be an arbitrarily small neighborhood of a. If a is an essential singularity of f, then $f(U)$ is dense in $\mathbb{C}$. This implies that the closure of $\exp(f(U))$ is equal to the closure of $\exp(\mathbb{C})$, i.e. it is the whole $\mathbb{C}$. If f has a pole, then there exists (Open Mapping Theorem) an $r > 0$, such that the region $B = \{ z ; |z| > r \}$ is contained in $f(U)$. The exponential function takes each of its values already in B.

7. Write f in the form $f(z) = (z - a)^k f_0(z)$. From the TAYLOR formula we have $f_0(a) = f^{(k)}(a)/k!$. The same considerations also apply for g.

8. The singularities are located in $1 + 4\mathbb{Z}$. Excepting $z = 1$ and $z = -3$, which are removable singularities, all other singularities are simple poles.

9. Apply partial integration for the functions $u = \sin^2(x)$ and $v = -1/x$. Because of $u' = \sin(2x)$, we are leaded to the well-known DIRICHLET integral $\int_0^\infty \frac{\sin x}{x} \, dx = \frac{\pi}{2}$.

10. Using the formula $\sin^2 2x = 4(\sin^2 x - \sin^4 x)$, we proceed as in the previous exercise.

Exercises in Sect. III.5

1. Using the formula for the geometric series, we have

$$\frac{z}{1+z^2} = \frac{1}{2}\frac{1}{z-\mathrm{i}} + \frac{1}{2}\sum_{n=0}^{\infty}(-1)^n\frac{(z-\mathrm{i})^n}{(2\mathrm{i})^{n+1}} \ .$$

The point $z = \mathrm{i}$ is a simple pole.

2. Using the partial fraction decomposition $f(z) = 1/(1-z) - 1/(2-z)$ we obtain

$$\frac{1}{(z-1)(z-2)} = \begin{cases} \sum_{n=0}^{\infty}\left(1 - 2^{-n-1}\right)z^n & \text{for } 0 < |z| < 1 \ , \\ -\sum_{n=1}^{\infty}z^{-n} - \sum_{n=0}^{\infty}\frac{z^n}{2^{n+1}} & \text{for } 1 < |z| < 2 \ , \\ \sum_{n=2}^{\infty}\left(2^{n-1} - 1\right)z^{-n} & \text{for } 2 < |z| < \infty \ . \end{cases}$$

The LAURENT series for the centers $a = 1$ and $a = 2$ are obtained analogously.

3. Make use of the partial fraction decomposition

$$\frac{1}{z(z-1)(z-2)} = \frac{1}{2z} + \frac{-1}{z-1} + \frac{1}{2(z-2)} \ .$$

4. The identity is valid only for $|z| < 1$ and $|1/z| < 1$, i.e. for the empty set.

5. The rational function $P(z) = 1/(1 - z - z^2)$ can be expanded into a power series around the center $a = 0$ in a suitable neighborhood. From the relation $(1 - z - z^2)P(z) = 1$ we see that the TAYLOR coefficients of P satisfy the same recursion as the FIBONACCI numbers.

   The explicit formula for f_n is obtained from the partial fraction decomposition

$$\frac{1}{1 - z - z^2} = \frac{1}{\sqrt{5}}\left(\frac{1}{z - \omega_2} - \frac{1}{z - \omega_1}\right) \qquad \text{with } \omega_{1|2} = \frac{-1 \pm \sqrt{5}}{2}$$

   by means of the geometric series.

6. The claim is clear for the polynomial $P(z) = z^m$, and follows then in general. Apply III.4.6.

7. (a) The function f is invariant under $z \mapsto -1/z$ and $z \mapsto -z$ on the one side, and $w \mapsto -w$ on the other side.

   (b) The formula is obtained from the integral representation of the LAURENT coefficients (Supplement of III.5.2), using the explicit parameter form of the involved line integral.

   (c) The substitution of the integration variable ζ in the integral giving the LAURENT coefficients by $2\zeta/w$, followed by integration along the unit circle gives rise to

$$J_n(w) = \frac{1}{2\pi\mathrm{i}}\left(\frac{z}{2}\right)^n \oint \zeta^{-n-1}\exp\left(\zeta - \frac{w^2}{4\zeta}\right)d\zeta \ .$$

   Using the exponential series we derive from this

$$J_n(w) = \frac{1}{2\pi\mathrm{i}}\sum_{m=0}^{\infty}\frac{(-1)^m}{m!}\left(\frac{w}{2}\right)^{n+2m}\oint\frac{e^\zeta}{\zeta^{n+m+1}}d\zeta \ .$$

   The integral has the value $2\pi\mathrm{i}/(n+m)!$. This can be seen by using the power series expansion of e^ζ, followed by termwise integration.

(d) To check the given differential equation use termwise differentiation of the series.

8.   The function $z \mapsto 1/z$ maps the open disk (centered at 0) of radius r onto the complement of the closed disk of radius $1/r$. This complement contains horizontal strips of height 2π. The exponential function has the same image when restricted to such strips.

9.   Else, the integral of the function $1/z$ along a circle around the origin would be 0.

10. In the upper half-plane, using the notation $q = e^{2\pi i z}$, $\operatorname{Im} z > 0$, we have:

$$\pi \cot \pi z = \pi \frac{\cos \pi z}{\sin \pi z} = \pi i \frac{q+1}{q-1} = \pi i - \frac{2\pi i}{1-q} = \pi i - 2\pi i \sum_{n=0}^{\infty} q^n .$$

Exercises in Appendix A to Sects. III.4 and III.5

1.   First, the essential point is to define in a strict sense the sum $f + g$ and the product fg of two meromorphic functions f, g. This was done for finite domains $D \subset \mathbb{C}$. If ∞ lies in the domain of definition D, then we best define $(f + g)(\infty)$ and $(f \cdot g)(\infty)$ by using the substitution $z \mapsto 1/z = w$ to reduce the case $z = \infty$ to the known case $w = 0$.

2.   By the way, this was tacitly used in the proof of A.2 (inversion of a meromorphic function). First, observe that D remains connected after eliminating a discrete set (the set of poles) from it. As a consequence of the Identity Theorem, the zero set has no accumulation point in the complement of the pole set. And a pole cannot be an accumulation point of the zero set, since the absolute values of a meromorphic function tend to ∞ near a pole.

3.   Removability means that $f(z)$ is bounded for $|z| \geq C$, where C is sufficiently large. A pole arises in the case of $\lim_{|z| \to \infty} |f(z)| = \infty$. An essential singularity is characterized by the fact that the regions $|z| \geq C$ $(z \neq \infty)$ are mapped onto dense subsets of $\mathbb{C}$.

4.   One verifies the formula by direct computation.

5. $\mathbb{E}$ is mapped (Open Mapping Theorem) onto an *open* subset of the plane. The complement of $\mathbb{E}$ is mapped by injectivity into the complement of this open set. Hence ∞ is not an essential singularity of f. This implies that f is a polynomial. The injectivity shows that its degree is 1.

6.   The only solutions are $f(z) = z$ and $f(z) = -z + b$ with a suitable constant b. The functional equation implies that f is injective. By the previous exercise f is a linear polynomial.

7.   Any meromorphic function on $\overline{\mathbb{C}}$ is rational (Proposition A.6). If a rational function is bijective, then it has exactly one zero and exactly one pole. In a reduced representation as a quotient of two polynomials the numerator and the denominator both have degree at most one. Now, (a) and (b) follow from Proposition A.9.

8.   The fixed point equation is a quadratic equation $az + b = (cz + d)z$.

9.   If a, b, c are not equal to ∞ then the cross ratio works. Else extend the cross ratio by an obvious limiting process.

10. The claim is clear for translations and rotation-dilations, i.e. for upper triangular matrices. The claim also follows for the substitution $z \mapsto 1/z$, as can be seen by writing a circle equation in the form $(z - a)\overline{(z - a)} = r^2$, and a line equation in the form $az + b\overline{z} = c$. An arbitrary matrix M can be written as a product of matrices of the above type. If M is not an upper triangular matrix, then $\alpha := M\infty$ is not ∞, and the matrix $N = \left(\begin{smallmatrix} 1 & \alpha \\ 0 & 1 \end{smallmatrix}\right)\left(\begin{smallmatrix} 0 & 1 \\ 1 & 0 \end{smallmatrix}\right)$ also has the property $N\infty = \alpha$. We then have $M = NP$ with a suitable upper triangular matrix P.

11. This is a consequence of Exercise 9, since any three different points uniquely determine a generalized circle through them.

12. There are two cases depending on the fact that M has one or two fixed points. In the first case choose A such that the fixed point is mapped to ∞. Then the matrix AMA^{-1} has ∞ as a fixed point, so its action is $z \mapsto az + b$. Since ∞ is the only fixed point, a must be 1. The corresponding matrix is a triangular matrix with equal diagonal entries. In the second case, the two fixed points can be mapped by a suitable matrix A to 0 and ∞. The transformed matrix is a diagonal matrix.

13. A triangular matrix with equal diagonal entries which is not diagonal, cannot have finite order. Then by Exercise 12 we can assume that M is a diagonal matrix.

Exercises in Sect. III.6

1. We exemplary consider part (e). At $z = 1$ there is a pole of second order. The residue is the first TAYLOR coefficient of $\exp(z)$ at $z = 1$, namely e.

2. The derivative of F is 0. Hence F is constant, in particular $F(1) = F(0)$. This implies that $G(1)$ is an integral multiple of $2\pi i$. The winding number is exactly $G(1)/2\pi i$.

3. (a) The function $\chi(\alpha, z)$ is continuous, and takes only integral values.

   (b) The formulas are direct consequences of the definition of the line integral.

   (c) The function

   $$h(z) = \chi(\alpha; 1/z) = \frac{1}{2\pi i} \int_\alpha \frac{z}{\zeta z - 1}\, d\zeta$$

   is first analytic on the set of all $z \neq 0$, such that $1/z$ does not lie in the image of α. It can be analytically extended into the zero point with the value 0. Since it is locally constant, it vanishes in a neighborhood of the zero point.

   (d) This was already proven in (c).

   (e) We deduce that there are two possibly different logarithms l_1 and l_2 of $\alpha(0) = \alpha(1)$, such that $2\pi i\chi(\alpha; a) = l_1 - l_2$.

4. (a) Substitute $\zeta \mapsto 1/\zeta$ in the integral representation.

   (b) The function $f(z) = z$ has a removable singularity at the origin, even more, it is a zero of f.

5. Choose R in Exercise 4 (a) large enough, such that all poles of f have modulus $< R$. The claim follows now from the Residue Theorem and 4(a).

6. The integrals can be computed by means of the Residue Theorem, combined with the relation in Exercise 5. For (a), the residue at ∞ is 0, and at $z = 3$ it is $(3^{13} - 1)^{-1}$. The mentioned relation implies the value $-2\pi i(3^{13} - 1)^{-1}$ for the integral.

7. Since fg has at most a simple pole we can apply III.5.4 (1).

8. A LAURENT series with vanishing coefficient a_{-1} can be integrated termwise.

9. Termwise differentiation of a LAURENT series delivers a LAURENT series with vanishing coefficient a_{-1}.

10. The transformation formula is obtained by using the parameter definition of the line integral, and the usual substitution rule. The residue formula is a special case.

Exercises in Sect. III.7

1. In the first example there is a zero in the interior of the unit circle, and no zero is located on its boundary. All other three zeros are in the complement. This can be shown by applying ROUCHÉ's Theorem III.7.7 to $f(z) = -5z$ and $g(z) = 2z^4 + 2$. For the second equation there are 3 solutions with $|z| > 1$. The third equation has 4 solutions in the annulus.
    The numeric locations of the solutions relative to the circles of radius one, resp. two are as follows:

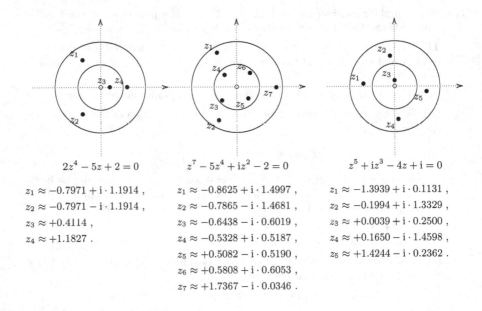

$$2z^4 - 5z + 2 = 0 \qquad z^7 - 5z^4 + iz^2 - 2 = 0 \qquad z^5 + iz^3 - 4z + i = 0$$

$z_1 \approx -0.7971 + i \cdot 1.1914$,	$z_1 \approx -0.8625 + i \cdot 1.4997$,	$z_1 \approx -1.3939 + i \cdot 0.1131$,
$z_2 \approx -0.7971 - i \cdot 1.1914$,	$z_2 \approx -0.7865 - i \cdot 1.4681$,	$z_2 \approx -0.1994 + i \cdot 1.3329$,
$z_3 \approx +0.4114$,	$z_3 \approx -0.6438 - i \cdot 0.6019$,	$z_3 \approx +0.0039 + i \cdot 0.2500$,
$z_4 \approx +1.1827$.	$z_4 \approx -0.5328 + i \cdot 0.5187$,	$z_4 \approx +0.1650 - i \cdot 1.4598$,
	$z_5 \approx +0.5082 - i \cdot 0.5190$,	$z_5 \approx +1.4244 - i \cdot 0.2362$.
	$z_6 \approx +0.5808 + i \cdot 0.6053$,	
	$z_7 \approx +1.7367 - i \cdot 0.0346$.	

2.   Example (2) on page 174 may serve as an orientation.

3.   Apply ROUCHÉ's Theorem III.7.7 with $f(z) = z - \lambda$ and $g(z) = \exp(-z)$. As path of integration choose the rectangle with vertices in $-iR$, $R - iR$, $R + iR$, iR for a sufficiently large R, such that the estimate $|g(z)| < |f(z)|$ holds on this path.

4.   The function $|\exp(z)|$ has on a given closed disk $|z| \leq R$ a positive minimum m. Since the exponential series converges uniformly on any compact set, there exists a natural number n_0 with

$$|e_n(z) - \exp(z)| < m \leq |\exp(z)| \quad \text{for } n \geq n_0 \text{ and } |z| \leq R .$$

In particular, $e_n(z)$ is not zero for $n \geq n_0$ and $|z| \leq R$.

5.   Apply ROUCHÉ's Theorem III.7.7 and reduce the claim to the trivial case $f \equiv 0$.

6.   The only singularity of the integrand in $\overline{U}_\varrho(a)$ is in $\zeta = f^{-1}(w)$. The residue is

$$\lim_{\zeta \to f^{-1}(w)} \left(\zeta - f^{-1}(w)\right) \frac{\zeta f'(\zeta)}{f(\zeta) - f\left(f^{-1}(w)\right)} = \frac{f^{-1}(w) f'\left(f^{-1}(w)\right)}{f'\left(f^{-1}(w)\right)} = f^{-1}(w) .$$

7.   The proof of the partial fraction decomposition of the cotangent III.7.13 may serve as an orientation. Consider the integral of g and respectively h along the contour Q_N. The limit of this integral vanishes for $N \to \infty$. The claim follows from the Residue Theorem. The singularities of g and h are located at integers $n \in \mathbb{Z}$. The residues are $f(n)$ and respectively $(-1)^n f(n)$.

8.   Apply Exercise 7 to the function $f(z) = 1/z^2$.

9.   For the first integral one must compute the residues of the rational function

$$\frac{z^6 + 1}{z^3 (2z - 1)(z - 2)}$$

in the unit disk, Proposition III.7.9. There is a pole of third order at the origin, which has the residue $21/8$. The point $1/2$ is a simple pole with residue $-65/24$. The pole at $z = 2$ is in the exterior of the unit circle. We obtain

$$\int_0^{2\pi} \frac{\cos 3t}{5 - 4\cos t} \, dt = \frac{\pi}{12} .$$

By the same method we compute

$$\int_0^\pi \frac{1}{(a + \cos t)^2} \, dt = \frac{1}{2} \int_0^{2\pi} \frac{1}{(a + \cos t)^2} \, dt = \frac{\pi a}{(a^2 - 1)\sqrt{a^2 - 1}} .$$

In the Exercises 10 to 13 one can use standard methods to verify the claimed results.

14. Let $\zeta = \exp(2\pi i/5)$. In the circle sector delimited by the half-lines $\{ t ; t \geq 0 \}$ and $\{ t\zeta ; t \geq 0 \}$ and the arc from r to $r\zeta$, $r > 1$, the integrand $(1 + z^5)^{-1}$ has exactly one singularity located at $\eta = \exp(\pi i/5)$. We have $\zeta = \eta^2$. The residue of $(1 + z^5)^{-1}$ at $z = \eta$ is $(5\eta^4)^{-1} = -\eta/5$. Since the integral along the arc converges to zero for $r \to \infty$, the Residue Theorem implies that the difference of the integrals along the two half-lines is $2\pi i$ times this residue. We thus obtain

$$\int_0^\infty \frac{dx}{1+x^5} - \zeta \int_0^\infty \frac{dx}{1+x^5} = -\frac{2\pi i}{5}\eta \ .$$

The above formula remains valid if we replace 5 by an arbitrary odd number > 1.

15. The integrals are improper at both limits. To apply the Residue Theorem we have to specify a branch of the logarithm. We choose the value $\log z = \log|z| + i\varphi$ with $-\pi/2 < \varphi < 3\pi/2$. This branch is analytic in the plane slit along the negative imaginary axis, and we define an integration path in this domain. Let ε be arbitrary with $0 < \varepsilon < r$. The path is a composition of the segment from $-r$ to $-\varepsilon$, the half-circle in the upper half-plane from $-\varepsilon$ to ε, and the segment from ε to r. From the Residue Theorem we derive using standard estimates

$$\lim_{r\to\infty} \int_\alpha \frac{(\log z)^2}{1+z^2}\,dz = -\frac{\pi^3}{4} \ .$$

Now we make use of the formula $\log(-x) = \log x + \pi i$ for $x > 0$. Passing to the limit $\varepsilon \to 0$, we obtain

$$2\int_0^\infty \frac{(\log x)^2}{1+x^2}\,dx + 2\pi i \int_0^\infty \frac{\log x}{1+x^2}\,dx - \pi^2 \int_0^\infty \frac{dx}{1+x^2} = -\frac{\pi^3}{4} \ .$$

The well-known value of the third above integral is $\pi/2$.

16. The integrand is an even function of x, so we can consider the integral from $-\infty$ to ∞ instead. Then take the imaginary part of the formula in Proposition III.7.1.

17. The function $f(z)$ has a simple pole at $z = a/2$, and no other singularities in the interior of the integration path. The residue is $\frac{1}{2}i\sqrt{\pi}$. The value of the line integral of f is $2\pi i$ times this residue, i.e. $\sqrt{\pi}$. The sum of the integrals along both horizontal lines gives $\int_{-R}^R \exp(-t^2)\,dt$. Both integrals from a to $R+a$, and from $-R$ to $-R+a$ tend to zero for $R \to \infty$.

18. Let us set $f(z) := \dfrac{\exp(2\pi i\, z^2/n\,)}{\exp(2\pi i\, z\,)-1}$. In both cases the residue theorem applied to f and α gives the GAUSS sum up to a trivial factor. We compute the integral in a different way taking the limit $R \to \infty$. The integrals along the horizontal segments converge to 0 for $R \to \infty$, since the integrand is rapidly decaying. The integral along the non-horizontal part can be reduced to one integral using

$$f(z+n) - f(z) \;=\; \exp\left(\frac{2\pi i}{n} z^2\right)\,[\,\exp(2\pi i z) + 1\,] \ .$$

Since the integrand has no singularity, one can integrate along the line at, $-\infty < t < \infty$ (i.e. one can take $\varepsilon = 0$) in the first case. In the second case one can replace the half circle by a straight segment. If one writes down explicitly the integrals in their parameter for, one is lead to the GAUSS integral from Exercise 17.

19. In the limit, the integral of f along the piecewise linear contour from r to $r+ir$ to $r + ir$ to $-r$ converges to zero as can be seen by standard estimates. Then using the Residue Theorem it is enough to understand the integrals of f along the half-circles of radius $\epsilon > 0$ around the simple poles p of f,

$$\delta_\epsilon : [0,1] \to \mathbb{C}, \quad \delta_\epsilon(t) := p + \epsilon \cdot \exp(\pi i(1-t)).$$

For this one writes

$$f(z) = \frac{c}{z-p} + h(z), \quad c := \mathrm{Res}(f;p),$$

where h is analytic near p. Now we obtain:

$$\lim_{\epsilon \to 0} \int_{\delta_\epsilon} f(z)\,dz = \lim_{\epsilon \to 0} \int_{\delta_\epsilon} \frac{c}{z-p}\,dz + \lim_{\epsilon \to 0} \int_{\delta_\epsilon} h(z)\,dz = -\pi i c + 0 \ .$$

VIII.4 Solutions to the Exercises of Chapter IV

Exercises in Sect. IV.1

1. The product in (a) diverges. The product in (b) converges. Its value is $1/2$, as it can be extracted from the partial products

$$\prod_{\nu=2}^{N}\left(1 - \frac{1}{\nu^2}\right) = \prod_{\nu=2}^{N} \frac{(\nu-1)(\nu+1)}{\nu^2} = \frac{1}{2}\frac{N+1}{N}\ .$$

The product in (c) converges, too. The Nth partial product is $\frac{1}{3}\left(1 + \frac{2}{N}\right)$. The value of the product is thus $1/3$. The last product also converges, its value is $2/3$. The Nth partial product is $\frac{2}{3}\left(1 + \frac{1}{N(N+1)}\right)$.

2. The corresponding series is a part of the geometric series, and thus it converges for $|z| < 1$. The value is obtained by the formula

$$(1-z)\prod_{\nu=0}^{n}\left(1 + z^{2^\nu}\right) = 1 - z^{2^{n+1}}\ .$$

3. The monotonicity follows from the trivial inequality $\log(1 + 1/n) > 1/(1+n)$. To show that 0 is a lower bound one uses the integral $\int_1^x \frac{dt}{t}$. (The sum over $1/\nu$ can be interpreted as the integral of a step function.)

4. The proof of IV.1.9 may serve as an orientation, use the formula for G_n given there.

5. From the STIRLING formula, the limit is equal to

$$\lim_{n \to \infty} \frac{(z+n)^{z+n-1/2}e^{-(z+n)}}{n^z n^{n-1/2}e^{-n}} = e^{-z}\lim_{n \to \infty}\left(1 + \frac{z}{n}\right)^n = 1\ .$$

6. From (a) we first deduce that $g = f/\Gamma$ is an entire function, which is periodic with the period 1. Because of (b) and Exercise 5, we have

$$\frac{g(z)}{g(1)} = \frac{g(z+n)}{g(n)} = \lim_{n \to \infty}\frac{g(z+n)}{g(n)} = 1\ .$$

7. The formula is a consequence of the LEGENDRE Doubling Formula IV.1.12, combined with the Completion Formula IV.1.11 specialized for $z = 1/3$.

8.   Both formulas follow from the Completion Formula IV.1.11

$$\Gamma(iy)\Gamma(1-iy) = -iy\Gamma(iy)\Gamma(-iy) \,,$$
$$\Gamma(-iy) = \overline{\Gamma(iy)} \,,$$
$$\Gamma(1/2+iy)\Gamma(1/2-iy) = \Gamma(1/2+iy)\Gamma\big(1-(1/2+iy)\big) \,.$$

9.   The fact that g is a polynomial of degree at most two, follows e.g. from the product formula IV.1.10 applied to $\Gamma(z), \Gamma(z+1/2)$ and $\Gamma(2z)$. The involved coefficients are determined by specializing $z = 1$ and $z = 1/2$.

10. Apply the auxiliary result to $g := f/\Gamma$. For the proof of this auxiliary result show the vanishing of the derivative of the logarithmic derivative $h(z) = (g'/g)'(z)$. It satisfies the functional equation $4h(2z) = h(z) + h(z+1/2)$. Its maximum $M \geq 0$ on the whole $\mathbb{R}$ exists for periodicity reasons, and satisfies the inequality $2M \leq M$. This gives $M = 0$, i.e. $h = 0$.

11. The functional equation and the boundedness in the vertical strip are evident. The normalizing constant can be determined by means of the STIRLING formula or by Exercise 19 in I.2.

12. (a) The integral is improper at both limits, so we need to show convergence and continuity. First, the proper integral

$$B_n(z,w) = \int_{1/n}^{1-1/n} t^{z-1}(1-t)^{w-1}\, dt$$

is continuous. Then proceed parallel to the investigation of the Γ-function at the lower integration point, and show that B_n converges in the specified region locally uniformly to B.

   (b) Use the argumentation from (a).

   (c) The functional equation is obtained by partial integration. In case of $z = 1$ the integrand admits a simple primitive.

   (d) The boundedness in a suitable vertical strip is evident. The normalization and the functional equation follow from (c).

   (c) Make use of the substitution $s = t/(1-t)$.

   (f) Make use of the substitution $t = \sin^2 \varphi$.

13. Let $\mu_n(r)$ be slightly more general the volume of the n-dimensional ball of radius r. A simple linear integral transformation gives $\mu_n(r) = r^n \mu_n(1)$. From FUBINI's Theorem for (iterated) integrals we obtain

$$\mu_n(1) = \int_{-1}^{1} \mu_{n-1}\big(\sqrt{1-t^2}\big)\, dt \,.$$

This gives the recursion formula. The involved integral reduces after the substitution $t = \sqrt{x}$ to a beta integral, and thus to a Gamma integral.

14. (a) The singularities of ψ are the zeros and the poles of the Γ-function. Then use the computation rule III.6.4 (3).

   It is now convenient to prove directly (e). For this, use the supplement of IV.1.7. The claims in (c), (f), (g) are then direct consequences. For (c) use the partial fraction decomposition of the cotangent. The last property (g) is clear since $\log \Gamma(x)' = \psi(x)$. For positive values of x the function value $\Gamma(x)$ is positive too, and we can take the logarithm.

15. We can assume $f(1) = 1$. Because of the functional equation it is sufficient to prove the identity $f(x) = \Gamma(x)$ for $0 < x < 1$. A double application of the logarithmic convexity leads to the estimate

$$n!(n + x)^{x-1} \le f(n + x) \le n!n^{x-1} .$$

Making use of the functional equation, this shows

$$\frac{n!n^x}{x(x + 1) \cdots (x + n)} \left(1 + \frac{x}{n}\right)^x \le f(x) \le \frac{n!n^x}{x(x + 1) \cdots (x + n)} \left(1 + \frac{x}{n}\right) .$$

The claim now follows by passing to the limit $n \to \infty$. It is easy to verify that Γ is indeed logarithmically convex (cf. Exercise 14(g)).

16. The identity is obtained by an iterated application of the functional equation needed to express $\Gamma(n - \alpha)$ in terms of $\Gamma(-\alpha)$. The asymptotic formula is delivered by Exercise 5.

17. First, we once more recall and emphasize that $w^{-z} := \exp(-z \log w)$ is defined by the principal value of the logarithm. The integrand is continuous on the whole integration path. Moreover, our chosen logarithmic branch leads to

$$\left|w^{-z}e^w\right| \le e^{\pi|y|} |w|^{-x} e^{\mathrm{Re}\, w} .$$

The integrand is rapidly decaying for $\mathrm{Re}\, w \to -\infty$, and the absolute convergence of the integral is then clear. An approximation of the integral, as in the case of the EULERian Gamma integral, by using proper integrals shows that it gives rise to an entire function. A particular case is $z = -n$, $n \in \mathbb{N}$. In this case the integrand is an entire function in w, and the CAUCHY Integral Theorem leads to the value zero for the HANKEL integral at $z = -n$, $n \in \mathbb{N}_0$. Using the Residue Theorem we obtain the value $2\pi i$ at $z = 1$.

It is convenient to prove a variant of the HANKEL formula, namely

$$\Gamma(z) = \frac{1}{2i \sin \pi z} \int_{\gamma_{r,\varepsilon}} w^{z-1} e^w \, dw .$$

The R.H.S. is analytic in $x > 0$, since the zeros of the sine are compensated by the zeros of the integral.

The two integral representations are equivalent. Now one proves the characterizing properties of the Gamma function for the second HANKEL integral:

The functional equation can be proved by partial integration. The boundedness in the strip $1 \le x \le 2$ follows from the mentioned estimate of the integrand correlated with standard estimates for the sine. The normalization is obtained from the first integral representation.

Exercises in Sect. IV.2

1. The formula (a) for the derivative is obtained by applying the product formula. The derivative of the exponent is $(z^k - 1)/(z - 1)$ by the geometric sum formula. From the formula for the derivative E_k' we see that its power series expansion consists of coefficients ≤ 0. The factor z^k implies that the first k coefficients vanish. The statement (b) for E_k follows by termwise integration.

For the proof of (c) we consider the entire function

$$f(z) = \frac{1 - E_k(z)}{z^{k+1}} = \sum_{n=0}^{\infty} c_n z^n \quad \text{with} \quad c_n \geq 0 .$$

We estimate this series by the sum over the moduli. Since $c_n \geq 0$ we get $|f(z)| \leq f(1) = 1$ for $|z| \leq 1$.

2. Specialize in the product expansion of the sine to $z = 1/2$, and consider the multiplicative inverse.

3. (a) Use the product expansion of the sine correlated with the equation

$$2 \cos \pi z \sin \pi z = \sin 2\pi z .$$

(b) Use (a) and the Addition Theorem

$$\cos \frac{\pi}{4} \left(\cos \frac{\pi}{4} z - \sin \frac{\pi}{4} z \right) = \cos \left(\frac{\pi}{4} z + \frac{\pi}{4} \right) .$$

4. First construct an entire function α, which has the zeros and respectively poles located exactly in the poles of f, depending on the fact whether the residues are positive and respectively negative. The multiplicities are given by the residues. Such a function can by obtained as a quotient of two WEIERSTRASS products. The function $f - \alpha'/\alpha$ is entire. If we could bring it into the form β'/β, then we would be done because of the representation $f = (\alpha\beta)'/(\alpha\beta)$. We have reduced the exercise to the case of an entire function f. In this case f admits a primitive F and $\exp F$ solves the problem.

5. (a) Indirect proof. We assume the contrary, namely the existence of finitely many functions $f_1, \ldots, f_n$ generating the ideal. Then there exists a natural number m, such that all f_j vanish in $m\mathbb{Z}$. Hence any function in the ideal vanishes in $m\mathbb{Z}$. But we can easily construct functions having zeros exactly in the set $2m\mathbb{Z}$.

(b) The functions with exactly one zero, which is moreover of first order are prime and irreducible.

(c) The functions without zeros are invertible.

(d) Only the functions with finitely many zeros are products of finitely many prime elements. The entire function $\sin \pi z$ cannot be written as a product of finitely many prime elements.

(e) By induction on the number of generators we can reduce the claim to the case of an ideal, which is generated by two elements f, g. By means of the WEIERSTRASS Product Theorem we will construct a function α, whose zero set is the union of the zero sets of f and g, and such that the order of any zero is the minimum of the orders for f and g at the same point. We have to show that f and g on the one side, and α on the other side generate the same ideal. Equivalently, $f/\alpha \in \mathcal{O}(\mathbb{C})$ and $g/\alpha \in \mathcal{O}(\mathbb{C})$ generate the unit ideal. We can thus assume from the beginning that f and g have no common zeros. Let us find an entire function h, such that the resulting function A has no poles. Then we are done since we get $Af + Bg = 1$ with $B = -h$. We have thus to construct h, such that $1 + hg$ vanishes at the zeros s of f with a sufficiently high order depending on s. This involves for each s equations for finitely many TAYLOR coefficients of h. Because of $g(s) \neq 0$ we can inductively solve these equations.

Exercises in Sect. IV.3

1. Let h be a solution of the given MITTAG-LEFFLER problem. For any natural number N we can determine in the disk $|z| < N$ the analytic function g_N, such that the logarithmic derivative of

$$f_N = \exp(g_N(z)) \prod_{s_n \le N} (z - s_n)^{m_n} \quad (|z| \le N)$$

is equal to h in the disk. This is a condition for the derivative of g_N. We can thus change g_N by an additive constant, such that all f_N coincide in a fixed point a, where f_1 does not vanish. Then, the functions f_{N+1} and f_N coincide in the disk $|z| < N$, so we can glue together the sequence of functions (f_N) to obtain an entire function.

2. Use the partial fraction decomposition of the cotangent in the form

$$\pi \cot \pi z = \frac{1}{z} + \sum_{n=1}^{\infty} \frac{2z}{z^2 - n^2} .$$

3. Use the partial fraction decomposition of the tangent and cotangent together with the formulas

$$\cot \pi z + \tan \frac{\pi}{2} z = \frac{1}{\sin z} , \qquad \cos \pi z = \sin \pi \left(\frac{1}{2} - z \right) .$$

The formula for $\pi/4$ is obtained by specializing $z = 0$.

4. A solution is the partial fraction series

$$\sum_{n=1}^{\infty} \left(\frac{\sqrt{n}}{z - \sqrt{n}} + 1 + \frac{z}{\sqrt{n}} + \frac{z^2}{n} \right) .$$

5. We seek for $f = gh$ with a WEIERSTRASS product g and a partial fraction series h. We choose some point $s \in S$, and assume for the sake of simplicity $s = 0$. We want to construct f, such that the first non-vanishing LAURENT coefficients are equal to $a_N, a_{N+1}, \ldots, a_M$. (All coefficients preceding a_N are zero.) We allow M to be positive. Of course, $M > N$. The function g is constructed such that its order M' in $s = 0$ is at least $M + 1$. This is a condition only in the case when M is non-negative. The function h is constructed to have a pole of order $N - M'$ in the origin. It is possible to prescribe its LAURENT coefficients $c_{N-M'}, \ldots, c_{-1}$. Let us denote by $b_{M'}, b_{M'+1}, \ldots$ the TAYLOR coefficients of g. Then we impose for the coefficients c_ν the conditions

$$\sum_{\mu+\nu=n} c_\mu b_\nu = a_n \quad \text{for} \quad N \le n \le M .$$

In the sum we isolate the term $b_{M'} c_n$. All other terms involve coefficients c_ν with $\nu < n - M'$. So we have a system of linear equations in triangular form, which can be solved inductively (since $b_{M'} \ne 0$).

Exercises in Sect. IV.4

1. The function $z \mapsto 1/z$ maps D onto a bounded domain. An analytic map from $\mathbb{C}^\bullet$ onto a bounded domain can be extended analytically to the whole $\mathbb{C}$ by the RIEMANN Removability Theorem and hence is constant by LIOUVILLE's Theorem. There is thus no conformal map with the required properties.

2. The conformal map is a dilatation $z \mapsto rz$.

3. By $z \mapsto (1-z)/(1+z)$ we map the unit disk onto the right half-plane. Then, composition with the map $w \mapsto w^2$ lands in the slit plane.

4. The map φ can be written as a composition of four conformal maps. The map $z \to w = z^2$ brings the quarter of the unit disk conformally onto the upper half of the unit disk. Then, applying $z \mapsto \frac{1+z}{1-z}$ we conformally switch to the first quadrant $\operatorname{Re} z > 0$, $\operatorname{Im} z > 0$. By $z \mapsto z^2$ the result is conformally mapped onto the upper half-plane, and we finally use $z \mapsto \frac{z-i}{z+i}$ to land in the unit disk $\mathbb{E}$.

5. The domain D is delimited by a branch of the hyperbola with equation $xy = 1$. The image of this hyperbola under the map $z \mapsto z^2 = x^2 - y^2 + 2ixy$ is the line $\operatorname{Im} w = 2$. The image of the point $2 + 2i \in D$ is $8i$. This implies, that D is conformally mapped by f onto the half-plane $\operatorname{Re} w > 2$.

6. If $\varphi : D \to D^*$ is some arbitrary conformal map, then $\gamma \mapsto \varphi\gamma\varphi^{-1}$ gives an isomorphism from $\operatorname{Aut} D$ to $\operatorname{Aut} D^*$.

7. If ψ is a second conformal map with the specified property, then $\psi\varphi^{-1}$ is a conformal self-map of the unit disk with the origin as a fixed point. So it is a multiplication map $z \to \zeta z$ for a suitable complex number ζ having modulus one. But if ζ is real and positive then $\zeta = 1$.

8. The function φ is analytic in the domain $\mathbb{C} \setminus \{1 \pm \sqrt{2}\}$, obtained from $\mathbb{C}$ by removing the two roots of the denominator. It is easy to check $\varphi(z_0) = 0$ and $\varphi'(z_0) > 0$. Similar to Exercise 4, we realize φ as a composition of simple known conformal maps,

$$z_1 = iz , \quad z_2 = \frac{1 + z_1}{1 - z_1} , \quad z_3 = z_2^2 , \quad \varphi(z) = -i\frac{z_3 - i}{z_3 + i} .$$

From this representation we see that D is conformally mapped onto the unit disk $\mathbb{E}$. The closure of D is then for continuity reasons mapped by φ onto the closed unit disk. The Maximum Principle now implies that the boundary is mapped onto the boundary. It remains to show that φ is injective on the boundary. Using the above splitting of φ we se that $\partial D \setminus \{-i\}$ is mapped topologically onto $\partial \mathbb{E} \setminus \{-i\}$. Because of $\varphi(-i) = -i$ the boundary of D is bijectively mapped onto the boundary of $\mathbb{E}$. Thus the map φ induces a bijective continuous map from $\overline{D}$ onto $\overline{\mathbb{E}}$. These spaces are compact, so the inverse map is also continuous.

9. Let $w_n = f(z_n)$. The claim is equivalent to the fact, that any accumulation value of the sequence (w_n) has modulus one. If this would not be the case, then there would exist an accumulation value $w \in \mathbb{E}$. Passing to a subsequence, we can assume that (w_n) converges to w. By continuity, the sequence with general term $z_n = f^{-1}(w_n)$ converges then to $z = f^{-1}(w)$. Contradiction.

    An example is the slit plane $D = \mathbb{C}_-$. We consider the function $w := (i\sqrt{z} + i)(i\sqrt{z} - i)^{-1}$. The image of $-1 + (-1)^n i/n$ has two accumulation values.

10. We consider the chain of transformations

$$z_1 = -iz \ , \quad z_2 = z_1^2 \ , \quad z_3 = z_2^2 - 1 \ , \quad z_4 = \sqrt{z_3} \ , \quad z_5 = iz_4 \ , \quad z_6 = \frac{z_5 - i}{z_5 + i} \ .$$

The conformal maps of D onto the upper half-plane $\mathbb{H}$ and onto the unit disk $\mathbb{E}$ are then given by $z \mapsto z_5$ and respectively $z \mapsto z_6$.

11. We only have to show that the transformation $z \mapsto (z - \lambda)(z - \bar\lambda)^{-1}$ is a conformal map of the upper half-plane onto the unit disk, since after proving this we can proceed as in the proof of III.3.10. The above transformation maps a real z into the unit circle. By the Open Mapping Theorem, this transformation maps the upper half-plane either onto the interior or onto the exterior of the unit circle. The second case is not possible, since $\lambda \in \mathbb{H}$ is mapped to $0 \in \mathbb{E}$.

Exercises to the Appendices A,B,C

1. Since two partitions have a common refinement, we only have to consider the case of adding one more point to a given partition. This point and the two left and right neighbors are contained in a disc, which is contained in the domain of definition D. The statement follows now from the CAUCHY Integral Theorem for triangular paths.

2. We can reduce the claim to the following statement. Let $\beta : [0, 1] \to \mathbb{C}^\bullet$ be a curve with starting point $a = \beta(0)$ and end point $b = \beta(1)$ both on the real axis. Assume that the image of the curve β has no other intersection points with the real axis. Then this image is contained either in the upper or in the lower half-plane. The value of the integral $\int_\beta dz/z$ is $\log b - \log a$, where we have to specify more precisely the values of the logarithm. If the curve is contained in the upper half-plane, then we take the principal value of the logarithm, since it is continuous in the closed upper half-plane. If the curve is contained in the lower half-plane, then we consider the continuous function on it which coincides with the principal branch on the open lower half-plane. The two logarithms coincide on the positive real axis, and differ by $2\pi i$ on the negative real axis. For the value of the integral we can now make a table that controls when β runs in the upper or lower half-plane, and where the corresponding intersection points a, b are located on the real or negative real axis. The integral along α is a finite sum of integrals of the specified type.

3. Let $\alpha, \beta : [0, 1] \to D$ be two not necessarily closed curves with the same starting point a and end point b, running in a simply connected domain. The closed curve

$$\gamma(t) = \begin{cases} \alpha(2t) & 0 \le 2t \le 1 \ , \\ \beta(2 - 2t) & 1 \le 2t \le 2 \ , \end{cases}$$

is then null-homotopic. So there exists a continuous family of curves γ_s, all of them starting and ending in a, which gives a deformation of $\gamma = \gamma_0$ into the constant curve $\gamma_1(t) = a$ inside the domain. We denote by $H(t, s) = \gamma_s(t)$ the corresponding homotopy. We now map the unit interval continuously onto the boundary of the homotopy square $[0, 1] \times [0, 1]$,

$$\varphi : [0, 1] \longrightarrow \partial([0, 1] \times [0, 1]) ,$$

$$\varphi(s) := \begin{cases} (0, 4s) & 0 \le 4s \le 1 , \\ (2s - 1/2, 1) & 1 \le 4s \le 3 , \\ (1, 4 - 4s) & 3 \le 4s \le 4 , \end{cases}$$

and use φ to construct a homotopy G, which deforms α into β (and fixes the starting point a and the end point b):

$$G(t, s) = H\big((1 - t)\varphi(s) + t(1/2, 0)\big) .$$

4. It is easy to reduce the claim to the case of a single curve by choosing a fixed point a in D, connecting it with the starting point of α_1, then running through α_1, then coming back to a on the reversal path, and repeating the same for α_2 and so on.

5. MORERA's Theorem is applied only to show the independence of a double integral of the order of integration. This argument is structurally simpler then the application of LEIBNIZ' criterion, which involves differentiation.

6. The proof steps can be found at different places in the text.

VIII.5 Solutions to the Exercises of Chapter V

Exercises in Sect. V.1

1. The claim follows directly from the well-known fact that for any real number x there exists an integer n with $0 \le x - n \le 1$.

2. The group property is clear. For the proof of the discreteness consider the function $g(z) = f(z + a) - f(a)$ for a fixed a, which is not a pole of f. The periods are zeros for this function. If there would be an accumulation point for this zero set in $\mathbb{C}$, then it is not a pole. The Identity Theorem then implies $g = 0$.

3. The first part $(L \cap \mathbb{R}\omega_1 = \mathbb{Z}\omega_1)$ easily follows from the following statement:
    If $a, b \ne 0$ are two real numbers with $|a| < |b|$, then there exists an integer n with $|b - na| < |a|$.
    In the second part we have to show that an arbitrary element $\omega \in L$ is a linear combination with integral coefficients of ω_1 and ω_2. At any rate, we have $\omega = t_1\omega_1 + t_2\omega_2$ with real t_1, t_2. After subtracting integers from t_1, t_2 we can assume $-1/2 \le t_1, t_2 \le 1/2$. Our proof is indirect. Supposing the converse, we can assume that both t_1, t_2 do not vanish. Then

$$|\omega| < |t_1\omega_1| + |t_2\omega_2| \le \frac{1}{2}(|\omega_1| + |\omega_2|) \le \frac{1}{2}(|\omega_2| + |\omega_2|) = |\omega_2| ,$$

a contradiction to the minimality of $|\omega_2|$.

4. The number of minimal vectors is always even, since together with a, the opposite $-a$ is also minimal. The problem is invariant with respect to rotation-dilations, so we can assume that 1 is a minimal vector. Let ω be a non-real vector of the given lattice L, which is minimal with this property. We already

know from the solution to Exercise 3, that L is generated by 1 and ω. In case of $|\omega| > 1$ the only minimal vectors are then ± 1, their number is 2. We further consider the case $|\omega| = 1$. In the subcase $\omega = \pm i$ there are exactly four minimal vectors, namely $\pm 1, \pm i$. So let's assume that there are strictly more than 4 minimal vectors. We can choose a minimal ω with non-vanishing real part. A further minimal vector is of the form $n + m\omega$ with non-zero integers n, m. From the triangle inequality we have $||n| - |m|| < |n + m\omega| = 1$, and thus $|n| = |m|$. Even more, we must have $|n| = |m| = 1$. This means that either $1 + \omega$ or $1 - \omega$ has modulus 1. Since ω is of modulus 1, the real part of ω is $\pm 1/2$. But then ω is a unit root of order six, which is not ± 1. The minimal vectors are in this case the sixth roots of unity.

Examples for the three types are

$$\mathbb{Z} + 2i\mathbb{Z}\,, \qquad \mathbb{Z} + i\mathbb{Z}\,, \qquad \mathbb{Z} + e^{\frac{2\pi i}{3}}\mathbb{Z}\,.$$

5. In (a) we consider the difference, and in (b) the quotient of f and g. We obtain entire elliptic functions, which have to be constants.

6. Assuming $L = L'$ we obtain linear equations

$$\begin{aligned} \omega_1' &= a\omega_1 + b\omega_2 \\ \omega_2' &= c\omega_1 + d\omega_2 \end{aligned} \qquad \text{and} \qquad \begin{aligned} \omega_1 &= \alpha\omega_1' + \beta\omega_2' \\ \omega_2 &= \gamma\omega_1' + \delta\omega_2' \end{aligned}$$

that admit integral solutions. Equivalently, the associated matrices

$$M = \begin{pmatrix} a & b \\ c & d \end{pmatrix}\,, \qquad N = \begin{pmatrix} \alpha & \beta \\ \gamma & \delta \end{pmatrix}$$

are inverse matrices. The product of their determinants is 1, so each determinant is invertible in $\mathbb{Z}$, i.e. ± 1. Conversely, if M is an integral matrix with determinant ± 1, then its inverse has integral entries.

7. We write $\omega_1 = x_1 + iy_1$, $\omega_2 = x_2 + iy_2$. The fundamental parallelogram $\mathcal{F}$ is the image of the unit square under a linear map. The volume of $\mathcal{F}$ is then the modulus of the determinant of the transformation matrix, namely $|x_1 y_2 - x_2 y_1|$, in concordance to the claimed formula. The invariance follows from Exercise 6.

8. A subgroup of $\mathbb{R}$, which is not dense, has no accumulation point. If the lattice $\mathbb{Z} + \sqrt{2}\mathbb{Z}$ were discrete, then there would exist a number a with $\mathbb{Z} + \sqrt{2}\mathbb{Z} = a\mathbb{Z}$, contradicting the irrationality of $\sqrt{2}$.

9. The degree of P_ω is independently of ω bounded from above, since the lattice is finitely generated. A sufficiently high derivative of f is then an entire elliptic function, and thus constant.

10. The function f'/f is elliptic. Its poles are the zeros of f. The residues are the corresponding zero orders, in particular they are positive. From the third LIOUVILLE Theorem we deduce that f'/f cannot have any pole, so it is a constant. Now we can apply Exercise 11 from I.5.

Exercises in Sect. V.2

1. Up to a constant factor we are dealing with the derivative of order $(n - 2)$ of the $\wp$-function.

2. If ω is a period of $\wp$, then we have in particular $\wp(0) = \wp(\omega)$, and thus ω is a pole of $\wp$.

3. This follows from $f(\omega/2) = f(\omega/2 - \omega) = f(-\omega/2) = -f(\omega/2)$.

4. Let l be the number of pairwise different poles modulo L of f (disregarding multiplicities), then we have $n = m + l$, since differentiation increases each pole order by one. The extremal values $l = 1$ and respectively $l = m$ are realized by the $\wp$-function and respectively by $\wp'^{-1}$.

5. The claimed bijectivity follows from Proposition V.2.10. The $\widehat{\Gamma}$-invariant meromorphic functions are exactly the even elliptic functions. In the next section (see V.3.2) we will prove that any such function can be written as a rational function of $\wp$. Another proof can be given by the specified bijection. The even elliptic functions can be transported to $\overline{\mathbb{C}}$. One can show that these functions are then meromorphic. But the meromorphic functions on $\overline{\mathbb{C}}$ are exactly the rational functions (A.6).

    By the definition of the quotient topology, the specified bijective map is continuous. A continuous bijective map between *compact* topological spaces is topological.

6. Let $L \subset \mathbb{C}$ be a lattice. Then

    $$L_{\mathbb{Q}} := \{ \ a\omega \ ; \quad a \in \mathbb{Q} \ , \ \omega \in L \ \}$$

    is a 2-dimensional $\mathbb{Q}$-vector space. If $L' \subset L$ are two lattices, then the corresponding generated $\mathbb{Q}$-vector spaces must coincide, since they have the same dimension and satisfy an inclusion. Both the condition (a) on the one side, and condition (b) on the other side, then imply that the $\mathbb{Q}$-vector spaces generated by L and respectively L' coincide. We conversely show now, that both (a) and (b) follow from this condition. For the proof we can assume that L and L' are contained in $\mathbb{Q}^2$. But in general, for any rational lattice $L \subset \mathbb{Q}^2$ there exists a natural number n, such that $n\mathbb{Z}^2 \subset L \subset (1/n)\mathbb{Z}^2$. This can be seen by choosing some rational basis, expressing it in terms of the canonical basis, and taking a multiple of the involved denominators. Then (a) and (b) become transparent.

    If the two fields have a non-constant elliptic function in common, then $L + L'$ is a lattice .

7. After subtracting a constant multiple of $\wp$ from an elliptic function with the properties specified in the Exercise, we can remove a possible pole of second order, obtaining an elliptic function of order ≤ 1. Such a function is constant.

Exercises in Sect. V.3

1. $\wp'^{-1} = \dfrac{\wp'}{4\wp^3 - g_2\wp - g_3}$, $\quad \wp'^{-2} = \dfrac{1}{4\wp^3 - g_2\wp - g_3}$, $\quad \wp'^{-3} = \dfrac{\wp'}{(4\wp^3 - g_2\wp - g_3)^2}$.

2. Make use of Proposition V.3.2.

3. Best, we first solve Exercise 5. By differentiation we obtain

    $$\wp''(z) = 2\big(\ (\wp(z) - e_1)(\wp(z) - e_2) + (\wp(z) - e_1)(\wp(z) - e_3) + (\wp(z) - e_2)(\wp(z) - e_3) \big) \ .$$

    Now we substitute $z = \omega_1/2$ and use $\wp(\omega_1/2) = e_1$.

4. We choose a point where f does not have a pole, and where the derivative f' does not vanish. In the image $f(U)$ of a small open neighborhood U of this point there exist an inverse function g. There exists a suitable open set V, which does not contain any lattice point, such that $\wp(V)$ is contained in $f(U)$. On this V the function $z \rightarrowtail h(z) = g(\wp(z))$ is well defined. From the equation $f(h(z)) = \wp(z)$ we deduce $f'(h(z))h'(z) = \wp'(z)$. Taking squares and using the differential equation satisfied by f and $\wp$, we are led to $h'(z)^2 = 1$. The only solutions of this differential equation are $h(z) = \pm z - a$. We obtain $f(\pm z - a) = \wp(z)$. Since $\wp$ is even, we can replace $\pm z$ by z (and correspondingly a by $\pm a$).

5. From the algebraic differential equation of the $\wp$-function we deduce that the polynomial $P(X) := 4X^3 - g_2 X - g_3$ has the roots e_1, e_2, e_3. This gives the factorization $P(X) = 4(X - e_1)(X - e_2)(X - e_3)$.

6. First differentiate twice termwise the LAURENT series of the $\wp$-function. Then take the square using the CAUCHY product of series, and make use of the equation $2\wp''(z) = 12\wp(z)^2 - g_2$.

7. We have successively:

    (a) $\Rightarrow$ (b) : Use Exercise 6.

    (b) $\Rightarrow$ (c) : The LAURENT coefficients are real.

    (c) $\Rightarrow$ (d) : Together with a pole ω, the conjugate $\overline{\omega}$ is also a pole of $\wp$, since $\wp$ is real. The lattice points are exactly the poles of $\wp$.

    (d) $\Rightarrow$ (a) : In the power series expansion of G_n appear only pairs of complex conjugated terms.

8. Rectangular and rhombic lattices are real by trivial reason. Let us show the converse. If ω is a lattice point of a real lattice, then $\omega + \overline{\omega}$ and $\omega - \overline{\omega}$ are also lattice points. Let L_0 be the sublattice of L generated by such lattice points. Then L_0 is generated by a real point ω_1 and a purely imaginary point ω_2. If L and L_0 coincide we are done. Else, there exists an element $\omega \in L - L_0$. We can assume that ω lies in the parallelogram with vertices $0, \omega_1, \omega_2, \omega_1 + \omega_2$. From the formula $2\omega = (\omega + \overline{\omega}) + (\omega - \overline{\omega})$ we deduce that 2ω lies in L_0. But since ω is neither real, nor purely imaginary, we deduce $2\omega = \omega_1 + \omega_2$. Then the lattice L is generated by

$$\omega = \frac{1}{2}(\omega_1 + \omega_2) \quad \text{and} \quad \omega - \omega_2 = \frac{1}{2}(\omega_1 - \omega_2) = \overline{\omega} \ .$$

    In this case L is rhombic.

9. If t is a real number, then

$$\wp(t\omega_j) = \overline{\wp(t\overline{\omega}_j)} = \overline{\wp(\pm t\omega_j)} = \overline{\wp(t\omega_j)} \ .$$

    Hence the values of $\wp$ are real on the boundary, where we of course except the lattice points. On the middle parallels we can proceed analogously, for instance we have for real t

$$\wp(\omega_1/2 + t\omega_2) = \overline{\wp(\omega_1/2 + t\omega_2)} = \overline{\wp(\omega_1/2 - t\omega_2)}$$
$$= \overline{\wp(-\omega_1/2 - t\omega_2)} = \overline{\wp(\omega_1/2 + t\omega_2)} \ .$$

10. The image of the closed fundamental parallelogram is the whole RIEMANN sphere. The pre-image of the real line including ∞ contains (Exercise 9) the boundary and the middle parallels of the fundamental parallelogram. Since the $\wp$-function takes any value exactly twice modulo L, the pre-image of $\mathbb{C}\backslash\mathbb{R}$ inside the fundamental parallelogram $\mathcal{F}$ consists of the four open sub-parallelograms obtained after removing from $\mathcal{F}$ the boundary edges and the middle lines. The $\wp$-function maps the union of the two left open sub-parallelograms bijectively onto $\mathbb{C}\backslash\mathbb{R}$. As a bijective, analytic map this restriction is conformal. A connectedness argument shows that the image of each sub-parallelogram is exactly a corresponding half-plane. From the LAURENT series of $\wp$ we see that $\wp\big(t(1+\mathrm{i})\big)$ has negative imaginary part for small positive values of t. The left lower sub-parallelogram is thus conformally mapped onto the *lower* half-plane.

11. The field of elliptic functions is algebraic over $\mathbb{C}(\wp)$.

12. If we could express $\wp$ and $\wp'$ rationally in terms of f, then there exists a discrete subset $S \subset \mathbb{C}$, such that from the equation $f(z) = f(w)$ we can deduce $z \equiv w$ mod L at least for $z, w \in \mathbb{C} \setminus S$. In the complement of the set $f(S)$ each point would have exactly one pre-image, so f would be an elliptic function of order 1.

Exercises in Sect. V.4

1. As has been explained at the beginning of Section V.4, we must consider the even and odd part of the function $f(z) = \wp(z+a)$ separately. We treat the hard case of the even part. (This is the hard case since the odd part is $\wp'(a)\wp'(z)$.) We start with

$$\frac{\wp(z+a) + \wp(z-a)}{2}\,[\wp(z) - \wp(a)]^2 = A + B\wp(z) + C\wp(z)^2 ,$$

and have to determine the constants A, B, C (which may depend on a). For this, we compare the LAURENT coefficients of z^{-4}, z^{-2}, z^0. (Odd powers do not occur.) For the computation we use the power series

$$\wp(z) = \frac{1}{z^2} + 3G_4 z^2 + \cdots$$

$$\wp(z)^2 = \frac{1}{z^4} + 6G_4 + \cdots$$

$$\wp(z) - \wp(a) = \frac{1}{z^2} - \wp(a) + 3G_4 z^2 + \cdots$$

$$[\wp(z) - \wp(a)]^2 = \frac{1}{z^4} - \frac{2\wp(a)}{z^2} + [\wp(a)^2 + 6G_4] + \cdots$$

$$\frac{\wp(z+a) + \wp(z-a)}{2} = \wp(a) + \frac{\wp''(a)}{2}z^2 + \frac{\wp^{(4)}(a)}{24}z^4 + \cdots \quad .$$

Using these first LAURENT coefficients we obtain

$$C = \wp(a) , \quad B = \frac{\wp''(a)}{2} - 2\wp(a)^2 , \quad A = \frac{\wp^{(4)}(a)}{24} - \wp(a)\wp''(a) + \wp(a)^3 .$$

Using the formulas for the derivatives of $\wp$ from Sect. V.3, we get simplified representations for these coefficients,

$$C = \wp(a) , \quad B = \wp(a)^2 - 15G_4 , \quad A = -15G_4\wp(a) - 70G_6 .$$

As final result we expect the representation

$$\frac{\wp(z+a) + \wp(z-a)}{2} = \frac{1}{4}\frac{\wp'(z)^2 + \wp'(a)^2}{(\wp(z) - \wp(a))^2} - \wp(z) - \wp(a) .$$

Using the algebraic differential equations we can see the equivalence of the two representations.

2. Substituting in the analytic form V.4.1 of the Addition Theorem the variable w by $-w$, and then z by $z + w$, we obtain the relation

$$\left(\frac{\wp'(z) - \wp'(w)}{\wp(z) - \wp(w)}\right)^2 = \left(\frac{-\wp'(z+w) - \wp'(w)}{\wp(z+w) - \wp(w)}\right)^2 .$$

This implies

$$\frac{\wp'(z) - \wp'(w)}{\wp(z) - \wp(w)} = \pm\frac{-\wp'(z+w) - \wp'(w)}{\wp(z+w) - \wp(w)} ,$$

where the sign $\pm$ does not depend on z and w. Specializing $w = -2z$, we see that the sign is in fact the $+$ sign. This formula is exactly the Addition Theorem in the geometric form V.4.4.

3. The equation of the tangent is $y = 5x - 6$. The equation $(5x - 6)^2 = 4x^3 - 8x$ has the solutions 2 (twice) and 9/4.

4. Develop the determinant from Proposition V.4.4 with respect to the third column.

5. The functions $f(z)$, $f(w)$ lie in the field $K = \mathbb{C}(\ \wp(z), \wp(w), \wp'(z), \wp'(w)\ )$. By the Addition Theorem for the $\wp$-function the function $\wp(z + w)$ also lies in this field. This implies that also $\wp'(z + w)$, and subsequently also $f(z + w)$ are elements of K. This field is algebraic over $\mathbb{C}(\wp(z), \wp(w))$. The three elements $f(z)$, $f(w)$ and $f(z + w)$ are thus algebraically dependent.

Exercises in Sect. V.5

1. If the zeros are real, then the coefficients are of course real, too. Now let P be a real polynomial of degree three. A figure for the real function $x \to P(x)$ shows that there are exactly three real zeros, iff there exists a zero between the two points with corresponding horizontal tangents. So let a, b be the roots of the derivative (in our case $\pm\sqrt{g_2/12}$), and the above condition becomes $P(a)P(b) \leq 0$. This leads to the condition $\Delta \geq 0$.

2. First, one must realize that it is possible to lift the curve α with respect to the map $\wp$ to a curve β with $\wp(\beta(z)) = \alpha(z)$. The argument is similar as in the case of exp during the proof of Proposition A.8. Let a be the starting point of β, and let b be its end point. Since α is closed, we have $b = \pm a + \omega$ with a suitable lattice element ω. From Theorem V.5.4 we deduce the relation

$$\int_0^x \frac{\alpha'(t)}{\sqrt{P(\alpha(t))}}\, dt = \beta(x) - \beta(0)$$

first for all x with $0 < x \le \varepsilon$ for a suitable ε, and then by the principle of analytic continuation for all x. In the case $x = 1$ we obtain in particular $b - a$. This element lies in the lattice L, if in the equation $b = \pm a + \omega$ the sign $\pm$ is the plus sign. This follows from the assumption $h(0) = h(1)$, which up to now has not yet been used.

3. The formula holds for $0 < x < 1$. The analytic nature of the formula allows us to restrict for a proof to "small" values of x. For the proof one can e.g. use the Remark V.5.2. It is simpler to substitute $t = s^{-1/2}$, which leads to

$$\int_0^x \frac{1}{\sqrt{1-t^4}}\, dt = \int_y^\infty \frac{1}{\sqrt{4t^3 - 4t}}\, dt \quad \text{with } y = x^{-2}\,.$$

The integral on the R.H.S. is in the normal form ($g_2 = 4$ and $g_3 = 0$). The claim now follows easily from the Addition Theorems for the $\wp$-function.

4. We use for the ellipse the parametrization

$$\alpha(t) = a\sin t + ib\cos t\,, \qquad 0 \le t \le 2\pi\,.$$

The arc length $l(\alpha)$ is given by the well-known formula

$$\int_0^{2\pi} |\alpha'(t)|\, dt = \int_0^{2\pi} \sqrt{a^2\cos^2 t + b^2\sin^2 t}\, dt = 4a \int_0^{\pi/2} \sqrt{1 - k^2\sin^2 t}\, dt\,.$$

Here, the parameter

$$k = \frac{\sqrt{a^2 - b^2}}{a}$$

is the so-called *eccentricity* of the ellipse. Substituting $x = \sin t$ one obtaines the other asserted formula for the integral.

Exercises in Sect. V.6

1. Slightly translate the period parallelogram, such that the origin lies in its interior. The integral of $\zeta(z)$ along the boundary, considered with the usual orientation, is equal to $2\pi i$ by the Residue Theorem. Compare now the integrals along opposite sides to obtain from the formula $\zeta(z + \omega_j) = \zeta(z) + \eta_j$ the claimed identity.

2. This function was already introduced in IV.1 in connection with WEIERSTRASS products. It can also be considered as a MITTAG-LEFFLER partial fraction series (IV.2).

3. Since there is exactly one zero in the fundamental region with respect to the lattice L_τ, it is enough to show that the theta series vanishes at $z = \frac{1+\tau}{2}$. This can be seen by substituting in the definition of the theta series the summation index n by $-1 - n$.

4. On both sides we have, for any arbitrary fixed $a \notin L$, an elliptic function in the variable z with the same zeros (namely $\pm a$) and poles. The terms of the claimed equality then coincide up to some constant factor. To see that this factor is one, we consider the limit $\lim_{z\to 0} z^2\big(\wp(z) - \wp(a)\big) = 1$, and on the R.H.S. we obtain the same corresponding limit because of $\sigma(a) = -\sigma(-a)$ and $\lim_{z\to 0}\big(\sigma(z)/z\big) = 1$. (These properties of σ are direct consequences of the definition.)

5. (a) Since the derivative of ζ is periodic, we have $\zeta(z+\omega) = \zeta(z) + \eta_\omega$ with a suitable number η_ω, which does not depend on z.

    (b) After subtracting from f a linear combination of derivatives $\wp^{(m)}(z-a)$, $m \geq 0$, we can reduce the claim to the case where all poles are simple. Using the function ζ from part (a), we can also remove these simple poles and obtain an elliptic function without poles, i.e. a constant.

6. The solution is $f(z) = 2\wp(z - b_1) + \zeta(z - b_1) - \zeta(z - b_2)$.

7. (a) Because of the $\mathbb{R}$-bilinearity, A is uniquely determined by the values $A(1,1), A(1,\mathrm{i}), A(\mathrm{i},1), A(\mathrm{i},\mathrm{i})$. Since A is alternating, we have $A(1,1) = A(\mathrm{i},\mathrm{i}) = 0$, and $A(1,\mathrm{i}) = -A(\mathrm{i},1)$. So A is in fact determined by $h = A(1,\mathrm{i})$.

    (b) To determine h we have to express 1 and i in terms of the basis,

    $$1 = t_1\omega_1 + t_2\omega_2 , \qquad \mathrm{i} = s_1\omega_1 + s_2\omega_2 .$$

    A simple computation gives

    $$\operatorname{Im} \frac{\omega_2}{\omega_1} = \frac{1}{s_2^2 + t_2^2} \det \begin{pmatrix} t_1 & s_1 \\ t_2 & s_2 \end{pmatrix} .$$

    (c) In the case of $\Theta = \zeta$ we have seen this in Exercise 1. The general case follows analogously.

Exercises in Sect. V.7

1. We first consider the case when one of the entries of the integral matrix M is 0. After multiplication of M with S from the right or from the left or from both sides if necessary, we can assume $c = 0$. In this case, either M or $-M$ is a power of T. Also use the fact that the square of S is the negative of the unit matrix.

    Let now M have all entries $\neq 0$. We take M such that μ is minimal. We can assume that $\mu = |c|$. Multiplication of M from the left by T^x has the effect of replacing a by $a + xc$. By means of the EUCLIDian algorithm we can find an $x \in \mathbb{Z}$, such that $|a + xc| < \mu$. This contradicts to the choice of μ.

2. $M = ST^{-3}ST^{-4}ST^2$. This representation is not unique.

3. Only the powers of S (and respectively ST) commute with S (and respectively ST).

4. The order is $n = 6$.

5. The lattice $L = \mathbb{Z} + \mathrm{i}\mathbb{Z}$ is invariant with respect to multiplication by i. This implies $G_k(L) = G_k(\mathrm{i}L) = \mathrm{i}^k G(L)$. In particular, $G_{2k}(\mathrm{i}) = 0$ for odd k, and thus $g_3(\mathrm{i}) = 0$. The lattice $L = \mathbb{Z} + e^{\frac{2\pi \mathrm{i}}{3}}\mathbb{Z}$ is invariant with respect to multiplication by $e^{\frac{2\pi \mathrm{i}}{3}}$. We deduce that $G_k(e^{\frac{2\pi \mathrm{i}}{3}}) = 0$, if k is not divisible by 6.

6. The last decisive assertion can be transported to the unit disk without any computation as follows. For any complex number ζ with modulus 1 there exists exactly one self-map of the unit disk, which fixes the origin and has the derivative equal to ζ. This property is transported to $(\mathbb{H}, \mathrm{i})$ from $(\mathbb{E}, 0)$, since

there exists a conformal map of $\mathbb{H}$ onto $\mathbb{E}$, which maps i to 0. The derivative of the map induced by the specified orthogonal matrix is $(\cos\varphi + i\sin\varphi)^{-2}$. Any complex number with modulus 1 is of this form. Since M and $-M$ define the same substitution, we only need to observe that the negative of the unit matrix is also orthogonal.

Exercises in Sect. V.8

1. The first point is equivalent with i, the second one with $1/2 + 2i$.
2. Recall that $\mathbb{C}$ is connected, in the sense that the only closed and open subsets are $\emptyset$ and $\mathbb{C}$. Moreover, we use the fact that being closed and being sequence closed are equivalent properties. (A set $A \subset \mathbb{C}$ is sequence closed, iff the limit of any convergent sequence contained in A lies in A.)
3. The claimed formulas can be directly extracted from the definition (V.8.2) of the EISENSTEIN series, and from the transformation formulas V.8.3. In particular it one has:
$$G_k(iy + 1/2) = \overline{G_k(iy - 1/2)} = \overline{G_k(iy + 1/2)} \ .$$
The EISENSTEIN series and j are thus real on $\operatorname{Re}\tau = 0$ and $\operatorname{Re}\tau = 1/2$. Since the j-function is also invariant with respect to the substitution $\tau \mapsto -1/\tau$, and since on the unit circle we have $-1/\tau = -\bar\tau$, we deduce that j is also real on the unit circle.
4. We have
$$\lim_{\operatorname{Im}\tau\to\infty} \Delta(\tau)/q = a_1 \ .$$
Because $\Delta(\tau)$ and q are real on the imaginary axis, the value of a_1 is real. If a_1 is positive, then
$$\lim_{y\to\infty} j(iy) = +\infty, \quad \lim_{y\to\infty} j(iy + 1/2) = -\infty \ .$$
The claim follows from the Mean Value Theorem for continuous functions. In case of $a_1 < 0$ we can proceed analogously. (In fact, the latter case does not occur.)
5. Make use of Exercise 5 from V.7.
6. The same proof works literally if we replace $\Gamma = \mathrm{SL}(2,\mathbb{Z})$ by the subgroup generated by the two specified matrices.

VIII.6 Solutions to the Exercises of Chapter VI

Exercises in Sect. VI.1

1. We have
$$M i = i \iff (ai + b) = i(ci + d) \iff a = d, \ b = -c \ .$$
Because of the formula
$$M^{-1} = \begin{pmatrix} d & -b \\ -c & a \end{pmatrix}$$
the above property [$a = d$, $b = -c$] is equivalent to $M' = M^{-1}$.

2.   (a) We can assume $w = i$, and use the formula

$$i = \begin{pmatrix} \sqrt{y}^{-1} & 0 \\ 0 & \sqrt{y} \end{pmatrix} \begin{pmatrix} 1 & -x \\ 0 & 1 \end{pmatrix} z \ .$$

(b) The given map is well-defined, the injectivity follows from Exercise 1, and the surjectivity from 2 (a).

The map is continuous by the definition of the quotient topology. To obtain that it is a homeomorphism, we show that it is open, which is enough. The group action is transitive, so it is enough to show that the image of a neighborhood U of $E \in \mathrm{SL}(2, \mathbb{R})$ by the map $M \mapsto Mi$ is a neighborhood of $i \in \mathbb{H}$. For this, we can even restrict to upper triangular matrices ($c = 0$) in U.

3.   One direction is trivial, namely, if M is elliptic, then M admits a fixed point, which is of course also a fixed point of any power of M. The converse is slightly more difficult. First we observe that the eigenvalues of elliptic matrices have always the modulus 1. This is so, because the characteristic polynomial

$$(a - \lambda)(d - \lambda) - bc = 0$$

has the free coefficient $ad - bc = 1$, and $|a + d| < 2$. Now let M^l be elliptic, and different from $\pm E$. We consider an eigenvalue ζ of M. Then ζ^l is an eigenvalue of M^l. It has modulus 1, and thus ζ has also modulus 1. Since the eigenvalues of the real matrix M come up as pairs of complex conjugated numbers, the numbers ζ and $\overline{\zeta} = \zeta^{-1}$ are the two eigenvalues of M. Because of

$$|\sigma(M^l)| = |\zeta^l + \overline{\zeta}^l| < 2 \ ,$$

ζ is not ± 1. This implies

$$|\sigma(M)| = |\zeta + \overline{\zeta}| < 2 \ ,$$

and M is elliptic.

4.   We can assume that i is the fixed point. The statement is. *Any finite subgroup of* $\mathrm{SO}(2, \mathbb{R})$ *is cyclic.* But this group is isomorphic to the group S^1 of all complex numbers of modulus 1, an isomorphism is given by

$$e^{i\phi} \longmapsto \begin{pmatrix} \cos\varphi & \sin\varphi \\ -\sin\varphi & \cos\varphi \end{pmatrix} \ .$$

The group S^1 is isomorphic to the group $\mathbb{R}/\mathbb{Z}$. If $G \subset S^1$ is a finite subgroup, then its pre-image in $\mathbb{R}$ is a discrete subgroup. The claim now follows from the fact that any discrete subgroup of $\mathbb{R}$ is cyclic.

Exercises in Sect. VI.2

1.   From $f(z) = f(Mz)$ we get by the chain rule

$$f'(z) = f'(Mz)M'(z) = f'(Mz)(cz + d)^{-2} \ .$$

2. We have $f'g - g'f = -f^2 \left(\dfrac{g}{f}\right)'$.

3. An analytic function $f : D \to \mathbb{C}$ is injective in a suitable neighborhood of a point $a \in D$, iff its derivative does not vanish at a (III.3). The j-function is injective modulo $\mathrm{SL}(2,\mathbb{Z})$. It is thus injective in a small neighborhood of a given point $a \in \mathbb{H}$, iff this point is not a fixed point of the elliptic modular group.

4. A bijective map between topological spaces is a homeomorphism, iff it is continuous and open. By the definition of the quotient topology, the j-function is a continuous map from $\mathbb{H}/\Gamma$ to $\mathbb{C}$. This map is open by the Open Mapping Theorem.

5. We call two points of the fundamental region *equivalent*, iff there exists a modular substitution mapping one point into the other. This gives an equivalence relation on $\mathcal{F}$. First it is clear that the quotient space of $\mathcal{F}$ with respect to this relation and $\mathbb{H}/\Gamma$ are topologically equivalent. We already know the equivalences inside the fundamental region $\mathcal{F}$, so we can map $\mathcal{F}$ onto a full square without one vertex. Two adjacent vertices of this square have to be identified. From the topology it is known that on this way one obtains from a closed square a sphere, removing one point we obtain a sphere without a point, i.e. a model of the real plane.

    Despite intuitive evidence, it is not easy to convert these arguments into a complete rigorous proof. This Exercise may seem unfair from this point of view. In the second volume we will develop, in connection with the topological classification of surfaces, instruments to attack such questions.

6. The most simple argument uses the j-function. Because of $\overline{j(z)} = j(-\bar{z})$ the quotient space is topologically equivalent to the quotient space of $\mathbb{C}$ with respect to the relation that identifies w and $\bar{w}$. The upper and the lower half-planes are thus folded together, and the result is the closed upper half-plane. (In contrast with Exercise 5, it is easy to prove it.)

Exercises in Sect. VI.3

1. Induction on the weight. The induction starts with the weight $k = 0$. The inductive step uses the fact that any modular form of positive weight without zeros in the upper half-plane is necessarily a cusp form, so we can divide by Δ.

2. We study the linear map $[\Gamma, k] \longrightarrow \mathbb{C}^{d_k}$, which associates to a modular form the tuple of the first d_k coefficients in its power series expansion. The Exercise claims that this map is bijective. Since both vector spaces have the same dimension, it is enough to show it is injective. Let f be in the kernel. Then f/Δ^{d_k} is an entire modular form of weight $k - 12 d_k$. From the formula for d_k we deduce, that in this weight there exists no entire non-zero modular form.

3. Since j takes infinitely many values, P has infinitely many zeros. The same applies (with the same argument, or as a consequence) for the function G_4^3/G_6^2. Let $\sum_{4\alpha + 6\beta = k} C_{\alpha\beta} G_4^\alpha G_6^\beta$ be a non-trivial linear relation. After multiplying with a suitable monomial we can assume that k is divisible by 6. We divide this relation by $G_6^{k/6}$, and obtain a linear relation between powers of G_4^3/G_6^2.

4. Start with $G_4^3 - C\Delta$.

5. Since a meromorphic modular form f has only finitely many poles in the fundamental region, there exists by Exercise 4 an entire modular form h, such that fh has no poles in the fundamental region, and hence in the whole $\mathbb{H}$. After multiplying with a suitable power of the discriminant, we obtain that it is also regular at $i\infty$.

6. We can represent a given modular function f as in Exercise 5 as a quotient $f = g/h$ of two entire modular forms g, h of the same weight. We can furthermore arrange that the common weight of g and h is divisible by 6, let's say it is $6k$. Because of the formula $g/h = (g/G_6^k)(h/G_6^k)^{-1}$ we can assume $h = G_6^k$. Expressing g as a polynomial in G_4 and G_6 we see that any modular function can be written as a rational function in G_4^3/G_6^2, and hence in j.

Exercises in Sect. VI.4

1. Apply the transformation behavior for the imaginary part, V.7.1.

2. If f is a cusp form, then $\exp(-2\pi i z)f(z)$ is bounded in domains of the form $y \geq \delta > 0$.

3. From the integral representation

$$a_n = \int_0^1 f(z)e^{-2\pi i n z}\,dx$$

we deduce concretely

$$|a_n| \leq C'e^{2\pi}n^{k/2}\ .$$

4. A simple estimate shows that the R.H.S. in the asserted asymptotic formula is greater than $\delta n^{m/2-1}$ for a suitable $\delta > 0$. The difference of the two sides has by Exercise 3 a smaller asymptotic behavior, namely $O(n^{m/4})$. Because of $m \geq 8$ we have $m/4 < m/2 - 1$.

5. If we subtract from a modular form a suitable constant multiple of the EISEN-STEIN series, then we obtain a cusp form. Using Exercise 3, we only have to prove the claim for the EISENSTEIN series. This follows easily from the formula mentioned in Exercise 4.

6. The rows of orthogonal matrices have the EUCLIDian length 1. If the matrix has integral entries, the rows are up to sign the canonical basis vectors. We can list all integral orthogonal matrices U, by writing all n canonical basis vectors in arbitrary order as rows of a matrix. We can modify the rows by s sign. There are $2^n n!$ possibilities.

7. Make use of the following fact. If A is an integral $n \times n$ matrix with non-zero determinant, then $L = A\mathbb{Z}^n$ is a sublattice of $\mathbb{Z}^n$ of index $|\det A|$. The square of this number is the GRAM determinant of the associated quadratic form.

    (a) We consider first $L \cap \mathbb{Z}^n$. This is the kernel of the homomorphism

$$\mathbb{Z}^n \longrightarrow \mathbb{Z}/2\mathbb{Z}\ , \qquad x \longmapsto x_1 + \cdots + x_n \quad \mathrm{mod}\ 2\ ,$$

    and thus a sublattice of index 2 of $\mathbb{Z}^n$. For odd n we have $L_n \subset \mathbb{Z}^n$, the determinant of a GRAM matrix is thus $2^2 = 4$. If n is even, then the vector $e = (1/2, -1/2, \ldots, 1/2, -1/2)$ lies in L_n, and any vector a of L_n is of the form $a = b$ or $a = e + b$ with $b \in L_n \cap \mathbb{Z}^n$. The index of $L_n \cap \mathbb{Z}^n$ in L_n is thus 2, and the determinant of a GRAM matrix is hence 1. The lattice L_n is thus of type II, iff n is even and $\langle a, a \rangle$ is even for all $a \in L_n$. Setting $a = e + b$, we have $\langle a, a \rangle \equiv n/4 \mod 2$. The lattice L_n is thus of type II, iff n is divisible by 8.

(b)   There are two types of minimal vectors:

The integral minimal vectors have twice the entry ± 1, and zeros else, the non-integer minimal vectors exist only in case $n = 8$. They contain only $\pm 1/2$ entries.

(c)   The following argument is simpler as the one given in the hint: L_8, and thus also $L_8 \times L_8$, is generated by minimal vectors, but this is false for L_{16}.

8.   Express $\vartheta_{a,b}$ in terms of the JACOBI theta function (V.6),

$$\vartheta_{a,b}(z) = e^{\pi i a^2} \vartheta(z, b + za) \; ,$$

and use the knowledge of the zeros (Exercise 3 in Sect. V.6). If the series vanishes, then we must have

$$b + za = \frac{\alpha}{2} + \frac{\beta}{2}z \; , \quad \alpha \equiv \beta \equiv 1 \mod 2 \; .$$

This is precisely the case which has been excluded.

9.   If we change b modulo 1, then the value of the theta series does not change. The substitution $a \mapsto a + \alpha$, $\alpha \in \mathbb{Z}$, changes the summation index as $n \mapsto n - \alpha$. Thus the series takes up the factor $\exp(-2\pi i a b)$.

As in Exercise 8, express the theta null value in terms of the JACOBI theta series, and apply the JACOBI theta transformation formula.

10.   Let $\mathcal{M}$ be the mentioned finite set of pairs (a, b). We first show, that for any modular substitution M there exists a bijective self-map $(a, b) \mapsto (\alpha, \beta)$ of $\mathcal{M}$ with the property

$$\vartheta_{a,b}(Mz) = v(M, a, b)\sqrt{cz + d}\,\vartheta_{\alpha,\beta}(z) \; ,$$

which involves a suitable root of unity of order eight $v(M, a, b)$. It is enough to show this only for the generators of the modular group. For the involution we obtain this property from (both parts of) Exercise 9. Together with (a, b) there is also a the coset of $(b, -a) \mod 2$ an element of $\mathcal{M}$. For the translation the claim is elementary. A suitable power of $\Delta_n(z)$ is a modular form without zeros, and hence by Exercise 1 in Sect. VI.3 it is a constant multiple of a power of the discriminant. Because of this a power of $f = \Delta_n^{24}/\Delta^{4n^2 - 1}$, and hence f itself, is constant,

$$\Delta_n(z)^{24} = C\Delta(z)^{4n^2 - 1} \; .$$

To determine the constant C we expand both sides as power series in $\exp\left(\frac{\pi i z}{4n}\right)$ and compare the lowest coefficients. Here, we use the fact that the coefficient of lowest order of a product of two power series is the product of the lowest coefficients of the involved power series. We obtain

$$C \cdot (2\pi)^{12(4n^2-1)} = \prod_{\substack{0 \leq b < 2n \\ b \neq n}} \left(1 + e^{-\pi i b/n}\right)^{24} = (2n)^{24} .$$

11. The evaluation of the square root uses the principal part, as required in the theta inversion formula. This means $\sqrt{1} = +1$, and

$$\sqrt{\frac{1}{1-i}} \cdot \sqrt{1+i} = \sqrt{\frac{1+i}{2}} \cdot \sqrt{1+i} = \frac{1+i}{\sqrt{2}} = e^{\pi i/4} .$$

The result is $e^{2\pi i n/8} = 1$, iff 8 divides n.

Exercises in Sect. VI.5

1. Let G_q be the subgroup generated by the two specified matrices. It is enough to show that for any matrix $M \in \mathrm{SL}(2, R)$ there exists a matrix $N \in G_q$, such that the first column of NM is the first column of the standard basis, because in this case NM is a translation matrix. We assume the knowledge of this fact in the case $R = \mathbb{Z}$ (i.e. $q = 0$). Let $a, c \in \mathbb{Z}$ be representatives of the first column of M. From the condition on the determinant we deduce that (a, c, q) are relatively prime. Using results from elementary number theory we can assume after a possible replacement of a, c modulo q that already a, c are relatively prime. Then we can extend a, c to a matrix in $\mathrm{SL}(2, \mathbb{Z})$, and we take for N its inverse.

2. Use Exercise 1.

3. The ring $\mathbb{Z}/p\mathbb{Z}$ is a field for prime numbers p. There exist $p^2 - 1$ non-zero vectors, and each such vector appears as the first column of an invertible 2×2 matrix. Fixing such a column vector v, we can extend it to an invertible matrix by choosing any vector w which is linearly independent of v as the second column, i.e. avoiding the p possible scalar multiples of it. For each v there are $(p^2 - p)$ possibilities for w.

    The group $\mathrm{SL}(2, R)$ is the kernel of the surjective homomorphism

    $$\mathrm{GL}(2, R) \longrightarrow R^\times , \qquad M \longmapsto \det M .$$

4. The kernel of the homomorphism

    $$\mathbb{Z}/p^{m+1}\mathbb{Z} \longrightarrow \mathbb{Z}/p^m\mathbb{Z} \qquad (m \geq 1)$$

    consists of all elements of the form ap^m, $a \in \mathbb{Z}/p^{m+1}\mathbb{Z}$. The assignment $ap^m \longmapsto \bar{a}$, where $\bar{a}$ is the coset of a modulo p, defines an isomorphism of the kernel onto $\mathbb{Z}/p\mathbb{Z}$. This argumentation also applies to the group $\mathrm{GL}(2)$, if we observe that a 2×2 matrix with coefficients in $\mathbb{Z}/p^{m+1}\mathbb{Z}$ is invertible, iff its image in $\mathbb{Z}/p^m\mathbb{Z}$ is invertible. Since we can test invertibility by means of the determinant, this observation follows from the fact that an element in $\mathbb{Z}/p^{m+1}\mathbb{Z}$ is a unit, iff its image in $\mathbb{Z}/p^m\mathbb{Z}$ is a unit.

    The given formula for the order of GL follows now by induction on m, the formula for SL then follows using the determinant homomorphism.

5. We have in general $\mathrm{GL}(n, R_1) \times \mathrm{GL}(n, R_2) = \mathrm{GL}(n, R_1 \times R_2)$.

6. Decompose q as a product of prime factors, and use Exercises 3 to 5.

7. From $\mathbb{H} = \bigcup_{M \in \Gamma} M\mathcal{F}$ we deduce

$$\mathbb{H} = \bigcup_{M \in \Gamma_0} \bigcup_{\nu=1}^{h} M M_\nu \mathcal{F} = \bigcup_{M \in \Gamma_0} M\mathcal{F}_0 .$$

Then S can be chosen to be the union of the boundaries of $M_\nu \mathcal{F}$.

8. Let f be a function on the upper half-plane, such that the integral

$$I(f) := \int_{\mathbb{H}} f(z) \frac{dxdy}{y^2}$$

exists as a LEBESGUE integral. From the transformation formula for double integrals we obtain

$$I(f) = I(f^M) \quad \text{with} \quad f^M(z) = f(Mz) \quad (M \in \mathrm{SL}(2, \mathbb{R})) . \quad (*)$$

The reason comes the fact that the real JACOBI functional determinant which appears in the general transformation formula for M is equal to $|cz + d|^{-4}$, and this factor is compensated by the factor which occurs form the transformation of y^{-2}. If f is the characteristic function of a set $A \subset \mathbb{H}$, then $v(A) = v(M(A))$. In particular, using the notations of Exercise 7 this implies

$$v(\mathcal{F}_0) = hv(\mathcal{F}) = [\Gamma : \Gamma_0]v(\mathcal{F}) .$$

An elementary computation shows that the integral of y^{-2} over the fundamental region $\mathcal{F}$ has the value $\pi/3$.

It remains to show the invariance of the integral. For this we need some basics of integration theory:

Let $\mathcal{F}_0$ and $\mathcal{F}_0'$ be two fundamental regions for Γ_0. The exceptional sets in the sense of Exercise 7(a) are denoted by S_0 and S_0'. We want to show the equality

$$\int_{\mathcal{F}_0} f(z) \frac{dxdy}{y^2} = \int_{\mathcal{F}_0'} f(z) \frac{dxdy}{y^2}$$

for a certain class of Γ_0-invariant functions f. This class consists of all continuous Γ_0-invariant functions with the following two properties:

(a) The support of the restriction of f to $\mathcal{F}_0 \setminus S_0$ is compact.

(b) The same is valid for $(\mathcal{F}_0', S_0')$ instead of $(\mathcal{F}_0, S_0)$.

Using a suitable partition we can reduce the claim to the situation where the support of f is "small"in the following sense. There exists a substitution $M \in \Gamma_0$, such that the image of the support of f under M is contained in $(\mathcal{F}_0', S_0')$. In this case we can apply $(*)$. The partition can be constructed using a net of squares as in the hint, or better by using a partition of unity.

9. First, the full modular group Γ acts on $K(\Gamma_0)$. Since the normal subgroup Γ_0 acts trivially, there is an induced action of the factor group. From algebra we know, that a field is always algebraic over the fixed field with respect to the action of a finite group of automorphisms.

10. Use the explanations for Exercise 11 in V.3.

11. The elements of the considered groups are triangular matrices when considered modulo q. The conjugation can be obtained by means of the involution S. The connection with the theta group in case of $q = 2$ follows from the result A.5 in the appendix.

12. We first observe that $\Gamma^0[p]$ has index $p + 1$ in the full modular group,

$$SL(2,\mathbb{Z}) = \Gamma^0[p] \begin{pmatrix} 0 & -1 \\ 1 & 0 \end{pmatrix} \cup \bigcup_{\nu=0}^{p-1} \Gamma^0[p] \begin{pmatrix} 1 & \nu \\ 0 & 1 \end{pmatrix} .$$

All translation matrices correspond to one cusp class.

Exercises in Sect. VI.6

1. We have to consider two cases, depending on the parity of q. If q is even, then we can define $\Gamma[q, 2q]$ by the conditions $a \equiv d \equiv 0 \mod q$ and $b \equiv c \equiv 0 \mod 2q$. An easy computation shows that these conditions define a group. If q is odd, then $\Gamma[q, 2q] = \Gamma[q] \cap \Gamma[1, 2]$. We only have to know that $\Gamma[1, 2]$ is a group. This is exactly the theta group.

2. If we transform the theta functions $\vartheta, \tilde{\vartheta}, \tilde{\tilde{\vartheta}}$ with a modular substitution $M \in SL(2,\mathbb{Z})$, then this triple is permuted up to some elementary factors. Encoding the initial order of the triple by the symbols $1, 2, 3$ we get a homomorphism $SL(2,\mathbb{Z}) \to S_3$. The involution corresponds to the transposition of 2 and 3. The translation corresponds to the transposition of 1 and 2. Since S_3 is generated by these transpositions, the homomorphism is surjective. The three theta series are modular forms with respect to $\Gamma[2]$, and hence the kernel of $SL(2,\mathbb{Z}) \to S_3$ contains $\Gamma[2]$. The induced homomorphism $SL(2,\mathbb{Z}/2\mathbb{Z}) \to S_3$ of groups of same order six is surjective, and thus an isomorphism.

3. Consider the pre-image of the alternating subgroup A_3 of S_3 with respect to the homomorphism described in Exercise 2.

4. The groups Γ with $\Gamma[2] \subset \Gamma \subset \Gamma[1]$ are in a bijective correspondence with the subgroups of $SL(2,\mathbb{Z}/2\mathbb{Z})$. The order of $SL(2,\mathbb{Z}/2\mathbb{Z})$ is six, so the order of any of its subgroups is 1, 2, 3 and 6. There exists exactly one subgroup of order 1, three conjugated subgroups of order 2, and one normal subgroup of order 3. There are thus exactly 6 possibilities for a group Γ with $\Gamma[2] \subset \Gamma \subset \Gamma[1]$, namely $\Gamma[2]$ itself, the three conjugates of the theta group, the normal subgroup of index 2, and the full modular group.

5. The assertion is a direct consequence of Theorem VI.6.3, since we know how the three basic theta functions transform under the action of the generators of Γ_ϑ.

6. Checking the transformation behavior and the constant power series coefficients one can verify the identities

$$G_4 = \zeta(4) \left(\vartheta^8 + \tilde{\vartheta}^8 + \tilde{\tilde{\vartheta}}^8 \right) ,$$

$$G_6 = \zeta(6) \left(\vartheta^4 + \tilde{\vartheta}^4 \right)\left(\vartheta^4 + \tilde{\tilde{\vartheta}}^4 \right)\left(\tilde{\vartheta}^4 - \tilde{\tilde{\vartheta}}^4 \right) .$$

7.   We consider

$$K := \{ \ (X,Y) \in \mathbb{C} \times \mathbb{C} \ ; \quad X^4 + Y^4 = 1 \ , \ XY \neq 0 \ \}$$

and $f = \tilde{\vartheta}/\vartheta$, $g = \tilde{\tilde{\vartheta}}/\vartheta$. It is convenient also to consider the function $h := f^8 g^8$. This function is invariant under the theta group, it has no zero, and hence it defines a map

$$h : \mathbb{H}/\Gamma_\vartheta \longrightarrow \mathbb{C}^\bullet \ .$$

One can show that h is bijective in an analogous way to the proof of the bijectivity of the j-function by using a substitute of the $k/12$-formula for the theta group. We don't give details and assume that the bijectivity of h has been proved. Consider the commutative diagram

$$
\begin{array}{ccccc}
\mathbb{H}/\Gamma[4,8] & \xrightarrow{\ (f,g)\ } & K & \ni & (X,Y) \\
\downarrow & & \downarrow & & \downarrow \\
\mathbb{H}/\Gamma_\vartheta & \xrightarrow[h]{\ \cong\ } & \mathbb{C}^\bullet & \ni & X^8 Y^8
\end{array}
$$

and compute the number of pre-images of points with respect to the vertical arrows. The claimed bijectivity is a consequence of this computation.

VIII.7 Solutions to the Exercises of Chapter VII

Exercises in Sect. VII.1

1.   The weight is 14. We must thus have $j'(z)\Delta(z) = C G_4(z)^2 G_6(z)$ with a suitable constant C. This constant can be determined by comparing the constant coefficients of the power series. In order to do this, we need the FOURIER expansion of the EISENSTEIN series VII.1.3, see also Exercise 5 below. We obtain

$$C = -\frac{(2\pi i)^{13}}{1728 \cdot 8\zeta(4)^2 \zeta(6)} \ .$$

2.   There must be a relation of the form

$$G'_{12}\Delta - G_{12}\Delta' = A G_4^2 G_6^3 + B G_4^5 G_6$$

with suitable constants A, B, which can be determined using the coefficients of the expansion of the EISENSTEIN series. Best, one proceeds as follows. Since we are dealing with a cusp form, we expect $G'_{12}\Delta - G_{12}\Delta' = C\Delta G_4^2 G_6$, and determine C from the knowledge of the formula for Δ in terms of G_4 and G_6. A comparison of the first coefficient gives

$$\frac{2(2\pi i)^{25}}{11!} - 2\zeta(12)(2\pi i)^{13} = C \, (2\pi)^{12} \cdot 8\zeta(4)^2 \zeta(6) \ .$$

3. Both the structure theorem in connection with the knowledge of special zeta values, and the recursion formula from Exercise 6 in V.3 lead to the formula

$$13 \cdot 11 G_{12} = 2 \cdot 3^2 G_4^3 + 5^2 G_6^2 \ .$$

4. Using the identity with the EISENSTEIN series we obtain the relation $240\sigma_3(5) = 30\,240$.

   The number of solutions of the equation $x_1^2 + \cdots + x_8^2 = 10$ with $x \in L_8$ (see VII.4) are

      7 168 solutions which contain once ± 2, six times ± 1, and once 0,
      6 720 solutions which contain twice ± 2, twice ± 1, and four times 0,
      224 solutions which contain once ± 3, once ± 1, and six times 0,
      7 168 solutions which contain once $\pm 5/2$, once $\pm 3/2$, and six times $\pm 1/2$,
   where one sign is determined by the remaining signs,
      8 960 solutions which contain four times $\pm 3/2$, and four times $\pm 1/2$, where one sign is determined by the remaining signs.

   We indeed can verify

   $$30\,240 = 7\,168 + 6\,720 + 224 + 7\,168 + 8\,960 \ .$$

5. Using the FOURIER expansion of the EISENSTEIN series, and the congruence

   $$5d^3 + 7d^5 \equiv 0 \mod 12$$

   we obtain the integrality of $\tau(n)$. The integrality of $c(n)$ is a consequence.

6. The logarithmic derivative of a normally convergent product is equal to the sum of the logarithmic derivatives of the factors,

   $$\frac{\eta'(z)}{\eta(z)} = \frac{\pi \mathrm{i}}{12} - \sum_{n=1}^{\infty} 2\pi i n \frac{\exp 2\pi i n z}{1 - \exp 2\pi i n z} \ .$$

   Expanding now $\left(1 - \exp(2\pi i n z)\right)^{-1}$ into a geometric series, we obtain $(4\pi)^{-1} G_2$ in the form VII.1.3.

   Both logarithmic derivatives are $\eta'(z)/\eta(z) + 1/(2z)$.

7. From Exercise 6 follows that $\eta(z)^{24}$ is a cusp form of weight 12.

8. At any rate, the 24th powers of both sides coincide. The quotient of both sides is an analytic function, whose values are roots of unity of order 24. By continuity, it is constant. This constant is 1 $(q \to 0)$.

9. First, we have

   $$\prod_{n=1}^{N}(1 - q^n) = \sum_{k=0}^{N} \sum_{1 \le n_1 < \cdots < n_k \le N} (-1)^k q^{n_1 + \cdots + n_k} \ .$$

   Sorting summands which correspond to fixed values of $n_1 + \cdots + n_k$, and taking the limit $N \to \infty$, we obtain the claimed formula.

Exercises in Sect. VII.2

1. For the proof of the Exercise, we need the estimate

$$\left|n^{-s} - m^{-s}\right| \le \left|\frac{s}{\sigma}\right| \left|n^{-\sigma} - m^{-\sigma}\right| ,$$

which follows directly from the integral representation

$$n^{-s} - m^{-s} = s \int_n^m t^{-(s+1)} \, dt$$

by estimating the factor s by $|s|$, the integrand by $t^{-(\sigma+1)}$ and evaluating then the integral.

We now assume that the DIRICHLET series converges at a point s_1. We must show the convergence in the half-plane $\sigma > \sigma_1$. After a translation of variables we can assume $s_1 = 0$. The series $\sum a_n$ converges in this case. We would like to apply the CAUCHY criterion for series, and in order to do this we must estimate for (large) $m > n$

$$S(n, m) = \sum_{\nu=n}^m a_\nu \nu^{-s} .$$

ABELian partial summation gives

$$S(n, m) = \sum_{\nu=n}^{m-1} A(n, \nu) \left(\nu^{-s} - (\nu + 1)^{-s}\right) + A(n, m) \, m^{-s} \qquad \text{with}$$

$$A(n, m) = \sum_{\nu=n}^m a_\nu .$$

For a given $\varepsilon > 0$, using the convergence of $\sum a_n$, we can find an N, such that $|A(n, m)| \le \varepsilon$ for $m > n \ge N$. Using the initial estimate we obtain

$$|S(n, m)| < \varepsilon \left(1 + \frac{|s|}{\sigma} \sum_{\nu=n}^{m-1} \left(\nu^{-\sigma} - (\nu + 1)^{-\sigma}\right)\right) = \varepsilon \left(1 + \frac{|s|}{\sigma} \left(n^{-\sigma} - m^{-\sigma}\right)\right) .$$

From this we deduce the uniform convergence in regions, where $|s|/\sigma$ is bounded from above.

Supplement. The inequality $\sigma_0 \ge \sigma_1$ is trivial. If the series converges in a point s, then the sequence $(a_n n^{-s})$ is bounded. Then we have that the DIRICHLET series converges absolutely at $s + 1 + \varepsilon$ for any positive ε. From this we deduce the second inequality.

2. We must prove these relations for the divisor sums $a(n) = \sigma_{k-1}(n)$. The relation in (a) follows from that the fact that any divisor of mn can be uniquely written as a product of a divisor of m and a divisor of n (m, n are relatively prime).

    For the relation (b) use the fact that the divisors of p^ν are exactly the p-powers p^j, $j \le \nu$.

    Expanding $(1 - p^{-s})^{-1}$ and $(1 - p^{k-1-s})^{-1}$ into geometric series, and multiplying them, we obtain a series of the form $\sum_{\nu=0}^\infty b(p^\nu) p^{-\nu s}$. A direct computation gives $b(p^\nu) = \sigma_{k-1}(p^\nu)$. The remaining part of the Exercise is proven analogously to the product expansion of the zeta function.

3.   The matrix is after a short computation

$$\begin{pmatrix} -\nu & \frac{\nu\mu+1}{p} \\ -p & \mu \end{pmatrix} \ .$$

4.   One checks the transformation behavior with respect to the generators. For the translation $z \mapsto z+1$ the function $f(pz)$ remains unchanged, and the terms of the sum are permuted. To see the behavior of the involution, we rewrite the formula as

$$(T(p)f)(z) = p^{k-1} f(pz) + \frac{1}{p} f\left(\frac{z}{p}\right) + \frac{1}{p} \sum_{\nu=1}^{p-1} f\left(\frac{z+\nu}{p}\right) \ .$$

Up to some necessary pre-factors, the involution interchanges the first two terms, and the terms of the sum are permuted by Exercise 3.

5.   For the proof, make use of the formula from Exercise 4 and of

$$\frac{1}{p} \sum_{\nu=0}^{p-1} e^{\frac{2\pi i n \nu}{p}} = \begin{cases} 1 & \text{if } n \equiv 0 \mod p \ , \\ 0 & \text{else} \ . \end{cases}$$

6.   The exercise claims that the coefficients of the normalized EISENSTEIN series satisfy a relation of the form

$$a(pn) + p^{k-1} a(n/p) = \lambda(p) a(n) \ .$$

By Exercise 2 such a relation occurs indeed, and the involved eigenvalue is $\lambda(p) = a(p)$. If p and n are relatively prime, then we use relation (a), else we have also to use relation (b).

7.   Apply Exercise 5 first for $n = 1$, then for an arbitrary n.

8.   The convergence follows from the estimate $|a(n)| \le Cn^{k-1}$ (Exercise 5 in V1.4). From the recursion formula for $a(p^\nu)$ in Exercise 7 we obtain after multiplication

$$\left(1 - a(p)x + p^{k-1}x^2\right)\left(1 + \sum_{n=1}^{\infty} a(p^n)x^n\right) = 1 \ .$$

The involved power series converges for $|x| < 1$. The product decomposition $D(s) = \prod D_p(s)$ follows from the relation $a(nm) = a(n)a(m)$ for relatively prime integers $n, m \ge 1$ by termwise multiplication. In full analogy to the case of the product expansion of the zeta function one has to prove that this formal multiplication of an infinite product is allowed.

9.   Directly from the definition of $T(p)$, and passing to the limit $y \to \infty$, we see that $T(p)$ transforms cusp forms into cusp forms.

10. From the formula for $T(p)$ (Exercise 4) follows

$$\left|\tilde{f}(z)\right| \le p^{k-1} |f(pz)| + \frac{1}{p} \sum_{\nu=0}^{p-1} \left|f\left(\frac{z+\nu}{p}\right)\right|$$

and from this we get

$$|\widetilde{g}(z)| \le p^{\frac{k}{2}-1}|g(pz)| + p^{\frac{k}{2}-1}\sum_{\nu=0}^{p-1}\left|g\left(\frac{z+\nu}{p}\right)\right| .$$

This gives $|\widetilde{g}(z)| \le p^{\frac{k}{2}-1}(1+p)m$, and the required estimate. If the eigenform f does not vanish identically, then $\widetilde{m} = \lambda(p)m$. This gives the desired estimate for $\lambda(p)$.

If there are given two non-cusp (modular) forms, then a suitable non-trivial linear combination delivers a cusp form. If the two non-cusp forms are eigenforms of an operator $T(p)$, then this cusp form is an eigenform with the same eigenvalue $1 + p^{k-1}$. Hence it must vanish, since for any $k \ge 4$, and any p we have $p^{\frac{k}{2}-1} < 1 + p^{k-1}$.

Exercises in Sect. VII.3

1. The series lies in the space $\{1, 2k, (-1)^k\}$. By Theorem VII.3.4 this space is isomorphic to $[1, 2k, (-1)^k]$, which is the space of modular forms of weight $2k$. In case of $k = 1$ this space is trivial, in the cases $k = 2, 3, 4$ it is 1-dimensional being generated by the EISENSTEIN series. Now one can make use of Exercise 2 in VII.2.

2. The proof follows along the lines in Exercise 1. Besides VII.3.9 one should also use a characterization of ϑ^k, $k < 8$, e.g. as it can be obtained from Exercise 5 in VI.6.

3. The discriminant is up to a constant factor the unique modular form of weight 12, whose coefficients are of the order $O(n^{11})$.

4. In the first series the sub-series from 0 to ∞, and from -1 to $-\infty$ coincide, as can be seen by means of the substitution $n \to -1 - n$. The terms with even $n = 2m$ of the second series lead to the terms from 0 to ∞ of the third series. Correspondingly, the terms with odd $n = 2m + 1$ lead to the terms of the third series from -1 to $-\infty$. The representation of f as a derivative of the JACOBI theta function evaluated at $w = 1/4$ is clear from the third formula for f. Now differentiate in the theta transformation formula with respect to w, and then specialize $w = 1/4$.

5. We write f as
$$f(z) = \sum_{n=0}^{\infty}(-1)^n(2n+1)e^{\frac{2\pi i(2n+1)^2}{8}}$$
and obtain the associated DIRICHLET series in the form
$$D(s) = \sum_{n=0}^{\infty}(-1)^n(2n+1)(2n+1)^{-2s} = \sum_{n=0}^{\infty}(-1)^n(2n+1)^{1-2s} = L(2s-1) .$$

    The functional equation for $D \in \{8, 3/2, 1\}$ according to VII.3.2 is identical with the desired functional equation for L.

6. The functional equation of the RIEMANN zeta function, and the functional equation for $L(s)$ from Exercise 5, together with LEGENDRE's Relation IV.1.12 for the Gamma function lead to the desired functional equation for $\zeta(s)L(s)$. The normalizing factor is obtained by passing to the limit $\sigma \to \infty$.

Exercises in Sect. VII.4

1. First we define $\mu(n)$ by the required formulas. The convergence of the DIRICH-
    LET series with coefficients $\mu(n)$ for $\sigma > 1$ is clear. Because of the uniqueness
    of the expansion as a DIRICHLET series the claim is equivalent to

    $$\sum_{n=1}^{\infty} \frac{1}{n^s} \cdot \sum_{n=1}^{\infty} \frac{\mu(n)}{n^s} = 1 .$$

    This means that for $C(N) := \sum_{n=1}^{N} \mu(n)$ we have $C(1) = 1$, and $C(N) = 0$
    for $N > 1$. Because of the obvious relations $\mu(nm) = \mu(n)\mu(m)$ and $C(nm) =$
    $C(n)C(m)$ for relatively prime m, n we can restrict ourselves to $N = p^m$. In
    case of $m > 0$ the defining sum for $C(N)$ has only two non-vanishing terms, 1
    and -1.

2. If we know the formula for two intervals $]x, y]$ and $]y, z]$, then we also know it
    for $]x, z]$. Therefore it is enough to prove the formula for those intervals, which
    do not contain any natural numbers in their interior. Then the function $A(t)$ is
    constant in the interior of such an interval, and the claim is easy to check now.

3. In Sect. VII.4 we showed that the first two formulas are equivalent, and the
    third one follows from them. If one goes through the proof carefully, one also
    gets the converse.

4. The convergence of the DIRICHLET series with the coefficients $\varphi(n)$ for $\sigma > 2$
    follows from the trivial estimate $\varphi(n) \le n$. The claimed identity

    $$\sum_{n=1}^{\infty} n n^{-s} = \sum_{n=1}^{\infty} n^{-s} \sum_{n=1}^{\infty} \varphi(n) n^{-s}$$

    is equivalent to the well-known relation

    $$\sum_{d \mid n} \varphi(d) = n .$$

5. A formal computation, which of course has to be rigorously verified, gives

    $$\sum \log(1 - p^{-s}) = \sum_{p} p^{-s} + \sum_{p} \sum_{\nu \ge 2} \frac{1}{\nu} p^{-\nu s} .$$

    The double series converges even in the region $\sigma > 1/2$, as it can be seen by a
    comparison with the zeta function. The first series on the R.H.S. is dominated
    by the series $\sum p^{-1}$, which converges by assumption. The whole R.H.S. remains
    thus bounded when we approach 1. Since the L.H.S. is a logarithm of the zeta
    function, we see that the zeta function $\zeta(s)$ remains bounded for $s \to 1$, which
    is a contradiction.

6. In the region $\sigma > 1$ we have the identity

    $$(1 - 2^{1-s})\zeta(s) = \sum_{n=1}^{\infty} \frac{(-1)^{n-1}}{n^s} .$$

    By LEIBNIZ' criterion of convergence for alternating series, the R.H.S. is con-
    vergent for any real $\sigma > 0$. From Exercise 1 of VII.2 we deduce that the R.H.S.

is even an analytic function in the region $\sigma > 0$. From the Principle of Analytic Continuation we obtain this identity in the whole half-plane.

The alternating series is always positive in the interval $]\,0,1[$ and the factor preceding the zeta function in the L.H.S. is negative.

7.   From the Prime Number Theorem we first easily deduce

$$\lim_{x \to \infty} \frac{\log \pi(x)}{\log x} = 1 \ .$$

Substituting instead of x the nth prime number p_n we deduce from $\pi(p_n) = n$

$$\lim_{n \to \infty} \frac{n \log n}{p_n} = 1 \ .$$

Conversely, let us assume that this relation is satisfied. For any fixed given $x > 2$ we then consider the biggest prime number p_n smaller then x, i.e. $p_n \le x < p_{n+1}$. From the assumption easily follows

$$\lim_{x \to \infty} \frac{x}{n \log n} = \lim_{x \to \infty} \frac{x}{\pi(x) \log \pi(x)} = 1 \ . \tag{$*$}$$

Taking logarithms this implies

$$\lim_{x \to \infty} \big(\log \pi(x) + \log \log \pi(x) - \log x\big) = 0 \ .$$

We divide by $\log \pi(x)$ and obtain

$$\lim_{x \to \infty} \frac{\log x}{\log \pi(x)} = 1 \ ,$$

and also using $(*)$ we get the Prime Number Theorem.

Exercises in Sect. VII.5

1.   The LAURENT series exists by the general Proposition III.5.2. It remains to show that $\gamma := \lim\limits_{s \to 1} \left(\zeta(s) - \dfrac{1}{s - 1} \right)$ is the EULER-MASCHERONI Constant γ (page 199). By Lemma VII.5.2 we have

$$\gamma = \frac{1}{2} - F(1) = \frac{1}{2} - \int_1^\infty \frac{\beta(t)}{t^2}\, dt \ .$$

The claim now follows from the formula

$$\sum_{n=1}^N \frac{1}{n} - \log N = \frac{1}{2} + \frac{1}{2N} - \int_1^N \frac{\beta(t)}{t^2}\, dt$$

after passing to the limit with $N \to \infty$. The applied formula can be proven by partial integration (compare with the proof of VII.5.2).

2. Both formulas are clear in the convergence domain $\sigma > 1$. The series $\sum(-1)^{n-1}n^{-s}$ converges by LEIBNIZ' criterion for alternating series first for all real $s > 0$. By Exercise 1 in VII.2 we obtain convergence in the half-plane $\sigma > 0$, and the represented function is analytic in this domain. The function $\zeta(s)$ can be thus analytically extended into the region $\sigma > 0$, excepting the zeros of $1 - 2^{1-s}$. Using $Q(s)$ we analogously can deduce the analytic continuation into the region $\sigma > 0$, excepting the zeros of $1 - 3^{1-s}$. The only common zero is $s = 1$. The residue is

$$\lim_{s \to 1} \frac{s-1}{1-2^{1-s}} \sum_{n=1}^{\infty} \frac{(-1)^{n-1}}{n} .$$

The value of the alternating series is known to be $\log 2$, the whole limit is thus equal to 1.

3. From the functional equation written in symmetric form

$$\pi^{-\frac{1-s}{2}} \Gamma\left(\frac{1-s}{2}\right)\zeta(1-s) = \pi^{-\frac{s}{2}}\Gamma\left(\frac{s}{2}\right)\zeta(s) ,$$

and from the Completion Formula for the Gamma function written as

$$\Gamma\left(\frac{1+s}{2}\right)\Gamma\left(\frac{1-s}{2}\right) = \frac{\pi}{\sin\left(\frac{\pi s}{2} + \frac{\pi}{2}\right)}$$

we deduce

$$\zeta(1-s) = \Gamma\left(\frac{s}{2}\right)\Gamma\left(\frac{s+1}{2}\right)\pi^{-s-\frac{1}{2}}\sin\left(\frac{\pi s}{2} + \frac{\pi}{2}\right)\zeta(s) .$$

The claim now follows from the Doubling Formula IV.1.12.

4. (a) The pole of $\zeta(s)$ is compensated by the pre-factor $s - 1$, the pole of $\Gamma(s/2)$ at 0 by the pre-factor s, and the remaining poles by the zeros of the zeta function (Exercise 3).

    (b) This is the functional equation of the zeta function, if we observe that the prefactor $s(s-1)$ satisfies the same functional equation.

    (c) Show $\overline{\Phi(\bar{s})} = \Phi(s)$, and use the functional equation.

    (d) Make use of $\zeta(0) = -1/2$, and of $\lim_{s \to 0} s\Gamma(s/2) = 2$.

    (e) None of the factors has a zero in the region $\sigma \geq 1$, $s \neq 1$. Because of (d), the function Φ has no zero in the closed half-plane $\sigma \geq 1$. From the functional equation it follows that there is also no zero in the region $\sigma \leq -1$. The symmetries follow from the functional equation and from $\overline{\Phi(\bar{s})} = \Phi(s)$.

5. The first part of the proof of Theorem VII.3.4 may serve as an orientation.

6. Since the function $t(1 - e^{-t})^{-1}$ is bounded from above we obtain the convergence of the integral from the convergence of the Gamma integral. For the proof of the formula we expand $(1 - e^{-t})^{-1}$ into a geometric series, and integrate termwise. The latter easily can be justified. The nth term gives exactly $\Gamma(s) n^{-s}$.

7. Make use of HANKEL's Integral Representation of the Gamma function (Exercise 17 in IV.1), and proceed as in Exercise 6.

Exercises in Sect. VII.6

1. In Exercise 1 from VII.4 we mentioned the MÖBIUS function, and its connection with the inverse of the zeta function. The assumptions of the TAUBERian Theorem are satisfied:

   First, the coefficients $a_n := \mu(n) + 1$ are indeed non-negative. For (I), we must use, besides the analytic continuation of the zeta function, the fact that $\zeta(s)$ has no zero on the vertical line Re $s = 1$. Then (II) follows from the estimates from above and below VII.5.1 for the zeta function. The residue ϱ is 1.

2. We would like to apply the Residue Theorem. In case of $0 < y \leq 1$ the integrand is rapidly decaying for $|\sigma| \to \infty$, $\sigma > 0$. Hence the integral vanishes, because the integrand is analytic in this region. In case of $y \geq 1$ the integrand is rapidly decaying for $\sigma \leq 2$. The integral is thus equal to the residue of the integrand at $s = 0$. This residue is equal to $\log y$, as it can be seen by using the power series $y^s = 1 + s \log y + \cdots$.

3. This is the special case $k = 0$ of VII.6.7. One can deduce it once again directly by means of the previous exercise.

4. The proof of VII.3.4 can be transposed without any problems.

5. As in Exercise 4, the proof of VII.3.4 may serve as an orientation. The needed transformation formula can be found in VI.4.8.

6. The proof is based on the formula

$$-\frac{\zeta'(s)}{\zeta(s)} = s \int_1^\infty \frac{\psi(x)}{x^{s+1}} \, dx \text{ for } \sigma > 1 \ ,$$

   which can be obtained from ABEL's identity in Exercise 2 of VII.4. A simple transformation gives

$$\Phi(s) := -\frac{\zeta'(s)}{s\zeta(s)} - \frac{1}{s-1} = \int_1^\infty \frac{\psi(x) - x}{x^{s+1}} \, dx \text{ for } \sigma > 1 \ .$$

   From the Prime Number Theorem written as $\psi(x) = x + o(x)$ one can deduce from this integral representation for a fixed t

$$\lim_{\sigma \to 0} (\sigma - 1)\Phi(\sigma + it) = 0 \ .$$

   If the zeta function would have a zero at $s = 1 + it$, then Φ would have a simple pole at this point, contradicting the value for the limit.

7. The summatoric function $S_r(n) := A_r(1) + \cdots + A_r(n)$ is equal to the number of lattice points $g \in \mathbb{Z}^r$, which lie in the (closed) ball of radius $\sqrt{n}$. For each of these lattice points we consider the unit cube $[g_1, g_1 + 1] \times \cdots \times [g_r, g_r + 1]$. Let $V_r(n)$ be the union of all these cubes. The volume of $V_r(n)$ is exactly $S_r(n)$. Obviously, $V_r(n)$ is contained in the ball of radius $\sqrt{n} + \sqrt{r}$, and contains the ball of radius $\sqrt{n} - \sqrt{r}$. Asymptotically, the volumes of these balls are of the order of the volume of the ball of radius $\sqrt{n}$, i.e. $V_r\sqrt{n}^r$.

   Now consider the EPSTEIN zeta function for the $r \times r$-unit matrix E,

$$\zeta_E(s) = \sum (g_1^2 + \cdots + g_r^2)^{-s} = \sum_{n=1}^\infty A_n n^{-s} \ .$$

The DIRICHLET series $D(s) := \zeta_E(s/2)$ satisfies the conditions of the TAUBERian Theorem with (see Exercise 5)

$$\varrho = \frac{\pi^{r/2}}{\Gamma(r/2)r/2} = \frac{\pi^{r/2}}{\Gamma(r/2+1)} \ .$$

References

Books on Complex Analysis

[Ah]   Ahlfors, L. V.: *Complex Analysis,* 3rd edn. McCraw-Hill, New York 1979

[As]   Ash, R. B.: *Complex Variables,* Academic Press, New York 1971

[BS]   Behnke, H., Sommer, F.: *Theorie der analytischen Funktionen einer komplexen Veränderlichen,* 3. Aufl. Grundlehren der mathematischen Wissenschaften, Bd. 77. Springer, Berlin Heidelberg New York 1965, Studienausgabe der 3. Aufl. 1976

[BG]   Berenstein, C. A., Gay, R.: *Complex Variables. An Introduction,* Graduate Texts in Mathematics, vol. 125. Springer, New York Berlin Heidelberg 1991

[Bi]   Bieberbach, L.: *Lehrbuch der Funktionentheorie,* Bd. I und II. Teubner, Leipzig 1930, 1931 — reprinted in Chelsea 1945, Johnson Reprint Corp. 1968

[Bu]   Burckel, R. B.: *An Introduction to Classical Complex Analysis, vol. I,* Birkhäuser, Basel Stuttgart 1979 (containing very detailed references)

[Cara]   Carathéodory, C.: *Theory of Functions of a Complex Variable,* (translated by F. Steinhardt) Vol. 1, Chelsea Publishing, New York, 1983.

[CH]   Cartan, H.: *Elementary Theory of Analytic Functions of One or Several Complex Variables.* Hermann, Paris and Addison-Wesley, Reading 1963

[Co1]   Conway, J. B.: *Functions of One Complex Variable,* 2nd edn. 7th printing Graduate Texts in Mathematics, vol. 11. Springer, New York Heidelberg Berlin 1995

[Co2]   Conway, J. B.: *Functions of One Complex Variable II,* corr. 2nd edn. Graduate Texts in Mathematics, vol. 159. Springer, New York Heidelberg Berlin 1995

[Din]   Dinghas, A.: *Vorlesungen über Funktionentheorie,* Grundlehren der mathematischen Wissenschaften, Bd. 110. Springer, Berlin Heidelberg New York 1961

[Ed]   Edwards, H. M.: *Riemann's Zeta-Function,* Academic Press, New York, London 1974

[FL]   Fischer, W., Lieb, I.: *Funktionentheorie,* 10. Aufl. Vieweg-Studium, Auf-
          baukurs Mathematik, Vieweg, Braunschweig Wiesbaden 2008

[Gam]   Gamelin, Theodore W.: *Complex Analysis,* 2nd corr. printing, Undergrad-
          uate Texts in Mathematics, Springer New York 2002

[Gre]   Greene, R.E., Krantz, St.G.: *Function Theory of One Complex Variable,*
          2nd edition, AMS, Graduate Studies in Mathematics, vol. 40, Providence,
          Rhode Island 2002

[Hei]   Heins, M.: *Complex Function Theory,* Academic Press, New York London
          1968

[HC]   Hurwitz, A., Courant, R.: *Funktionentheorie. Mit einem Anhang von H.
          Röhrl,* 4.Aufl. Grundlehren der mathematischen Wissenschaften, Bd. 3.
          Springer, Berlin Heidelberg New York 1964

[How]   Howie, J.H.: *Complex Analysis,* Springer, London 2003

[Iv]   Ivic, A.: *The Riemann Zeta-Function,* Wiley, New York 1985

[Jä]   Jänich, K.: *Funktionentheorie. Eine Einführung,* 6. Aufl. Springer-
          Lehrbuch, Springer, Berlin Heidelberg New York 2004

[Kne]   Kneser, H.: *Funktionentheorie,* Vandenhoeck & Ruprecht, Göttingen 1966

[Kno]   Knopp, K.: *Theory of Functions,* (5 Volumes), Dover, New York, 1989.

[La1]   Lang, S.: *Complex Analysis,* 3rd corr. printing, Graduate Texts in Mathe-
          matics 103. Springer, New York Berlin Heidelberg 2003

[LR]   Levinson, N., Redheffer, R.N.: *Complex Variables,* Holden-Day, Inc. San
          Francisco 1970

[LZL]   Lu, J-K. L., Zhong, S-G., Liu, S-Q.: *Introduction to the Theory of Complex
          Functions,* Series in Pure Mathematics, vol 25, World Scientific, New Jersey,
          London, Singapore, Hong Kong, 2002

[Ma1]   Maaß, H.: *Funktionentheorie I,* Vorlesungsskript, Mathematisches Institut
          der Universität Heidelberg 1949

[Mar1]   Markoushevich, A.I.: *Theory of Functions of a Complex Variable,* Prenti-
          ce-Hall, Englewood Cliffs 1965/1967

[MH]   Marsden, J.E., Hoffmann, M.J.: *Basic Complex Analysis,* third edn., W.M.
          Freeman and Company, New York 1998

[McG]   McGehee, O. Carruth: *An Introduction to Complex Analysis,* John Wiley
          & Sons, New York 2000

[Mo]   Moskowitz, M.A.: *A Course in Complex Analysis in One Variable,* World
          Scientific, New Jersey, London, Singapore, Hong Kong, 2002

[Na]   Narasimhan, R.: *Complex Analysis in One Variable,* Birkhäuser, Boston
          Basel Stuttgart 1985

[NP]   Nevanlinna, R., Paatero, V.: *Einführung in die Funktionentheorie,* Birk-
          häuser, Basel Stuttgart 1965

[Os1]   Osgood, W.F.: *Lehrbuch der Funktionentheorie I, II_1, II_2,* Teubner, Leipzig
          1925, 1929, 1932

[Pal]   Palka, B.P.: *An Introduction to Complex Function Theory,* Undergraduate
          Texts in Mathematics. 2nd corr. printing, Springer, New York 1995

[Pat]   Patterson, S.T.: *An Introduction to the Theory of Riemann's Zeta–Function*, Cambridge University Press, Cambridge 1988

[Re1]   Remmert, R.: *Theory of Complex Functions*, Graduate Texts in Mathematics, Readings in Mathematics, vol. 120, 1st. edn. 1991. Corr. 4th printing, Springer New York 1999

[ReS1]   Remmert, R., Schumacher, G.: *Funktionentheorie I*, 5. Aufl. Springer-Lehrbuch, Springer, Berlin Heidelberg New York 2002

[ReS2]   Remmert, R., Schumacher, G.: *Funktionentheorie II*, 3rd edn., Springer-Lehrbuch, Springer, Berlin Heidelberg New York 2005

[Ru]   Rudin, W.: *Real and Complex Analysis*, 3rd edn. Mc Graw-Hill, New York 1987

[SZ]   Saks, S., Zygmund, A.: *Analytic Functions*, PWN, Warschau 1965

[Tit1]   Titchmarsh, E.C.: *The Zeta–Function of Riemann*, Cambridge Tracts in Mathematical Physics, No. 26, Cambridge, University Press 1930, second revised (Heath–Brown) edition, Oxford University Press 1986

[Ve]   Veech, W.A.: *A Second Course in Complex Analysis*, Benjamin, New York 1967

Supplementary and Completing Literature

[AS]   Ahlfors, L., Sario, L.: *Riemann Surfaces*, Princeton University Press, Princeton NJ 1960

[Ap1]   Apostol, T.M.: *Modular Functions and Dirichlet Series in Number Theoric*, 2nd edn. Graduate Texts in Mathematics, vol. 41. Springer, New York Berlin Heidelberg 1992. Corr. 2nd printing 1997

[Ap2]   Apostol, T.M.: *Introduction to Analytic Number Theory*, 2nd edn. Undergraduate Texts in Mathematics, Springer, New York Heidelberg Berlin 1984. Corr. 5th printing 1998

[Apé]   Apéry, R.: *Irrationalité de $\zeta(2)$ et $\zeta(3)$*, Astérisque 61, pp. 11-13, 1979

[Ch1]   Chandrasekharan, K.: *Introduction to Analytic Number Theory*, Grundlehren der mathematischen Wissenschaften, Bd. 148. Springer, Berlin Heidelberg New York 1968

[Ch2]   Chandrasekharan, K.: *Elliptic Functions*, Grundlehren der mathematischen Wissenschaften, Bd. 281. Springer, Berlin Heidelberg New York 1985

[Ch3]   Chandrasekharan, K.: *Arithmetical Functions*, Grundlehren der mathematischen Wissenschaften, Bd. 167. Springer, Berlin Heidelberg New York 1970

[CS]   Conway, J.H., Sloane, N.J.A.: *Sphere Packings, Lattices and Groups*. 2nd edn. Grundlehren der mathematischen Wissenschaften 290. Springer, New York Berlin Heidelberg 1999

[DS]   Diamond, F., Shurman, J.: *A First Course in Modular Forms*, Graduate Texts in Mathematics, vol. 228, Springer 2005

[Die1]   Dieudonné, J.: *Calcul infinitésimal*, 2ième édn. Collection Méthodes, Hermann, Paris 1980

[Fr1]   Fricke, R.: *Die elliptischen Funktionen und ihre Anwendungen,* first part: Teubner, Leipzig 1916, second part: Teubner, Leipzig 1922. Reprinted by Johnson Reprint Corporation, New York London 1972

[Fo]   Forster, O.: *Lectures on Riemann Surfaces.* Graduate Texts in Mathematics, vol. 81, Springer, Berlin Heidelberg New York 1981 (2nd corr. printing 1991)

[Ga]   Gaier, D.: *Konstruktive Methoden der konformen Abbildung.* Springer Tracts in Natural Philosophy, vol. 3. Springer, Berlin Heidelberg New York 1964

[Gu]   Gunning, R. C.: *Lectures on Modular Forms.* Annals of Mathematics Studies, No 48. Princeton University Press, Princeton, N. J., 1962

[He1]   Hecke, E.: *Lectures on Dirichlet Series, Modular Functions and Quadratic Forms,* Vandenhoeck & Ruprecht, Göttingen 1983

[Hen]   Henrici, P.: *Applied and computational complex analysis, vol. I, II, III.* Wiley, New York 1974, 1977, 1986

[Iw]   Iwaniec, H.: *Topics in Classical Automorphic Forms.* AMS, Graduate Studies in Mathematics, vol. 17, 1997

[JS]   Jones, G. A., Singerman, D.: *Complex Functions, an Algebraic and Geometric Viewpoint.* Cambridge University Press, Cambridge 1987

[KK]   Koecher, M., Krieg, A.: *Elliptische Funktionen und Modulformen,* 2. Aufl. Springer, Berlin Heidelberg 2007

[Ko]   Koblitz, N.: *Introduction to Elliptic Curves and Modular Forms,* 2nd edn. Graduate Texts in Mathematics, vol. 97. Springer, New York Berlin Heidelberg 1993

[Lan]   Landau, E.: *Handbuch der Lehre von der Verteilung der Primzahlen, Bd. I, Bd. II.* Teubner, Leipzig 1909; 3rd edn. Chelsea Publishing Company, New York 1974

[La2]   Lang, S.: *Algebra,* 4rd corr. printing, Graduate Texts in Mathematics 211, Springer New York 2004

[Le]   Leutbecher, A.: *Vorlesungen zur Funktionentheorie I und II,* Mathematisches Institut der Technischen Universität München (TUM) 1990, 1991

[Ma2]   Maaß, H.: *Funktionentheorie II, III,* Vorlesungsskript, Mathematisches Institut der Universität Heidelberg 1949

[Ma3]   Maaß, H.: *Modular Functions of one Complex Variable,* Tata Institute of Fundamental Research, Bombay 1964. Revised edn.: Springer, Berlin Heidelberg New York 1983

[Mi]   Miyake, T.: *Modular forms,* Springer, Berlin Heidelberg New York 1989

[Mu]   Mumford, D.: *Tata Lectures on Theta I,* Progress in Mathematics, vol. 28. Birkhäuser, Boston Basel Stuttgart 1983

[Ne1]   Nevanlinna, R.: *Uniformisierung,* 2. Aufl. Grundlehren der mathematischen Wissenschaften, Bd. 64. Springer, Berlin Heidelberg New York 1967

[Ne2]   Nevanlinna, R.: *Eindeutige analytische Funktionen,* 2. Aufl. Grundlehren der mathematischen Wissenschaften, Bd. 46. Springer, Berlin Heidelberg New York 1974 (reprint)

[Pf]   Pfluger, A.: *Theorie der Riemannschen Flächen,* Grundlehren der mathe-
         matischen Wissenschaften, Bd. 89. Springer, Berlin Göttingen Heidelberg
         1957

[Po]   Pommerenke, Ch.: *Boundary Behaviour of Conformal Maps,* Springer
         Berlin 1992

[Pr]   Prachar, K.: *Primzahlverteilung,* 2. Aufl. Grundlehren der mathematischen
         Wissenschaften, Bd. 91. Springer, Berlin Heidelberg New York 1978

[Ra]   Rankin, R. A.: *Modular Forms and Functions.* Cambridge University Press,
         Cambridge, Mass., 1977

[Ro]   Robert, A.: *Elliptic Curves.* Lecture Notes in Mathematics, vol. 326 (2nd
         corr. printing). Springer, Berlin Heidelberg New York, 1986

[Sb]   Schoeneberg, B.: *Elliptic Modular Functions.* Grundlehren der mathemati-
         schen Wissenschaften, Bd. 203. Springer, Berlin Heidelberg New York 1974

[Sch]   Schwarz, W.: *Einführung in die Methoden und Ergebnisse der Primzahltheo-
         rie,* B I-Hochschultaschenbücher, Bd. 278/278a. Bibliographisches Institut,
         Mannheim Wien Zürich 1969

[Se]   Serre, J. P.: *A Course in Arithmetic,* Graduate Texts in Mathematics, vol.
         7. Springer, New York Heidelberg Berlin 1973 (4th printing 1993)

[Sh]   Shimura, G.: *Introduction to Arithmetic Theory of Automorphic Functions,*
         Publications of the Mathematical Society of Japan 11. Iwanami Shoten,
         Publishers and Princeton University Press 1971

[Si1]   Siegel, C. L.: *Topics in Complex Function Theory, vol. I, II, III.* Intersc.
         Tracts in Pure and Applied Math., No 25. Wiley-Interscience, New York
         1969, 1971, 1973

[Sil]   Silverman, J. H.: *Advanced Topics in the Arithmetic of Elliptic Curves,*
         Graduate Texts in Mathematics, vol. 151 Springer, New York Berlin Hei-
         delberg 1994

[ST]   Silverman, J., Tate, J.: *Rational Points on Elliptic Curves,* Undergraduate
         Texts in Mathematics, Springer, New York Berlin Heidelberg 1992

[Sp]   Springer, G.: *Introduction to Riemann Surfaces,* Addison-Wesley, Reading,
         Massachusetts, USA 1957

[Tit2]   Titchmarsh, E. C.: *The Theory of the Riemann Zeta-Function,* Clarendon
         Press, Oxford 1951, reprinted 1967

[We]   Weil, A.: *Elliptic Functions according to Eisenstein and Kronecker,* Ergeb-
         nisse der Mathematik und ihrer Grenzgebiete, Bd. 88. Springer, Berlin Hei-
         delberg New York 1976

[WK]   Weierstraß, K.: *Einleitung in die Theorie der analytischen Funktionen,* Vor-
         lesung, Berlin 1878. Vieweg, Braunschweig Wiesbaden 1988

[WH]   Weyl, H.: *Die Idee der Riemannschen Fläche,* 4. Aufl. Teubner, Stuttgart
         1964, new edition 1997, editor R. Remmert

History of the Complex Numbers and Complex Functions

[Bel]   Belhoste, B.: *Augustin-Louis Cauchy. A Biography,* Springer, New York
         Berlin Heidelberg 1991

[CE]   Cartan, É.: *Nombres complexes*, Exposé, d'après l'article allemand de E. Study (Bonn). Encyclop. Sci. Math. édition francaise I 5, p. 329–468. Gauthier-Villars, Paris; Teubner, Leipzig 1909; see also E. Cartan: Œvres complètes II.1, p. 107–246, Gauthier-Villars, Paris 1953

[Die2]   Dieudonné, J. (Ed.): *Abrégé d'histoire des mathematiques* I & II, Hermann Paris 1978

[Eb]   Ebbinghaus, H.-D. et al.: *Numbers*, 3rd. corr. printing, Graduate Texts in Mathematics 123, Springer New York 1996 Springer-Lehrbuch, Springer, Berlin Heidelberg New York 1992

[Fr2]   Fricke, R.: *IIB3. Elliptische Funktionen*. Encyklopädie der mathematischen Wissenschaften mit Einschluß ihrer Anwendungen, Bd. II 2, Heft 2/3, S. 177–348. Teubner, Leipzig 1913

[Fr3]   Fricke, R.: *IIB4. Automorphe Funktionen mit Einschluß der elliptischen Funktionen*. Encyklopädie der mathematischen Wissenschaften mit Einschluß ihrer Anwendungen, Bd. II 2, Heft 2/3, S. 349–470. Teubner, Leipzig 1913

[Hi]   Hirzebruch, F.: chapter 11 in [Eb]

[Hou]   Houzel, C.: *Fonctions elliptiques et intégrales abéliennes*, chap. VII, pp. 1–113 in [Die2], vol. II

[Kl]   Klein, F.: *Vorlesungen über die Entwicklung der Mathematik im 19. Jahrhundert, Teil 1 und 2*, Grundlehren der mathematischen Wissenschaften, Bd. 24 und 25. Springer, Berlin Heidelberg 1926. Nachdruck in einem Band 1979

[Mar2]   Markouschevitsch, A. I.: *Skizzen zur Geschichte der analytischen Funktionen*, Hochschultaschenbücher für Mathematik, Bd. 16. Deutscher Verlag der Wissenschaften, Berlin 1955

[Neu]   Neuenschwander, E.: *Über die Wechselwirkung zwischen der französischen Schule, Riemann und Weierstraß. Eine Übersicht mit zwei Quellenstudien.* Arch. Hist. Exact Sciences **24** (1981), 221–255

[Os2]   Osgood, W. F.: *Allgemeine Theorie der analytischen Funktionen a) einer und b) mehrerer komplexer Größen*, Enzyklopädie der Mathematischen Wissenschaften, Bd. II 2, S. 1–114. Teubner, Leipzig 1901–1921

[Re2]   Remmert, R.: *Complex Numbers*, Chap. 3 in [Eb]

[St]   Study, E.: *Theorie der gemeinen und höheren complexen Grössen*, Enzyklopädie der Mathematischen Wissenschaften, Bd. I 1, S. 147–183. Teubner, Leipzig 1898–1904

[Ver]   Verley, J. L.: *Les fonctions analytiques*, Chap IV, pp. 129–163 in [Die2], vol. I

In [ReS1] and [ReS2] one can find many facts related to the history of the theory of complex functions.

Original Papers

[Ab1]   Abel, N. H.: *Mémoire sur une propriété générale d'une classe très étendue de fonctions transcendantes* (submitted at 30. 10. 1826, published in 1841).

Œvres complètes de Niels Henrik Abel, tome premier, XII, p. 145–211. Grondahl, Christiania M DCCC LXXXI, Johnson Reprint Corporation 1973

[Ab2]   Abel, N. H.: *Recherches sur les fonctions elliptiques,* Journal für die reine und angewandte Mathematik **2** (1827), 101–181 und **3** (1828), 160–190; see also Œvres complètes de Niels Henrik Abel, tome premier, XVI, p. 263–388. Grondahl, Christiania M DCCC LXXXI, Johnson Reprint Corporation 1973

[Ab3]   Abel, N. H.: *Précis d'une théorie des fonctions elliptiques,* Journal für die reine und angewandte Mathematik **4** (1829), 236–277 und 309–370; see also Œvres complètes de Niels Henrik Abel, tome premier, XXVIII, p. 518–617. Grondahl, Christiania M DCCC LXXXI, Johnson Reprint Corporation 1973

[BFK]   Busam, R., Freitag, E., Karcher, W.: *Ein Ring elliptischer Modulformen,* Arch. Math. **59** (1992), 157–164

[Cau]   Cauchy, A.-L.: *Abhandlungen über bestimmte Integrale zwischen imaginären Grenzen.* Ostwald's Klassiker der exakten Wissenschaften Nr. 112, Wilhelm Engelmann, Leipzig 1900; see also A.-L. Cauchy: Œuvres complètes 15, 2. Ser., p. 41–89, Gauthier-Villars, Paris 1882–1974
The source appeared as *"Mémoire sur les intégrales définies, prises entre des limites imaginaires"* in 1825.

[Dix]   Dixon, J. D.: *A brief proof of Cauchy's integral theorem,* Proc. Am. Math. Soc. **29** (1971), 635–636

[Eis]   Eisenstein, G.: *Genaue Untersuchung der unendlichen Doppelproducte, aus welchen die elliptischen Functionen als Quotienten zusammengesetzt sind, und der mit ihnen zusammenhängenden Doppelreihen (als eine neue Begründungsweise der Theorie der elliptischen Functionen, mit besonderer Berücksichtigung ihrer Analogie zu den Kreisfunctionen).* Journal für die reine und angewandte Mathematik (Crelle's Journal) **35** (1847), 153–274; see also G. Eisenstein: Mathematische Werke, Bd. I. Chelsea Publishing Company, New York, N. Y., 1975, S. 357–478

[El]   Elstrodt, J.: *Eine Charakterisierung der Eisenstein-Reihe zur Siegelschen Modulgruppe,* Math. Ann. **268** (1984), 473–474

[He2]   Hecke, E.: *Über die Bestimmung Dirichletscher Reihen durch ihre Funktionalgleichung,* Math. Ann. **112** (1936), 664–699; see also E. Hecke: Mathematische Werke, 3. Aufl., S. 591–626. Vandenhoeck & Ruprecht, Göttingen 1983

[He3]   Hecke, E.: *Die Primzahlen in der Theorie der elliptischen Modulfunktionen,* Kgl. Danske Videnskabernes Selskab. Mathematisk-fysiske Medelelser XIII, 10, 1935; see also E. Hecke: Mathematische Werke, S. 577–590. Vandenhoeck & Ruprecht, Göttingen 1983

[Hu1]   Hurwitz, A.: *Grundlagen einer independenten Theorie der elliptischen Modulfunktionen und Theorie der Multiplikator-Gleichungen erster Stufe,* Inauguraldissertation, Leipzig 1881; Math. Ann. **18** (1881), 528–592; see also A. Hurwitz: Mathematische Werke, Band I Funktionentheorie, S. 1–66, Birkhäuser, Basel Stuttgart 1962

[Hu2]   Hurwitz, A.: *Über die Theorie der elliptischen Modulfunktionen,* Math. Ann. **58** (1904), 343–460; see also A. Hurwitz: Mathematische Werke, Band I Funktionentheorie, S. 577–595, Birkhäuser, Basel Stuttgart 1962

[Ig1]   Igusa, J.: *On the graded ring of theta constants,* Amer. J. Math. **86** (1964), 219–246

[Ig2]   Igusa, J.: *On the graded ring of theta constants II,* Amer. J. Math. **88** (1966), 221–236

[Ja1]   Jacobi, C. G. J.: *Suite des notices sur les fonctions elliptiques,* Journal für die reine und angewandte Mathematik **3** (1828), 303–310 und 403–404; see also C. G. J. Jacobi's Gesammelte Werke, I, S. 255–265, G. Reimer, Berlin 1881

[Ja2]   Jacobi, C. G. J.: *Fundamenta Nova Theoriae Functionum Ellipticarum,* Sumptibus fratrum Bornträger, Regiomonti 1829; see also C. G. J. Jacobi's Gesammelte Werke, I, S. 49–239, G. Reimer, Berlin 1881

[Ja3]   Jacobi, C. G. J.: *Note sur la décomposition d'un nombre donné en quatre quarrés,* C. G. J. Jacobi's Gesammelte Werke, I, S. 274, G. Reimer, Berlin 1881

[Ja4]   Jacobi, C. G. J.: *Theorie der elliptischen Funktionen, aus den Eigenschaften der Thetareihen abgeleitet,* after a lecture of Jacobi, revised at his request by C. Borchardt. C. G. J. Jacobi's Gesammelte Werke, I, S. 497–538, G. Reimer, Berlin 1881

[Re3]   Remmert, R.: *Wielandt's Characterization of the Γ-function,* pp. 265–268 in [Wi]

[Ri1]   Riemann, B.: *Grundlagen für eine allgemeine Theorie der Functionen einer veränderlichen complexen Grösse,* Inauguraldissertation, Göttingen 1851; see also B. Riemann: Gesammelte mathematische Werke, wissenschaftlicher Nachlaß und Nachträge, collected papers, S. 35–77. Springer, Berlin Heidelberg New York; Teubner, Leipzig 1990

[Ri2]   Riemann, B.: *Ueher die Anzahl der Primzahlen unterhalb einer gegebenen Grösse,* Monatsberichte der Berliner Akademie, November 1859, S. 671–680; see also B. Riemann: Gesammelte mathematische Werke, wissenschaftlicher Nachlaß und Nachträge, collected papers, S. 177–185. Springer, Berlin Heidelberg New York, Teubner, Leipzig 1990

[Si2]   Siegel, C. L.: *Über die analytische Theorie der quadratischen Formen,* Ann. Math. **36** (1935), 527–606; see also C. L. Siegel: Gesammelte Abhandlungen, Band I, S. 326–405. Springer, Berlin Heidelberg New York 1966

[Wi]   Wielandt, H.: Mathematische Werke, vol 2, de Gruyter, Berlin New York 1996

Collections of Exercises

Parallely to the problem books among the Knopp [Kno] editions we especially recommend:

[He]   Herz, A.: *Repetitorium Funktionentheorie,* Vieweg, Lehrbuch Mathematik 1996

[Kr]   Krzyz, J.G.: *Problems in Complex Variable Theory,* Elsevier, New York London Amsterdam 1971

[Sha]   Shakarchi, R.: *Problems and Solutions for Complex Analysis.* Springer, New York Berlin Heidelberg 1995
(In this book one can find solutions to all exercises in LANG's book [La1].)

[Tim]   Timmann, S.: *Repetitorium der Funktionentheorie,* Verlag Binomi, Springe 1998

and also the classical

[PS]   Polya, G. Szegö, G.: *Problems and Theorems in Analysis II,* Theory of Functions, Zeros, Polynomials, Determinants, Number Theory, Geometry, Classics in Mathematics, Springer 1998, Reprint of the 1st ed. Berlin, Heidelberg, New York 1976

Symbolic Notations

iff	if and only if
L.H.S.	left hand side
R.H.S.	right hand side
$\mathbb{N} = \{\, 1, 2, \dots \,\}$	set of natural numbers
$\mathbb{N}_0 = \{\, 0, 1, 2, \dots \,\}$	set of natural numbers including zero
$\mathbb{Z}$	ring of integers
$\mathbb{R}$	field of real numbers, real axis
$\mathbb{C}$	field of complex numbers, complex plane
$\mathbb{C}_- = \mathbb{C} \setminus \{\, x \in \mathbb{R} \,;\, x \le 0 \,\}$	slit plane along the negative real half-line
$\mathbb{C}^\bullet = \mathbb{C} \setminus \{0\}$	punctured plane
$\overline{\mathbb{C}} = \mathbb{C} \cup \{\infty\}$	RIEMANN sphere
$P^n(\mathbb{C})$	n-dimensional projective space
$\mathbb{H}$	upper half-plane
$\mathbb{E}$	open unit disk
S^1	unit circle
$\mathcal{H}$	HAMILTONian quaternions
$\operatorname{Re} z$, $\operatorname{Im} z$	real and imaginary part of a number z
$\operatorname{Re} f$, $\operatorname{Im} f$	real and imaginary part of a function f
$\overline{z}$	complex conjugate of z
$\lvert z \rvert$	modulus, absolute value of z
$\operatorname{Arg} z \; (-\pi < \operatorname{Arg} z \le \pi)$	principal value of the argument
$\operatorname{Log} z = \log \lvert z \rvert + \mathrm{i} \operatorname{Arg} z$	principal value of the logarithm
$\overset{\circ}{D}$	set of interior points in D
$\overline{A}$	closure of A
$J(f, a) : \mathbb{C} \to \mathbb{C}$	JACOBI map of f in a
$\Delta = \partial_1^2 + \partial_2^2$	LAPLACE operator
$\int_\alpha f$	Line integral of f along the curve α
$l(\alpha)$	length of the piecewise smooth curve α

$\alpha \oplus \beta$	composition of two curves
α^-	inverse (reciprocal) curve
$\langle z_1, z_2, z_3 \rangle$	triangular path
$U_r(a)$, $\overline{U}_r(a)$	open resp. closed disk centered at a with radius r
$\oint f$	integral along a circle
$\mathcal{O}(D)$	ring of analytic functions on D
$\mathcal{A}$	annular domain
$\mathcal{A}(a; r, R)$	annular domain with center a and radii r, R
$\chi(\alpha; a)$	index of the closed curve α around a
$\mathrm{Res}(f; a)$	residue of f in a
$\mathrm{Int}(\alpha)$	interior of the closed curve α
$\mathrm{Ext}(\alpha)$	exterior of the closed curve α
S^2	unit sphere in $\mathbb{R}^3$
$\mathfrak{M}$	group of MÖBIUS transformations
$\mathrm{Aut}(D)$	group of conformal self-maps of D
$\mathcal{M}(D)$	field of meromorphic functions on a domain D
$CR(z, a, b, c)$	cross ratio
$\Gamma(z)$, $\Gamma(s)$	gamma function
$B(z, w)$	beta function
$\wp$	$\wp$-function of WEIERSTRASS
G_k	EISENSTEIN series of weight k
g_2 , g_3	$g_2 = 60\,G_4$, $g_3 = 140\,G_6$,
$K(L)$	field of elliptic function for the lattice L
$K(\Gamma)$	field of elliptic modular functions for the modular group Γ
$\sigma(z)$	WEIERSTRASS' σ-function
$\vartheta(\tau, z)$, $\vartheta(z, w)$	JACOBI theta function
$j(\tau)$	absolute invariant
$\Delta(\tau)$	discriminant
$\mathrm{SL}(2, \mathbb{R})$	group of real 2×2 matrices with determinant 1
$\Gamma = \mathrm{SL}(2, \mathbb{Z})$	elliptic modular group
$[\Gamma, k]$	vector space of all modular forms of weight k
$[\Gamma, k]_0 \subset [\Gamma, k]$	vector space of all cusp forms of weight k
$\mathcal{F}$	fundamental region of the modular group
Γ_ϑ	theta group
$\mathcal{F}_\vartheta$	fundamental region of the theta group
$\Gamma[q]$	principal congruence group of level q
$\Theta(x)$, $\psi(x)$	TSCHEBYSCHEFF functions
$\pi(x)$	prime number function
$\mathrm{Li}(x)$	integral logarithm
$\zeta(s)$	RIEMANN zeta function

Index